MARINE FUNGI OF INDIA

(MONOGRAPH)

By

B.D. BORSE

Principal, Uttamrao Patil Arts & Science College; Dadasaheb Manikrao Patil Jr. College, Dahivel, Dhule, Maharashtra

(Ex Reader, Dr Ghogrey Science College, Dhule, Maharashtra)

D.J. BHAT

Department of Botany, Goa University, Goa

K.N. BORSE

S S V P S's L K Dr P R Ghogrey Science College, Dhule, Maharashtra

A.R. TUWAR

M.E.S's Arts, Commerce & Science College, Sonai, Maharashtra

N.S. PAWAR

S S V P S's Arts, Commerce & Science College, Shindkheda, Maharashtra

Published by

MARINE FUNGI OF INDIA

(MONOGRAPH)

© Author:

B.D. BORSE

D.J. BHAT

K.N. BORSE

A.R. TUWAR

N.S. PAWAR

First Edition: Febryary , 2012

1st Floor, Ashirwad Bldg., Next to Rizvi Towers,
Near Caculo Island, 18th June Road, Panjim, Goa
Tel: 2420677, 6647037, 6647038, Fax: 6647038
Website: www.bbcbooks.net
Email: bbcbooks@rediffmail.com

ISBN : 9789380837383

Printed by : Rama Harmalkar 9326102225

CONTENTS

Preface

Publications on freshwater and terrestrial mycota of India far exceed those dealing with fungi in marine habitats. Though a sizeable body of information is presently available on marine fungi, details on their activities, ecology, physiology, relationship with other organisms and taxonomy are very few. Herein lies the first purpose of this monographic work, a detailed summation of all the available data and observations on marine fungi of India. Although this work attempts to be a consolidation of information, including some of previously unpublished data, we make no claim that our treatment and evaluations are final.

Fifty years have elapsed since the first comprehensive treatise on '*Fungi in oceans and estuaries*' by Johnson and Sparrow (1961) and it is now thirty years after the monumental work '*Marine Mycology: The Higher Marine fungi*' by Kohlmeyer and Kohlmeyer (1979). Both these great writings are partly out-of-date in view of several newer developments, continued flow of additional information and consequent controversy arouse in relation to this ecological group of fungi. Late Dr. Sam Meyers (1995) had reviewed the first five decades of marine mycology. Dr. E.B.G. Jones (2011a) recently made a comprehensive review of development in marine mycology since 1995.

The newer developments are, among others, i) renewed study of fungal species in marine habitats, particularly in mangroves, ii) use of newer isolation techniques and growth of species in axenic cultures, iii) studies on ecology and distribution, iv) discovery of newer taxa in the levels of order, family, genus and species, v) ultrastructural studies using TEM and SEM and vi) use of molecular techniques and study of phylogeny of marine fungi which have reformed many of the older treatments including origin and evolution. Jones et al. (2009a) provided a comprehensive classification and descriptions of genera and key to the major taxa, i.e. genera and species of filamentous marine fungi described up to 2008.

This monograph on '*Marine fungi of India*' is the outcome of work carried out by various researchers on this ecological group of fungi. Marine fungi have been the subject of study in our country for almost six decades, but no attempt has been made to consolidate and summarize the widely scattered data. Johnson and Sparrow (1961) and Kohlmeyer and Kohlmeyer (1979) have discussed the species known until that time. Karling (1981) and Dick (2001) discussed the lower marine fungi. During the past four decades several new genera and species have been discovered and combinations were made. Additional intensive studies using molecular techniques on long known species have modified our understanding within the group. In the light of these discoveries and the fact that these fungi are significant from the standpoint of phylogeny, evolution, decomposition of organic matter, drug productivity, as pathogens of algae and animals and in the energy budget of marine and estuarine ecosystems, we felt that a comprehensive treatment on marine fungi is very essential, particularly to stimulate further researches on marine fungi of India.

This monographic work deals with 208 species of marine fungi [14 Labyrinthulomycota, 4 Chytridiomycota, 4 Oomycota, 140 Ascomycota, 3 Basidiomycota and 43 Anamorphic (mitosporic) fungi] reported so far from marine waters of various coasts and Islands of India. The taxonomic part contains complete descriptions of each genus and species and distribution along the coasts of India. The descriptions of taxa have been arranged in alphabetical order, first the genus followed by species within the genus. They are numbered accordingly. There are keys for all higher marine species reported from India within the genus and a general key leading to individual species of higher marine fungi which will be useful for further research and understanding of their ecology. Some fungi collected from marine habitats are incorrectly identified and therefore, they are not considered while keying out within the genus and in the general key. Most of these fungi are listed in the ecological studies of mangrove marine fungi.

Description, distribution and discussion of a few doubtfully recorded species (25 Ascomycota and 2 Mitosporic fungi) are also presented in this monograph with the purpose of making these data available to future research workers. In order to provide a complete data on marine fungi of India, wherever we could not collect and study on our own only descriptions of the taxa are provided. In such instances, we urge the readers to refer the original papers. Original descriptions of species provided by numerous authors have been used in the text and grateful thanks are herein recorded for their courtesy. These contributors are too numerous for individual mention but due reference and credits are given in the text for their contributions.

The study of marine mycota has always been challenging because of their exiting nature. However, the literature was too scattered in numerous journals and many of them were difficult to obtain. We hope that students of mycology in future can use this monograph as a ready reference and enjoy the delight of studying marine fungi. Professor E.B.G. Jones (2011a) rightly stated that "some may regard the study of marine fungi as somewhat esoteric but they do play a vital role in ecology of marine ecosystems and in the food web of the oceans".

Foreword

Foreword

By Prof. E. B. Gareth Jones

It is my very great pleasure to write the forward to this monograph on the marine fungi of India as it is well conceived and authoritative and I believe it will be most useful to those interested in marine fungi, especially to a new generation of Indian mycologists. I personally have a connection to the study of marine fungi in India as C.V Subramanian visited me in Portsmouth to learn about them at first hand. This lead to the first study of marine fungi in India by Sheshagiri Raghukumar with the results published in Kavaka (1973). He subsequently went on to study the zoosporic Thrausochytriales with Dr Alwyn Gaertner at the Institute für Meeresforschung, Bremerhaven, Germany. There I also met his wife Dr Chandralata Raghukumar, who went on to make a singular contribution to Indian marine mycology, specialising in bioremediation. Another colleague who has contributed much to our knowledge of marine mycology is Professor Bhat.

I first met the senior author of this volume in 1985 when he attended the International Marine Mycology Symposium at the University of Portsmouth, England. There he and Dr Kevin Hyde (then my student) compared mangrove fungi they had collected in India and the Seychelles, resulting in the description of three mangrove ascomycetes: *Acrocordiopsis patilii*,

Institut Sains Samudera dan Bumi
C308, Bangunan IPS, Universiti Malaya,
50603 Kuala Lumpur, MALAYSIA
Tel: (+603) 7967-4640 / 6995
Faks: (+603) 7967-6994 • http://ioes.um.edu.my

Biatriospora marina and *Massarina velatospora* (now *Morosphaera velatospora*). In subsequent years many Indian mycologists have been attracted into the study of marine fungi, thus making a substantial contribution to our knowledge of fungal diversity in different oceans of the world.

This monograph provides the reader with a comprehensive introduction to marine fungi: an extensive introduction and literature review, with an up to date reference list. The bulk of the volume details 205 species in 139 genera of Indian marine fungi in the taxonomic groups: Chytridiomycota, Ascomycota, Basidiomycota, mitosporic fungi and fungal-like organisms, such as, the Thraustochytriales. Each species is provided with a detailed description, collection data and substrates they grown on. Some genera include key to species, while the volume concludes with a key to the identification of all taxa recorded for India and a table listing the distribution and substrata on which these fungi grown on.

This is a masterly account of the marine fungi of India and it has taken a sound knowledge of the group and the current literature in its preparation. I commend the authors for their diligence; endeavour and foresight in the production of what I believe will be a valuable contribution to marine mycology. I look forward to seeing the printed volume on my bookshelf.

15th September 2011

Dedicated
To

Late Prof. S.D. Patil

Ex. Professor & Head, Department of Botany, University of Pune

Late Prof. S.D. Patil (14/5/1926 –29/12/1988) or Shamrao Patil as he was known among his colleagues at Pune University campus and affiliated Colleges of the University, one of the well known mycologists of India who made significant contribution on Myxomycetes, Aquatic fungi, Ascomycetes, Uredinales, Ustilaginles and Hyphomycetes with notable contribution to *Beltrania* and *Beltraniella*. He joined the Department of Botany, University of Pune, as a herbarium Curator and, by his hard work, research, devotion, dedication and human values, later became the Professor and Head of the Department. His deep knowledge on fungi, excellent teaching ability and pleasant personality attracted many students to his laboratory. The Department of Microbiogy, University of Pune, was established under his leadership in 1977. He was a friend, philosopher and guide to many institutions of higher learning. He was great in action and thought. After retirement from the University of Pune in 1985, he worked as Director of Shahu Institute of Ecology at Kolhapur, Maharashtra. Prof. Patil was a Member of Mycological Society of Japan. He also served as President of Indian Mycological Society (1985-86). It is a pleasure to dedicate this monograph to his memory as a mark of profound respect for his original research and administrative and academic calibers.

B.D. Borse
D.J. Bhat
K.N. Borse
A.R. Tuwar
N.S. Pawar

Acknowledgement

We are grateful to Prof. E.B.G. Jones, BIOTEC Bioresources & Technology Unit, NSTDA, 113 Thailand Science Park, Phaholyothin Road, Klong Luang, Pathum Thani, 12120, Thailand; and Visiting Professor, University of Malaya, Malaysia, for kindly reading an early draft of the manuscript and generous comments which assisted in the organization of this treatment. We thank Prof. Jones for writing a Foreword to this monograph.

We are particularly grateful to Drs. T.W. Johnson, Jr (USA), J. Kohlmeyer (USA), late Erika Kohlmeyer (USA), A.R. Cavaliere (USA), E.B.G. Jones UK), C.A. Shearer (USA), L.M. Zelezinskaya (Rassia), D. Porter (USA), J.D. Miller (Canada), late S.T. Moss (UK), J. Koch (Denmark), L.E. Leightley (Australia), K.D. Hyde (Hong Kong, China), P.W. Kirk, Jr (USA), A. Nakagiri (Japan), B. Volkamann-Kohlmeyer (USA), L.P. Vrijmoed (Hong Kong, China), K.R.L. Petersen (Denmark), M.A. Abdel-Wahab (Egypt), T.K. Tan (Singapore), M. Barata (Portugal), N.I. Pena (Argentina), Ka-Lal Pang (China) and Huzefa Raja (USA), among others who all have generously supplied reprints and photocopies of articles published in journals/books which were otherwise not available with us.

Mycologists from India, Drs. S. Raghukumar (NIO, Goa), C. Raghukumar (NIO, Goa), B.P.R. Vittal (Chennai), K.R. Sridhar (Mangalore), Ch. Ramesh (Dharwad), P. Manimohan (Calicut), D.R. Ravikumar (Chennai), K.B. Patil (Songir, Dhule), V.V. Sarma (Pondicherry), K. Raveendran (Kerala) and G. Nambiar (Kerala), among others who have generously provided us reprints and photocopies of articles in journals/books which were lacking with us. They also assisted us with information on personal communications.

One of us (BDB) wishes to acknowledge the generous cooperation of Dr. Bhaidas C. Patil, President, SSVP Sanstha, Dhule; Shri Aravind M. Patil, President; Ex Principal L.A. Nandre, Secretary; Shri. Rajendra M. Patil and Shri. Vasantroa Bachhav, Director, Navodaya Shaikshanik Sanstha, Dahivel (Dist. Dhule,

Maharastra); Prin. B.M. Patil and Prin. S.N. Nandan; Head of the PG Department of Botany, Shri. V.C. Vaghamare, Dr. S.D. Deorey, Shri. A.C. Patil and Dr. Sandhya S. Patil. SSVPS's LK Dr PR Ghogrey Science College, Dhule (M S) is specially thanked for providing the numerous facilities which enabled the authors (BDB, KNB) to complete researches and writings that formed the foundation of this monograph.

We (NSP, KNB & BDB) wish to acknowledge the generous cooperation of Shri Bapusaheb P.M. Sisode, Vice-President, SSVP Sanstha, Dhule and Chaiman SSVPS's ACS College, Sindkheda (Dist. Dhule, MS). We also avail this opportunity to express our sincere thanks to former Prin. Dr B.V. Kamble, ACS college, Sindkheda (Dist.-Dhule, MS) for kind co-operation and encouragements.

One of us (ART) is extremely indebted to Shri Bhausaheb Prashanta Patil, Secretary, Mula Education Society, Sonai (Dist.-Ahmednagar, MS) for his unparalleled support, inspiration and encouragements.

The senior author also wishes to acknowledge the generous cooperation of Profs. M.S. Patil (Ex Head, Dept. of Botany, Shivaji University, Kolhapur, MS), B.P. Kapadnis (Dept of Microbiology, Pune University, Pune), R.V. Gandhe, Mrs. Kanchanganga V. Gandhe (Dept of Botany, Modern College, Pune), his colleagues, Drs./Profs. P.R. Patil, M.T. Bhadane, D.A. Patil (Ex Deen of Science Faculty, North Maharashtra University, Jalgaon), V.B. Salunkhe, S.T. Kottawar, Mrs. Lata S. Kottawar, D.J. Kelkar, M.R. Kumavat (PG Dept of Botany, SSVP S's LK Dr .PR Ghogrey Science College, Dhule, MS); his various Ph.D./M.Phil. students viz., R.V. Patil, N.B. Pawar, C.M. Pawara, V.R. Patil, R.S. Patil, V.M. Kamble, S.Y. Patil, L.C. Nemade, P.K. Ahire, S.N. Wagh, C.S. Jadhav, R.V. Karanda, Mrs. Vandana S. Patil, D.S. Borade and S.A. Gosavi who all have assisted with data from their projects on marine and freshwater fungi.

The senior author is grateful to his teacher and Ph.D. supervisor late Prof. S.D. Patil (Ex Prof and Head, Department of Botany, University of Pune), who introduced him to this fascinating ecological group of fungi and aptly therefore this monograph is dedicated to him. He records his deep sense of gratitude to his wife, Mrs. Asha Borse, sons Amar and Amit for patience, dedicated support, unfailing understanding, and further faithfully following him along the coast of India in search of marine fungi during the many field trips.

One of us (DJB) places on record the continued assistance provided by Goa University, in many ways. Thanks are also due to various national funding agencies, viz. Department of Science and Technology, Ministry of Environment and Forests, Council of Scientific and Industrial Research, and the University Grants Commission, New Delhi, for their generous support.

This monograph would not have possible without the financial support of UGC (WRO), Pune. We are thankful to M/s Broadway Book Centre, Panjim, Goa, for printing and publishing this volume very elegantly.

INTRODUCTION

Until late 1930s, marine mycology was not thought as a discrete branch of science. Following the works of Sparrow (1936), Barghoorn and Linder (1944), Wilson (1951), Hohnk (1952), Meyers (1953), Johnson (1956), Cribb and Cribb (1955), Kohlmeyer (1959) and Jones (1962), marine fungi were considered as a recognizable group in biological sciences. Of all these, investigations by Barghoorn and Linder (1944) also made us to realize that fungi in sea waters greatly contribute to the decomposition of organic matter.

The first facultative marine fungus, *Phaeosphaeria typharum*, was described by Desmazieres (1849) as *Sphaeria scirpicola* var. *typharum* on *Typha* found growing in freshwater. Durieu and Montagne (1869) discovered the first obligate marine fungus on the rhizomes of the sea grass, *Posidonia oceanica*. During the period of 1930-1939, there was only sporadic interest in marine fungi. Descriptions of marine species in the first hundred years were provided by researchers working on a wide variety of substrates with no particular interest in marine habitats. Exceptions were Crown and Crown (1867) who described five marine fungi in their '*Florule du Finistere*' and Sutherland (1915, 1916a, b) who published a series of papers dealing exclusively on marine fungi. There has been a sudden increase of description of newer marine fungi after 1940 which is attributed primarily to the publication of Barghoorn and Linder (1944) and again after 1980 with the publication of Kohlmeyer and Kohlmeyer (1979, 1991), all of which further stimulated interest on marine mycology worldwide.

Besides descriptions of numerous species, there have been several monographs and proceedings published on the morphology, taxonomy, ultrastructure and culturing methods on lower marine fungi (Sparrow, 1960; Johnson and Sparrow, 1961; Pokorny, 1967; Olive, 1975; Moss, 1985, 1986; Fuller and Jaworski, 1987; Porter, 1990; Ulken, 1990; Raghukumar, 1996a; Bremer, 2000; Dick, 2001). Keys to the species of labyrinthulids and thraustochytrids have been presented by Dick (2001).

There have been a number of monographic work and edited books on higher marine fungi, viz. Johnson and Sparrow (1961), Kohlmeyer and Kohlmeyer (1979), Jones (1976), Moss (1986), Hyde and pointing (2000), Hyde (2002), Raveendran and Manimohan (2007), Jones et al. (2009a), Alias and Jones (2009) and Pang et al. (2010a). A number of newer generic names have been introduced and many of the classical species were moved here and there as new combinations to one or other existing or new genera made, on the basis of ultrastructural studies and molecular phylogenetic analysis. The study of marine fungi continued to attract mycologists, ecologists, physiologists and those interested in drug discovery (Jones, 2008; Jones et al., 2008a)

Excellent keys aimed at identification of higher marine fungi have been published by Kohlmeyer and Kohlmeyer (1991), Hyde and Sarma (2000), Sarma and Hyde (2000a) and Jones et al. (2009a). Key to the species by Kohlmeyer and Kohlmeyer (1979) included 209 species (149 ascomycetes, 4 basidiomycetes, and 56 anamorphic fungi) of filamentous higher marine fungi. The taxonomic key by Kohlmeyer and Kohlmeyer (1991) included 321 species (255 ascomycetes, 6 basidiomycetes, and 60 anamorphic fungi) of filamentous higher marine fungi. The key by Hyde and Sarma (2000) accommodated 444 species (360 ascomycetes, 10 basidiomycetes, and 74 anamorphic fungi) of filamentous higher marine fungi. The treatment by Jones et al. (2009a) embodied 530 species (424 ascomycetes, 12 basidiomycetes, and 94 anamorphic fungi) of filamentous higher marine fungi. Kohlmeyer and Kohlmeyer (1979) and Jones et al. (2009a) also presented classifications of higher marine fungi.

Suetrong et al. (2009) published a major review of marine Dothideomycetes in a four-gene study of 50 taxa. In the last decade, there has been a steady decline in the description of newer taxa. The decadal figure of marine fungi for 1990-1999 was 156, with only 43 in 2000-2009 (Jones et al., 2009a). Work in the 1990s included description of newer taxa from mangrove substrata and the salt marsh sedge *Juncus roemerianus* (Kohlmeyer and Volkmann-Kohlmeyer, 2001a-d; Kohlmeyer et al., 2005). Since recently, mycologists have been focusing studies on the phylogeny of marine fungi and introduced many new orders and families (Kohlmeyer et al., 2000; Campbell et al., 2009; Thongkantha et al., 2009; Suetrong et al., 2009, 2011; Zhang et al., 2009a, b).

Biodiversity is defined as "the variety of organisms at all levels, from genetic variants of a single species through arrays of species to genera, families and still higher taxonomic levels; it includes the variety of ecosystems which comprise both the communities of organisms within the particular habitats and physical conditions under which they live" (Lincoln et al., 1998). The biodiversity is nothing but variability among the living organisms including terrestrial, marine and all aquatic ecosystems and the ecological complexes of which they are part. This includes diversity of forms right from the molecular unit to individual organism and onto the population, community landscape and biosphere levels. It may be the diversity within species, between species and of ecosystems. The biodiversity has assumed more significance since the United Nations Convention on Biological Diversity which was formulated during the Earth Summit held at Rio de Janeiro (Brazil) in 1992. It now serves as the focal point for documenting knowledge on the biological resources of our earth.

The biodiversity may be genetic (genes), species, and ecosystem diversity. The genetic diversity refers to the variation of genes within the species and covers distinct population of the same species or genetic variation within population. The species diversity refers to the species within a region. The earth biodiversity comprises at least 1.7 million of species that have been described and reported. Recent estimates suggested that there are million of species of organisms on earth and of which insects and microorganisms account for major proportion (Stork, 1993).

The ecosystem diversity may be aquatic (fresh and marine waters) and terrestrial (forest, desert and man-made). Of these, most important are the tropical forests (both dry and moist), wetlands, mangroves, freshwater lakes, marine coasts and coral reefs. Amongst these, the tropical rain forest is said to contain maximum diversity.

The wide variety of living organisms on earth today is the outcome of hundreds of millions of years evolutionary processes. In the course of time, human societies have evolved and adapted to the environment of particular regions. Continued uses and alterations of the local biotic resources led to further shaping of the biodiversity. Systematic methods of identifying the biodiversity require two separate but interdependent activities: (i) compiling good data on the distribution and abundance of the species to be conserved and (ii) development of appropriate procedures for using these data to determine priorities (Margules et al., 2002; Williams et al., 2002).

Aquatic Biodiversity of the World

About 3047 taxa of fungi have so far been reported from aquatic habitats around the world (Table 1), the largest taxonomic groups being the meiosporic and mitosporic Ascomycota, followed by the Chrytridiomycata.

Table 1- Fungi from aquatic habitats from around the world

Taxonomic group	Number of species
Chytridiomycota	576
Freahwater meiosporic ascomycota	450
Mangrove meiosporic ascomycota	612
Marine meiosporic and mitosporic ascomycota	465
Ingoldian mitosporic fungi	290
Aeroaquatic mitosporic fungi	90
Miscellaneous mitosporic fungi	405
Oomycota (Saprolegniales)	138
Basidiomycota from freshwater habitats	11
Basidiomycota from brackish and marine habitats	10
Total number of taxa	**3047**

(**Source:** Shearer et al., 2007)

Biodiversity of India

India is very rich in its biological diversity, with huge endemic flora and fauna. This richness is due to the impact of varied climatic and latitudinal conditions coupled by a variety of habitats which range from humid evergreen forests of the Western Ghats to the hot desert of Rajasthan, from the cold desert of Ladakh to the icy mountain of Himalayas and the warm marine coasts of peninsular India (Agrawal, 1998).

India is a tropical country and recognized as one of the 12 mega-biodiversity nations in the world. Today, of the 1.7 million species of identified organisms 0.2 million (13%) are known from India alone. This is truly remarkable since India covers only 2% of the earth's land area. As with rest of earth's surface, the biodiversity is largely clumped within the warm and humid regions of the Indian subcontinent. Accordingly, the wet-evergreen forests of Western Ghats and Eastern Himalayas are said to be the richest repositories of biological species (Daniels, 1993).

Biodiversity of Fungi of India

Of the nearly 100000 species of fungi known from around the world (Kirk et al., 2008), about 27500 have been reported from India (Bhat, 2010). Over 100 species of coral fungi (Calvariales), about 165 species of cup fungi (Pezizales), about 350 species of slime moulds (Myxomycetes), about 52 species of *Phellinus*, 18 species of *Inonotus,* over 700 species of *Cercospora* and about 330 species of *Puccinia* - all add to the richness of Indian fungal flora (Sharma et al., 1997).

About 3500 species of fungi (about 12%) are endemic to India. Mitosporic fungi constitute about 40% of the total Indian mycoflora. The fungi of India are the largest biotic community next to insects (Sarbhoy et al., 1996). According to the 9th edition of "*Dictionary of the fungi*", the true fungi belong to super kingdom Eukaryota which has four phyla, 103 orders, 484 families and 4979 genera (Kirk et al., 2001). The Deuteromycotina is not accepted as a formal taxonomic category. Following Hawksworth et al. (1995), the number of genera of fungi reported from world and India between 1905 and 1995 are as shown in Table 2. Approximately 27,500 species are reported to colonize the diverse habits and habitats in India. According to Manoharachary et al. (2005), about 205 genera have been newly described from India.

Table-2: Fungal genera reported from world over and India

Phyla	World	India
Myxomycotina	450	380
Mastigomycotina	308	205
Zygomycotina	55	50
Ascomycotina	2000	745
Basidiomycotina	357	232
Deuteromycotin	4100	468
Total	**7270**	**2080**

(**Source:** Manoharachary et al., 2005)

Definition of Marine Fungi

Marine fungi are not a taxonomic and physiologically defined group (Hyde et al., 2000). Several workers have tried to define the marine fungi. Johnson and Sparrow (1961) and Tubaki (1969) defined marine fungi based on their ability to grow at certain seawater concentrations. Meyers (1968) and Jones and Jennings (1964) have tried to determine the physiological requirements, especially of sodium chloride for the growth of fungi in seawater.

According to Kohlmeyer and Kohlmeyer (1979), "obligate marine fungi" are those that grow and sporulate exclusively in marine or estuarine habitat; "facultative marine fungi" are those from freshwater or terrestrial milieu able to grow (and possibly also sporulate) in the marine environment. This is the most widely accepted definition of marine fungi (Hyde et al., 2000). They further suggested that a valid criterion for defining a marine fungus could be its ability to germinate and to form mycelium under natural marine conditions.

As pointed out by Raghukumar (1996b) "the marine system is as complex as its terrestrial counterpart and it incorporates the deep-sea coral reefs, mangroves and intertidal beaches". In addition, estuarine habitats such as river-mouths, lagoons and backwaters containing brackish water also frequently support the growth of these fungi. Some of these habitats display a fascinating gradient of saline conditions where obligate marine fungi, facultative marine fungi and purely terrestrial once co exist. Aerial parts of mangroves support the growth of terrestrial fungi while the inundated portion of the same supports growth of obligate marine fungi (Hyde et al., 2000).

Biodiversity of Marine Fungi:

Nearly 100000 fungal species are known from all over the world but, in contrast, less than 1500 species have been described from oceans and estuaries (Jones and Mitchell, 1996; Jones et al., 2009a). The higher filamentous marine fungi include 530 species in 321 genera which included 424 Ascomycota, (251 genera), 94 Mitosporic fungi (61 genera) and 12 Basidiomycota (9 genera). Most species were described between 1980 and 1999 and only 54 species added during 2000-2011. Current figure of higher filamentous marine fungi is 549 with 16 new genera and 19 new species described since 2009 (Jones, 2011a) and the lower marine fungi comprise probably less than 1000 species (Jones et al., 2009a).

Marine mycology was established as a discrete science in the late 1930's following the works of Sparrow (1937) and Barghoorn and Linder (1944). The first obligate marine fungus *Sphaeria posidoniae* was described on *Posidonia*. Sparrow (1937) reported occurrence of a novel group of zoosporic fungi, the thraustochytrids on marine algae. Barghoorn and Linder (1944) described 25 species from submerged wood in New England and California. These two pioneering works gave the impetus to the study of marine fungi of all over the world. About 100 new species were described by 1970, mainly by Hohnk (Germany), Johnson (United States), Kohlmeyer (Germany) and Meyers (United States). Johnson and Sparrow (1961) published a treatise on "Fungi in oceans and estuaries".

The monograph *"Marine Mycology: the higher fungi"* by Kohlmeyer and Kohlmeyer (1979) lists 209 filamentous species. Twelve years later the number increased to 321 (Kohlmeyer and Volkmann-Kohlmeyer, 1991). Another 10 years, the total number of marine fungi known in literature was 444 species (Hyde et al., 2000). By 2011, as shown in Table 3, 549 filamentous marine fungi have been described (Jones et al., 2009a; Jones, 2011a).

Table-3: Number of higher Marine Fungi described at various times

Group	Kohlmeyer and Kohlmeyer, 1979		Kohlmeyer and Volkmann-Kohlmeyer, 1991		Hyde et al., 2000		Jones et al., 2009a	
	Genera	Species	Genera	Species	Genera	Species	Genera	Sp.
Ascomycetes	62	149	115	255	177	360	251	424
Basidiomycetes	04	04	05	06	07	10	09	12
Hyphomycetes	25	34	25	39	28	46	61	94
Coelomycetes	15	22	16	21	23	28		
Total	**106**	**209**	**161**	**321**	**235**	**444**	**321**	**530**

Habitats for Marine Fungi:

A wide variety of substrates are available in marine and estuarine waters for colonization by fungi. These include seaweeds, decaying leaves, rhizomes and roots of marine angiosperms; driftwood, leaf petioles, prop roots and fruits of mangroves; dead animals and their faecal pellets; shells of various mollusks, sewage effluent and man made structures such as polythene and plastics (Jones, 1988). Some marine fungi are hydrocarbonoclastic (Kirk and Gordon, 1988). Most marine fungi have been recorded from submerged, intertidal or driftwood, each forming a niche that supported a characteristic mycota (Jones, 1988). For example, the so called arenicolous fungi colonize wood buried in the sand with their ascocarps often developing on sand grains (Jones, 1988).

Biodiversity of Mangrove Marine Fungi:

Fungi have been reported from mangrove forests since 1920's (Stevens, 1920). Extensive fungal surveys have been carried out in the mangroves of Caribbean and Pacific Islands (Kohlmeyer, 1984; Kohlmeyer and Volkmann-Kohlmeyer, 1993a, b) and in South East Asia (Sarma and Hyde, 2001).

Hawksworth (1991) made a conservative estimate of world's fungal resources, based on plant/fungus ratio, as about 1.5 million species. However, there are now many postulations, from 0.5 million to 9.9 million species (Cannon, 1997). Mangroves are considered as a major niche of fungi. The fungal diversity is dependent on the age and diversity of the mangrove plants and the physico-chemical features of mangrove habitat, i.e. temperature, salinity and tidal range (Jones, 2000; Hyde and Jones, 1988). Twenty eight mangrove tree species yielded 120 higher marine fungi (Hyde and Jones, 1988). Fifty five mangroves and their associates yielded about 200 higher marine fungi (Jones and Alias, 1997). Among the mangrove tree species, *Rhizophora apiculata* alone harbored 63 higher marine fungi (Sarma and Vittal, 2000).

Ecological studies have focused on the succession of fungi on exposed mangrove substrate test blocks (Alias and Jones, 2000a) and their vertical distribution in mangroves (Hyde, 1988a 1990a; Alias and Jones, 2000b). Numerous studies on the occurrence and biodiversity of fungi on natural occurring substrata, especially mangrove wood (Vittal and Sarma, 2006; Raveendran and Manimohan, 2007; Sridhar, 2009a). Schmit and Shearer (2003, 2005) reviewed published results on the occurrence of mangrove fungi and showed that fewer taxa were available from the Atlantic ocean than from the Pacific Ocean. More recently, Alias and Jones (2009) published a monograph on the *'Mangrove fungi of Malaysia'* and Pang et al. (2010a) published another monograph on *'Marine mangrove fungi of Taiwan'*. These authors recognised a core group of mangrove fungi.

Tomilson (1986) listed 116 woody mangrove plant species. However, fungi have been reported from only 72 of them (Schmit and Shearer, 2003). Schmit and Shearer, (2003) listed 625 mangrove associated fungi which included 278 ascomycetes, 277 mitosporic fungi, 30 basidiomycetes and 14 oomycetes. They listed for each fungus, their known geographic distribution in mangrove forest, host association, as well as known habitat and type substrate. They further stated that Ascomycetes and Mitosporic mangrove associated fungi differed in their substrate affinity and habitats. Terrestrial and sediment fungi have more restricted geographical distribution compared to intertidal fungi.

Justifications:

In India, several taxonomic and ecological studies of marine fungi have been undertaken from various coasts and most of them were aimed at compilation of information on biodiversity, taxonomy and ecology. Running through several provinces along the west and east coasts, India has a long coastline (7516.6 km) which embraces both tropical and subtropical climates (Map 1&2). Besides sandy, rocky and muddy beaches and small to big islands, these coasts exhibits varied habits and habitats, including mangroves and river mouths. In comparison to the voluminous work carried out on terrestrial and freshwater fungi by a galaxy of eminent mycologists (Subramanian, 1971, 1983; Thind, 1977; Lakhanpal and Mukerji, 1981; Bilgrami et al., 1979, 1981, 1991; Bilgrami, 1995; Sarbhoy et al., 1986, 1996; Dayal, 1996, 1997, 2000; Sati et al., 1997 a, b; Khulbe, 2001; Janardhanan et al., 1997; Rao et al., 2003; Rao et al., 2004; Jamaluddin et al., 2004, Rai and Deshmukh, 2005; Dargan et al., 2005; Pande, 2008; Bhat, 2010), hitherto studies on fungi of Indian oceans and seas are very little.

Thus, this monographic work makes an effort to provide a detailed compilation of information on taxonomy, ecology, distribution and substrata range of fungi from Indian marine waters.

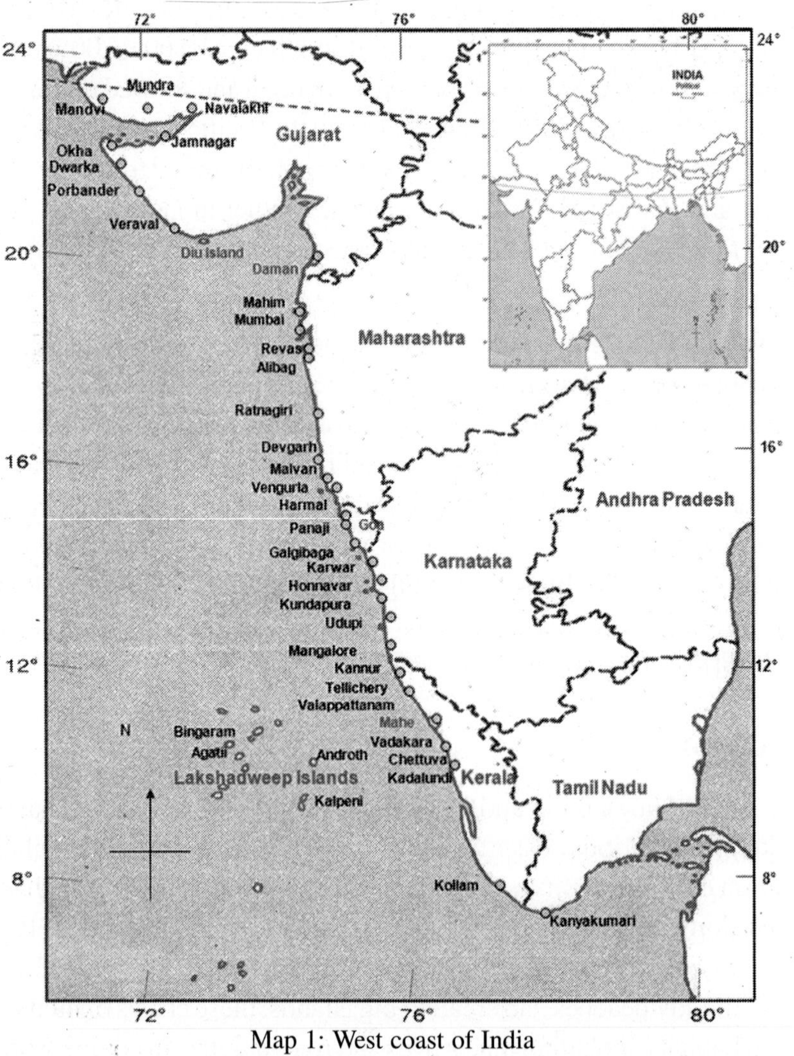

Map 1: West coast of India

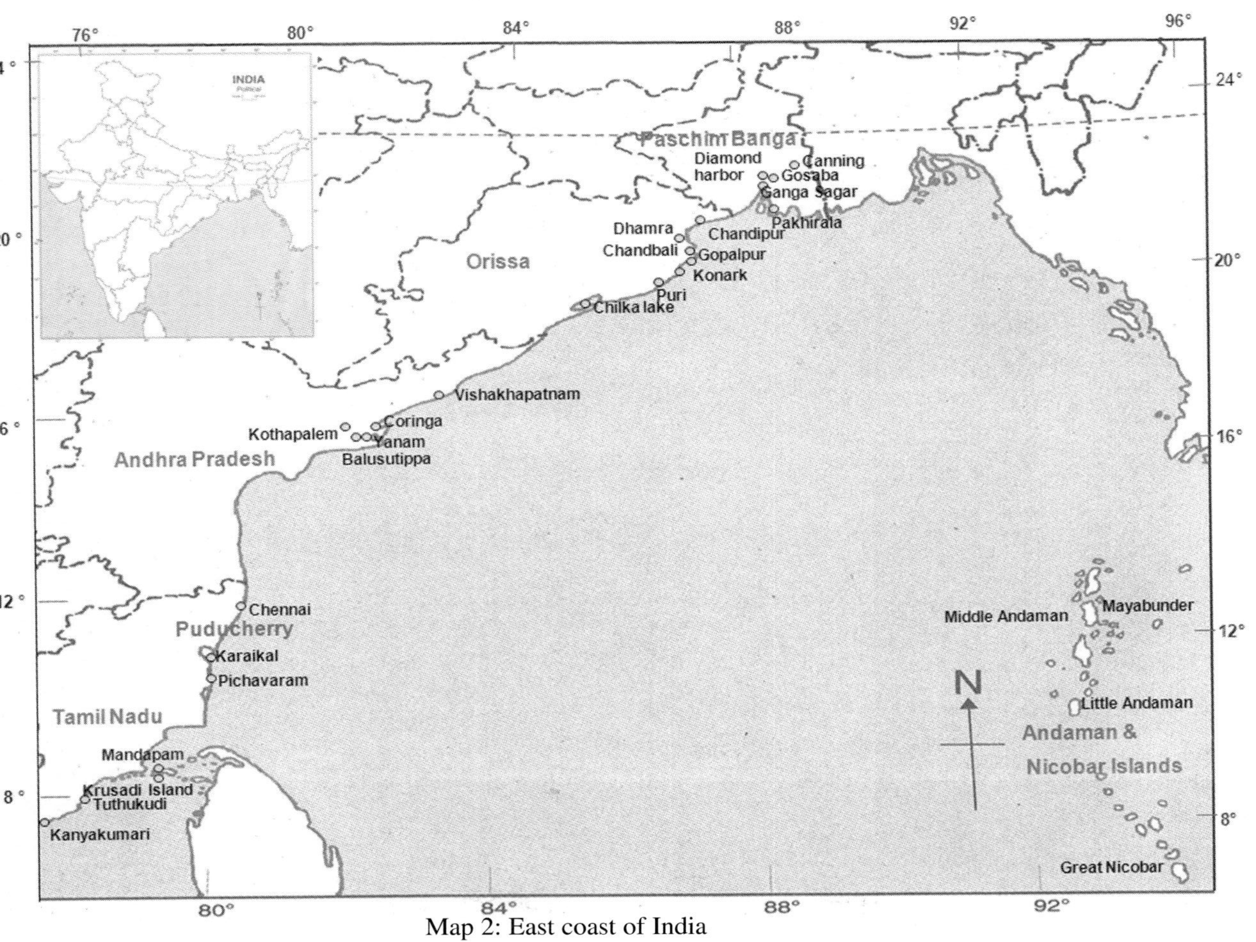

Map 2: East coast of India

Habits and habitats of marine and mangrove fungi

Rhizophora appiculata

Acrostichum aureum at Bhitarkanica (Orissa)

Avicennia alba

Salicornia brachiata

Sonneratia apetala

Suaeda maritime

Avicennia marina

Avicennia officinalis

Excoecaria agallocha

Heritiera fomes

Kandelia candel

REVIEW OF LITERATURE

The study of marine fungi has continuously attracted mycologists, ecologists, physiologists and those interested in drug discovery (Jones, 2008; Jones et al., 2008a). Work on marine fungi in the 1990s included largely description of new taxa from the mangrove substrata and salt marsh sedge *Juncus roemerianus* Scheele (Kohlmeyer and Volkmann-Kohlmeyer, 2001a-d; Kohlmeyer et al., 2005). Since 2000, a number of studies has been devoted to the taxonomy and phylogeny of marine fungi. In the last decade, two edited volumes and four monographs have been published (Hyde and Pointing, 2000; Hyde, 2002; Raveendran and Manimohan 2007; Jones et al., 2009a; Alias and Jones 2009; Pang et al., 2010a).

Suetrong et al. (2009) published a major review of the marine Dothideomycetes in a four-gene study of 51 taxa. In the last decade however there has been a declining rate of description of new taxa, the decadal figure for 1990-1999 being 156 and mere 43 for 2000-2009 (Jones et al., 2009a). Most of the marine fungal species were described between 1980 and 1999 and only 54 species during 2000-2011. Current figure of higher filamentous marine fungi is 549 with 16 new genera and 19 new species described since 2009 (Jones, 2011a). The lower marine fungi comprise less than 1000 species (Jones et al., 2009a).

In the recent years, mycologists have focused on the phylogeny of marine fungi and introduced newer orders, viz. Lulworthales (Kohlmeyer et al., 2000) and Koralionastetales (Campbell et al., 2009), and the new families, Magnoporthaceae (Thongkantha et al., 2009), Aigialaceae and Morosphaeriaceae (Suetrong et al., 2009) and Trematosphaeriaceae (Suetrong et al., 2011).

Fungi are the less known group of marine organisms, although first fungus was reported long ago by Desmazieres (1849) as *Sphaeria scirpicola* var. *typharum* (now *Phaeosphaeria typharum*). Chromistan or Straminipilan organisms and chytrids (lower marine fungi) were studied as early as 1865 with the description of *Chytridium polysiphoniae* by Cohn (1865). Most of these early observations were on fungi parasitic on either microalgae or marine animals, e.g. *Lagendium oophilum* on rotifers (Sparrow, 1937), *Rhizophydium coicola* on *Codium mucronatum* (Zellers, 1918) and so on. The momentum for these early studies arose from the work of Sparrow on Chytrids and Thraustochytriales in the early 1930's and the discovery of diverse marine mycota on driftwoods by Barghoorn and Linder (1944).

A number of years lapsed before further publications appeared on these organisms (Wilson, 1951; Meyers, 1953; Hohnk, 1954; Johnson, 1956; Becker and Kohlmeyer, 1958a, b). Since then, there has been steady progress in the discovery of marine fungi (Johnson and Sparrow, 1961; Jones, 1976; Kohlmeyer and Kohlmeyer, 1979; Hyde and Ponting, 2000; Hyde, 2002; Raveendran and Manimohan, 2007; Jones et al., 2009a; Alias and Jones, 2009; Pang et al., 2010a).

About 100000 species of fungi are known from all over the world (Kirk et al., 2008). In contrast, 549 species of higher marine fungi have been described from oceans and estuaries. These included 424 species in 251 genera of Ascomycota, 12 species in 9 genera of Basidiomycota and 94 species in 61 genera belonging to mitosporic fungi (Jones et al., 2009a). Sixteen new genera and 19 new species were described since 2009 (Jones, 2011a).

Lower Marine Fungi:

Detailed information on the morphology, taxonomy, ultra-structure and culturing methods of marine fungi is available in the following literature: Johnson and Sparrow, 1961; Pokorny, 1967; Olive, 1975; Karling, 1977, 1981; Moss, 1985, 1986; Fuller and Jaworski, 1987; Porter, 1990; Ulken, 1990; Raghukumar, S. 1996a; Bremer, 2000; Dick, 2001. Labyrinthulids and Thraustochytrids became hot topic of research and studies have traced their phylogeny (Yokoyama and Honda, 2007; Yokoyama et al., 2007), resulting with establishment of a number of new genera based on sequence data.

Some of the early studies of marine chytrids were by Cohn (1865) and Sparrow (1934, 1936). Intensive researches during the period 1950-1970 were by Hohnk (1955a, b, 1961), Hohnk and Aleem (1953), Harder and Uebelmesser (1955), Ulken (1967, 1968, 1969, 1974), Chakravarty (1974), Booth (1969, 1971) and Sparrow (1969). Some of these studies dealt with parasitic chytrids, while others perused the saprobic species (Patersen, 1958). Catastrophic collapse of chytrids was well documented (Walsh, 1983; Tillmann et al., 1999; Kagami et al., 2007).

Chromistan organisms have been isolated by sampling water at various depths and most of them belonged to Thraustochytriales and Labyrinthulales (Hohnk, 1955b; Goldstein, 1963; Gaertner, 1972, 1974). Thraustochytrids are rarely seen growing on recently recovered substrata. Pollen baiting technique has been used for their isolation and growth in culture (Gaertner, 1968; Clokie and Dickinson, 1972; Bahnweg and Sparrow, 1974; Clokie, 1974). In the recent years, this group has aroused considerable interest for their ability to produce high yields of polyunsaturated omega-3-fatty acids (Bajpai et al., 1991; Singh et al., 1996; Yaguchi et al., 1997; Bowles et al., 1999; Fan et al., 2000). Thus, more than 38 years after Goldstein (1973) wrote his review on marine zoosporic fungi, one could still repeat his statement that 'these organisms are swimming in an ocean of benign neglect'.

Marine Yeasts:

Yeasts are abundant in the coastal waters. They form extensive communities in open seawaters and said to be more numerous than filamentous fungi (Lachance and Starmer, 1998). Kohlmeyer and Kohlmeyer (1979) listed some 140 facultative marine yeasts. Marine yeasts were intensively studied by Fell (1967, 1974, 1976), Fell et al. (1960), Meyers et al. (1967), Ahearn et al. (1968), van Uden and Casttelo-Branco (1963), van Uden and Fell (1968), van Uden and Zobell (1962), Ahearn and Crow (1986). Recent studies of marine yeasts have been by Kutty and Philip (2008), Chen et al. (2009) and Fell and his co-workers (Fell et al., 2004, 2010).

Currently there are about 1500 marine yeasts, particularly in mangrove habitats (Fell et al., 2010). Statzell-Tallman et al. (2008) reported 55 and 58 species respectively of ascomycetous and basidiomycetous yeasts from three mangrove habitats and 50% of which were undescribed. Fell et al. (2010) and Statzell-Tallman et al. (2010) described marine yeasts from Everglades, coral reefs in the Florida Keys and mangrove regions in Belize and Bahamas. In the

Everglades' study, 74 previously described species were documented with an equal number of new taxa (Fell et al., 2010).

Marine Ascomycota:

From marine habitats, majority of the fungi recorded so far were Ascomycetous forms with new genera and species being constantly added to the list. Jones et al. (2009a) recognized 251 genera of Ascomycota from marine habitats, with a total of 424 species. The Halosphaeriales is the largest order of marine fungi with 126 species in 53 genera and of which 35 are monotypic (Sakayaroj et al., 2010a). Using data from molecular studies, a number of new families have been added to the list of marine taxa. Within the Dothideomycetes, eight new families have been proposed: Lautosporaceae (Kohlmeyer et al., 1995); Aliquandostipitaceae (Inderbitzin et al., 2001) (Jahnulales); Hypostromataceae (Mugambi and Huhndorf, 2009); Aigialaceae, Morosphaeriaceae (Suetrong et al., 2009), Amniculicolaceae, Lentitheciaceae (Zhang et al., 2009a,b) and Trematosphaeriaceae (Suetrong et al., 2011) (Pleosporales).

The recent update of novel Ascomycota described from marine habitats was that of Jones et al. (2009a). Besides, Al-Saadoon (2006) described a new species of *Corollospora*, *C. mesopotamica*. Suetrong et al. (2009) described the new genera, *Halomassarina*, *Morosphaeria* and *Rimora* in Pleosporales. Zhang et al. (2009b) described *Neomassariosphaeria* (a freshwater entity) as a new genus of Amniculiolaceae. Sakayaroj et al. (2010a) described the new genera, *Kochiella*, *Toriella* and *Tubakiella* in the Halosphaeriaceae. Abdel-Wahab and Nagahama (2011a) described the new species, *Halosarpheia japonica,* from Japan. Abdel-Wahab and Nagahama (2011b) described the new genus *Gesasha* (Halosphaeriaceae), with three new species from Gasashi mangroves in Japan. Pang et al. (2010b) described the monotypic new genus *Sedesimiella taiwanensis* from Taiwan. Boonmee and Hyde (2011) described the new genus *Halokirschsteinionthelia* in the Mytilinidiaceae.

Marine Basidiomycota:

Filamentous Basidiomycota as a group are rare in the marine habitats, with only 12 species belonging to 9 genera known (Jones et al., 2009a). Basidiomata and basidiospores of the marine basidiomycetes are quite distinct and fungi can easily be separated using their characteristics. *Halocyphina* and *Calathella* are tropical mangrove genera, superficially similar with white to cream disc or cup-shaped basidiomata. They are visible even to naked eye on wood. The spores of both genera are hyaline, one-celled and subglobose (Jones and Agerer, 1992).

Nia is a saprobe, with an enclosed basidiocarp. The basidiospores are distinct being ovoid, one-celled, with five slender appendages. *Digitatispora* is also a saprobe on wood; it is a temperate genus with an exposed hymenium. Basidiospores are distinct with a cylindrical basal and three radiating arms. *Flamingomyces* (Urocystaceae) is parasitic on *Ruppia maritime* (Ruppiaceae), producing dark sori containing globose or ovoid, dark spores on stem and leaves of the host (Jones et al., 2009a). *Parvulago marina* (Ustilaginaceae) is sporulating at the base of the culms of *Eleocharis parvula* (Cyperaceae). *Mycaureola dilseae* is the only basidiomycete known to occur on the red algae, *Dilsea carnosa,* and its basidiospores are filiform (Porter and Farnham, 1986). *Physalacria maipoensis* produces a capitate basidocarp with a globose head and a short stalk (Inderbitzin and Desjardin, 1999). *Haloalerodiscus mangrovei* was described from decaying and decorticated branch of a living *Sonneratia alba* tree (Maekawa et al., 2005).

Marine Mitosporic (Anamorphic) Fungi:

Since the description of the first conidial fungus, *Camarosporium roumeguerii* Sacc. on salt marsh plant *Salicornia herbacea* L. by Saccardo (1880), more than 50 species of marine conidial fungi have been recorded (Kohlmeyer and Kohlmeyer, 1979). These fungi also occur as parasites, saprophytes and symbionts on a wide variety of substrates in oceans and estuaries.

Hosts or substrates of marine anamorphic fungi include marine algae, submerged parts of tropical mangrove trees and temperate salt marsh halophytes, dead woods, other cellulosic substances and animal chitin and conchyolin. Conidial fungi constitute approximately one forth of all described marine filamentous fungi.

The manner of sampling is most important for deciding whether a fungus found in a marine milieu is active in a particular habitat or whether it occurs there only in a dormant or inactive state. According to Kohlmeyer and Kohlmeyer (1979), pour plate techniques of marine sediments or water performed in the laboratory generally resulted in the isolation of terrestrial fungi derived from dormant propagules. The latter are prevented from germinating by a mycostatic principle that occurs under natural conditions in seawater.

Jones et al. (2009a) recognized 94 mitosporic fungal species. Majority of them belonged to well-known terrestrial genera such as *Camarosporium,*

Trichocladium and *Zalerion*. Genera such as *Asteromyces, Cirrenalia, Orbiomyces* and *Varicasporina* appeared to have only marine species. More recently, Abdel-Wahab et al. (2010) on the basis of phylogenetic evaluation of anamorphic species of *Cirrenalia* and *Cumulospora* described seven new marine helicoid genera, viz. *Glomerulispora, Halazoon, Hydea, Matsusporium, Moheitospora, Moleospora* and *Moromyces,* with four new marine species.

Molecular systematics:

The advent of molecular systematics has enhanced our understanding of the origin and evolution of marine fungi. The first paper on molecular studies of marine fungi was by Spatafora and Blackwell (1994) where they included a sequence data of *Halospheriopsis medisetigera* belonging to ophiostomid ascomycetes. Subsequently Spatafora et al. (1998) established the independent terrestrial origin of the Halosphaeriales, while the genera *Lindra* and *Lulworthia* formed a separate clade of marine perithecial ascomycetes. Thereafter followed extensive researches on the phylogeny and taxonomy of marine fungi, which proceeded along three separate lines:

1. Resolution in the delineation of species and genera (Pang, 2002; Sakayaroj et al., 2005a, b, 2010a; Campbell et al., 2003, 2005; Schoch et al., 2006a, b; Jones et al., 2006, 2008b, 2009a, b; Abdel-Wahab et al., 2010; Maekawa et al., 2005; Koch et al., 2007; Pang et al., 2010b; Abdel-Wahab and Nagahama, 1011a, b; Zhang et al., 2009a, b; Statzell-Tallman et al., 2008, 2010; Fell et al., 2010),
2. Higher order taxonomic placement of marine fungi (Spatafora et al., 1998; Kohlmeyer et al., 2000; Sakayaroj et al., 2010a; Campbell et al., 2009; Suetrong et al., 2009, 2011; Zhang et al., 2009a, b; Inderbitzin et al., 2001; Huhndorf, 1994; Mugambi and Huhndorf, 2009; Kohlmeyer et al., 1995; Pang et al., 2002; Schoch et al., 2006a) and
3. Application of ecological studies, especially in the identification of culturable sterile endophytes and unculturable isolates (Zuccaro et al., 2003, 2004; Sakayaroj et al., 2010b; Pivikin, 2000; Pivikin et al., 1999; Jones, 2011a).

Marine lignicolous fungi:

Biodivrsity studies have been the main domain of researches on fungi over the past 50 years. Initially, the focus was on temperate-water fungi, especially wood-inhabiting taxa (Jones, 1968a; Schaumann, 1968; 1969, 1975; Grasso et al, 1985, 1990; Cuomo et al., 1985, 1988; Shearer and Burgos, 1987) with additional studies in more recent years (Petersen and Koch, 1997; Panebianco et al., 2002; Abdel-Wahab, 2011a, b). The more intensive collections were made by Koch and Petersen (1996), Linttot and Linttot, (2002), Jones et al. (1998), Jones (2010) and Abdel-Wahab et al. (2008) who reported 75 from Denmark, 38 fromNew Zealand, 92 from San Juan Islands and 41 taxa from Italy, respectively. Jones (2011a) listed 25 most common temperate marine fungi from various studies. Majority of these are members of Halosphaeriales, while very few bitunicate ascomycetes found on submerged wood (Jones, 1985).

Marine fungi associated with mangroves:

Study on marine fungi inhabiting the mangroves was initiated 50 years ago in Australia (Cribb and Cribb, 1955, 1956). In the following 25 years, majority of the data collected were from the temperate and tropical regions of Atlantic Ocean, with few reports from the tropical Indian and pacific oceans (Kohlmeyer and Kohlmeyer, 1979; Kohlmeyer, 1984). Extensive ecological and taxonomic studies have been conducted since 1980s in the Pacific and Indian oceans. These studies focused on succession of fungi on exposed mangrove wood test-blocks (Alias and Jones, 2000a) and their vertical distribution in the mangroves (Hyde, 1988a, 1990a; Alias and Jones, 2000b). Several studies dealt on the occurrence and distribution of fungi on natural substrata, especially the mangrove woods (Vittal and Sarma, 2006; Raveendran and Manimohan, 2007; Alias and Jones, 2009; Sridhar, 2009a; Nambiar and Raveendran, 2009i; Pang et al., 2010b; Basitulo et al., 2010). All these together contributed to the increase in the recorded number of mangrove fungal species.

Numerous studies of tropical marine fungi were undertaken over the past 25 years, largly those occurring on mangrove substrata. Kohlmeyer and Kohlmeyer (1979) listed 42 mangrove fungi. Hyde and Jones (1988) recognized that mangrove fungi constituted the second largest ecological group of marine fungi which are mostly distributed in the Old and New world mangroves. Jones and Alias (1997) estimated that there were 269 species of higher marine fungi in the mangroves. Schmit and Shearer (2003) listed 625 taxa but this figure included those growing on terrestrial parts of mangrove trees also. Alias and Jones (2009) identified a core group of mangrove fungi which included *Hydea pygmea*, *Dactylospora haliotrepha*, *Halorosellinia ocenica*, *Lulworthia grandispora* and *Verruculina enalia*. Alias et al. (2010) listed 287 species on submerged mangrove substrata. Schmit and Shearer (2003, 2005) reviewed published results on the occurrence of mangrove fungi and showed that fewer taxa were known from the Atlantic Ocean than the Pacific Ocean. Illustrated monographs of mangrove fungi have been published for Kerala (India) (80 species: Raveendran and Manimohan, 2007), Malaysia (140 species: Alias and Jones, 2009) and Taiwan (69 species: Pang et al., 2010a).

The Ascomycota is the most numerous and common taxonomic group in mangroves, with the Basidiomycota least frequently collected (Alias and Jones, 2009; Pang et al., 2010a). Biodiversity studies have showcased a wide variation in the frequency and abundance of mangrove fungi: 154 from Thailand (Sakayaroj et al., 2004), 139 from Malaysia (Alias et al., 2010), 131 from east-coast of India (Vittal and Sarma, 2006), 128 from Hong Kong (Jones and Vrijmoed, 2003) and 91 fron the west-coast of India (Maria and Sridhar, 2003a). Alias and Jones (2009) and Jones (2011a) identified a core group of mangrove fungi (14 species) which included *Dactylospora haliotrepha*, *Halocyphina villosa*, *Halorosellinia ocenica*, *Kallichroma tethys*, *Leptosphaeria australiensis*, *Lulwirthia grandispora*, *Verruculina enalia*, *Hydea pygmea*, *Lignincola laevis*, *Phoma* sp. *Halosarpheia marina*, *Dictyosporium pelagicum*, *Matsusporium tropicalis* and *Trichocladium alopallonella*.

Fungi of tidal salt marshes and other substrata:

Kohlmeyer and Kohlmeyer (1979) listed 33 species of teleomorphic Ascomycota, 1 Basidiomycota and 23 species of anamorphic Ascomycota from tidal salt marshes. Many of them were found to be host specific or restricted to host plants belonging to one family. The most intensively surveyed salt marsh plant is *Spartina* (Gessner and Kohlmeyer, 1976; Barata, 2002). In a review of fungi growing on *Spartina* spp., Barata (2002) listed 123 species from temperate and tropical salt marshes. The most diverse taxonomic group was the bitunicate ascomycetes with the genera *Leptosphaeria* (13 sp.), *Phaeosphaeria* (4 sp.), and *Pleospora* (6 sp.), as well-represented. In her study of the marine fungi on *Spartina maritima* in Portugal, the most frequently collected species were: *Natantispora retorquens*, *Phialophorophoma litoralis*, *Sphaerulina oraemaris*, *Phoma* sp., *Dictyosporium pelagicum* and *Byssothecium obiones*.

Kohlmeyer et al. (1995, 1997, 2005) and Kohlmeyer and Volkmann-Kohlmeyer (2000, 2001a-c) have undertaken detailed studies of the fungi colonizing the salt marsh sedge, *Juncus roemerianus*. Their studies indicated that 107 species (44 obligate, 25 facultative and 38 halotolerent/terristrial) are growing on the salt marsh plant, with 48 new species. These taxa belonged to 7 orders, 20 families and 44 genera. Sridhar et al. (2010) have undertaken a study of fungi colonizing the mangrove sedge *Cyperus malaccernsis*, in Netravathi river delta, Karnataka (India). Their studies confirmed that initially terrestrial fungi are dominant, followed by typical mangrove and marine fungi (*Acrocordiopsis patilii*, *Cumulospora* sp., *Okeanomyces cucullata*, *Leptosphaeria australiensis*, *Lignincola laevis*, *Lulworthia* sp., and *Periconia prolifica*).

Sadaba et al. (1995) surveyed *Acanthus ilicifolius* for fungi. Forty four species were found on the decaying standing parts of *A. ilicifolius* in the intertidal zone at Mai Po Marshes, Hong Kong. Poon and Hyde (1998a, b) recorded 61 (17 ascomycetes, 44 mitosporic fungi) species on the grass, *Phragmites australis*. Wong et al. (1998) reported 31 (6 ascomycetes, 25 mitosporic fungi) species on the sedge, *Schoenoplectus litoralis,* in Mai Po Marshes, Hong Kong. When new substrata were surveyed for marine fungi, invariably new genera and species encountered. This was well illustrated by studies of the brackish water palm *Nypa fruiticans* (Hyde et al., 1999). Pilantanapak et al. (2005) collected 81 species of marine fungi on *N. fruticans* in Thailand. Hyde and Sarma (2006) documented 46 species on *N. fruiticans* from the Tutong River, Brunei. Many of the fungi, *Aniptodera nypae*, *Helicascus nypae*, *Linocarpon nypae*, *Tirisporella beccariana* and *Vibrissea nypicola*, known only from *Nypa* are host specific or host recurrent.

Marine fungi on Seagrasses:

Early studies on seagrasses listed the basidiomycetes, *Falmingomyces ruppiae* on *Ruppia maritima* and *Halotthia posidoniae* and *Ponteporeia biturbina* on *Posidonia oceanica* (Kohlmeyer and Kohlmeyer, 1979). The seagrasses *Enhalus acoroides*, *Halodule bermudensis*, *Halophila ovalis*, *Syringodium filiforme*, *Thalassia testudinum*, *Zostera japonica*, *Z. marina*, and *Z. muellleri* (mostly tropical) and the temperate-water species *Cymodocea nodosa* and *Posidonia oceanica*, have been surveyed for saprobic and endophytic fungi (Newell and Fell, 1980; Wilson, 1998; Cuomo et al., 1982, 1985; Alva et al., 2002; Devarajan et al., 2002; Sakayaroj et al., 2010b). Jones (2011a) listed 11 marine saprobic species on seagrasses.

Fungi associated with Algae:

Representatives of all fungal groups have been found associated with marine algae. They included parasites, saprophytes and edophytes of seaweeds and planktonic taxa, and most of them are members of Ascomycota. Kohlmeyer and Volkmann-Kohlmeyer (1991) listed 45 species of Ascomycota, 1 of Basidiomycota and 6 of mitosporic fungi on marine algae. Among the Ascomycota, 31 species have been collected from living algae (Kohlmeyer and Kohlmeyer, 1979). Kohlmeyer and Volkmann-Kohlmeyer (2003) provided illustrations and full descriptions of 8 marine ascomycetes, which included 6 from algae and 2 from animal hosts. The algicolous species belonged to the genera *Didymella, Haloguignardia, Spathulospora* and *Turgidosculum*; those from animals were members of *Abyssomyces* and *Laboulbenia*.

Major compilations of marine fungi (Johnson and Sparrow, 1961; Kohlmeyer and Kohlmeyer, 1964-1969, 1979; Kohlmeyer and Volkmann-Kohlmeyer, 1991; Hyde and Ponting, 2000; Jones et al., 2009a) show that researches during the past six decades have concentrated mostly on lignicolous and rarely on algicolous fungi. Among the 465 described filamentous higher marine fungi, 79 are associated with algae as parasites or symbionts, and 18 with animal hosts (Kohlmeyer and Kohlmeyer, 1979, Kohlmeyer and Volkmann-Kohlmeyer, 1991).

The first group of marine fungi was those collected on algae (Sutherland, 1915, 1916a, b). Bugni and Ireland (2004) commented that fungi isolated or growing on algae were the second largest group of marine fungi. Algal genera that have shown to support fungi include, *Ascophylum* (Sutherland, 1915; Webber, 1967), *Ballia* (Kohlmeyer, 1967), *Chondrus* (Schatz, 1980a, b), *Dilsea* (Stanley, 1992), *Fucus* (Zaccaro et al., 2008), *Laminaria* (Kohlmeyer, 1967) and *Sargassum* (Kohlmeyer, 1972a, b). Jones (2011b) drew attention to the large number, i.e. 9200 to 12500 algal species that are yet to be studied for occurrence of the fungi.

Recently, there has been a resurgent of interest in fungi growing on marine algae. Loque et al. (2009) studied filamentous fungi (27 spp.) and yeasts (48 spp.) associated with marine algae *Adenocystis utricularis*, *Desmarestia anceps* and *Palmaria decipiens* from Antarctica. Chytrids and the Chromistan (Straminipiles) Oomycetes parasitic on algae have also attracted interest (Sekimoto et al., 2008a, b; Strittmatter et al., 2009; Gachon et al., 2006, 2009, 2010).

Many of these studies have been concerned with commercially important algae, e. g. *Porphyra* in the production of 'nori', and the resultant losses (Gachon et al., 2010). Filamentous marine alga, e.g. *Pylaiella*, has also been examined for marine Oomycetes (Sekimoto et al., 2008b).

Fungi on calcareous substrate and Coral reef:

Calcium carbonate, produced by marine algae and animals, is deposited in large amount along the sea shores and with the sediments in the ocean. The occurrence of filamentous higher marine fungi in and on submerged carbonaceous substances was not thought of earlier. However, occurrence of filamentous higher fungi on coral bases has been discovered by Kohlmeyer and Volkmann-Kohlmeyer (1992 a, b). They reported 22 species of teleomorphic and anamorphic Ascomycota on calcium carbonate in marine habitats.

Few studies examined shells, calcareous algae, coral and soft corals in any detail to determine the role marine fungi in their biology (Kohlmeyer, 1969a). Le Campion-Alsumard et al. (1995) showed endolithic septate fungal hyphae in coral skeletons and soft coral tissue. Porter and Lingle (1992) found thraustochytrids boring into mollusk shells. Kendrick et al. (1982) found evidence of 'microborings' in aragonite of coral skeletons. Vast quantities of mollusk shells are cast up on shores and undergo decomposition and fungi utilize the conchyolin which makes up part of the shells (Alderman and Jones, 1967).

The mycota of coral reef has been little explored and so far only 9 taxa of Ascomycota have been collected *in situ*. It included 5 species of *Koralionates* (Kohlmeyer and Volkmann-Kohlmeyer, 1987a, 1990a), two species and one verity of *Lulworthia* (Kohlmeyer and Volkmann- Kohlmeyer, 1989a); one species of *Halograhis* (Kohlmeyer and Volkmann- Kohlmeyer, 1988a) and *Xenus lithaphylli* (Kohlmeyer and Volkmann- Kohlmeyer, 1992b).

Arenicolous fungi:

Sandy beaches harbour interesting organisms that are adapted to living in water-filled spaces between grains of sand. The fungi inhabiting in these habitats are known as arenicolous fungi. Kohlmeyer and Kohlmeyer (1979) defined theses fungi as living among or on grains of sand without implying that they obtain any nutrients from the sand itself. These fungi degrade organic matter of plant and animal origin. So far, about 40 arenicolous fungi have been documented, with members of the genera *Arenariomyces*, *Carbosphaerella* and *Corollospora*, and the mitosporic species *Varicosporium ramulosa*, as most commonly found (Kohlmeyer, 1966a; Koch, 1974; Rees et al., 1979; Tokura, 1982; Koch et al., 1983; Farrant et al., 1985; Nakagiri and Tokura, 1987) primarily on temperate water seashores. Accounts of arenicolous fungi from the tropics include those of Steinke and Jones (1993) from South Africa, Sundari et al. (1996a, b) for Malysia and Singapore, and Ananda et al. (1998) for India.

It is now known that spores of these fungi are trapped and accumulated in sea foam on sandy beaches and that the sea foam acts as source of the areniculous fungi on the beach (Kirk, 1983; Kohlmeyer, 1966; Nakagiri, 1989; Nakagiri et al., 1999; Sundari et al., 1996a, b; Tokura, 1982). Gonzales and Hanlin (2010) suggested that arenicolous ascomycetes might be used as indicators of ecosystem disturbances on sandy beaches.

Pathogenic marine fungi:

Marine fungi cause diseases of marine animals and plants and it a relatively unexplored topic of research (Kohlmeyer, 1973, 1979; Kohlmeyer and Demoulin, 1981; Gachon et al., 2010; Jones, 2011a). Crustacean species, fish and algae are the most frequently cited hosts of pathogenic marine fungi. Species of *Fusarium* have been reported to cause diseases on prawns (Khao et al., 2005), tiger prawn (Khao et al., 2004) and infection of the eggs of loggerhead sea turtle (Sarmiento-Ramirez et al., 2010); *Aphyanomyces sinensis* infections of juvenile soft-shelled turtle (Tokuma et al., 2011) and various fungi causing skin infections of southern right whale (Reeb et al., 2011). Many new fungi have been described as pathogens: *Trichomaris invadens* in tanner crab (Sparks, 1982), *Labyrinthuloidea haliotidis* of juvenile abalone (Bower, 1987), *Haliphthoros milfordensis* juvenile stages of lobster (Fisher et al., 1975), *Atkinsiella awabi* from stocked abalone (Kitancharoen et al., 1994), and *Plectosporium oratosquillae* in marine shrimp (Duc et al., 2009).

Marine algae susceptible to fungal infections include: the red algae *Bangia*, *Palmaria*, *Polysiphonia* and *Porphyra* (Pueschel and Vandermee, 1985; Muller et al., 1999; Sekimoto et al., 2008a, b) and the brown algae *Cystoseira*, *Halidrys* and *Pylaiella* (Alongi et al., 1999; Gachon et al., 2006; Harvey and Golf, 2010). The role of marine zoosporic fungi in parasitizing marine plants and animals is known only partly (Gleason et al., 2011). A list of pathogenic marine fungi is provided by Zuccaro et al. (2004).

Deep-sea fungi:

Seawater moulds show variations in distribution as per the depth of water. Although some authors restrict the definition of the deep-sea to waters below 3800 m (Zobell, 1968), Kohlmeyer and Kohlmeyer (1979) drew an arbitrary line at 500 m and included the mycota growing and reproducing below this depth at hydrostatic pressure of 50 ftm and more.

Early records of deep-sea fungi were by Jones and Le Camion-Alsumard (1970a, b) and Kohlmeyer (1977) on wood and polyurethane covered panels. Recently, Dupont et al. (2009) recovered two fungi (*Alisea longicola* and *Oceanitis scuticella*) from a depth of 1000 m in the Pacific Ocean, off the Vanuatu Islands. All these reports were based on the fungi sporulating on retrieved substrata. Only 5 species of filamentous higher marine fungi have been described from the deep-sea, these included 4 ascomycetes viz., *Abyssomyces hydrozoicus* (631-641 m), *Alisea longicola* (1000 m), *Bathyascus vermisporus* (1615-1720 m), *Oceanitis scuticella* (1000-3975 m) and 2 mitosporic fungi, viz. *Allescheriella bathygena* (1720 m) and *Periconia abyssa* (3975-5315 m).

Recent studies indicated utilization of a combination of culturing and molecular techniques to characterize the diversity of fungi in the deep-sea, especially those present in deep-sea sediments (Nagano et al., 2010; Raghukumar et al., 2010). Damare et al. (2006a) and Singh et al. (2010) have recovered 163 cultivable fungi from the Central Indian Basin deep-sea sediments, most filamentous strains belonged to Ascomycota and most yeasts belonged to Basidiomycota. Common genera were *Aspergillus*, *Cladosporium* and *Penicillium*, thus complementing the data for 'marine derived fungi' isolated from various substrata in marine environment.

The fungal diversity reported from deep-sea has been low with few new species recovered (Bass et al., 2007). However, Le Calvez et al. (2009) recovered new species using DNA extracts from hyphothermal vent samples. Lopez-Garcia et al. (2001, 2003) studied diversity of small eukaryotes in hyphotermal sediments in deep-sea. Lai et al. (2007) and Takishita et al. (2007) studied fungal communities from methane hydrate-bearing deep-sea sediments. The study of deep-sea fungi has been hampered by the cost of such researched and rarity of sampling methods and equipments (Raghukumar et al., 2010).

Endophytic marine fungi:

Considerable progress has been made in the documentation of endophytes from marine hosts, including seagrasses (Sakayaroj et al., 2010b), the brown alga *Fucus serratus* L. (Zuccaro et al., 2003, 2008), saltmarsh plants and mangrove trees (Suryanarayanan and Kumaresan, 2000; Okane et al., 2001; Pang et al., 2008b). Fungi isolated have been predominately typical terrestrial, although marine fungi were reported by Ananda and Sridhar (2002) from the roots of *Avicennia, Rhizophora* and *Sonneratia* species. Zuccaro et al. (2008) noted that many marine species of Dothideomycetes and Halosphaeriaceae are common endophytes of *Fucus serratus* L.

Marine and maritime "borderline" Lichens:

Rocky seashores harbour a large number of lichens, often in typical band-like zones. While 450 species of lichens have been reported from midlittoral to superlittoral rocks around the coasts of the British Isles (Fletcher, 1980), the number of permanently or periodically submerged marine lichen is small. Kohlmeyer and Kohlmeyer (1979) listed five species of submarine lichenoid associations and the permanently submerged *Halographis runica* has been found on coral slab and snail shells in Belize and Australia (Kohlmeyer and Volkmann-Kohlmeyer, 1988a, 1992a). These have been referred to as "primitive lichens", i.e. loose symbioses in which the algal or cyanobacterial partners occur also in a free-living state. In these cases, the fungal partner does not develop a well-differentiated cortical layer as expected in true lichen associations and as generally defined (Hawksworth, 1988; Hawksworth and Hop-Negger, 1994), and neither the fungal partner is immersed in unmodified algae or cyanobacteria as in mycophycobioses (Kohlmeyer and Kohlmeyer, 1979; Hawksworth, 1987b).

Often, the morphology of algal or cyanobacterial parner is modified and the associations show a high degree of specilization involving fungi derived largely from ascomycete orders, but without the formation of well-differentiated fungal tissues generally considered characteristic of lichens (except in the ascomata and conidiomata). Therefore, these associations in this book are referred as "borderline" lichens.

Kovacik and Pereira (2001) and Lud et al. (2001) examined the biology and developmental morphology of one of these borderline lichens, *Mastodia tessellata* (formerly listed as *Turgidosculum complicatulum*). Both these groups of workers independently concluded that the association could be interpreted as a lichen, albeit one "— with simple organizational level" (Lud et al., 2001), but did not fully address the nomenclatural implications of their conclusions. A second borderline species, *Collemopsidium pelvetiae* (Syn. *Leiophloea pelvetiae, pyrenocollema pelvetiae*), treated as "primitive lichen" by Kohlmeyer and Kohlmeyer (1979), has also received little attention, both morphologically and nomenclaturally.

Marine-derived fungi:

The terrestrial-like fungi isolated from marine environment have been frequently reffered to as marine-derived fungi which do not show morphological affinities with the so called obligate or facultative marine fungi recovered from substrata such as drift-wood and attached mangroves, intertidal seagrasses and seaweeds. Marine-derived fungi have been isolated from a broad range of substrata: saprobic fungi on marine algae; woods (but not sporulating on the wood); sediments and sand; in the water colums; as parasitic on algae and marine animals; as endophytes (Udea, 1980; Udea and Udagawa, 1983; Mantel et al., 2006; Phongpaichit et al., 2006; Jones et al., 2008a; Duc et al., 2009). Many of these fungi are known from mangroves, marine algae, marine sediments and coastal soils (Swart, 1958, 1963; Chawdhery and Rai, 1980; Pawar et al., 1967; Janson et al., 2005).

Many marine-derived fungi are mitosporic fungi belonging to the genera such as *Apergillus*, *Cladosporium*, *Fusarium*, *Penicillium* and *Phoma*. They have been isolated in high numbers from various sources: 617 from coral reefs (Morrison-Gardiner, 2002), 1000 from marine sediments (Pivikin et al., 1999), 800 from mangrove endophytes (Pang et al., 2008b) and 1743 from diverse marine organisms (Schulz et al., 2008). Many of these strains remained sterile and others could only be identified to genus level. According to Sakayaroj et al. (2010b), sequence data has greatly aided in the identification of these strains, especially those isolated as endophytes and their true origin still remains unclear. Jones (2011a) asked following questions concerning marine-derived fungi: What roles do they play in the ecology of oceans? Are they truly adapted to life in the marine environment? According to Alker et al. (2001) and Zuccaro et al. (2004), preliminary studies indicate that the terrestrial species might have evolved into marine forms and this aspect warrants further study.

Wood decay by marine fungi:

Wood decay researches in the 1960-70's focused on the great losses of timber structures in the marine environment by wood boring animals and the role of fungi in the preconditioning of wood prior to larval settlement. Some of the early studies of physiology and biochemistry of marine fungi dealt on their ability to cause decay of wood (Jones, 1971; Meyers, 1971a, b; Leightly, 1980), in particular the attack on wood cellwalls resulting with soft rot (Mouzouras, 1986; Mouzouras et al., 1988). Jones (1971) examined the ability of marine fungi to cause soft rot attack of wood. Mouzouras (1986) screened 24 species for soft rot attack and cell-wall penetration.

Mouzouras (1989) and Mouzouras et al. (1988) showed that temperature affected the degree of weight loss caused. The cold water basidiomycete, *Digitatispora marina,* caused 14% weight loss of balsa wood at 10°C but only 5% at 22°C., while the opposite was noted for the mangrove basidiomycete *Halocyphina villosa*, as 0% at 10°C. and 23% at 22°C. The results for *Nia vibrissa*, a cosmopolitan marine basidiomycete, were 13°C and 28% loss at 10°C and 22°C, respectively. Butcher et al. (2004) also examined the wood mass loss under exposed and submerged conditions by 48 marine strains and observed that weight losses were much lower under submerged conditions.

Lightly and Eaton (1979) demonstrated white rot decay of wood by *Nia vibrissa*. Mouzouras (1989) studied white rot decay of wood by *Halocyphina villosa*. The mechanism of soft rot attack has been only speculated until Hale and Eaton (1984, 1985a, b) observed cavity formation by two marine fungi, using a combination of continuous photomicrography and scanning and transmission electron microscopy. They were able to observe the oscillatory growth of the fine probosccis hyphae in the wood cellwall leading to the formation of cavities.

Bioactive compounds from marine fungi:

Since the discovery of the first bioactive compound 'Siccayne' from the marine Basidiomycete *Halocyphina villosa* Kohlmeyer (Kupta et al., 1981), there has been an increase in the number of newer chemical structures and bioactive compounds reported from marine fungi. Currently, about 800 compounds have been discovered from marine or marine-derived fungi (Ebel, 2006; Jones, 2008; Jones et al., 2008a; Pang et al., 2008a; Proksch et al., 2008; Schulz et al., 2008).

Marine fungi produce a wide spectrum of secondary metabolites, i.e. lipids: ceramide was obtained from *Lignicola laevis* (Abraham et al., 1994); heterocycles: penicilazine from a marine derived *Penicillium* sp. (Lin et al., 2000); allenolic series: xyloallenolide A, isolated from a mangrove *Xalaria* sp. strain #2508 (Lin et al., 2001a); ketal series: xyloketals 9-16 were isolated from a mangrove *Xylaria* strain #2508 (Lin et al., 2001b); despipeptides: enniatin G was isolated from mangrove fungus *Halosarpheia* sp. (Lin et al., 2002); alkaloids: two new diletopiperazines and roquefortine H, J and I were obtained from a *Penicillium* sp. isolated from a deep ocean sediment (Du et al., 2010); cyclic peptides: azonazine was from the marine derived fungus *Aspergillus insulicola* (Wu et al., 2010); isocumarins: isoculmorin was obtained from the mangrove fungus *Kallichroma tethys* (Alam et al., 1996); terpanes: dendryphielins A-D obtained from *Dendriphyiella salina* (Guerriero et al., 1988); diketopiperazines: an *Aspergillus* sp. isolated from *Mytilus edulis* (mussel) yielded a series of prenylated Diketopiperazines, notoamides A-D and F-K (Kato et al., 2007); lactones: *Helicascus kanaolans*, a mangrove ascomycete, yielded helicascolides A and B (Poch and Gloer, 1989); anthraquinones: many marine fungi have been shown to produce known and new antraquinones: *Halorosellina* sp. (Jiang et al., 2000).

Marine fungi have been shown to have activity against a broad range of microoraganisms in pharmacological conditions. While the terrestrial fungi are more active against gram negative bacteria, marine fungi are in addition active against fungi also (e.g. *Candida albicans*). Biological activity range from: antibacterial, antifungal, broad antimicrobial activity, antimalarial, cytotoxic and platelet activating factor (PAF) antagonist. Other uses of metabolites from marine fungi are discussed by Verbist et al. (2000). Pan et al. (2008) screened more than 100 fungal strains isolated from various substrata collected in the South China Sea, with 25 yielding 40 new compounds, of which 20 were new bioactive compounds. This is an area of marine mycology that will continue to yield fascinating results in the future (Jones, 2011a).

Marine fungi in Bioremediation:

The use of fungi in bioremediation has attracted interest, especially with those taxa able to produce lignin peroxidase, manganese peroxidase and laccase (Pointing et al., 1998; Raghukumar, 2008). Marine fungi are able to degrade polycyclic aromatic hydrocarbons (Raghukumar et al., 2006) and coloured effluents and synthetic dyes (D'Souza et al., 2006).

The ability of marine fungi to produce lignin degrading enzymes has stimulated research into the decolourazation of bleach plant effluent from pulp and paper mills, effluent from textile and dye making industries and molasses spent wash from alcohol distilleries (Raghukumar, 2008). Various researchers have screened marine fungi for their ability to decolourize a range of dyes or produce lignin degrading enzymes (Pointing et al., 1998; Raghukumar et al., 1999; Raghukumar, 2002, 2008).

Raghukumar (2002) screened 11 marine fungi for lignin degrading enzymes and their ability to decolourize dyes, with 70% showing laccase activity, and 82% cellulose activity. Three of these showed manganese peroxidase and one peroxidase activity. *Flavodon flavus,* a white rot basidiomycete, isolated from a decaying seagrass in a coral lagoon of Lakshadeep Islands, India, produced all three major classes of degrading enzymes: manganese-dependent peroxidase, lignin peroxidase and laccase, and turned out to be the most effective in the decolourization of dyes. For example, Cango red, remazol brilliant blue, and 80% of pigments in spent molasses were decolourized.

Raghukumar (2002) has discussed some of the problems related to bioremediation of waterborne pollutants and dyes by marine fungi. Pointing (2001) reviewed the wide range of pollutants in the environment, the ability of white rot fungi to bring about their transformation and minerlization, and set out some of the conditions that have to be met for their use in bioremidation. Sadaba and Sarinas (2010) pointed out that marine fungi may have a greater potential in the bioremidation of oil spills in coastal waters.

Distribution of marine fungi:

Studies on distribution of marine fungi was first attempted by Hughes (1974). A detailed analysis of distribution of marine fungi have been made by Hughes (1986) and Booth and Kenkel (1986). These studies depicted that water temperature is the most important factor which controls the geographic distribution.

Marine fungi have been grouped into three categories: cosmopolitan in distribution, confined to tropics or sub tropics and restricted to temperate or arctic locations (Jones, 1993). *Antennospora quadricornuta, Halosarpheia abonnis* and *Halorosselinia oceanicum* are restricted to the tropics/subtropics. *Ceriosporopsis halima, lignincola laevis* and *Trichocladium achrasporum* are cosmopolitan in distribution. *Ondiniella torquata* and *Ceriosporopsis trullifera* are confined to temperate region (Hughes, 1974; Kohlmeyer and Voklmann-Kohlmeyer, 1993a; Jones, 1993).

Physiology of marine fungi:

Marine fungi can be defined, based on their morphology, physiology and ecology and their ability to reproduce in the marine environment. Early physiological studies of marine fungi focused on their salinity tolerance (Jones and Jennings, 1964; Jennings, 1986a, b), temperarure requirement (Jones et al., 1971), nutrient requirement (Amon, 1986), enzyme production (Molitories and Schaumann, 1986; Schaumann et al., 1986), aspects of wood decay (Jones, 1971; Meyers, 1971a, b; Leightly, 1980) and ability to grow on different polysaccharides (Barghoorn and Linder, 1944; Meyers, 1971b).

a) Salinity requirement:

Studies of the effect of salinity on growth and morphology were undertaken in order to try and define them based on their physiology and need of seawater for growth and reproduction. It soon became apparent that marine fungi showed a broad response to growth in seawater, while vegetative growth occurred at all salinities, maximum growth was often at 20-60% seawater (Jones and Jennings, 1964; Meyers, 1971b). Similar variation was noted for sporulation: most required salinities of 40-60% seawater, others formed perithecia in media made up with distilled water.

Byrne and Jones (1975a) investigated the effect of salinity and temperature on spore germination of terrestrial, freshwater and marine fungi. Increasing salinity inhibited germination in terrestrial species, while marine species exhibited a broad tolerant to salinity and temperature stresses. The second major factor is salinity, which controls occurrence and distribution of marine fungi. The fluctuation in salinity affects the population of marine fungi. Allem (1980), Kirk and Brandt (1980), Kirk and Schatz (1980) and Vrijmoed et al. (1982) have reported on occurrence of marine fungi with reference to salinity levels. The studies so far conducted suggested that salinity is an important factor affecting the distribution of marine fungi. The influence of salinity on reproduction of higher marine fungi was also examined by Byrne and Jones (1975b) who showed that five ascomycete species produced ascocarps on balsa wood in yeast extract broth in seawater concentration. Zoosporic fungi were shown to be more sensitive to changes in salinity (Harrison and Jones, 1974; Tsui et al., 2011). Marine yeasts have also been studied for their ability to withstand high salinities (Norkrans, 1966; Norkrans and Kylin, 1969).

b) Oxygen requirements:

The influence of dissolved oxygen in seawater on marine fungi has not been tested, although field observations indicate that the higher marine fungi are unable to grow under low oxygen condition. A deep-sea area off the California coast is known as "Minimum Oxygen Zone", having content of 0.30 ml/lit, did not allow growth of soft- rot fungi. It was also found that when such fungi occurred on wood at neighboring site with a content of dissolved oxygen of 1.26 ml/lit they did not grow further (Kohlmeyer, 1968).

Another indication for the high oxygen requirement of higher marine fungi is the fact that wood inhabiting soft-rot fungi usually penetrate only a few mm into the wood substrate. A deeper penetration of an Ascomycota species observed only in light wood containing vessels with wide lumen which permits adequate aeration of the tissue (Becker and Kohlmeyer, 1958 a, b).

c) Light requirement:

Kohlmeyer (1968) attributed the lack of fruiting structures of higher marine fungi at a depth of 1615m, off the coast of California, to the constant darkness of the environment. Jones and Le Campion-Alsumard (1970) reported abundant perithecia of *Haligena unicaudata* at 437 m depth. Kirk (1969) stated that the illumination had no apparent effect on the fruiting of marine fungi.

Indian Work

The biodiversity of marine fungi was studied by various researchers (Jamaluddin et al., 2004). Borse (2002) listed 166 species (13 Labyrinthulomycota, 4 Chytridiomycota, 3 Oomycota, 120 Ascomycota, 3 Basidiomycota and 23 Mitosporic fungi) of marine fungi from India.

a) Lower marine fungi:

Chakravarty (1979) reported *Schizochytrium aggregatum*, a marine fungus from Tamilnadu. Gopalan et al. (1980) described *Lagenidium marina* and *Sirolpidium parasitica* fungal infection in the tiger prawn *Penaeus monodon* and other crustaceans from the Cochin backwaters (Kerala). However, both names are now invalid (E.B.G. Jones: personal communications). Lower marine fungi have been studied by Drs. Seshagiri Raghukumar and Chandralata Raghukumar (National Institute Oceanography, Goa, India) and their students since 1986.

Raghukumar (1987) described the new genus of thraustochytrid *Corallochytrium* with *C. limacisporum* as its type species from the coral reef lagoons of Lakshadweep islands in the Arabian Sea. However, this fungus now belongs to Protozooa. Raghukumar and Raghukumar (1988) have given an account of nine species of Thraustochytrids from the coastal waters of Goa, Tamilnadu, Lakshadweep islands and the Arabian Sea which included *Labyrinthuloides minuta, L. yorkensis, Thraustochytrium aggregatum, T. motivum, T. maltirudimentale, T.striatum,* and *Ulkenia visurgensis.*

Raghukumar (1986a) gave an account of fungal parasites of marine green algae, *Cladophora* and *Rhizoclonium* and reported *Sirolpidium bryopsidis, Olpidium rostriferum, Coenomyces* sp., and *Labyrinthula* sp. from the beaches of Goa and Lakshadweep islands. Raghukumar (1986b) recorded *Chytridium polysiphoniae* a fungal parasite on the red alga, *Centroceras clavulatum,* from Goa, as new report from India. Raghukumar (1987a) gave an account of fungal parasites of marine algae from Mandapam and reported *Pontisma lagenidioides* parasite of *Chaetomorpha media* and *Chytridium polysiphoniae* on *Sphacelaria* sp. as fungal parasites of marine algae, *Cladophora* and *Rhizoclonium.*

Raghukumar (1988a) described *Schizochytrium mangrovei,* a thraustochytrid from mangroves in India. Raghukumar (1988b) detected the thraustochytrid protests *Ulkenia visurgensis* in a hydroid, using immunofuorescence. Raghukumar and Lande (1988) reported shell disease of rock oyster *Crassostrea cucullata* caused by *Ostracoblabe implexa.* Raghukumar and Balasubramanian (1991) reported occurrence of thraustochytrid fungi in corals and coral mucus. Raghukumar et al. (1992a) studied associations of thraustochytrids and fungi living with marine algae. Raghukumar (1996) has given an account of 8 zoosporic fungal parasites of marine biota from India, including *Ulkenia amoebiodea.* Raghukumar (1996a) has given an account of morphology, taxonomy and ecology of Traustochytrids and Labyrinthuloids.

Raghukumar and Raghukumar (1994) conducted a survey of the Indian coast to explore the presence of parasitic fungi in marine algae. They examined 35 filamentous algae and found that 15 harbored fungal parasites. The percentage of algal cells infected in nature varied from 5 to 60% for the different species. A total of 7 fungal genera belonging to Mastigomycotina, labyrinthulales and Thraustochytriales were found regularly as parasitic in various algae. Ramasamy et al. (1996) reported *Lagenidium callinectes* infection and its control in cultured larval Indian tiger prawn *Penaeus monodon* Fabricus from Madras (Tamilnadu). Raghukumar (2002) studied ecology of marine protists, the *Labyrinthulomycetes* (Thraustochytrids and Labyrinthulids). Bongiorni et al. (2005) described a new species namely *Thraustochytrium gaertnerium* from mangroves of Goa. Damare and Raghukumar (2006) studied morphology and physiology of *Aplanochytrium yorkensis* isolated from the equatorial Indian Ocean.

b) Higher marine fungi:

Becker and Kohlmeyer (1958b) recorded the presence of soft rot fungi on small fishing craft in India. The only species reported was *Halosphaeria quadricornuta* (Now *Antennospora quadricornuta*). Kohlmeyer (1959) described new genus *Paraliomyces,* with *P. lentiferus* as its type species from Madras coast.

Kohlmeyer et al. (1967) described a new species of *Corollospora* (Ascomycota), *C. pulchella,* with its imperfect state *Clavariopsis bulbosa* (Now *Clavatospora bulbosa*) from Kerala coast. Jones (1968) reported five higher marine fungi from test blocks of wood submerged at Bombay and Cochin coast. Nair (1970) reported some higher marine fungi from Kerala.

Raghukumar (1973) reported 12 species Ascomycota and 6 Mitosporic fungi from Madras coast. Patil and Borse (1982, 1983 a, b, 1985, 1986) published a series of papers dealing exclusively higher marine fungi from the coast of Maharashtra. They described two new marine species of Ascomycota viz., *Didymella avicenniae* and *Halosarpheia ratnagiriensis* (Now *Saagaromyces ratnagiriensis*). Kohlmeyer and Vittal (1986) described *Lophiostoma mangrovei* (Now *Rimora mangrovei*) a new marine ascomycete from India.

Borse (1984, 1985a, 1987a-e) published a series of papers dealing with higher marine fungi from Maharashtra coast and described three new species of Ascomycota namely *Aigialus mangrovei*, *A. rhizophorae* and *Pleospora avicenniae* (Now *Julella avicenniae*) from mangroves of Maharashtra. Borse (1985b) reported two new marine Loculoascomycetes from Maharashtra, India. Venkatesan and Natarajan (1985) reported *Cirrenalia pygmea* along with several marine-derived fungi from Pichavaram mangroves.

Hyde and Borse (1986a) described the new genus *Biatriospora* (Ascomycota) with *B. marina* as its type species. Hyde and Borse (1986b) described a new species, *Massarina velataspora* (Now *Morosphaeria velataspora*) from Maharashtra coast and Seychells.

Ravikumar and Vittal (1987) studied marine fungi from Tamilnadu coast and reported ten species of Ascomycota occurring on intertidal woods of various mangroves. Raghukumar (1988c) reported four lignicolous, marine fungi from the coast of Goa and Lakshadweep Islands. Raghukumar et al. (1988a) described a new species of Deuteromycotina, *Cirrenalia basiminuta,* on material collected from Goa and Kuwait coast. Borse (1988) studied the frequency of occurrence of higher marine fungi from Maharashtra coast and observed that *Massarina velatospora* was the most dominant species in mangrove habitats, while *Antennospora quadricornuta* was the most dominant species from beach habitats.

Borse et al. (1988) reported some marine fungi from Maharashtra coast. Borse and Shrivastava (1988) reported *Halosarpheia abonnis*, and *Zopfiella latipes* as additions to the fungi of India. Ravikumar and Purushothaman (1988a) reported *Cirrenalia tropicalis* from Tamilnadu coast. Ravikumar and Purushothaman (1988b) recorded *Corollospora intermedia* from Tamilnadu coast. Sridhar and Kaveriappa (1988) studied occurrence and survival of freshwater Hyphomycetes in brackish and seawater.

Shrivastava (1989) reported 7 species of Ascomycotina from Bombay coast. Borse and Hyde (1989) described a new genus of Ascomycotina viz., *Acrocordiopsis* with *A. patilii* as its type species, occurring on mangrove wood from Maharashtra and Seychelles Islands. Further, Ramesh and Borse (1989) reported some higher marine fungi from Maharashtra coast.

Ravikumar and Vittal (1991a) described a new species of *Bathyascus* (*B. mangrovei*) from Pichavaram estuary, Tamilnadu. Ravikumar and Vittal (1991b) reported two species of *Corollospora* viz., *C. angusta* and *C. filiformis* from Tamilnadu. Sridhar and Kaveriappa (1991) reported five Ascomycetes and four Deuteromycetes from Manglore coast. Hyde et al. (1992) described a new species of Ascomycota viz., *Massarina armatispora* occurring on intertidal mangrove wood from Andaman Islands, Goa and Karnataka coast. Chinnaraj and Untawale (1992) reported nine species of Ascomycotina occurring on intertidal wood of mangroves from Andaman and Nicobar Islands, Goa and Karnataka coast.

Chinnaraj (1992) recorded 32 species of higher marine fungi from Lakshadweep Islands. Chinnaraj (1993) recorded 63 species of higher marine fungi from six mangrove tree species from Andaman and Nicobar Islands. Chinnaraj (1994) reported higher marine fungi from mangroves. Borse and Shrivastava (1994) reported *Ascocratera manglicola* Kohlmeyer and *Massarina thalassiae* (Now *Halomassarina thalassiae*) from Maharashtra coast. Nandan et al. (1993) recorded 14 higher marine fungi from Goa coast with *Aniptodera mangrovei* Hyde and *Ocostaspora apilongissima* Jones, Johnson and Moss as new reports to the fungi of India.

Sridhar and Prassanarai (1993) reported 4 ascomycetes from Mangalore coast, in which *Halosarphiea viscosa* (Now *Panorbis viscosus*) and *Trematospaeria mangrovei* as new reports for Indian marine fungi. Shrivastava (1994) recorded 10 species of Deuteromycotina from the coast of Bombay (Maharashtra). Srivastava (1995) reported 14 marine ascomycetes from different localities around Bombay (M.S.).

Ravikumar and Vittal (1996) studied fungal diversity on *Rhizophora* and reported 48 fungal species, where most common and abundant were *Lophiostoma mangrovei, Verruculina enalia, Halocyphina villosa,* and *Monodictys pelagica.* Raghkumar (1996b) has given an account of fungi in the marine realm with status, challenges and prospects.

Prasannarai and Sridhar (1997) studied the effect of incubation on occurrence of marine fungi from west coast of India. Ananda et al. (1998) studied occurrence of higher marine fungi on marine animal substrates from Mangalore coast. Raghukumar (1998) recorded 3 Ascomycetes viz., *Abysomyces hydrozoicus, Bathyascus vermisporus, Oceanitis scuticella* and two Mitosporic species viz, *Allecheriella bathygena, Periconia abysa* for the first time from Indian marine waters.

Borse et al. (1999a) recorded 15 marine ascomycetes from Goa. Borse et al, (1999b) reported 8 marine fungi from Diu Island with *Cryptovalsa halosarceicola* Hyde as new to India. Prasnnarai et al. (1999) studied seasonal occurrence and colonization of marine fungi from mangalore harbour in which *Didymosphaeria lignomaris* and *Nimbospora effusa* are new records for the Indian coast. Sarma and Vittal (1998-99) studied seasonal occurrence of manglicolous fungi from Godavari and Krishna deltas, east coast of India.

Borse et al. (2000a) studied frequency of occurrence of marine fungi from Pirotan Island (Gujarat) and collected 29 species of higher marine fungi on mangroves. Borse et al. (2000b) reported 10 ascomycetes and 3 dueteromycetes along with frequency of occurrence from Daman coast. Borse (2000a) reported 3 species of *Savoryella* from Maharashtra, Orissa and West Bengal with *Savoryella appendiculata* Hyde and Jones as a new record to India. Borse (2000b) listed 83 species (62 Ascomycota, 3 Basidiomycota, and 18 Deuteromycotina) from the coast of Maharashtra. Prasannarai et al. (2000) described new species of *Corollospora* viz., *C. indica* from west coast of India.

Sarma and Vittal (2000) studied the biodiversity of mangrove fungi on *Rhizophora apiculata* and *Avicennia* sp. from Godavari and Krishna deltas, east coast of India, and recorded 63 species. Sarma and Hyde (2000b) described a new species of *Tirispora* viz., *T. mandoviana* from Goa on *Rhizophora mucronata*. Prasanarai and Sridhar (2000-2001) recorded 11 species of higher marine fungi as new records for fungi of India.

Sarma and Vittal (2001) collected 88 species of manglicolous fungi from Godavari and Krishna delta (Andhra Pradesh), east coast of India. Sarma et al. (2001) studied frequency of occurrence of mangrove fungi from the east coast of India. Patil and Borse (2001) reported 50 spp. of higher marine fungi along with frequency of occurrence from Gujarat coast. Borse and Borse (2001) reported *Remispora* as a new generic record and *Trematosphaeria lineolatispora* a new record for India. Borse and Pawar (2001) reported *Carbosphaerella* and *Dryosphaera* as new generic records to the fungi of India from West Bengal.

Ananda and Sridhar (2001a) described *Aniptodera indica,* (synonym of *Tirispora unicaudata* Jones and Vrijmoed) from West coast of India. Ananda and Sridhar (2001b) reported mycoflora of dead animal substrates in mangrove habitats of Karnataka coast, India. Vishwakiran et al. (2001) recorded 33 species of marine fungi (20 Ascomycota, 1 Basidiomycota and 12 Mitosporic fungi) on mango panels submerged in the Mandovi and Zuari estuarine systems of Goa.

Prasannarai and Sridhar (2001) studied the diversity and abundance of higher marine fungi from west coast of India and reported 88 species belonging to 48 genera including 22 species reported as new records for the Indian penninsula. Sridhar and Prasannarai (2001) reviewed biogeography and biodiversity of higher marine fungi in tropics. Borse et al. (2001a) reported propagules of 7 Ascomycota and 2 Deuteromycetes in foam samples from sandy beaches of Orissa. Borse et al. (2001b) studied propagules of 7 Ascomycota and 1 Mitosporic fungus from foam samples of Sunderbans. Sarma et al. (2001) studied frequency of occurrence of mangrove fungi from the east coast of India.

Borse et al. (2002a) reported nine species of *Corollospora* from the coast of Orissa. Borse et al. (2002b) reported 3 species of *Arenariomyces* from Sunderban region. Shindikar and Borse (2002) reported 21 species of higher marine fungi (14 Ascomycota, 1 Basidiomycota and 6 Mitosporic fungi) from mangrove swamps of Vikhroli (Mumbai, M.S.). Maria and Sridhar (2002a) reported a total of 78 species belonging to 45 genera comprising 46 Ascomycota, 1 Basidiomycota and 31 Mitosporic fungi on woody litter of mangroves along with west coast of India.

Maria and Sridhar (2002b) described a new species of Ascomycete, *Passeriniella mangrovei* from the mangrove forest of India. Ananda and Sridhar (2002) studied the diversity of endophytic fungi in the roots of mangrove species on the west coast of India. Sarma and Vittal (2002) studied vertical distribution of manglicolous fungi on prop roots of *Rhizophora apiculata* at Krishna delta, east coast of India.

Maria and Sridhar (2003a) studied the diversity of filamentous fungi on woody litter of five mangrove plant species from the south-west coast of India. Maria and Sridhar (2003b) studied endophytic fungal assemblage of two halophytes from west coast mangrove habitats, India. Ananda and Sridhar (2003) studied on filamentous fungal assemblage of mangrove fungi from Coconut Island and Minicoy Island of the Arabian Sea.

Ananda and Sridhar (2004) studied diversity of filamentous fungi on decomposing leaf and woody litter of mangrove forest in the south-west coast India. Maria and Sridhar (20C4) studied fungal colonization of immersed wood in mangroves of the southwest coast of India. Sarma and Vitthal (2004) recorded 65 Ascomycota, 1 Basidiomycota and 22 Mitosporic fungi (6 Coelomycetes and 16 Hyphomycetes) from mangroves of Godavary and Krishna deltas of east coast of India along with dichotomous key notes on some taxa.

Kumaran et al. (2004a) studied fossil record of marine manglicolous fungi from Malvan (Konkan), west coast of India. Kumaran et al. (2004b) studied mangrove associated lignite beds of Malvan, Konkan as evidence for higher sea-level during the late tertiary (Neogene) along the west coast of India. These authors observed spores of mangrove fern (*Acrostichum aureum*) and an estuarine fungus *Cirrenalia* indicate that these lignites are autochthous and deposited in a near-shore environment. Pawar and Borse (2004) reported 12 species of higher marine fungi collected on samples of mangroves and salt marsh plants from the coast of Sunderbans.

Borse et al. (2005) have given review of diversity and taxonomy of marine fungi from Maharashtra. Borse and Borse (2005) recorded 18 mitosporic fungi from coast of Orissa. Pawar and Borse (2005a) recorded some marine fungi from Sunderbans. Pawar and Borse (2005b) recorded 15 Mitosporic species from Sunderbans. While Borse et al. (2005) recorded 5 Ascomycota, Pawar et al. (2005) also reported 5 Ascomycota for the first time from east-coast of India. Maria et al. (2005) studied antimicrobial and enzyme activities of mangrove endophytic fungi of southwest coast of India.

Vittal and Sarma (2005) studied fungal diversity associated with decaying substrates of nine mangrove plant species in Godavari and Krishna deltas (Andhra Pradesh) on the east coast of India. A total of 88 species of fungi were recorded which included 65 Ascomycota, 1 Basidiomycota and 22 Mitosporic fungi. Borse and Pawar (2005) reported 13 species of marine fungi (12 Ascomycota and one Mitosporic fungus) in the sea water foam samples collected from Mahe (Pondicherry) coast.

Borse and Tuwar (2006) reported 8 species of marine fungi (7 Ascomycota and 1 Mitosporic fungus) in the seawater foam samples collected from Goa coast. Vittal and Sarma (2006) reviewed diversity and ecology of fungi on mangroves of Bay of Bengal region and listed a total number of 131 species belonging to 77 genera. Sridhar and Maria (2006) reported 66 fungal species on mangrove woody litter *Rhizophora mucronata* (Rhizophoraceae). Nambiar and Raveendran (2006) gave a comparative account of mangrove associated marine fungi of Valapattanam estuary in Kannur district (Kerala). Nambiar et al. (2006) studied marine fungi from north Malabar region of Kerala coastal waters recorded 31 marine fungi from Valapattanam estuary, Kannur District (Kerala).

Raveendran and Manimohan (2007) wrote a book entitled "*Marine Fungi of Kerala*". They described 80 species of marine fungi from Kerala Coast. These included 61 Ascomycota, 1 Basidiomycota and 18 Mitosporic fungi. Nambiar and Raveendran (2007) encountered 52 marine fungi from three estuaries of North Malabar region of Kerala. These included 35 Ascomycetes, 2 Basidiomycetes and 15 Mitosporic fungi. Nambiar and Raveendran, (2007b) carried out an exploratory study and elucidated marine fungal diversity in the coastal wetlands of Kerala.

Kamble et al. (2008) recorded five species of marine ascomycets from Kutccha (Gujarat). Nambiar and Raveendran (2008a) recorded 22 marine fungi from Kerala. These included 16 Ascomycota, 1 Basidiomycota and 6 Mitosporic fungi. Among these, 11 species reported from coir-retting field. Nambiar and Raveendran (2008b) reported 61 marine fungi from coastal wetlands of Kerala. These included 42 Ascomycota, 2 Basidiomycota and 17 Mitosporic fungi. Nambiar and Raveendran, (2008c) recorded 18 marine fungi from Pondicherry and Mahe. These included 10 Ascomycota and 8 Mitosporic fungi.

Nambiar and Raveendran (2008d) recorded 23 marine fungi from estuaries of North Malabar region of Kerala. These included 16 Ascomycota, 1 Basidiomycota and 6 Mitosporic fungi. Nambiar and Raveendran (2008e) studied biodiversity of marine mangrove fungi of Valapattanam and Pichavaram mangrove forests (South India). Nambiar and Raveendran (2008f) have given a checklist of marine fungi from Kerala state, India.

Nambiar et al. (2008a) recorded 30 marine fungi from backwater and brackish water of North Malabar region of Kerala. These included 19 Ascomycota, 1 Basidiomycota and 10 Mitosporic fungi. Nambiar et al. (2008b) reported 51 marine fungi from four localities of Tamil Nadu (Tuthukudi, Chennai, Kanyakumari and Pichavaram). These included 28 Ascomycota, 1 Basidiomycota and 22 Mitosporic fungi. Sridhar (2009a) reviewed published reports of mangrove fungi of the Indian Peninsula. Besides, Sridhar (2009b) studied fungal diversity of Pichavaram mangroves, southeast coast of India.

Nambiar and Raveendran (2009a) studied influence of coir retting on mangrove ecosystem and recorded 19 marine fungi from mangrove field (15 Ascomycota, 1 Basidiomycota and 3 Mitosporic fungi) of Kerala. Nambiar and Raveendran (2009b) studied lignicolous marine fungi in selected wetlands of north Malabar (Kerala). Nambiar and Raveendran (2009c) also studied frequency of arenicolous marine fungi on sand buried wood substrates along Kerala beaches. Nambiar and Raveendran (2009d) recorded manglicolous marine fungi on *Avicennia* and *Rhizophora* along Kerala coast (India). Nambiar and Raveendran (2009e) exploited untapped potentiality of coastal paddy fields of Kerala (India). Nambiar and Raveendran (2009f) studied manglicolous marine fungi of Kerala.

Nambiar and Raveendran (2009g) studied frequency and abundance of marine mycoflora in mangrove ecosystem of North Malabar, Kerala (India). Nambiar and Raveendran (2009h) studied marine mycoflora on *Avicennia* and *Bruguiera* along Kerala coast. Nambiar and Raveendran (2009i) studied marine mycoflora in backwater ecosystem of Kerala. Nambiar and Raveendran (2009j) also studied marine mycoflora of Andhra Pradesh (India). Nambiar and Raveendran (2009k) gave an account of marine mycoflora of south India with special emphasis to lignicolous marine fungi. Nambiar and Raveendran (2009l) studied eco-biodiversity of marine mycoflora of Gulf of Mannar biosphere reserve (India). Nambiar and Raveendran (2009m) detailed an account of manglicolous marine fungi of Kerala.

Shini et al. (2009-10) studied assemblage and diversity of higher marine fungi from Thalassery (Kerala) and Bakkhali (West Bengal) under explored mangroves. Sridhar et al. (2010) studied fungal colonization and breakdown of sedge (*Cyperus malaccensis* Lam.) in a mangal located along the coast of Karnataka. Nambiar and Raveendran (2010a) studied frequency and abundance of arenicolous marine fungi along South Indian Beaches. Nambiar and Raveendran (2010b) studied impact of hospital waste on mangrove fungi of Kerala. Nambiar and Raveendran (2010c) studied frequency and abundance of marine fungi of South India. Nambiar and Raveendran (2010d) also studied arenicolous marine fungi of Kerala.

Kamble et al. (2011) recorded six speicies of marine ascomycetes from Gujarat coast. Nambiar and Raveendran (2010e) provided frequency and abundance of algicolous marine fungi along Indian coast. Nambiar and Raveendran (2011a) reported *Arenariomycec triseptatus* and *Halospheria appendiculata* as new record of marine fungi from India. Nambiar and Raveendran (2011b) reported *Anthostomella nypensis* and *Biconiosporella corniculata* as an addition to the marine mycoflora of India. Khan and Manimohan (2011) reported *Etheirophora bijubata*, *Remispora quadric-remis* and *Robillarda rhizophorae* as new records for India. Manimohan et al. (2011) studied a comparison of diversity of marine fungi on three co-habiting mangrove plants from Kerala coast. Borse (2011) recorded biodiversity of marine and freshwater fungi from India. Tuwar (2011) recorded biodiversity of marine fungi from Goa coast (India).

c) Other related studies:

Durve and Bal (1960) reported a shell disease of *Crassostrea gryphoides* Schlotheim from Bombay, the etiology of which is yet unknown (Ramaiah, 2006). Raghukumar (1985) enumerated thraustochytrids (heterotrophic microorganisms) from the Arabian Sea. Salique et al. (1985) studied ecology of fungal populations in the Vellar estuary and Pichavaram mangroves, South India. Raghukumar et al.

(1988b) studied colonization of marine fungi on seven kinds of wood panels treated with wood preservatives (copper-chrome-arsenate; Copper-chrome-borate and creosote) in Mandovi estuary, Goa.

Raghukumar et al. (1990) noticed abundance of thraustochytid fungi in the Arabian Sea. Raghukumar (1990) speculated on niches occupied by fungi in the sea with relation to bacteria. Raghukumar and Raghukumar (1991) studied fungal invasion of massive corals. Raghukumar et al. (1992b) studied on endolithic fungi from deep-sea calcareous substrata, isolation and laboratory studies. Sathe-Pathak et al. (1993) and Sharma et al. (1994) studied thraustochytrid and fungal component of marine detritus of the brown alga *Sargassum cinereum*.

Raghukumar et al. (1994) found out laccase and other lignocellulose modifying enzymes of marine fungi isolated from the coast of India. Raghukumar et al. (1994) studied role of thraustochytrids and fungi in decomposition of leaves of the mangrove *Rhizophora apiculata* Blume in the laboratory. Raghukumar et al. (1995) further studied role of thraustochytrids and fungi in decomposition of leaves of the mangrove *Rhizophora apiculata* in the field conditions. Raghukumar et al. (1996) studied degradation of lignin and decolourization of paper mill bleach plant effluent by marine fungi. Ravikumar and Suryanarayanan (1998) showed that increasing salinity elevated the activities of polyoldehydrogenase (NADP+), D-Manital dehydrogenase and Peroxide of the mangrove fungus *Cirrenalia pygmea*. Beena et al. (1997) studied association of arbuscular mycorrhizal fungi with *Launaea sarmentosa* on maritime sand dunes of west coast of India.

Raghukumar and Raghukumar (1998) studied barotolerence of fungi isolated from deep-sea sediments of the Indian Ocean. Raghukumar and Raghukumar (1999) recorded thraustochytrid fungoid protests in faecal pellets of the tunicate *Pegea confoederata* and their tolerance to deep-sea conditions and implications in degradation processes. Raghukumar et al. (1999) studies lignin modifying enzymes of *Flavodon flavus*, a basidiomycete isolated from coastal marine environment.

Raghukumar (2000) studied decolourization of beach plant effluents and synthetic dyes using fungi from marine habitats. Suryanarayana and Kumaresan (2000) studied endophytic fungi of some halophytes from an estuarine mangrove forest from Tamilnadu. Raghukumar et al. (2001) observed thraustochytrid protists as a component of marine microbial films. Beena et al. (2000a) studied fungal endophytes of three sand dune plant species of west coast of India. Beena et al.

(2000b) studied diversity of arbuscular mycorrhizal fungi on the coastal sand dunes of the west coast of India.

Raghukumar et al. (2001) studied dynamics of thraustochytrid protists in the water column of the Arabian Sea. Ravindran et al. (2001) studied fungi in association with healthy and deseased corals, *Porites lutea.* Raikar et al. (2001) studied degradation of hydrocarbons by thraustochytrids. Kumaresen and Suryanarayanan (2001) studied occurrence and distribution of endophytic fungi in a mangrove community. Kumaresen and Suryanarayanan (2002) studied endophytes in young, mature and senescent leaves of *Rhiziphora apiculata* and provided evidence for the role of endophytes in mangrove litter degradation. Raghukumar (2002) worked on bioremediation of colored pollutants by terrestrial versus facultative marine fungi. Raghukumar et al. (2002) reviewed ecology of the marine protists, the Labyrinthulomycetes (thraustochytrids and labyrinthulids). Devarajan et al. (2002) studied endophytic fungi associated with the tropical seagrass *Halophila ovalis* (Hydrocharitaceae).

Jain et al. (2004) studied on enhancement of production of polyunsaturated fatty acid, docosahexaenoic acid in thraustchytrid protists. Karunasagar et al. (2004) reviewed microbial diseases in shrimp aquaculture. Raghukumar (2004) reviewed the role of fungi in marine detrital processes. Raghukumar. (2004) reviewed enzymes of marine fungi in decolorization of coloured effluents. Raghukumar et al. (2004a) studied on xylanases of marine fungi as potential use for biobleaching of paper pulp. Raghukumar et al. (2004b) studied on culturable fungi in a deep-sea sediment core from the Chagos Trench, Indian Ocean.

Jain et al. (2005) worked on extra-cellular polysaccharide production by thraustchytrid protists. Raghukumar (2006) studied algal-fungal interactions (symbiosis to parasitism) in the marine ecosystem. D'Souza et al. (2006) studied on enhanced production of laccase by a marine fungus during treatment of colored effluents and synthetic dyes. Damare et al. (2006a) studied fungi in deep-sea sediments of the Central Indian Basin. Damare et al. (2006b) investigated deep-sea fungi as a source of alkaline and cold-tolerent proteases. Raghukumar (2006) reviewed marine eukaryotic diversity, with particular reference to fungi. Ramaiah (2006) reviewed on fungal diseases of algae, marine fishes, shrimps and corals. Ravishanker et al. (2006) studied strategies for osmoregulations in the marine fungus *Cirrenalia pygmea* Kohlm.

Raghukumar (2008) reviewed marine fungal biotechnology from an ecological perspective. Raghukumar (2008) reviewed the thraustochytrids for production of PUFAs. Kamat et al. (2008) studied marine-derived fungi as a source of proteases. Purvatkar et al. (2009) screened marine-derived fungus *Apergillus terreus* for Aspernoilides A and B and Butenolides. Suryanarayanan et al. (2010) studied distribution, diversity and biotechnology potential of internal mycobiota of marine macroalgae from Tamilnadu coast. Raghukumar et al. (2010) provided review of occurrence, diversity and adaptations of deep sea fungi. Singh et al. (2010) provided phylogenetic diversity of culturable fungi from the deep sea sediments of the Central Indian Basin and their growth characteristics. Raghukumar and Damare (2011) provided evidences of role of Labyrinthulomycetes in the functioning of marine ecosystems.

Table- 4: New genera and species of marine fungi described from India

SN	Taxon	Subsrate	Location	Reference
	Labyrinthulomycota			
1	*Schizochytrium mangrovei* Raghukumar	*R. mucronata, A. officinalis*	Goa	Raghukumar, 1988a
2	*Thraustochytrium gaertnerium* Bong. et al.	Mangroves	Goa	Bongiorni et al., 2005
	Ascomycotina			
3	**Acrocardiopsis patilii* Borse & K.D. Hyde	*Avicennia alba, R. mucronata*	Maharashtra	Borse and Hyde, 1989
4	*Aigilus mangrove* Borse	*R. mucronata*	Maharashtra	Borse, 1987e
5	*A. rhizophorae* Borse	*R. mucronata*	Maharashtra	Borse, 1987e
6	*Bathyascus mangrovei* Ravikumar & Vittal	*R. apiculata*	Tamilnadu	Ravikumar and Vittal, 1991a
7	**Biatriospora marina* K.D. Hyde & Borse	*A. marina*	Maharashtra	Hyde and Borse, 1986a
8	*Corollospora indica* Prasannarai et al.	Calcarious substrates	Karnataka	Prasannarai et al., 2000
9	*C. pulchella* Kohlm. et al.	Intertidal wood	Kerala	Kohlm. et al.,1967
10	*Didymella avicenniae* S.D. Patil & Borse	*Avicennia alba*	Maharashtra	Patil and Borse, 1985
11	*Saagaromyces (Halosarpheia) ratnagirinsis* S.D. Patil & Borse	Mangrove wood	Maharashtra	Patil and Borse, 1982
12	*Julella avicenniae* (Borse) K.D. Hyde	Avicennia alba	Maharashtra	Borse, 1987b
13	*Rimora (Lophiostoma) mangrovei* Kohlm. & Vittal	*R. mucronata*	Tamilnadu & Maharashtra	Kohlmeyer and Vittal,1986

14	*Massarina armatispora* K.D. Hyde et al.,	Intertidal mangrove wood	And..Islands, Karn. & Goa	Hyde et al.,1992
15	*Morosphaeria (Massarina) velataspora* K.D. Hyde & Borse	*R. mucronata*	Maharashtra	Hyde and Borse, 1986b
16	**Paraliomyces lentiferus* Kohlm.	Intertidal wood	Tamilnadu	Kohlmeyer, 1959
17	*Tirispora mandoviana* V.V. Sarma & K.D. Hyde	*Rhizophora mucronata*	Goa	Sarma and Hyde, 2000 b
18	*Passeriniella mangrovei* Maria & K.R. Sridhar	*Rhizophora mucronata*	Karnataka	Maria and Sridhar, 2002 a
	Mitosporic Fungi			
19	*Cirrenalia basiminuta* Raghukumar & Zainal	Mangrove wood	Karnataka	Raghukumar et al., 1988 b

MATERIALS AND METHODS

ISOLATIONS OF LOWER MARINE FUNGI:

Detailed information on the morphology, taxonomy, ultra-structure and culturing methods of these organisms is available in the following literature: Johnson and Sparrow, 1961; Pokorny, 1967; Olive, 1975; Karling, 1981; Moss, 1985,1986; Fuller and Jaworski,1987; Porter,1990; Ulken,1990; Raghukumar, 1996; Bremer, 2000; Dick, 2001; Yokoyama and Honda, 2007; Yokoyama et al., 2007. The basic methods are detailed below.

ISOLATIONS OF HIGHER MARINE FUNGI:

From Intertidal and Driftwood:

Samples of various substrata supporting higher marine fungi are collected randomly during low tide period from different coastal areas along the coast line of India. The samples are placed in plastic bags and sealed well in order to avoid moisture loss. On returning to the laboratory, samples are washed thoroughly under running tap water to remove the adhering debris. Surface fouling organisms are scrapped off, following rinsing in tap water. The samples are examined under a stereoscope for fungal fruiting structures. The surface of the wood is sliced away in order to locate the buried ascomata or pycnidia. Spore mass contents are scooped out onto a slide in a drop of sterile seawater for examination under a compound microscope. Superficial fruiting bodies are picked up with a fine-tipped needle and transferred to a drop of sterile seawater on a slide. These are then crushed to release the spores for observations.

After initial observations, samples are incubated in plastic boxes and kept moist by spraying of sterile seawater and periodically examined for developing fungal fruiting bodies. The structure and morphology of ascomata, ascomata wall, the presence or absence of catenophyses, paraphyses, pseudoparaphyses and periphyses, ascus and ascospores are important criteria for identification of the ascomycetous fungi. For Basidiomycota, the morphology of the basidiomata, basidia and basidiospores serves as an essential feature for identification. For diagnosis of mitosporic fungi, the morphology of conidiomata and the mode of conidiation are vital, in addition to the conidial characteristics.

Sea foam samples are collected during high tide from sandy beaches with the help of a ladle and placed in clean wide mouth plastic bottles and kept for 24 h to enable the foam to dissolve. It is preserved by adding Formalin (45%) to yield 5% foam solution. Then, the samples are brought to the laboratory and scanned under low and high power of a microscope for presence of fungal spores.

Sporulating material is mounted first in seawater, so that any appendages present can dialate and their true morphology be determined. In some cases, ascospores are allowed to stand in seawater for up to 30 min, e.g. species of *Natantispora, Panorbis, Aniptodera* etc. If spores are mounted directly in lactophenol, appeandages are often not seen and lead to misidentification of the species. For some fungi, Delafield's Haematoxylin stain or India ink are used to detect the presence or absence of gelatinous or mucilaginous sheaths or appendages of the spores. Congo red and Melzer's reagent are also used to study the ascus morphology.

For some of the Ascomycota, if the material is very sparse, no asci could be observed. The features of ascospores are so characteristic of each species in marine Ascomycetes that, in the absence of all information, there is difficulty in assigning them to their respective species.

Permanent mounts of the fungi are done by replacing the seawater with lactophenol (with or without cotton blue). Surplus mountant is cleaned away by a blotting paper. Free-hand sections are taken for forms embedded in the substrates. The sections are stained with cotton blue and, if necessary, mounted in lactophenol. The cover glass is sealed with D.P.X. mountant.

The slides are made semi-permanent by using 'double cover glass method' of Volkmann-Kohlmeyer and Kohlmeyer (1996). For initial examination of the fungus, a square cover glass (22 x 22 mm) is placed on a slide (76 x 26 mm) and tiny droplets of distilled water are introduced with a dissecting needle on two sides of the cover glass to make it to adhere to the slide. A larger droplet of distilled water is dropped in the center of the cover glass and the specimen placed in it and covered with a small cover glass (18 x 18 mm). After necessary microscopic observations and drawings or photomicrographs are made, the measurements are taken. A droplet of lactophenol (with or without cotton blue) is added to the water from the side. The slide is stored horizontally in a dust proof container for several days to permit the water to evaporate. After cleaning the edges of cover glass, the mount is sealed with D.P.X. mountant. The large cover glass is detached from the slide and a drop of D.P.X. or Canada Balsum placed on the small cover glass. The preparation is turned over with the drop of D.P.X. or Canada balsam hanging down and then placed on a fresh slide. The drop of D.P.X. or Canada balsam flattens out around the edges of the small cover glass, thereby permanently covering the dried ring of D.P.X.

From Test Blocks:

Blocks of size 5 x 2 x 1 cm of teak or bamboo wood are made for baiting in the sea. Teak and bamboo wood are preferred because they are generally used in the building of small boats, jetty construction and various fishing activities in marine waters.

These blocks are drilled centrally and strung into ladders with blocks separated by 2 cm apart knots on a nylon string. For each wood, replicates are maintained. These are sterilized by autoclaving at 120^{0}C for 20 min, transported to the study sites and tied to the stems or roots of mangrove trees standing 5 to 10 m apart at the mid-tide level where they remain submerged twice a day by seawater.

The test blocks are removed once in every four months over a period of one year. The blocks are scrapped with a point scrapper at every collection throughout the investigation in order to remove surface fouling organisms.

During every sampling of blocks, they are placed in polythene bags with little seawater, tied by a rubber band and brought to the laboratory for isolation of growing fungi. In the laboratory, the blocks are washed with autoclaved seawater, first examined for the presence of fungi and then incubated at 28-30^{0}C under diurnal light conditions in plastic boxes lined with a layer of moist sterile sea sand or filter paper. The blocks are examined for over 6 months incubation period.

STAINS AND REAGENTS:

Lactophenol is a standard and commonly used mounting medium for routine examination. This has a high refractive index (1.45). This acts as a clearing, swelling and preservative agent; also can be used for making semi-permanent mounts because glycerin, due to its very slow drying nature keeps the preparation wet for longer period. However, care should be taken to avoid its contact with skin or inhaling as it has been reported to be carcinogenic. The mounting medium is prepared by adding Phenol (pure crystals) 20 g + Lactic acid 20 ml + Glycerol 40 ml + Water 20 ml.

Cotton blue in Lactophenol: It is prepared by dissolving 0.05-0.01 g of Cotton blue/Aniline blue powder in 100 ml Lacto-phenol and filtered to remove sediments. It is the most common mounting cum clearing medium. This rapidly stains the cytoplasm leaving the cell-wall unstained; thus pseudoseptate nature of the ascospores or conidia becomes clear. By applying nail varnish or D.P.X. mountant, permanent preparations can be made in this mounting medium.

It is an acidic stain used to study the fungal structures. It is also a cytoplasmic stain, leaving the hyphal or spore walls and septa hyaline. Thus, the size, shape, septation and structures of ascospores/ conidia can be studied by using this stain. Another aspect of this stain is to study the nature of ascus wall. For critical study of ascopsore/conidia ornamentations, a mixture of cotton blue and lactic acid is used. This helps to study warts, spines and reticulate thickenings on the outer surface of ascospores/conidia.

Congo-red: This is an important stain used to differentiate the two layers of bitunicate ascus by colour differentiation. It is an amyloid stain which stains the inner layer of the ascus wall as pink or red with outer wall remaining unstained. 12% Congo-red is prepared by dissolving 1 g of Congo red powder in 100 ml. of distilled water. One drop of stain is applied in the slide preparation. After a few minutes, it gives good result. Material, if dipped in 1 % acidic alcohol, better differentiation can be achieved.

Delafield's Haematoxylin (Mordant combined): It is prepared by dissolving Haematoxylin powder (8.5 g) in alcohol (100 ml) and ammonium alum (6.25 g) in 320 ml distilled water. Glycerol (80 ml) is added first to the Haematoxylin solution, followed by alum solution. The contents are mixed and stored for 3 months for ripening. The solution is filtered before every use. The stain is used to detect the presence of gelatinous or mucilaginous sheath or appendages of ascospores.

Melzer's Reagent: Melzer's reagent containing Iodine is prepared by dissolving Potassium iodide (1.5 g), Iodine 0.5 g and Choloral hydrate (20 ml) in 20 ml distilled water (This stain is to be prepared in a closed cupboard or fume cupboard: Dissolve Potassium iodide in distilled water in a glass beaker and then add iodine. Cover the container and leave for over 24 h to dissolve). It imparts blue colour to the hymenium. Also due to its clearing properties, it can be used for general examination of the structure. The ascal structures such as amyloid apical ring and apical plug or appartus of the ascus tip, stain blue in this stain. The positive reaction of Iodine test (J +ve) confirms the unitunicate / inopreculate nature of asci. The blue reaction of the pore plug of ascus is a useful feature in separating Sphaeriales from Loculoascomycetes.

Choloral hydrate acts as a clearing agent. Iodine also imparts a yellowish brown colour to the non-gelatinous hyphal walls in general and may help to decide the presence of septa in ascospores. Pre-treatment of material with 5 % or 10 % KOH is generally recommended. Pre-treatment in 0.1 ml of NaOH, heated gently and then treated with 0.1 ml of HNO3, gives better reaction. The carotenoid pigments turn green in iodine while dextrinoid (pseudoa.myloid) stain reddish brown. Non-gelatinized hyphal walls appear yellow-brown. After drawing out the reagent from one side and replacing with other mountants like lactophenol, permanent mounts can be prepared after sealing with nail varnish or D.P.X.

India ink: India ink is used to make visible thin hyaline appendages on ascospores/ conidia. These appendages appear white against black background of ink.

Measurements: The measurements of various parts of fungi are taken and used in the identification and classification of different taxa. Metric units mm and µm are used for measurements of microstructures. Measurements are taken with the help of a calibrated ocular micrometer.

ILLUSTRATIONS:

Drawings: All illustrations are made with the help of an Erma Camera Lucida or an Olympus make drawing tube, fitted to the microscope and using 5x, 10x, 45x, 100x objectives and 5x, 10x, 15x eyepieces. The choice of the objective and eyepiece is made according to the size of fungal structure. Sketches are drawn on plain paper with the help of black, water proof ink. The thickness of each line is maintained uniform throughout.

Photomicrographs: Photomicrographs of the fungi are taken using Olympus photomicrographis unit fitted to a trinocular microscope at P.G. Department of Botany, S.S.V.P.Sanstha's L.K. Dr. P.R. Ghogrey Science College, Dhule, Maharashtra and the Department of Botany, Goa University, Goa.

IDENTIFICATIONS:

Monographs and illustrated taxonomic keys for the identification of lignicolous marine fungi, viz., Kohlmeyer and Kohlmeyer (1979), Kohlmeyer and Volkmann-Kohlmeyer (1991), Hyde et al., (2000), Hyde (2000), Jones et al. (2009a) and other relevant published literature are consulted for identification fo the fungi.

In the present monograph, the nomenclature of fungi adopted by Kohlmeyer and Kohlmeyer (1979), Kohlmeyer and Volkmann-Kohlmeyer (1991), Hyde et al. (2000), Hyde (2002) and Jones et al. (2009a) and other relevant literature are followed. In this work, each genus is described to include all taxonomic features. Further, details of type species, number of hitherto described species and detailed description of each species are given.

Lower Marine Fungi from India

Phylum: Labyrinthulomycota

1. ***Aplanochytrium*** G. Bahnweg & F.K. Sparrow, 1972. *Arch. Mikrobiol.*, **81:** 46.

= *Labyrinthuloides* Perkins, *Arch. Mikrobiol.*, **90:** 5 (1973).

Sporangia are globose or subglobose, sessile or free. Walls are made up of scales. Bothrosomes are present. Ectoplasmic net system is endobiotic or exobiotic, and does not enrode the cells. Cells move independently, and may reverse direction. Applanospores always formed. Zoospores may be formed. Plasmodia and amoeboid stages may be formed. The genus is represented by 8 species (Leander and Porter, 2000; Moro et al., 2003).

Type species: *Aplanochytrium kerguelensis* G. Bahnweg & F.K. Sparrow
Description: Based on Leander and Porter (2000).

1.1. ***Aplanochytrium minuta*** (Watson & Raper) Leander & D. Porter, 2000. *Mycotaxon*, **76**: 439-444. [For Fig. refer original paper]

Basionym: *Labyrinthula minuta* Watson & Raper, *J. Gen. Microbiol.*, **17:** 368 (1957).

≡ *Labyrinthuloides minuta* (Watson & Raper) Perkins, *Mycologia,* **66:** 697 (1974).

Cells ovoid or discoid, hyaline or pale orange, 5-15 x 3-5 μm, dividing into tetrads by two successive divisions at right angles, at colony periphery, cells may enlarge to ovoid or irregular, multinucleate units fragmenting into uni- or multinucleate, irregular bodies. Slimeways filamentous, interconnected, hyaline, 0.5-1μm wide. Sori not generally observed.

Habitat: On marine algae and sea grasses.

Description:- Based on Johnson and Sparrow (1961).

Distribution: West Coast:-Goa: On marine algae *Lyngbya* sp. and *Caulerpa* sp. (Raghukumar and Raghukumar, 1988; Raghukumar et al., 1992), on brown alga *Sargassum cinereum* (Sathe-Pathak et al., 1993); **Lakshadweep Islands:** On sea grass *Thalassia hemprichii* (Sathe and Raghukumar, 1991).

1.2. *Aplanochytrium yorkensis* (Perkins) Leander & D. Porter, 2000. *Mycotaxon*, **76**: 439-444. [For Fig. refer original paper]

Basionym: *Labyrinthuloides yorkensis* Perkins, *Arch. Mikrobiol.*, **90:** 1-17 (1973).

This species is characterized by the ectoplasmic net elements or pseudorhizoids extend from the thallus body or cell and do not enrobe it as in most species of the genus (except *A. minuta*). Sporulation is by cleavage of the globose thallus into aplanospores which have a slow gliding movement associated with the net elements, filaments or pseudorhizoids. Such filaments are membrane bound and arise from sagenogenetosomes. Interphase centrioles are present, and the zoospores do not appear to have an eyespot.

Habitat: On marine algae, corals and coral mucus

Description: -Based on Karling (1981).

Distribution: West Coast:-Goa: On marine algae (Raghukumar and Raghukumar, 1988); **Lakshadweep Islands:** On corals and coral mucus (Raghukumar and Balasubramanian, 1991).

2. *Aurantiochytrium* R. Yokayama & D. Honda, 2007. *Mycoscience,* **48**: 203.

Thallus is thin-walled, globose, and orange. Cells contain astaxanthin phoenicoxanthin, canthaxanthic acid ß-carotene and possessing minor arachidonic acids. Colonies are small by continuous binary cell divisions. Ectoplasmic nets not well developed. Zoospores are biflagellate hetrokont; reniform to ovoid. 18S rRNA gene sequence is distinct. Resting spore not observed.

Type species: *Aurantiochytrium limacinum* (Honda & Yokochi) Yokoyama & Honda

Description:- Based on Yokoyama and Honda (2007).

2.1. ***Aurantiochytriu mangrovei*** (Raghukumar) R. Yokoyama & D. Honda, 2007. *Mycoscience,* **48**: 203. [For Fig. refer original paper]

Basionym: *Schizochytrium mangrovei* Raghukumar, *Trans. Br. Mycol. Soc.,* **90**: 627 (1988).

Isolated in seawater pine pollen culture, the thallus develops epibiotically on pine pollen, interbitically with only the ectoplasmic net element formed from a single site touching the pine pollen. The young thallus is 37 x 44 μm diam., globose and formed by encystment of the zoospore. The mature thallus is at a stage prior to division 6.3-9.2 μm diam., globose or subglobose. The mature thallus divides by a binary division to form a diad, then tetrad and form cluster of 12 cells. At maturity each cell of the cluster gives rise to biflagellate zoospore. The escape of the zoospore is successive. The zoospores are 4.1 x 2.6-3.3 μm, elliptic with slight depression at the anterior end. The long anterior flagellum and shorter posterior flagellum laterally attached.

Habitat: Saprophytic on leaves of mangroves.

Description:-Based on Raghukumar (1988a).

Distribution: West Coast:-Goa: On leaves of *Rhizophora apicualata* (Raghukumar, 1988a, Raghukumar et al., 1994), on brown alga *Sargassum cinereum* (Sharma et al., 1994).

3. ***Labyrinthula*** Cienkowski, 1867. In: *Schultze, Arkiv mikro. Anat.,* **3:** 274.

Thallus an intricate network of delicate, hyaline, filamentous slimeways connecting single or colonial, fusiform, ovoid or discoid "spindle cells". Spindle cells are hyaline or faintly pigmented, with or without contractile or non-contractile vacuoles; uninucleate; aggregating into masses or mounds; reproducing by transverse, diagonal or longitudinal division, or by endogenous production of four small irregular or ovoid cells. Slimeways elaborated by spindle cells as tube-like or track-like structures; cells moving on the slimeways. Reproduction varied: fragmentation of multinucleate amoeboid or non-amoeboid cells; formation of sori, production of pseudosori and pseudospores; discharge of postriorly biflagellate swarmers; spindle cell encystment followed by liberation of amoeboid cells or production of giant cells and tetrasporocysts; various cell types formed from fructifications. Cysts occasionally developed, elongating into spindles or cleaving endogenously into small spore-like cells.

Type species: *Labyrinthula macrocystis* Cienkowski
Habitat: On algae and sea grasses.
Description:-Based on Johnson and Sparrow (1961).

3.1. *Labyrinthula* sp.

The infected algae (*Chaetomorpha media*) appeared pale green. The host cells filled with a labyrinth of spindle-shaped cells in an ectoplasmic network. Chloroplasts aggregated into a sphere in the centre of the host cells. Sori are golden brown.

Habitat: On marine algae and sea grass.
Description:-Based on Raghukumar (1987a).
Distribution: West Coast:-Goa: On green algae *Chaetomorpha media, Cladophora frescatii, C. repens, Rhizoclonium* sp. and blue green alga *Lyngbya* sp., (Raghukumar, 1986a, Raghukumar and Raghukumar, 1994); **Lakshadweep Islands:** On green algae *Chaetomorpha media, Cladophora frascatii, and Rhizoclonium* sp. (Raghukumar, 1986a, Raghukumar and Raghukumar, 1994), on sea grass *Thalassia hemprichii* (Raghukumar, 1996); **Gujarat** (Veraval)**:** *Valoniopsis pachynema*, (Raghukumar and Raghukumar, 1994); **Karanataka:** On brown alga *Lobophora* sp., red alga *Laurencia* sp. (Raghukumar and Raghukumar, 1994).
East Coast:- Tamil Nadu (Mandapum)**:** On green algae *Chaetomorpha media, Cladophora* sp. *Rhizoclonium* sp., blue green alga *Lyngbya*, brown alga *Lobophora variegate*, red alga *Laurencia* sp., (Raghukumar, 1987a, b; Raghukumar and Raghukumar, 1994).

4. *Schizochytrium* S. Goldstein & Belsky emend. R. Yokoyama & D. Honda, 1964. *Am. J. Bot.*, **51:** 72; *Mycoscience,* 48: 202 (2007).

Thallus is thin-walled, globose, pale yellow. Cells possensing only ß-carotene as carotenoid pigment and possessing ca. 20 % arachidonic acid as the fatty acids. Colonies are large by continuous binary cell divisions. Zoospores are biflagellate heterokont; reniform to ovoid. 18S rRNA gene is sequence distinct. Resting spores not observed.

Type species: *Schizochytrium aggregatum* Goldstein & Belsky
Habitat: Saprophytic in sea water, pollen grains, viable and heat-killed marine and freshwater algae, and grass.
Description:-Based on Yokoyama and Honda (2007).

4.1. *Schizochytrium aggregatum* S. Goldstein & Belsky, 1964. *Am. J. Bot.*, **51:** 72. [For Fig. refer original paper]

Globose stage of thallus, 4-12 µm diam., tetrad stage, 13-29 x 14-27 µm, amorphous stage, 30-140 µm diam. Rhizoids or ectoplasmic filaments formed at any stage but usually on individual segments during and after the tetrad stage; extending through the water and contacting other pollen grains, branched with spindle-shaped swellings, and sometimes anastomosing. Zoospores are motile in the sporangium, reniform, ovoid, or subglobose, 2.5-5 x 4-7.5 µm, laterally biflagellate, heterocont; movement primarily by the undulations of the longer anteriorly directed flagellum.

Habitat: Saprophytic in sea water, pollen grains, viable and heat-killed marine and freshwater algae, and grass.
Description:-Based on Karling (1981).
Distribution: East Coast:-West Bengal: Isolated from sea water and sediments (Chakravarty, 1979).

4.2. *Schizochytrium* sp.

Habitat: On corals and marine algae.
Distribution: West Coast:-Goa: On corals and coral mucus (Raghukumar & Balasubramanian, 1991); **Lakshadweep Islands:** On corals and coral mucus (Raghukumar and Balasubramanian,1991).
East Coast:-Tamil Nadu (Mandapum)**:** On marine algae (Raghukumar, 1996).

5. *Thraustochytrium* F. K. Sparrow, 1936. *Biol. Bull.*, **70:** 259; Emend. Johnson & Sparrow, 1961. *Fungi in Oceans and Estuaries*, pp. 331.

Thallus is monocentric, eucarpic. Sporangia usually globose, with or without internal proliferation, thin or thick-walled, smooth. Rhizoids (Pseudorhizoids) are usually endobiotic, occasionally interbiotic, branched and extensive or simple and unbranched; apophysis present or absent. Zoospores are laterally biflagellate, usually hetrocont, motile or nonmotile when liberated through irregular fissure in the sporangial wall or by its deliquescence. Resting spores where present epibiotic, smooth, spherical, and containing a single centric or eccentric oil globule. Zoospore discharge is thraustotheccoid, that is, results from the partial dissolution of the sporangial wall. The genus is represented by 14 species in marine waters (Karling, 1981).

Type species: *Thraustochytrium proliferum* Sparrow
Habitat: Saprobic on organic matter in marine habitats.
Description:- Based on Johnson and Sparrow (1961) and Karling (1981).

5.1. *Thraustochytrium aggregatum* A. Ulken, 1965. *Veroff. Inst. Meeresforsh. Bremerh.*, **9:** 291. [For Fig. refer original paper]

Sporangia aggregated in clusters, sessile, globose, small, 10-15 μm diam., with a smooth, hyaline, thin wall. Rhizoids (Pseudorhizoids) stout, branched, endo- or ectobiotic between pollen grains and clumping them together; on seawater agar and silica gel, spreading in the substrate and becoming extensive up to 30 μm long; sporangium wall dissolving and disintegrating, releasing polyhedral spore initials which remain quiescent ½ to 1 hr. before becoming motile. Zoospores are globose, ovoid, reniform, 2.5-3.5 x 4-5.5 μm, heterocont, with 2 lateral flagella.

Habitat: On pollen grains in sea water and marine algae.
Description:-Based on Karling (1981).
Distribution: West Coast:-Goa: On marine algae (Raghukumar and Raghukumar, 1988).

5.2. *Thraustochytrium gaertnerium* Bongiorni, G. Jain, R., Raghukumar, & R. K. Aggarwal, *2005*. Protist, 156: 303-315. [For Fig. refer original paper]

The organism is characterized by large zoosporangia with two distinct development cycles. In one, typical thalli with ectoplasmic net elements mature into zoosporangia that divide to form heterokont biflagellate zoospores, leaving behind a proliferation body. In the second type, the thalli develop into amoeboid cells, reminiscent of the genus *Ulkenia* Gaertner. Unlike *Ulkenia*, however, the 'amoebae' do not immediately produce zoospores, but round up prior to division into zoospores. The two types of development occur simultaneously in single cell-derived in-vitro cultures.

Habitat: On mangrove leaves.
Description:- Based on Bongiorni et al. (2005).
Distribution: West Coast:- Goa: On mangrove leaves (Bongiorni et al., 2005).

5.3. *Thraustochytrium motivum* S. Goldstein, 1963. *Am. J. Bot.*, 50: 273. [For Fig. refer original paper]

Sporangia gregarious, up to 60 thalli on a pollen grain, proliferating from 1 basal rudiment, sessile, obpyriform, subglobose, globose, fusiform to ellipsoidal, 95-35 µm long by 7-28 µm diam., smooth, hyaline, thin-walled, with a basal endobiotic, branched pseudorhizoid. Apophysis is present or absent, when present intra- or extrametrical. Zoospores angular at first, then becoming reniform, ellipsoid or spherical, 2-4 x 3.5-6 µm when motile, laterally biflagellate, heterocont, anterior flagellum 7-16 µm long; posterior one 5-9 µm long; containing 1-4 refractive osmophylic globules; becoming flagellate and churning in the sporangium prior to discharge, emerging singly or in groups through a fissure in the wall, separating and swimming away. Basal rudiment or incipient secondary sporangium delimited before zoospore formation, occasionally so large that only a single layer of spore initial surrounds it in the primary sporangium. Resting bodies (spores ?) in Booth and Miller's (1968) isolate TMA-5, spherical, 10-25 µm diam. with a 2-4.5 µm wall, thicker on one side, and dense, faintly orange red. Germination is unknown. (Description is Based on Karling, 1981).

The sporangia are obpyriform to globose and granular throughout. Towards zoospore formation, the granules are clustered in the middle of the sporangium. The proliferation body recognized at the base of the sporangium as a clear zone of cytoplasm free of granules. Zoospores are formed in the upper remaining part of the sporangium and by centripetal invagination starting at the peripheral zone of the cytoplasm which is free from granules. By the time the centripetal invagination is completed, the freshly delimited zoospores can be seen to be arranged radially and are coniform. Later they round up and are irregularly scattered in the sporangium. The zoospores are 3.5-5.6 x 5.6-7.7 µm and elliptical. The flagella are lateral and almost equal in length. The zoospores exhibit very rapid, darting movement. (Description based on Raghukumar and Raghukumar, 1988).

Habitat: On marine algae, Corals, sea grass and mangroves.
Description:-Based on Karling (1981) and Raghukumar and Raghukumar, (1988).

Disribution: West Coast:- Goa: Isolated from decaying leaves of *Rhizophora mucronata* (Raghukumar and Raghukumar, 1988); **Lakshadweep Islands**: On corals and coral mucus (Raghukumar and Balasubramanian, 1991); on sea grass *Thalassia hemprichii* (Sathe and Raghukumar, 1992).
East Coast:-Tamil Nadu (Mandapum)**:** On red alga *Lourencia* sp. (Raghukumar and Raghukumar, 1988).

5.4. *Thraustochytrium multirudimentale* S. Goldstein, 1963. *Am. J. Bot.*, **50:** 273. [For Fig. refer original paper]

Sporangia proliferating with 2-4 rudiments, sessile, globose, subglobose, 11-28 high by 14-37 µm diam., smooth- and thin-walled, with a branched endobiotic basal pseudorhizoid. Zoospores elongate, subfusiform, or fusiform, ellipsoid, 2-3.5 x 4.5-7-5 µm, with 1-3 refractive osmophilic globules and 2 apically or laterally inserted flagella; heterocont , anterior flagellum 6-14 µm long, posterior one 3-7 µm long and inconspicuously appressed to spore body; becoming motile in the sporangium; liberated through 1, seldom 2 apical ruptures in the sporangial wall. Basal rudiments delimited when zoospores are formed, globose, 6-17 µm diam. Resting spores are unknown. (Description based on Karling, 1981).

Mature thallus converted into single sporangium and produces up to 100 zoospores. Zoospore formation is initiated by the appearance of zoospore contours all over the sporangium. The zoospores are 2.8 x 5.5-7 µm and obclavate. The anterior flagellum 7.2-14 µm long and appears to be apical. The posterior flagellum 5.5-11 µm long and appears to be subapical. The swimming movement of the zoospores is slow and steady. (Description is based on Raghukumar and Raghukumar, 1988).

Habitat: On pollen in sea water and on mangrove leaves.
Description:-Based on Karling (1981) and Raghukumar and Raghukumar, (1988).
Distribution: West Coast:-Goa: Isolated on decaying leaves of *Avicennia officinalis* (Raghukumar and Raghukumar, 1988).

5.5. *Thraustochytrium striatum* Schneider, 1968. *Kieler Meeresforsch.*, **23:** 19. [For Fig. refer original paper]

Sporangia globose, 14.6-22 µm diam. with a thin or fairly thick, hyaline, smooth wall, a large central vacuole and striate plasm which cleaves into radiate sectors. Pseudorhizoid branched. Zoospores delimited in the sporangium and arranged around the central vacuole which enlarges to form a hollow sphere of spores;

becoming motile in the sporangium as the wall dissolves, ovate, ellipsoid, 2 x 4 µm, laterally biflagellate. Resting spores are unknown.

Description:-Based on Karling (1981).

Habitat: On pollen in sea water and mangroves.

Distribution: West Coast:-Goa: On mangroves (Raghukumar and Raghukumar, 1988).

East Coast:-Tamil Nadu (Mandapum)**:** On mangroves (Raghukumar and Raghukumar, 1988).

5.6. *Thraustochytrium* sp.

Habitat: On corals, algae and diatoms.

Distribution: West Coast:-Goa: On diatoms *Navicula* sp. and *Grammatophora* sp. (Raghukumar and Raghukumar, 1994); **Karnataka:** On *Bryopsis* sp. (Raghukumar and Raghukumar, 1994); **Lakshadweep Islands:** On corals and coral mucus (Raghukumar and Balasubramanian, 1991), on green alga *Cladophora frascatii* and red alga *Polysiphonia* sp. (Raghukumar and Raghukumar, 1994); **Central Arabian Sea:** On diatoms *Coscinodiscus* sp, *Navicula* sp., *Nitzschia* sp., *Grammatophora* sp. and *Melosira* sp. (Raghukumar and Raghukumar, 1994). **East Coast:- Tamil Nadu** (Mandapum)**:** On marine algae *Caulerpa cupressoides* (Raghukumar, 1978a), on green alga *Caulerpa cupressoides* (Raghukumar and Raghukumar, 1994).

6. *Ulkenia* A. Gaertner, 1977. *Veroff. Inst. Meeresforsch. Bremerh.*, **16**: 141; emend. Yokoyama, Salleh &Honda, 2007. Mycoscience, **48**: 323.

Thallus is epi- and endobiotic, thin-walled during active growth, with a rhizoidal system (syn. Ectoplasmatic net) endobiotic or interbiotic. Thallus is globose, sunglobose, or pear shaped, very variable in diameter with varying and differing cell content during development. Colonies are small. Ectoplasmic nets not well developed. At maturity, cell wall disappears yotally, and the protoplast becomes fully naked or creeps like an amoeba out of the sporangium wall by a small opening at the apical part of the sporangium. After settlement, the protoplast becomes a zoosporangium, which released heterokont zoospores. Cells possess astaxanthin, phenicoxanthin, echinenon, and ß-carotene. The genus is represented by 3 species in marine waters (Yokoyama & Honda, 2007).

Type species: *Ulkenia visurgensis* (Ulken) Gaertner
Habitat: Saprobic on organic matter in marine habitats.
Description:-Based on Yokoyama et al. (2007).

6.1. *Ulkenia visurgensis* (A. Ulken) A. Gaertner, 1977. *Veroff. Inst. Meeresforsch. Bremerh.*, **16:** 141. [For Fig. refer original paper]

Basionym: *Thraustochytrium visurgensis* Ulken, *Veroff. Inst. Meeresforsch. Bremerh.*, **9:** 289 (1965).

Sporangia are sessile, globose, 15-20 µm diam., with a 2 µm thick wall which swells and loosens, whereby the sporangia become irregular in shape. Zoospores becoming active in the sporangium and discharged as a swelling glutinous mass and swimming away; subglobose to reniform, 2.5-3.5 x 4-4.5 µm, heterocont, coming to rest and germinating directly.

Habitat: Isolated on pine pollen from marine algae and Corals.
Description:-Based on Karling (1981).
Distribution: West Coast:-Goa: On marine algae (Raghukumar and Raghukumar, 1988), on brown alga *Sargassum cinereum* (Sathe-Pathak et al., 1993); **Lakshadweep Islands**: On corals and coral mucus (Raghukumar and Balasubramanian, 1991).
East Coast:-Tamil Nadu (Mandapum)**:** On marine algae (Raghukumar and Raghukumar, 1988).

6.2. *Ulkenia amoeboidea* (G. Bahnweg & F. K. Sparrow) A. Gaertner, 1977. *Veroff. Inst. Meeresforsch. Bremerh.,* **16:** 141. [For Fig. refer original paper]

Basionym: *Thraustochytrium amoeboidea* Bahnweg & Sparrow, *Am. J. Bot.,* **61:** 764 (1974).

Sporangia are sessile, globose, subglobose, to broadly elliptical, 20-25 µm diam., with a thin hyaline wall. Pseudorhizoidal system monoaxial, branched; amoeboid content of sporangium emerging through an apical pore as an amoeboid mass to the outside where it becomes more rigid and undergoes euglenoid movement; later becoming quiescent, rounding up, and dividing successively into numerous (up to 64) daughter cells which at first remain connected by protoplasmic strands; later separating into amoeboid spores, 3.5-5 µm diam., which after a period of crawling encyst and develop rhizoids to become new secondary thalli and sporangia.

Zoospores biflagellate, heterocont, ellipsoid to ovoid, 2.5-3.5 x 5-6.5 µm, possibly formed from such above noted sporangia; anteriorly directed flagellum, thick, 8-12 µm long; posterior one shorter and thin.

Habitat: Isolated on brine shrimp in sea water and on Diatoms.

Description:-Based on Karling (1981).

Distribution: West Coast:-Goa: On Diatoms, *Coscinodiscus, Navicula, Nitzschia, Grammatophora, Melosira* from Central Arabian Sea and Mandovi estuary, Goa (Raghukumar, 1996).

Phylum: Chytridiomycota

7. *Chytridium* Braun, 1851. *Betrachtungen uber die Erscheinung der Verjungung in der Natur,* Leipzig, p.198.

Thallus is epi-endobiotic, eucarpic, monocentric; sporangia formed from epibiotic portion; operculate; sessile. Rhizoidal system is developing from endobiotic portion; variable in extent; apophysis present or absent. Resting spores, when known, endibiotic, sexually or asexually formed; thick-walled; on germination, producing an epibiotic, operculate sporangium. A rhizoidal portion of the thallus may be a simple, unbranched, peg-like tube, or a delicate, extensive, branched system that may arise from an apophysis. Because of its tenuos nature, the rhizoidal system is often difficult to observe.

Type species: *Chytridium olla* Braun

Habitat: Occurring in aquatic environment.

Description:-Based on Johnson and Sparrow (1961).

7.1. *Chytridium polysiphoniae* F. Cohn, 1865. *Hedwigia*, **4:** 169. [For Fig. refer original paper]

Basionym: *Rhizophydium polysiphoniae* (Cohn) Peterson, *Oversigt Kgl. Danske Vidensk. Selskabs Forhandl.*, p. 486 (1905).

= *Rhizophydium olla* Peterson, *Oversigt Kgl. Danske Vidensk. Selskabs Forhandl.*, p. 485 (1905).

Sporangia subspherical, subangular, pyriform, unceolate or spherical, sessile; wall smooth or densely punctuate, hyaline, yellowish or dark; 12-33 x 12-42 μm. Rhiziods branched or unbranched; apophysis absent. Zoospores 2.5- 4 μm in diam., postriorly uniflaggelate; spherical; escaping, on dehiscence of an operculum, through a terminal pore 7-13 μm in diam.; Resting spores not observed.

Habitat: On marine algae.

Description:-Based on Johnson and Sparrow (1961).

Distribution: West Coast:-Goa: On red alga *Centroceras clavulatum* (Raghukumar, 1986b; Raghukumar and Raghukumar, 1994); **Lakshadweep Islands:** On red alga *Centroceras clavulatum* (Raghukumar, 1996; Raghukumar and Raghukumar, 1994).

East Coast:-Tamil Nadu (Mandapum)**:** On red alga *Centroceras clavulatum* and brown alga *Sphacelaria* sp. (Raghukumar, 1987a; Raghukumar and Raghukumar, 1994).

8. ***Coenomyces*** Deckenbach, 1902-03. *Scripta Bot. Horti. Univ. Imper. Petro.,* **19:** 15.

= *Deckenbachia* Jaczewski, *Opredelitel gribov. I. Fikomitsety*, p.32 (1931).

Thallus is epi-endobiotic; hyphae slender, branched, distantly septate. Sporangia are extrametrical; borne at tips of hyphae, or lateral, sometimes sessile; with one (rarely two) long slender discharge tube. Zoospores are posteriorly uniflaggelate. Resting spores not observed.

Type species: *Coenomyces cobsuens* Deckenbach

Habitat: Occurring in gelatinous sheath of algae.

Description:-Based on Johnson and Sparrow (1961).

8.1. ***Coenomyces*** sp.

In *Cladophora repens* (J. Agardh) Harvey, the thallus epibiotic and endobitic, eucarpic consisting of a filamentous, branched, segmented loose hypha-like complex of tubes bearing terminal and intercalary turbinate swellings. Sporangia are sessile and inoperculate. The intercalary turbinate swellings (sporangis?) 15-17 μm long, with a 65-70 μm long discharge tube.

Habitat: Occurring in gelatinous sheath of algae.

Description:-Based on Raghukumar (1986a).

Distribution: West Coast:-Goa: On green algae *Cladophora repens* (Raghukumar, 1986a; Raghukumar and Raghukumar, 1994); **Gujarat** (Veraval): On green algae *Cladophora repens* and *C.fascicularis* (Raghukumar, 1996; Raghukumar and Raghukumar, 1994); **Lakshadweep Islands:** On green alga *Rhizoclonium* sp. (Raghukumar, 1986a).

9. *Olpidium* (Braun) Rabenhorst, 1868. *Flora Europaea algarum,* **3:** 288, *Sensu,* Schroeter, *Kryptogamen-Fl. Schlesien*, **3:** 180 (1985).

Thallus at first without discrete wall, then becoming walled; endobiotic, holocarpic, monocentric; converted entirely into a sporangium or resting spore. Spornagia are predominantly spherical or ellipsoidal; wall smooth; inoperculate; provided with one, sometimes several, discharge tubes of variable length. Zoospores posteriorly uniflaggelate; formed endogenously; escaping on apical rupture or deliquescence of discharge tube. Resting spores thick-walled, endobiotic; formed in some instances from a planozygote resulting from isoplanogametic fusion; producing zoospores on germination.

Type species: *Olpidium endophytum* (Braun) Rabenhorst

Habitat: Occurring in aquatic habitats.

Description:-Based on Johnson and Sparrow (1961).

9.1. *Olpidium rostriferum* Tokunaga, 1933. *Trans. Sapporo Nat. Hist. Soc.*, **13:** 80. [For Fig. refer original paper]

In *Cladophora frascatii* Collins and Harvey, most of the cells of infected algal filament become brown. Sporangia are 8-15 per host cell, spherical to ovoid, 14-16 µm wide with thin smooth wall. Discharge tubes single per sporangium, 25-30 µm in length, strongly constricted when passing through the host cell wall and extend to various distances outside the host cell wall. Zoospores are 3-4 µm wide, spherical, with a single flagellum.

Habitat: Parasitic on green algae.

Description:-Based on Raghukumar (1986a).

Distribution: West Coast:-Goa: On green algae *Cladophora frascatii* (Raghukumar, 1986a; Raghukumar and Raghukumar, 1994); **Gujarat** (Veraval): On green alga *Rhizoclonium* sp. (Raghukumar,1996; Raghukumar and Raghukumar, 1994); **Lakshadweep Islands**: On green algae *Cladophora frascatii* (Raghukumar, 1986a; Raghukumar and Raghukumar, 1994).

East Coast:-Tamil Nadu (Mandapum)**:** On green algae *Cladophora expansa, C. fascicularis* and *Rhizoclonium* sp. (Raghukumar, 1987a, 1996; Raghukumar and Raghukumar, 1994).

9.2. ***Olpidium*** sp.

Habitat: On green algae.

Distribution:-East Coast:-Tamil Nadu (Mandapum)**:** On green algae *Rhizoclonium* sp (Raghukumar, 1987a).

Phylum: Oomycota

10. ***Lagenidium*** Schenk, 1859. *Verhandl. Phys. – Med. Gesell. Wurzburg, A. F.,* **9:** 27.

= *Reticularia* Dangeard, *Le Botaniste*, **Ser. 2**, p. 96 (1890-91).

= *Lagenidiopsis* de Wildeman *Ann. Soc. Belge Micro.* (Mem.), **20:** 109 (1896).

Thallus usually endobiotic or extensively ectobiotic, 1-celled and olpidioid and occupying a single host cell, or extensive, tubular, cylindrical, filamentous, contorted, irregular, branched or un-branched and extending through several host cells; becoming transversely septate into segments which may remain attached or disarticulate and separate; segments transformed into sporangia or gamentangia. Sporangia usually of the same size and shape as the segments or unicellular thalli, with 1 or 2, or up to 20 or more, short, long, curved, tortuous, branched or un-branched exit tubes whose wall may be thick-ended at the base to form a spreading or forcing apparatus, or inflated at the apex, or lacking and replaced by a pore in the sporangium; exit tubes ending at the surface of host cell or extending short or long distances outside; content often emerging to form a globular mass at the exit orifice. Zoospores bean-shaped, reniform, or somewhat pyriform, biflagellate with several small refractive globules; primary zoospores when isocont, secondary ones heterocont. Zoospores formed (1) extramatrically by cleavage of the emerged mass of protoplasm which may be naked or surrounded by a vesicular membrane as in *Pythium*; (2) delimited in the sporangium and completing maturation at the exit orifice; (3) completely formed in the sporangium, emerging and swimming away; (4) primary zoospores emerging and encysting at the exit orifice and germinating to form secondary ones as in *Achlya*. Gamentangia more differentiated than those of *Myzocytium*. Antheridia borne on the same thalli with the oogonia, usually elongate, cylindrical, and developing the conjugation canal; oogonia terminal or intercalary; differentiation of an oosphere and periplasm lacking; fertilization through a canal; oospores sexually formed or parthenogenetic, in the former case lying free in the oogonia, usually spherical or ellipsoidal, thick-walled, smooth or with one to several refractive globules; functioning as a sporangium in germinating.

Type species: *Lagenidium giganteum* Couch (Dick, 2001).
Habitat: Species of this genus occur as parasites of algae, fungi, rotifers and their eggs, nematodes, eggs of barnacles, crabs, oysters, copepods, Daphne and mosquito larvae. Few species are saprophytic in soil and water.
Description:-Based on Johnson & Sparrow (1961) and Karling (1981).

10.1. *Lagenidium callinectes* Couch, 1942. *J. Elisha Mitchell Sci. Soc.*, **58:** 158. [For Fig. refer original paper]

Thalli predominantly endobiotic, mycelioid, coarse, irregularly branched, septate at maturity, thin-walled, 5.4-12.6 µm, or sometimes 5-50 µm diam., each segment becoming a sporangium with an ectobiotic tubular discharge tube, 29 x 25-27 µm, or sometimes 80-1500 µm; protoplasm emerging as an irregular, subspherical or spherical mass, 14-200 µm diam., enveloped by a persistent thick gelatinous vesicoospores. Zoospores partly formed in the sporangium in and completing their maturation in and swarming in the vesicle and liberated by its rupture; tapering at the anterior and rounded at the posterior end, 10.9 (7.2-12) x 13.4 (10.8-14.4) µm, with a diagonal groove, containing several oil globules; laterally biflagellate; isocont, anteriorly directed flagellum tinsillate, posterior one of the whip-lash type; encysting, monoplanetic; cystospores oblong or subspherical, 10-11.3 µm diam. Resting spores are asexually formed, spherical, subspherical, or ovoid, 18-30 µm diam., with a 3 um thick walls; containg pale whitish cytoplasm and an eccentric mass of oil globules; germination unknown.

Habitat: Parasitic in eggs and larvae of *Callinectes sapindus*, *Pinnotheres ostreum* and *Neopenope texiana*; ova of *Chelinibia patula*; on *Ectocarpus* sp. *Chordaria* sp. and *Ceramium* sp.
Description:-Based Johnson and Sparrow (1961) and Karling (1981).
Distribution: East Coast:-Tamil Nadu: Parasitic in larvae of Indian tiger prawn *Penaeus monodon* (Ramasamy et al., 1996; Karunasagar et al., 2004; Ramaiah, 2006).

11. *Sirolpidium* Petersen, *Oversigt Kgl. Danske Videnskab. Selskab. Forhandl.*, **5:** 478.

Thallus is predominantly endobiotic but becoming partially ectobiotic under certain conditions; olpidiod or elongate and filamentous, transformed into a sporangia or becoming septate and fragmenting to form a row of septate sporangia. Zoosporangia usually several in a host cell, variable in size and shape with one simple or branched exit tube which varies markedly in length and may extend considerably beyond the host wall. Zoospores isocont, with two flagella attached at or near the anterior end oppositely directed; fully formed and swarming in the sporangium; emerging fully developed and swimming directly away; occassinally liberated in the host cell. Sexuality unknown; resting spores apparently asexually formed.

Type species: *Sirolpidium bryopsidis* (de Bruyne) Petersen
Habitat: Occurring in marine algae and animals.
Description:-Based on Johnson and Sparrow (1961) and Karling (1981).

11.1. *Sirolpidium bryopsidis* (de Bruyne) Petersen, 1905. *Oversigt Kgl. Danske Vidensk. Selskabs Forhandl.,* **5:** 480. [For Fig. refer original paper]

Basionym: *Olpidium bryopsidis* de Bruyne, *Arch.de Biol.*, **10:** 85 (1890).

Thallus is endobiotic, at first plasmodial in shape, elongate, irregular, septate; disarticulating into elongate, oblong, ovate and olpidiod subthalli 7-16 x 18-90 µm; olpidiod ones 12-17 x 18-38 µm, with a thin hyaline wall. Sporangia of the same size and shape as thalli and disarticulated segments, with a narrow, cylindrical exit tube up to 300 µm long and extending for varying distances beyond the host. Zoospores fully formed and swarming in the sporangium, narrowly pyriform, somewhat arched or nephroid, 2-3 x 4-6 µm, with two laterally inserted flagella, isocont. Resting spores uncertain (Description based on Karling, 1981).

In *Cladophora frascatii* Collins and Harvey, the terminal cell of the infected algal filament is golden brown and swollen (Hypertrophied). Golden brown patches alternate with hyaline patches of fungal sporangia with a discharge tube penetrating the host cell wall. Five to seven sporangia are in a row per host cell. Adjacent sporangia produce discharge tubes on alternate sides. Sporangia are 15-25 x 10-15 µm, smooth, hyaline and appressed to the host cell wall. The discharge tubes are simple or branched, 10-12 x 2.5 µm, and extend beyond the host cell wall.

Zoospores 2-4 μm long, with two flagella (Description based on Raghukumar, 1986a). .

Habitat: Parasitic on green algae *Bryopsis* and *Cladophora* sp.

Description:-Based on Karling (1981) and Raghukumar, (1986a).

Distribution: West Coast:-Goa: On green algae *Cladophora frascatii* (Raghukumar, 1986a; Raghukumar and Raghukumar, 1994); **Lakshadweep Islands:** On green algae *Cladophora frascatii* (Raghukumar, 1986a, 1996; Raghukumar and Raghukumar, 1994).

East Coast:-Tamil Nadu (Tuticorin)**:** On green algae *Cladophora gracilis* (Raghukumar, 1987a, 1996; Raghukumar and Raghukumar, 1994).

12. *Pontisma* Petersen, 1905. *Oversigt Kgl. Danske Vidensk. Selskabs Forhandl.,* p. 482.

Thallus is holocarpic, endobitic; either a branched or unbranched tube that becomes segmented at maturity, or an olpidioid or tubular, nonseptate cell; segments typically not disarticulating. Sporangia formed by direct conversion of segments; provided with one (or sometimes several) discharge tube. Zoospores are biflaggelate; pyriform; endogenously formed. Resting spores are spherical, thick-walled.

Type species: *Pontisma lagenidioides* Petersen

Habitat: Occurring in marine algae.

Description:-Based on Johnson and Sparrow (1961).

12.1. *Pontisma lagenidioides* Petersen, *Oversigt Kgl. Danske Vidensk. Selskabs Forhandl.,* p. 482. [For Fig. refer original paper]

Basionym: *Sirolpidium lagenidioides* (Petersen) Karling, In: *Simple Holocarpic Biflaggelate Phycomycetes,* New York, p. 66 (1942).

Thallus endobiotic, occasionally unicellular, lobed, curved and olpidioid; usually elongate, irregular, frequently diarticulating into isolated segments, 11-30 x 20-60 μm, olpidioid segments 14-16 x 13-15 μm high; tubular, non-septate ones 130-200 μm long. Sporangia of same size and shape as thalli segments, or subthalli, with 1 to several cylindrical, tapering exit tubes, up to 145 μm long, and extending a short distance through the host wall. Zoospores fully formed and swarming in the sporangium, and emerging by flagellar action; bent, rod-shaped, reniform, or pyriform, 2.5-3 x 4.5-7 μm, with a strongly refractive region at either end, with two short laterally attached flagella; motion erratic, tumbling. Resting spores are spherical, 27 um diam., somewhat thick-walled with several refractive oil drops; germination is unknown. (Desription is based on Karling, 1981).

Infected filaments of alga (*Chaetomorpha media*) show a gradual browning from the tip downward. The young fungal thallus at early stages of development appears knob-like inside the host cell. The thallus further develops into a tubular filament 15-20 µm long, becomes septate and then fragment into a number of segments. The segments 30-60 µm long, give rise to sporangia. Sporangia olpidioid (14-20 µm wide) or elongate (30-60 µm long) or lobed. Discharge tubes simple or branched (40-100 µm long), within the host cell or penetrate the host cell and projected out for a great distance (100-120 µm long). Zoospores come out singly from the tip of the discharge tube, and swam away directly or encysting inside the host cell. The zoospores are with two oppositely directed flagella. (Description based on Raghukumar, 1987a).

Habitat: Weakly parasitic or saprophytic in marine green algae.

Description:-Based on Karling (1981) and Raghukumar (1987a).

Distribution: West Coast:-Goa: On green algae. *Chaetomorpha media* (Raghukumar and Raghukumar, 1994); **Gujarat** (Veraval)**:** On green alga *Valoniopsis pachenema* (Raghukumar, 1996; Raghukumar and Raghukumar, 1994).

East Caost:-Tamil Nadu (Mandapum)**:** On green algae. *Chaetomorpha media* (Raghukumar, 1987a; Raghukumar and Raghukumar, 1994).

Phylum: Oomycota *incertae sedis*

13. *Ostracoblabe implexa* Bornet & Flahault, 1889. *Bull. Soc. Bot. France, Ser.* 2, 11: CXLVII-CXLXXVI. [For Fig. refer original paper]

Ostracoblabe implexa Bornet & Flahault causes a shell disease in oysters, and since its discovery in 1889 it has been reported several times in marine waters of Western Europe and Canada. It consists of a straight mycelium, 2 um diam., which bears at intervals of 40-100 um. Swellings, 4 x 6 um, called prochlamydospores by Alderman & Jones (1971). So far no fruiting structures have been shown that they contain a nucleus closely appressed to one of its walls. The mycelium becomes septate in senescent cultures, and chlamydospores are formed from the prochlamydospores at low temperature. *O. implexa* has been regarded as the hyphal component of a lichen or as having pythiaceous affinities. So far as it is known there are no cogent grounds for includings for including this fungus in the biflagellate phycomycetes (Karling, 1981).

Habitat: Parasitic on oysters,

Description:-Based on Karling (1981)

Distribution:-West Coast:-Lakshadweep Islands: On Rock Oyster *Crassostrea cucullata* (Raghukumar and Lande, 1988).

Note: Eckblad and Kristinsen (1989) studied the distribution and culture of the causal agent of oyster disease *Ostracoblabe implexa*, and stated that this fungus should be considered a distinct genus and species, not a species of *Arthopyrenia*.

Phylum: Ascomycota

Classification of the marine Ascomycota of India

The following classification is based on Hibbett et al. (2007), Cannon and Kirk (2007) and Jones et al. (2009a).

		Page No.
Phylum:	**ASCOMYCOTA**	
Subphylum:	**Pezizomycotina**	
Class:	**Dothideomycets**	
Sub-class:	**1) Dothideomycetidae**	
	2) Pleosporomycetidae	
Sub-class:	**Dothideomycetidae**	
	1. Capnodiales	
	2. Dothideales	
Order	**1. Capnodiales**	
	Mycosphaerellaceae	
	Mycosphaerella	229
	Pharcidia	259
	2. Dothideales *incertae sedis*	
	Botryosphaeriaceae	
	Belizeana	117
	Passeriniella	252
	Dothideomycetidae family *incertae sedis*	
	Lautosporaceae	
	Lautospora	202
	Zopfiaceae	
	Coronopapilla	151
Sub-class:	**Pleosporomycetidae**	
	Pleosporales	

	Pleosporales	
	Aigialaceae	
	Aigialus	82
	Ascocratera	108
	Rimora	266
	Amniculicolaceae	
	Neomassariosphaeria	234
	Didymosphaeriaceae	
	Didymosphaeria	163
	Testudinaceae	
	Verruculinia	296
	Leptosphaeriaceae	
	Leptosphaeria	203
	Lophiostomataceae	
	Lophiostoma	217
	Paraliomyces	251
	Quintaria	261
	Melanommataceae	
	Acrocordiopsis	81
	Astrosphaeriella	110
	Bicrouania	121
	Caryosporella	126
	Morosphaeriaceae	
	Helicascus	189
	Morosphaeria	227
	Phaeosphaeriaceae	
	Lautitia	200
	Phaeosphaeria	256
	Pleosporaceae	
	Pleospora	260
	Teichosporaceae	
	Byssothecium	123
	Trematosphaeriaceae	
	Halomassarina	176
	Trematosphaeria	293
	Pleosporales incertae sedis	
	Biatriospora	118
	Didymella	161
	Halothia	187
	Julella	190
	Kirschsteiniothelia	196
	Lineolata	215
	Salsuginea	277
	Dothideomycetidae *incertiae sedis*	
	Patellariaceae	

	Patellaria	250
	Jahnulales	
	Hypostromataceae	
	Manglicola	225
Class:	**Lecanoromycetes**	
Sub-class:	**Lecanoromycetidae**	
	Lacanorales	
	Dactylosporaceae	
	Dactylospora	158
Class:	**Leotiomycetes**	
Sub-class	**Lecanoromycetidae**	
	Helotiates	
	Helotiaceae	
	Amylocarpus	89
Class:	**Arthoniomycetes**	
	Arthoniales Roccellaceae	
	Roccellaceae	
	Halographis	174
Class:	**Sordariomycetes**	
Sub-class:	**1) Hyphocreomycetidae**	
	2) Sordariomycetidae	
	3) Xylariomycetidae	
Sub-class:	**Hyphocreomycetidae**	
	1. Hyphocreales	
	2. Halosphaeriales	
	1. Hyphocreales	
	Bionectriaceae	
	Heleococcum	188
	Kallicroma	192
	Hyphocreaceae	
	Payosphaeria	254
	Hyphocreales *incertiae sedis*	
	Torpedospora	291
	Swampomyces	285
	Etheirophora	166
	2. Halosphaeriales	
	Halosphaeriaceae	
	Aniptodera	90
	Antennospora	98
	Arenariomyces	101
	Bathyascus	112
	Carbosphaerella	125
	Ceriosporopsis	128
	Corollospora	133
	Haiyanga	172

	Halosarpheia	180
	Halosphaeria	186
	Kochiella	198
	Lignincola	208
	Limacospora	211
	Natantispora	232
	Neptunella	236
	Nimbospora	238
	Oceanitis	241
	Ocostaspora	244
	Okeanomyces	246
	Ophiodeira	247
	Panorbis	248
	Remispora	263
	Saagaromyces	269
	Thalassogena	288
	Tirispora	289
Sub-class:	**Sordariomycetidae**	
	1. Diaporthales	
	2. Chaetosphaeriales	
	3. Sordariales	
	4. Ophiostomatales	
	1. Diaporthales	
	Valsaceae	
	Cryptovalsa	155
	2. Sordariales	
	Lasiosphaeriaceae	
	Biconiosporella	120
	Chaetomiaceae	
	Zopfiella	299
	Sordariales *incertae sedis*	
	Abyssomyces	80
	Savoryella	279
Sub-class:	**Xylariomycetidae**	
	Xylariales	
	Diatrypaceae	
	Cryptosphaeria	154
	Eutypa	169
	Eutypella	171
	Pedumispora	255
	Xylariaceae	
	Anthostomella	101
	Halorosellinia	177
	Sordariomycetes *incertae sedis*	
	Lulworthales	

	Koralionasteles	
	Phyllachorales	
	1. Lulworthales	
	Lulworthaceae	
	Koralionasteles	
	Phyllachorales	
	1. Lulworthales	
	Lulworthaceae	
	Lindra	213
	Lulworthia	218
	2. Koralionastetales	
	Koralionastetaceae	
	Koralionastes	199
	Phyllachorales *incertae sedis*	
	Marinosphaeria	226
	Unitunicate Ascomycetes genera *incertae sedis*	
	Biflua	122
	Crinigera	153
	Dryosphaera	164
	Rhizophila	264
	Saccardoella	274

Phylum : Ascomycota

14. ***Abyssomyces*** Kohlm., 1970. *Ber. Dtsch. Bot. Ges.*, **83**: 505.

Ascomata: solitary, subglobose or pyriform, superficial, ostiolate, periphysate, papillate, coriaceous, brown, setose, paraphyses absent. *Asci*: 8-spored, subcylindrical or fusiform, pedunculate, unitunicate, thin-walled, persistent, developing at the base of the ascocarp venter. *Ascospores*: subcylindrical, 3-septate, hyaline, with a semiglobose appendage at each end. The genus is monotypic (Jones et al., 2009a).

Type species: *Abyssomyces hydrozoicus* Kohlm.
Habitat: Parasitic or saprobic on marine invertebrates (Hydrozoa).
Description: Based on Kohlmeyer and Kohlmeyer (1979).

14.1. ***Abyssomyces hydrozoicus*** Kohlm., 1970. *Ber. Dtsch. Bot. Ges.*, **83**: 505.
[For Fig. refer original paper]

Ascomata: 130-155 µm high, excluding necks, 135-155 µm in diam, solitary, subglobose to pyriform, superficial, ostiolate, periphysate, papillate, coriaceous, light brown, setose, paraphyses absent. Necks: 95-120 x 40-55 µm, subconical to cylindraical, periphysate. *Peridium*: 12-20 µm thick, 3-layered; on the outside a crustose, brown, non-cellular layer,; middle layer of truncate to ellipsoidal cells, hyaline, forming setae; inner layer of flattened cells, hyaline. Setae: 8-60 µm long, 3.5-5 µm in diam at the base, 2-2.5 µm at the apex, subuliform, non or one-septate. *Paraphyses*: absent. *Asci*: 60-90 x 6-9 µm, 8-spored, subcylindrical to fusiform, short pedunculate, unitunicate, thin-walled, developing at the base of the ascomata venter. *Ascospores*: 18-20 x 3.5-4 µm, biseriate, subcylindrical, 3-septate, straight or slightly curved, not constricted at the septa, hyaline, appendages at each end, sometimes surrounded by a mucous cover; appendages 0.5-1 µm in diam, hyaline, semiglobose, cap-like, terminal.

Habitat: Parasitic or saprobic on chitinous hydrorhiza and hydrocaulon of hydrozoa attached to stony rocks.
Description: Based on Kohlmeyer and Kohlmeyer (1979) and Kohlmeyer and Volkmann-Kohlmeyer (2003).
Distribution: West Coast: On deep sea sediments in the Arabian sea (Raghukumar, and Raghukumar, 1998).

15. *Acrocordiopsis* Borse & K.D. Hyde, 1989. *Mycotaxon,* **34**: 535.

Ascomata: conical or semiglobose, seated on a thin black stroma that covers the wood surface; also next to the large ascomata are similar smaller fruiting bodies , containing hyaline, one-celled spores, most probably spermatia of the same species; superficial, solitary or gregarious, black, ostiolate, carbonaceous. *Peridium*: thick, composed of 2-3 layers. *Pseudoparaphyses*: numerous, filiform, persistent, anastomosing, septate. *Asci*: 8-spored, bitunicate with an apical thickening and an ocular chamber, cylindrical, pedunculate. *Ascospores*: hyaline or slightly yellow, 1-septate, constricted at the septum, obovoidal or ellipsoidal, guttulate, lacking a sheath or appendages. The genus is represented by two species in marine habitats (Jones et al., 2009a).

Type species: *Acrocordiopsis patilii* Borse & K.D. Hyde
Habitat: Saprobic on intertidal mangrove wood.
Description: Based on Borse and Hyde (1989) and Jones et al. (2009a).

15.1. *Acrocordiopsis patilii* Borse & K.D. Hyde, 1989. *Mycotaxon,* **34**: 535.

[Fig. 15.1]

Ascomata: 1-3 mm in diam, 1-2 mm high, conical or semiglobose, seated on a thin black stroma that covers the wood surface; also next to the large ascomata are similar smaller fruiting bodies , containing hyaline, one-celled spores, most probably spermatia of the same species; superficial, solitary or gregarious, black, ostiolate, carbonaceous. *Peridium*: 250-350 µm thick, 2-3 layered; outer layer 240-338 µm, thick, black, pseudoparenchymatous, composed of thick-walled black cells; inner layer 9-14 µm thick, hyaline light-brown, merging with pseudoparaphyses, composed of irregular elongate thin-walled cells. A third layer of rectangular thick-walled cells, brown in colour, fills the angle between the cover and basis of the ascomata. *Pseudoparaphyses*: 0.5-1.5 µm, hyaline, numerous, filiform, persistent, anastomosing and sparsely septate. *Asci*: 150-250 x 12-16 µm, 8-spored, bitunicate, with an apical thickened refractive region and an ocular chamber, cylindrical, pedunculate, developing from base and corners of the ascomata. *Ascospores*: 16-25 x 10-16 µm, hyaline or slightly yellow, 1-septate, constricted at the septum, ovoidal or ellipsoidal, one large guttule per cell, lacking a sheath or appendages.

Habitat: Saprobic on intertidal mangrove wood.

Description: Based on Borse and Hyde (1989) and Jones et al. (2009a).

Distribution: West Coast:-Maharashtra: On intertidal wood of *Avicennia alba, A, officinalis* and *Rhizophora mucronata* (Borse and Hyde, 1989: Borse et al., 2005a).

East Coast:-Andhra Pradesh: On intertidal wood (Nambiar and Raveendran, 2009j); **West Bengal:** On intertidal wood of *Avicennia marina* and *A. alba* (Pawar et al., 2005); **Andaman & Nicobar Islands:** On intertidal wood of *Avicennia marina, A. officinalis, Rhizophora apiculata, R. mucronata* and *Sonneratia alba* (Chinnaraj, 1993a).

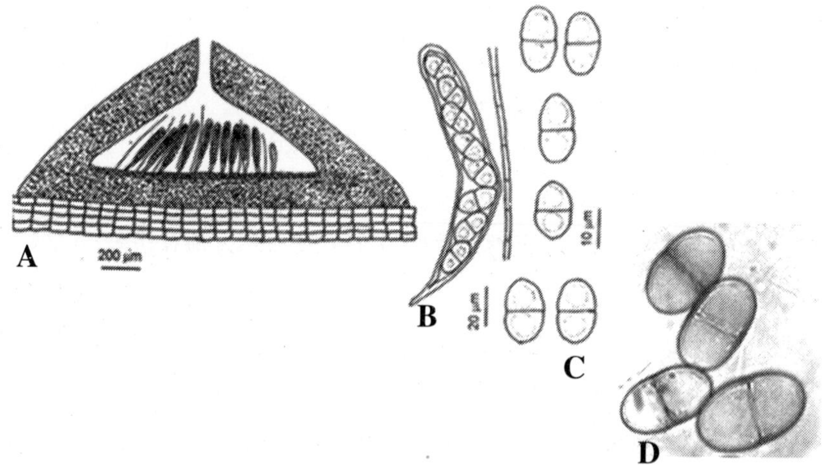

Fig. 15.1. ***Acrocordiopsis patilii,*** A-D; A- v. s. of ascocarp, B- ascus and paraphysis, C- ascospores, D- ascospores (1000x).

16. ***Aigialus*** Kohlm. & S. Schatz, 1985. *Trans. Br. Mycol. Soc.,* **85**: 669.

Ascomata: subglobose in frontal view, fusiform in sagittal section, laterally compressed, completely or ¾ immersed in a black stroma, with a longitudinal furrow at the top, ostiolate, apapillate, carbonaceous to coriaceous, black, gregarious. *Peridium*: 2-layered; outer stratum composed of elongate more or less irregular cells, encrusted with melanin particles, interspersed with cells of host; inner stratum composed of smaller, hyaline cells that merge with pseudoparaphyses; the peridium extends at the base into the locule with brown pseudoparaphysate strands. *Ostiole*: depressed or slightly projecting, in the centre of the apical furrow, circular; ostiolar canal subglobose, filled with a network of branched or forked septate periphyses. *Pseudoparaphyses*: trabeculate, unbranched at the base, becoming branched and anastomosing above the asci, merging with the periphyses in the ostiolar canal, embedded in a gelatinous matrix. *Asci*: 8-spored, cylindrical, pedunculate, thick-walled, fissitunicate, with a refractive apical plate in the ectoascus and a refractive apical ring in the endoascus, not bluing in IKI, arising from a basal ascogenous tissue. *Ascospores*: biseriate, ellipsoidal to broadly fusiform in side view (flat on one side, convex on the other), muriform, slightly constricted at the septa, dark brown except for hyaline to light brown apical cells, glabrous, apical and subapical cells covered by a gelatinous sheath or cap. The genus is represented by five species (Suetrong et al. 2009).

Type species: *Aigialus grandis* Kohlm. & S. Schatz
Habitat: Saprobic on intertidal mangrove wood.
Description: Based on Kohlmeyer and Schatz (1985).

Key for *Aigialus* species from India:

1 Ascospopres with 6 (rarely 8)-transverse-septate, 35-55 x 10-16 µm .... *Aigialus mangrovei*

1' Ascospores with more than 8-transverse-septate 2

2 (1') Ascospores with (9)-10-11-(12)-transverse-septate, 49-72 x 19-26 µm .... *Aigialus parvus*

2' (1') Ascospores with more than 13 transverse-septate 3

3 (2') Ascospores rarely longi-septate in subapical cells, 67-101 x 18-29 µm, ... *Aigialus grandis*

3' (2') Ascospores always long-septate in subapical cells, 70-95 x 18-25 µm, ... *Aigialus rhizophorae*

16.1. ***Aigialus grandis*** Kohlm. & S. Schatz, 1985. *Trans. Br. Mycol. Soc.*, **85**: 669. [Fig. 16.1]

Ascomata: 950-1215 µm high, 945-1500 µm wide, 241-476 µm thick, subglobose in frontal view, fusiform in sagittal section, laterally compressed, completely or ¾ immersed in a black stroma, with a longitudinal furrow at the top, ostiolate, apapillate, carbonaceous to coriaceous, black, gregarious, often close together.

Peridium: 2-layered; outer stratum 50-100 µm thick and clypeoid near the ostiole, 30-50 um at the sides, composed of elongate more or less irregular cells, encrusted with melanin particles, interspersed with cells of host; inner stratum 10-15 µm thick, composed of smaller, hyaline cells that merge with pseudoparaphyses; the peridium extends at the base into the locule with brown strand composed of pseudoparaphyses; that separate above, become hyaline and are surrounded by a gelatinous matrix. *Ostiole*: 30-90 µm diam, depressed or slightly projecting, in the centre of the apical furrow, circular; ostiolar canal subglobose, filled with a network of branched or forked septate periphyses 3-5 µm thick, filling the ostiole. *Pseudoparaphyses*: 1.5-2 µm diam, trabeculate, unbranched at the base, becoming branched and anastomosing above the asci, merging with the periphyses in the ostiolar canal, embedded in a gelatinous matrix. *Asci*: 380-520 x 26-36 µm, 8-spored, cylindrical, long-pedunculate, thick-walled, fissitunicate, with a refractive apical plate, 6-12 µm diam, in the ectoascus and a refractive apical ring in the endoascus; ring not bluing in IKI, 12-13 µm diam, 2.5-5 µm thick, tear-shaped in optical section; asci arising from a basal ascogenous tissue. *Ascospores*: 67-97 µm long, 18-28 µm broad, 14-20 µm thick, biseriate, ellipsoidal to broadly fusiform in frontal view, in side view flat on one side, convex on the other, muriform, with 14-17 trans-septa and 1-3 longisepta in all but the end cells, slightly constricted at the septa, yellow-brown except for hyaline to light brown apical cells, glabrous, with a gelatinous cap, 2.5-3 µm thick, around the apical and subapical cells.

Habitat: Saprobic on intertidal roots and branches of mangroves.

Description: Based on Kohlmeyer and Schatz (1985).

Distribution: West Coast:-Gujarat: On intertidal wood of *Rhizophora mucronat* (Borse et al., 2000a), on intertidal wood and mangrove wood (Patil and Borse, 2001); **Maharashtra:** On intertidal wood of *Rhizophora mucronata, Sonneratia acida* and *S. apetala* (Borse, 1987c, 1988, Srivastava, 1989); **Goa:** On intertidal prop roots of *Rhizophora mucronata* (Borse et al., 1999a); **Kerala:** On intertidal mangrove wood of *Avicennia* sp. and *Rhizophora* sp. (Raveendran and Manimohan, 2007; Nambiar et al., 2008a, Nambiar and Raveendran, 2009d, e, f, g, h, i).

East Coast:-Tamil Nadu: On submerged parts of *Rhizophora* spp. (Ravikumar and Vittal, 1996); **Andhara Pradesh:** On intertidal wood of *Avicennia marina, Rhizophora apicula,* and *R. mucronata* (Sarma and Vittal, 1998-99; Sarma and Vittal, 2000; Sarma and Vittal, 2001; Sarma et al., 2001; Sarma and Vittal, 2002; Sarma and Vittal, 2004; Vittal and Sarma, 2005); **West Bengal:** On intertidal wood of *Ceriops tagal; Excoecaria agallocha* and *Sonneratia apetala* (Pawar et al., 2005); **Andaman & Nicobar Islands:** On intertidal wood of *Avicennia marina, Rhizophora apiculata,* and *R. mucronata* (Chinnaraj, 1993a).

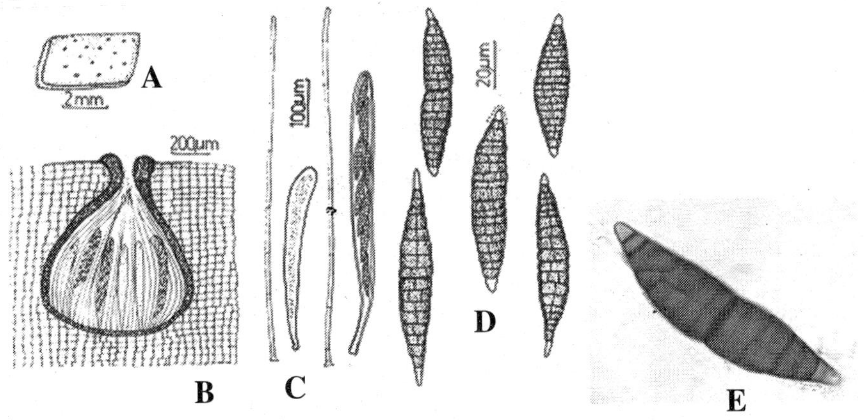

Fig. 16.1. ***Aigialus grandis,*** A-E; A- Habit, B- v. s. of ascocarp, C- asci and paraphyses, D- ascospores, E- ascospore (1000x).

16.2. ***Aigialus mangrovei*** Borse, 1987c. *Trans. Br. Mycol. Soc.,* **88**: 424. [Fig. 16.2]

Ascomata: 600-850 µm high, 600-800 µm wide, 150-200 µm thick. *Ostiole*: 40-50 µm diam. *Pseudoparaphyses*: 1.5-2 µm diam. Asci: 300-425 x 20-30 µm. *Ascospores*: 35-55 x 10-16 µm, muriform, with 6-7 transverse septa and 1-2 longitudinal septa in all but the end cells, slightly constricted at the septa, yellow-brown except for the hyaline or light brown apical cells, glabrous, with a gelatinous cap, around the apical and subapical cells may present.

Habitat: Saprobic on intertidal prop roots of *Rhizophora mucronata*.

Description: Based on Borse (1987c).

Distribution: West Coast:-Maharashtra: On proproots of *Avicennia marina* and *Rhizophora mucronata* (Borse, 1987c, 1988, Borse et al., 2005a); **Karnataka:** on intertidal wood of *Avicennia officinals* and *Rhizophora mucronata* (Maria and Sridhar, 2004); **Pondicherry (Mahe):** On intertidal wood (Nambiar and Raveendran, 2007, 2008a, c); **Kerala:** On intertidal wood (Raveendran and Manimohan, 2007), on intertidal mangrove wood (Nambiar and Raveendran, 2006, 2007, 2008a, d, 2009a, b, e, g, 2010c; Nambiar et al., 2008a).

East Coast:-Tamil Nadu: On submerged parts of *Rhizophora* spp. (Ravikumar and Vittal, 1996),), on mangrove wood (Nambiar and Raveendran, 2008e, 2009l); **Andhara Pradesh:** On intertidal wood of *Avicennia marina* and *A. officinails* (Sarma and Vittal, 1998-99; Sarma and Vittal, 2000; Sarma and Vittal, 2001; Sarma et al., 2001; Sarma and Vittal, 2004; Vittal and Sarma, 2005), on intertidal wood (Nambiar and Raveendran, 2009j); **West Bengal:** On intertidal wood of *Avicennia marina* and *A. alba* (Pawar et al., 2005); on mangrove woody litter (Shini et al., 2009-10); **Andaman & Nicobar Islands:** On intertidal wood of *Avicennia marina, A. officinalis, Rhizophora apiculata* and *Sonneratia alba* (Chinnaraj, 1993a).

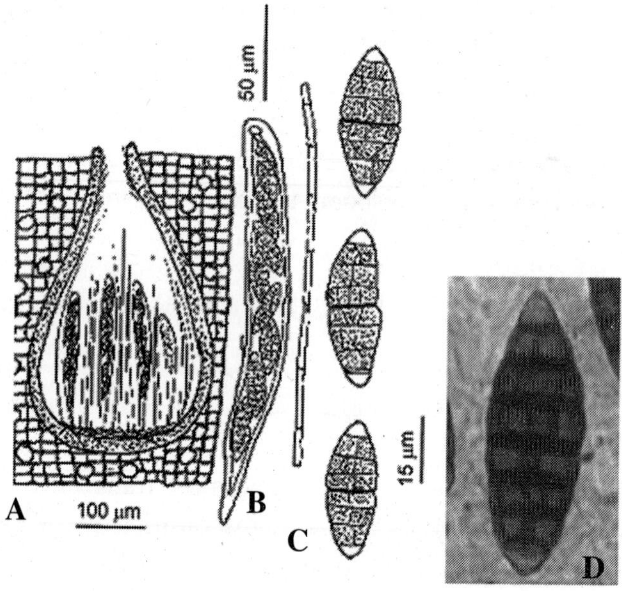

Fig. 16.2. ***Aigialus mangrovei***, A-D; A- v. s. of ascocarp, B- ascus and paraphysis, C- ascospores, D- ascospore (1000x).

16.3. ***Aigialus parvus*** S. Schatz & Kohlm., 1985. *Trans. Br. Mycol. Soc.,* **85**: 774. [Fig. 16.3]

Ascomata: 788-1257 µm high, 954-1333 µm wide, 305-405 µm thick, subglobose in frontal view, fusiform in sagittal section, laterally compressed, completely or ¾ immersed in a black stroma, with a longitudinal furrow at the top, ostiolate, carbonaceous to coriaceous, black, gregarious, often close together. *Peridium*: 2-layered; outer stratum 40-90 µm thick and clypeoid near the ostiole, 15-45 µm at the sides, composed of elongate more or less irregular cells, encrusted with melanin particles, interspersed with cells of host; inner stratum 7-15 µm thick, composed of smaller, hyaline cells that merge with pseudoparaphyses; the peridium extends at the base into the locule with brown strand composed of pseudoparaphyses that separate above, become hyaline and are surrounded by a gelatinous matrix. *Ostiole*: 50-70 µm diam, depressed or slightly projecting, in the centre of the apical furrow, circular; ostiolar canal subglobose, filled with a network of branched or forked septate periphyses 3-4 µm thick, filling the ostiole. *Pseudoparaphyses*: 1.5-2 µm diam, trabeculate, unbranched at the base, becoming branched and anastomosing above the asci, merging with the periphyses in the ostiolar canal, embedded in a gelatinous matrix. *Asci*: 300-420 x 30-35 µm, 8-spored, cylindrical, pedunculate, thick-walled, fissitunicate, with a refractive apical plate in the ectoascus and a refractive apical ring in the endoascus, not bluing in IKI, asci arising from a basal ascogenous tissue. *Ascospores*: 49-72 µm long, 19-26 µm broad, 16-21 µm thick, biseriate, ellipsoidal to broadly fusiform in frontal view, in side view flat on one side, convex on the other, muriform, with 10-11 trans-septa and 1-3 longisepta in all but the end cells, slightly constricted at the septa, yellow-brown except for hyaline to light brown apical cells, glabrous, with a gelatinous cap, around the apical and subapical cells.

Habitat: Saprobic on intertidal roots and branches of mangroves.

Description: Based on Kohlmeyer and Schatz (1985).

Distribution: West Coast:-Gujarat: On intertidal wood of *Avicennia marina* (Borse et al., 2000a) on intertidal mangrove wood (Patil and Borse, 2001); **Maharashtra:** On intertidal wood and roots of *Avicennia alba* and *A. marina* (Borse, 1987c, 1988, Borse et al., 2005a), on submerged wood of Sonneratia acida (Srivastava, 1989), on intertidal mangrove wood (Maria and Sridhar, 2002a); **Kerala:** On intertidal wood (Raveendran and Manimohan, 2007), on intertidal mangrove wood (Nambiar and Raveendran, 2006, 2008d; Nambiar et al., 2008a). **East Coast:-Tamil Nadu:** On submerged parts of *Rhizophora* spp. (Ravikumar and Vittal, 1996),), on mangrove wood (Nambiar and Raveendran, 2008e, 2010c); **Andhara Pradesh:** On intertidal wood of *Avicennia marina, A. officinalis* and *Rhizophora apiculata* (Sarma and Vittal, 1998-99; Sarma and Vittal, 2000; Sarma and Vittal, 2001; Sarma et al., 2001; Sarma and Vittal, 2004; Vittal and Sarma, 2005); **West Bengal:** On intertidal wood of *Avicennia marina, A alba* and *Excoecaria agallocha* (Pawar et al., 2005); **Andaman & Nicobar Islands:** On intertidal wood of *Rhizophora mucronata* and *Sonneratia alba* (Chinnaraj, 1993a).

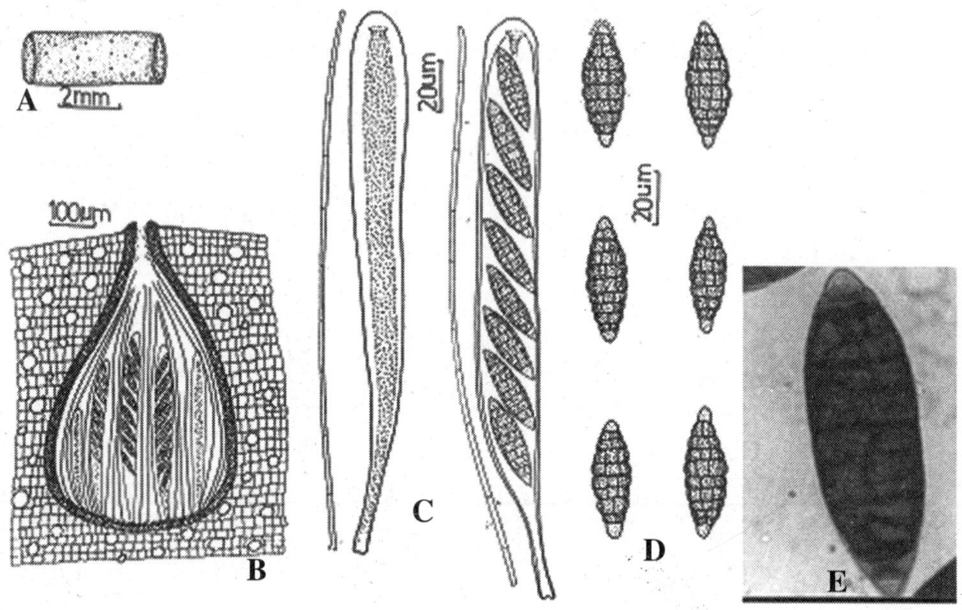

Fig. 16.3. ***Aigialus parvus***, A-E; A- Habit, B- v. s. of ascocarp, C- asci and paraphyses, D- ascospores, E- ascospore (1000x).

16.4. ***Aigialus rhizophorae*** Borse, 1987c. *Trans. Br. Mycol. Soc.,* **88**: 425. [Fig. 16.4.]

Ascomata: 990-1200 μm high, 950-1100 μm wide, 300-400 μm thick. *Ostiole*: 50-75 μm diam. *Pseudoparaphyses*: 1.5-2 μm diam. *Asci*: 350-475 x 30-40 μm. *Ascospores*: 70-95 x 18-25 μm, muriform, with 13-15 transverse septa and 1-4 longitudinal septa in all but the end cells, slightly constricted at the septa, yellow-brown except for the hyaline or light brown apical cells, glabrous, with a gelatinous cap, around the apical and subapical cells may present.

Habitat: Saprobic on intertidal prop roots of *Rhizophora mucronata*.

Description: Based on Borse (1987c).

Distribution: West Coast:-Maharashtra: On intertidal prop roots of *Rhizophora mucronata* (Borse, 1987c, 1988, Borse et al., 2005a).

East Coast:-West Bengal: On intertidal wood of *Rhizophora mucronata* (Pawar *et al.*, 2005).

Note: Kohlmeyer and Volkmann-Kohlmeyer (1991) and Jones et al. (2009a) rejected *Aigialus rhizophorae* as it shared a number of features with *Aigialus grandis*, but only different in the vertical septation in the subapical cells of ascospores. Recent collections of *Aigialus rhizophorae* made in Thailand had enabled to sequence this species and it is clearly distinct species from *Aigialus grandis* (Suetrong et al., 2009).

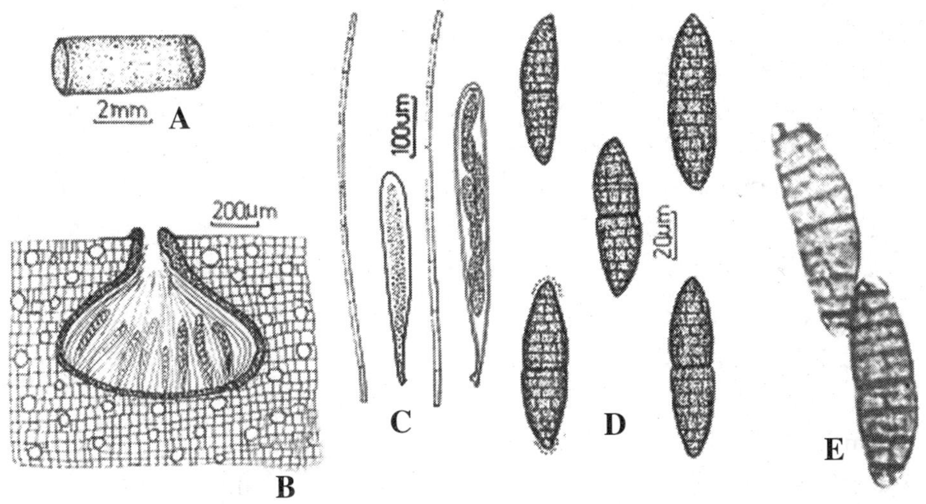

Fig. 16.4. ***Aigialus rhizophorae,*** A-E; A- Habit, B- v. s. of ascocarp, C- asci and paraphyses, D- ascospores, E- ascospores (bar- 10 µm).

17. *Amylocarpus* Curr., 1857. *Proc. R. Soc. Lond.,* 9: 119.

Ascomata: solitary or gregarious, cleistothecial, globose or subglobose, erumpent to superficial, coriaceous, variously coloured: cream-yellow, yellow or reddish yellow, dissolving irregularly above. Paraphyses: absent. *Asci*: 8-spored, clavate or ellipsoidal, apiculate, pedunculate, unitunicate, thin-walled with thickened apical wall, without an apical apparatus, and deliquescing early; ascogenous hyphae and asci scattered irregularly throughtout the ascomata center between sterile hyphae. *Ascospores*: hyaline, subglobose to ovoidal, unicellular, with 10-25 awl-shaped appendages distributed over the surface. The genus is monotypic (Jones *et al.*, 2009a).

Type species: *Amylocarpus encephaloides* Curr.
Habitat: Saprobic on intertidal wood.
Description: Based on Kohlmeyer and Kohlmeyer (1979).

17.1. ***Amylocarpus encephaloides*** Curr., *Proc. R. Soc. Lond.,* **9**: 119.

[For Fig. refer original paper]

= *Plectolitus acanthosporum* Kolhm., Nova Hedwigia, 2: 329 (1960).

Ascomata: 0.3-1.4 mm high, 0.3-3 mm in diam, solitary or gregarious, cleistothecial, ellipsoid or subglobose, erumpent and finally superficial, rarely partly immersed, seated in the substratum with broad subicles, coriaceous or cartilaginous, variously coloured: cream-yellow, yellow or reddish yellow, amber coloured and corny when dried, wall dissolving irregularly above. *Peridium*: 32-80 µm thick, 2-layered, pseudoparenchymatous on the outside, prosenchymatous on the inside. *Paraphyses*: absent. *Asci*: 27-44 x 18-27 µm in the sporogenous part, 8-16 x 3-5.5 µm in the peduncle, 8-spored, broadly clavate or ellipsoidal, apiculate, pedunculate, unitunicate, thin-walled with thickened apical wall, aphyloclastic, without an apical apparatuses, and deliquescing early; ascogenous hyphae and asci scattered irregularly throughtout the ascomata center between sterile hyphae. *Ascospores*: 8-16 µm in diam (excluding appendages), hyaline, subglobose to ovoidal, unicellular, appendaged, containing one large oil globule; 10-25 awl-shaped appendages, 5.5-10 x 1 µm, slender, acuminate, rigid, distributed more or less evenly over the ascospore surface.

Habitat: Saprobic on intertidal wood.
Description: Based on Kohlmeyer and Kohlmeyer (1979).
Distribution: West Coast:-Karnataka: On intertidal wood (Prasannarai and Sridhar, 2001, 2000-2001).

18. ***Aniptodera*** Shearer & M.A. Mill., 1977. *Mycologia,* **69**: 893.

Ascomata: peritheciod, immersed, semi-immerssed or superficial, globose or subglobose, coriaceous, hyaline to brown, papillate, ostiolate, solitary or gregarious. *Ostiole*: central, or ascomata lying horizontal to the host surface, and then neck at one end and curving upwards, cylindrical to conical, hyaline to brown, periphysate. *Peridium*: relatively thin, comprising several layers of hyaline or brown, thick-walled, compressed or angular cells. *Catenophyses*: wide, slightly constricted at the septa. *Asci*: 8-spored, clavate, or becoming balloon-shaped or swollen, short pedunculate, unitunicate, apically truncate, with a J- apical thickening which has a central pore, and subapical cytoplasm retraction, mostly persistent. *Ascospores*: 2-3-seriate, ellipsoidal or fusiform, 1-3-euseptate, not constricted at the septum, hyaline, smooth, thick-walled, guttulate, with or without polar appendages; appendages filamentous, unfurling in water, long or short. The genus is represented by 11 freshwater (Hyde and Tsui, 2006) and 8 marine water species (Jones et al., 2009a).

Type species: *Aniptodera chesapeakensis* Shearer & Miller
Habitat: Saprobic on submerged wood in aquatic environment.
Description: Based on Cai et al. (2006).

Key to *Aniptodera* species from India:

1 Ascospores with polar appendages 3

1' Ascospoes without polar appendages 2

2 Ascospores 10.5-13 x 7-8 µm .... *A. intermedia*

2' Ascospores 20-25 x 14-18 µm .... *A. haispora*

2" Ascospores 32-54 x 9-13 µm .... *A. longispora*

2"' Ascospores 21-37 x 7-15 µm .... *A. chesapeakensis*

2 "" Ascospores 16-21.5 x 6.5-7.5 µm .... *A. lignicola*

3 Ascospores with very thick wall, 21-37 x 7-15 µm .... *A. chesapeakensis*

3' Ascospores wall less thick 4

4 Ascospores 24-31 x 8-12 µm .... *A.juncicola*

4' Ascospores 14-20 x 4-7 µm .... *A. salsuginosa*

4" Ascospores 37-45 x 12-14 µm .... *A. mangrovei*

18.1. *Aniptodera chesapeakensis* Shearer & M.A. Mill., 1977. *Mycologia*, **69**: 894. [Fig. 18.1]

Ascomata: 130-300 µm high, 170-325 µm in diam, peritheciod, immersed, semi-immerssed or superficial, globose or subglobose, hyaline to brown, papillate, ostiolate, solitary or gregarious, membranaceous. *Ostiole*: central, or ascomata lying horizontal to the host surface, and then neck at one end and curving upwards, cylindrical to conical, hyaline to brown, periphysate. *Necks*: 81-326 μm long, 36-80 μm in diam, cylindrical, periphysate, brown at or below the tip. *Peridium*: 12-16 μm thick, composed of elongated, thin-walled cells with large lumina, forming a textura angularis, merging into the pseudoparenchyma of the venter. *Catenophyses*: wide, slightly constricted at the septa. *Asci*: 64-116 x 14-38 μm, 8-spored, clavate, or becoming ballon-shaped or swollen, short pedunculate, unitunicate, apically truncate, with a J- apical thickening which has a central pore, and subapical cytoplasm retraction, mostly persistent, slightly constricted at below the apex. *Ascospores*: 21-37 x 7-15 μm, 2-3-seriate, ellipsoidal or fusiform, 1-euseptate, not constricted at the septum, hyaline, smooth, thick-walled, guttulate, with or without polar appendages; appendages filamentous, unfurling in water, long or short.

Habitat: Saprobic on submerged wood in aquatic environment.

Description: Based on Shearer and Miller (1977) and Kohlmeyer and Kohlmeyer (1979).

Distribution: West Coast:-Maharashtra: On intertidal wood and prop roots of *Rhizophora mucronata* (Borse, 1987a, 1988), on intertidal mangrove wood (Maria and Sridhar, 2002a); **Gujarat:** On intertidal wood (Kamble, et al., 2011); **Goa:** On intertidal prop roots of *Rhizophora mucronata* (Nandan et al., 1993), on intertidal wood (Prasannarai and Sridhar, 2001), on intertidal mangrove wood (Maria and Sridhar, 2002a); **Karnataka:** On intertidal wood of *Bambusa* sp. (Prasannarai and Sridhar, 1997, 2001), on intertidal wood (Prasannarai et al., 1999), on intertidal mangrove wood of *Acanthus ilicifolius* and *Avicennia officinalis* (Maria and Sridhar, 2002a, 2003a), on intertidal leaf and woody litter of mangroves (Ananda and Sridhar, 2004); **Pondecherry (Mahe):** On intertidal wood (Nambiar and Raveendran, 2007): **Kerala:** on intertidal wood (Prasannarai and Sridhar, 2001, 2000-2001), on intertidal mangrove wood (Maria and Sridhar, 2002a), on intertidal wood (Raveendran and Manimohan, 2007), on intertidal wood, mangrove wood of wood of *Avicennia* sp. and *Bruguiera* sp. (Nambiar and Raveendran, 2006, 2007, 2008d, 2009a, b, d, e, f, g, h, I, 2010b, c; Nambiar et al., 2006, 2008a).

East Coast:-Tamil Nadu: On intertidal wood ((Prasannarai and Sridhar, 2001; Nambiar et al., 2008b), on mangrove wood (Nambiar and Raveendran, 2008e, 2009k, l, 2010c); **Andhara Pradesh:** On intertidal wood of *Avicennia marina,*

A. officinali, Ceriops decandras and *Rhizophora apiculata* (Sarma and Vittal, 1998-99; Sarma and Vittal, 2000; Sarma and Vittal, 2001; Sarma et al., 2001; Sarma and Vittal, 2004; Vittal and Sarma, 2005), on intertidal wood (Nambiar and Raveendran, 2009j); **West Bengal:** On intertidal wood of *Avicennia marina* (Pawar and Borse, 2004); **Andaman-Nicobar Islands:** On intertidal wood of *Rhizophora mucronata* and *Sonneratia alba* (Chinnaraj, 1993a).

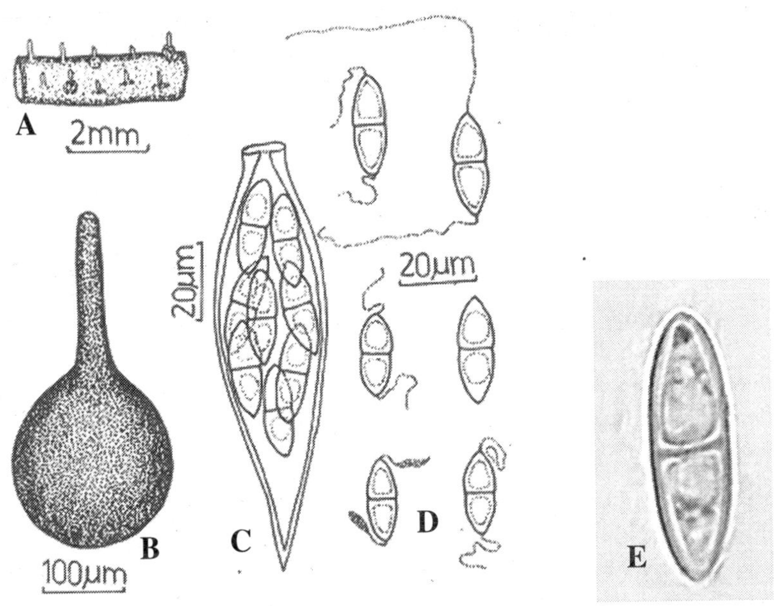

Fig. 18.1. ***Aniptodera chesapeakensis,*** A-E; A- Habit, B- ascocarp, C- ascus, D- ascospores, E- ascospore (1000x).

18.2. ***Aniptodera haispora*** Vrijmoed, K.D. Hyde, & E.B.G. Jones, 1994. *Mycol. Res.*, **98**: 701. [For Fig. refer original paper]

Ascomata: 202-283 x 161-282 µm, immersed, semi-immersed or superficial, axis horizontal or vertical to the host surface, light-brown to pale grey, ostiolate, papillate, solitary or gregarious. *Necks*: central or lateral, 172-303 x 50-91 µm, terminating in a number of hair-like structures. *Peridium*: up to 25 µm thick, light brown, comprising 5-7 layers of thick-walled elongate cells, which are inwardly hyaline and outwardly brown. *Catenophyses*: absent. *Asci*: 81-113 x 16-28 µm, 8-spored, clavate, thin-walled, unitunicate, pedunculate, with thickened and flattened apex containing a refractice pore, asci developing from a pad at the base of the ascoma. *Ascospores*: 14-22 x 10-12 µm, hyaline, 2-celled, ellipsoidal, weakly constricted at the septa and thick-walled.

Habitat: Saprobic on intertidal mangrove driftwood.

Description: Based on Vrijmode et al. (1994).

Distribution: West Coast:-Pondecherry (Mahe): On intertidal mangrove wood (Nambiar and Raveendran, 2008a); **Kerala:** On intertidal wood (Raveendran and Manimohan, 2007; Nambiar and Raveendran, 2009d, e, f); on mangrove woody litter (Shini et al., 2009-10).

East Coast:-Andhara Pradesh: On intertidal wood of *Avicennia officinali,* and *Rhizophora apiculata* (Sarma and Vittal, 1998-99; Sarma and Vittal, 2000; Sarma and Vittal, 2001; Sarma et al., 2001; Sarma and Vittal, 2004; Vittal and Sarma, 2005).

18.3. *Aniptodera intermedia* K.D. Hyde & Alias, 1999. *Mycol. Res.*, **103**: 1409. [For Fig. refer original paper]

Ascomata: 210-330 µm in diam., subglobose, hyaline, membraneous, solitary in old fruiting bodies of other ascomycetes, with a periphysate papilla, which is erumpent through the host surface. *Peridium*: 12.5 -15 µm wide, composed of several layers of hyaline cells in the form of *textura epidermoidea. Catenophyses*: present. *Asci*: 46-62 x 15.5-18.5 µm, 8-spored, clavate, unitunicate, pedunculate, apically truncate, with a refractice apical thickening and pore. *Ascospores*: 10.5-13 x 7-8 µm, hyaline, 2-3-septate, ellipsoidal, unicellular or occasionally bicelled, smooth-walled, lacking any sheath or appendages.

Habitat: Saprobic on intertidal petioles of *Nypa fruticans.*

Description: Based on Hyde et al. (1999).

Distribution: East Coast:-West Bengal: On intertidal mangrove woody litter (Shini et al., 2009-10).

18.4. *Aniptodera juncicola* Volkm.-Kohlm. & Kohlm., 1994. *Bot. Mar.*, **37**: 109. [For Fig. refer original paper]

Habitat: Saprobic on dead culms of salt marsh plant *Juncus roemerianus* Scheele

Distribution: West Coast:-Kerala: On intertidal mangrove wood (Shini, et al., 2009-10).

East Coast:-Tamil Nadu:- on intertidal woody debris (Ravikumar, et al., 2009); **West Bengal:** On mangrove woody litter (*Shini* et al., 2009-10).

Note: This is a substrate (host) specific ascomycete collected on dead culms of salt marsh plant *Juncus roemerianus* Scheele in North Carolina, USA (Volkmann-Kohlmeyer and Kohlmeyer, 1994).

18.5. ***Aniptodera lignicola*** Hyde, Ho & Tsui, 1999. Mycoscience, 40: 171. [For Fig. refer original paper]

Ascomata: 160-240 µm in diam., globose to subglobose, partly immersed or superficial, hyaline, becoming black at maturity, membraneous, ostiolate, papillate, periphysate. *Peridium*: of comptressed cells. *Catenophyses*: present. *Asci*: 50-64 x 14-18 µm, 8-spored, cylindrical to clavate, pedicillate, thin-walled, apically truncate and thickened, with a apical pore and cytoplasm retracted below the ascus apex, persitent. *Ascospores*: 16-21.5 (-26) x 6.5- 7.5 (-)8 µm, hyaline, ellipsoidal, bicelled, not constricted at the septum, relatively thin-walled (< µm 1 thick).

Habitat: Saprobic on submerged woody material in freshwater habitats.

Description: Based on Hyde et al. (1999).

Distribution: East Coast:-Tamil Nadu: On intertidal woody debris (Ravikumar et al., 2009).

Note: This is a freshwater ascomycete collected on submerged woody material on submerged woody material from Australia, Brunei, Malyasia and Philippines. This species is somewhat similar to *A. limnetica* Shearer, but differs as ascospores in *A. lignicola* are never allantoid and have thin walls (Hyde et al., 1999).

18.6. ***Aniptodera longispora*** K.D. Hyde, 1990. *Bot. Mar.*, **33**: 335.

[For Fig. refer original paper]

Ascomata: 270-390 x 180-260 µm, pyriform to subglobose in fronta view, ellipsoidal in end view, lying horizontally immersed in the substrate, membranous, light-brown to black above, hyaline below, ostiolate , papillate, solitary or gregarious, *Peridium*: 15-28 µm thick, composed of an outer layer of melanized thick-walled angular cells and an inner layer of hyaline thin-walled elongate cells. *Necks*: periphysate. Catenophyses: present. Paraphyses: absent. *Asci*: 145-201 x 24-31 µm, 8-spored, clavate, pedunculate, unitunicate, persistent, thin-walled, but thickened apically, and with a pore, which is refractive, developing at the base of the ascomata venter. *Ascospores*: 39-51 x 9-14 µm, cylindrical with rounded apices, 2-3 seriate, hyaline, 2-celled with a central septum, slightly constricted at the septum, a single large and several smaller oil globules in each cell, appendages not observed.

Habitat: Saprobic on dead roots of *Rhizophora apiculata*.

Description: Based on Hyde (1990b).

Distribution: West Coast:-Goa: On intertidal wood ((Prasannarai and Sridhar, 2001); **Karnataka:** On intertidal wood ((Prasannarai and Sridhar, 2001); **Kerala:** On intertidal wood (Prasannarai and Sridhar, 2001), on intertidal wood (Raveendran and Manimohan, 2007), on intertidal wood, and wood of *Avicennia* sp. (Nambiar and Raveendran, 2009e, h, 2010c); on mangrove woody litter (Shini et al., 2009-10).

18.7. ***Aniptodera mangrovei*** K.D. Hyde, 1986. *Can. J. Bot.,* **64**: 2989.

[Fig. 18.7]

Ascomata: 240-400 x 200-366 µm, globose to subglobose, light-brown to black, immersed, ostiolate (periphysate), papillate, solitary or gregarious, membranous. *Peridium*: pseudoparenchymatous. Catenophyses: present. *Asci*: 106-144 x 28-39 µm, 8-spored, clavate, thin-walled, unitunicate, short pedunculate, with an apical apparatus (composed of a ring and plug), deliquescing before or at spore maturity, developing at the base of the ascomata. *Ascospores*: 37-45 x 12-14 µm, hyaline, 2-celled, ellipsoidal to fusiform, nor or slightly constricted at the septum, appendaged. *Appendages*: 5-30 µm in length (prior to uncoiling), bipolar, appressed closely to one side of the spore and composed of a single filament which is tightly coiled and appears hamate. On contact with water the appendages uncoil to become a long thread still attached at the polar region.

Habitat: Saprobic on driftwood, dead mangrove wood and twigs.

Description: Based on Hyde et al. (1986).

Distribution: West Coast:-Maharashtra: On intertidal mangrove wood (Maria and Sridhar, 2002a); **Goa:** On intertidal wood (Nanadan et al., 1993), on intertidal mangrove wood (Maria and Sridhar, 2002a); **Karnataka:** On intertidal mangrove wood of *Acanthus ilicifolius, Avicennia officinalis, Brugueiria gymnorrhiza, Rhizophora mucronata* and *Sonneratia caseolaris* (Maria and Sridhar, 2002a, 2003a, 2004), on intertidal mangrove wood (Ananda and Sridhar, 2003);

Pondecherry (Mahe): On intertidal mangrove wood (Nambiar and Raveendran, 2008a); **Kerala:** On intertidal mangrove wood (Maria and Sridhar, 2002a), On intertidal wood (Raveendran and Manimohan, 2007), on intertidal mangrove wood, wood of *Avicennia* sp. and *Bruguiera* sp. (Nambiar and Raveendran, 2006, 2007, 2008a, d, 2009a, e, f, g, h, 2010c); **Lakshadweep Islands:** On mangrove wood (Ananda and Sridhar, 2003).

East Coast:-Tamil Nadu: On intertidal wood (Nambiar et al., 2008b),), on mangrove wood (Nambiar and Raveendran, 2008e, 2009l, 2010c); **Andhara Pradesh:** On intertidal wood of *Avicennia officinalis* and *Rhizophora apiculata* (Sarma and Vittal, 1998-99; Sarma and Vittal, 2000; Sarma and Vittal, 2001; Sarma et al., 2001; Sarma and Vittal, 2002; Sarma and Vittal, 2004; Vittal and Sarma, 2005); **West Bengal:** On intertidal wood of *Avicennia alba* (Pawar and Borse, 2004); **Andaman & Nicobar Islands:** On intertidal wood of *Rhizophora mucronata* (Chinnaraj, 1993a).

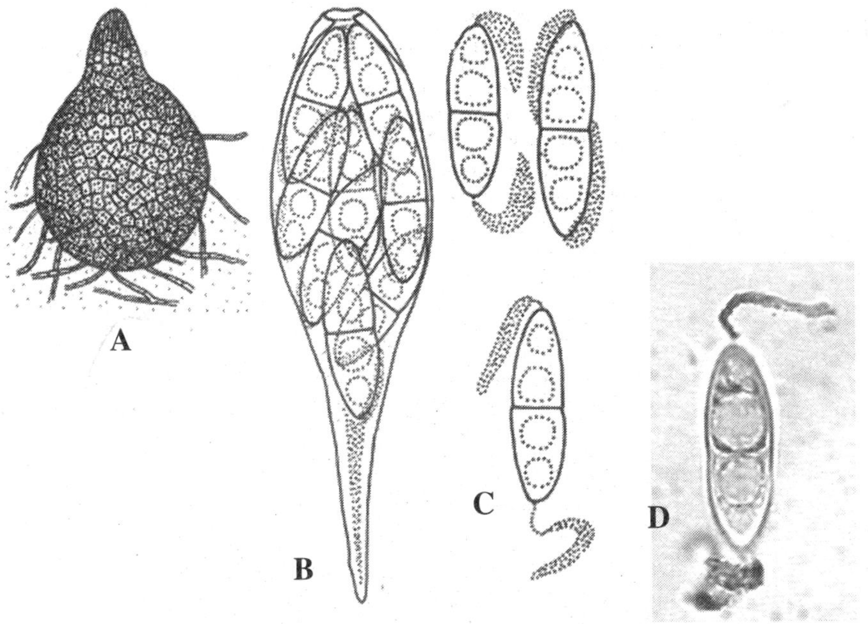

Fig. 18.7. ***Aniptodera mangrovei,*** A-D; A- ascocarp (100x), B- ascus (1000x), C- ascospores (1000x), D- ascospore (1000x).

18.8. *Aniptodera salsuginosa* Nakagiri & Tad Ito, 1994. *Mycol. Res.*, **98**: 931. [For Fig. refer original paper]

Ascomata: 240-280 µm diam., 92-112 µm high, ellipsoidal to flattened cortex, main axis horizontal, black upper side, hyaline bottom, solitary, immersed just below the epidermis of bark, ostiolate, papillate. *Peridium*: 6-10 µm thick, membranous, composed in the upper region of black elongate cells, interspersed among host cells; in the lower region of hyaline flat cells. Necks: lateral, bending upward, 280-520 µm long above the wood surface, 100-160 µm long in the wood, 35-48 um diam., cylindrical, hyaline to subhyaline in the upper region, black in the lower region, with periphyses. *Paraphyses*: absent. Catenophyses: present. *Asci*: 76-90 x 12-14 µm, 8-spored, cylindrical to clavate, pedunculate, unitunicate, thin-walled, persistent, apex truncate with an apical thickening (ring) with a pore, with cytoplasm retracted in the subapical region. Asci develop in a hymenium at the base of the ascomata and are easily detached from the base and released into the ascomata venter. *Ascospores*: 14-20 x 4-7 µm, ellipsoidal to fusiform, hyaline, 2-celled, not constricted at the septum, with apical appendages. *Appendages*: attached apically at both ends, initially stiff, 9-14 µm long, unfurling in seawater into a membranous structure.

Habitat: Saprobic on intertidal bark of Rhiziphora stylosa and Bruguiera gymnorrhiza.

Description: Based on Nakagiri and Ito (1994).

Distribution: West Coast:-Pondecherry (Mahe): On intertidal mangrove wood (Nambiar and Raveendran, 2008a, c); **Kerala:** On intertidal wood and wood of *Avicennia* sp. (Raveendran and Manimohan, 2007; Nambiar and Raveendran, 2007, 2008a, c, 2009d, f, g, h, 2010c; Nambiar et al., 2008a).

East Coast:-Tamil Nadu: On mangrove wood (Nambiar and Raveendran, 2008e).

19. *Antennospora* Meyers, 1957. *Mycologia*, **49**: 501.

Ascomata: subglobose or ellipsoidal, immersed, superficial when growing on the calcareous tubes of mollusks, ostiolate, papillate, coriaceous or subcabonaceous, dark beown, solitary or gregarious. *Catenophyses*: deliquescing. *Necks*: periphysate. *Asci*: clavate, pedunculate, unitunicate, thin-walled, no apical apparatus, deliquescing early. *Ascospores*: ellipsoidal, 1-septate, slightely constricted at the septum, hyaline with polar appendages. Appendages: subterminal, at each end of the spore, at right angles to each other, cylindrical, and attenuate. The genus is monotypic (Jones et al., 2009a).

Type species: *Antennospora quadricornuta* (Cribb & J.W. Cribb) T.W. Johnson
Habitat: Saprobic on intertidal wood.
Description: Based on Meyers (1957) and Jones et al. (2009a).

19.1. ***Antennospora quadricornuta*** (Cribb & J.W. Cribb) T.W. Johnson, 1958. *J. Elisha Mitchella Sci.* Soc., **74**: 46. [Fig. 19.1]

Basionym: *Halosphaeria quadricornuta* Cribb & J.W. Cribb, *Univ. Queensl. Pap. Dept. Bot.*, **3**: 99 (1965).

= *Antennospora caribbea* Meyers, *Mycologia*, **49**: 503 (1957).

Ascomata: 130-260 µm high, 140-285 µm diam., subglobose or ellipsoidal, immersed or becoming exposed, superficial when growing on the calcareous tubes of mollusks, ostiolate, papillate, coriaceous or subcabonaceous, dark brown to black, often surrounded by dark brown hyphae, solitary or gregarious. *Peridium*: 9-13 µm thick, composed of 3 or 4 layers of small, irregular, polygonal, thick-walled cells, forming textura angularis, merging more or less abruptly into the large-celled catenophyses of the venter. *Necks*: 70-560 µm long, 20-70 µm in diam., subconical or cylindrical, centric or eccentric. *Catenophyses*: thin-walled, large, polygonal or ellipsoidal, deliquescing. *Asci*: 8-spored, clavate, pedunculate, unitunicate, thin-walled, aphysoclastic, no apical apparatus, deliquescing early; developing at the base of the ascomata venter on a small-celled ascogenous tissue. *Ascospores*: 20-37 x 8-12 µm (excluding appendages), ellipsoidal, 1-septate, slightly constricted at the septum, hyaline with polar appendages. Appendages: 20-37 µm long, 1-2 µm diam., at each end two subterminal, cylindrical, attenuate, stiff, pair of appendages at right angles to each other..

Habitat: Saprobic on intertidal wood.
Description: Based on Kohlmeyer and Kohlmeyer (1979).
Distribution: West Coast:-Diu Island: On intertidal wood (Borse et al., 1999a); **Daman:** On intertidal wood and *Avicennia marina* (Borse et al., 2000a); **Gujarat:** On intertidal wood and mangrove wood (Patil and Borse, 2001); **Maharashtra:** On intertidal wood panel (Jones, 1968), on decaying intertidal pneumatophores of *Sonneratia acida* (Patil and Borse, 1983a, 1986; Borse, 1988), intertidal wood of *Rhizophora mucronata* and *Sonneratia acida* (Patil and Borse, 1986, Borse et al., 2005a), on intertidal wood (Shrivastava, 1995); **Goa:** On submerged wood panels of *Anogeissus latifolia, Carapa molluscensis, Acacia*

auriculiformis, and *Hevea braziliensis* (Raghukumar et al., 1988a), on intertidal wood (Borse et al., 1999b); **Karnataka:** On intertidal wood of *Bambusa* sp. (Prasannarai and Sridhar, 1997, 2001, 2000-2001), on intertidal wood and teak panels (Prasannarai et al., 1999), on intertidal wood of *Acanthus ilicifolius* and *Rhizophora mucronata* (Maria and Sridhar, 2003a); **Kerala:** On intertidal wood of fishing crafts (Becker and Kohlmeyer, 1958), on intertidal wood panel (Jones, 1968), on intertidal wood (Prasannarai and Sridhar, 2001), on intertidal wood (Raveendran and Manimohan, 2007; Nambiar and Raveendran, 2007, 2010a, c); **Lakshadweep Islands:** On intertidal wood of *Cocos nucifera* and *Pemphis acidula* (Chinnaraj, 1992). **East Coast:-Tamil Nadu:** On intertidal wood of fishing crafts (Becker and Kohlmeyer, 1958), on drift wood, intertidal wood, logs of *Casuarina* sp. (Raghukumar, 1973), on submerged parts of *Rhizophora* spp. (Ravikumar and Vittal, 1996), on intertidal wood (Prasannarai and Sridhar, 2001), on intertidal wood (Nambiar and Raveendran, 2009l); on intertidal woody debris (Ravikumar et al., 2009); **Pondecherry (Karaikkal):** on intertidal woody debris (Ravikumar et al., 2009); **West Bengal:** On intertidal wood of *Avicennia marina* (Pawar and Borse, 2004); **Andaman & Nicobar Islands:** On intertidal wood of *Rhizophora mucronata* (Chinnaraj, 1993a).

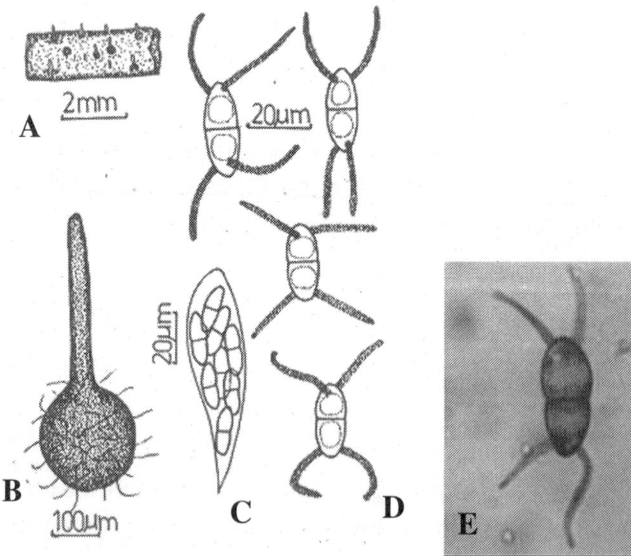

Fig. 19.1. ***Antennospora quadricornuta,*** A-E; A- Habit, B- ascocarp, C- ascus, D- ascospores, E- ascospore (1000x).

20. ***Anthostomella*** Sacc., 1875. *Atti Soc. Veneto-Trent. Sci. Nat., Padova*, **4:** 84.

Ascomata: globose to subglobose, immersed in the host, pale brown to dark brown to black, ostiolate, epappilate clypeate, coriaceous, periphysate, solitary or gregarious, paraphyses numerous, simple, septate. *Asci*: cylindrical, short pedunculate, unitunicate, thick-walled with J+ apical apparatus. *Ascospores*: ellipsoidal, unicellular, with germ-slits extending over the whole length of the spore, hyaline with gelatinous caps or sheaths. Most of the marine species of this genus have been collected on *Juncus roemerianus*, *Nypa fruiticans*, *Spartina* spp. and *Pandanus* sp., but many also occur on mangrove wood (Jones *et al.*, 2009a).

Type species: *Anthostomella limitata* Sacc.
Habitat: Marine species saprobic on *Juncus roemerianus*, *Nypa fruiticans*, *Spartina* spp. and *Pandanus* sp., mangrove wood.
Description: Based on Jones et al. (2009a).

20.1. ***Anthostomella nypensis*** K.D. Hyde, Alias & B.S. Lu, 1999. Mycol. Res., 103: 1440. [For Fig. refer original paper]

Ascomata: globose to subglobose, immersed in the host, pale brown to dark brown to black, ostiolate, epappilate clypeate, coriaceous, periphysate, solitary or gregarious, paraphyses numerous, simple, septate. *Asci*: cylindrical, short pedunculate, unitunicate, thick-walled with J+ apical apparatus. *Ascospores*: 12-18 x 5.5-7.5 µm, inequilaterely ellipsoidal, unicellular, with germ-slits extending over the whole length of the spore, hyaline with a thin gelatinous sheath.

Habitat: Saprobic on intertidal wood.
Description: Based on Hyde et al. (1999).
Distribution: West Coast:-Kerala: On intertidal wood (Namiar and Raveendran, 2011b).

21. ***Arenariomyces*** Hohnk, 1954. *Veroff. Inst. Merresforchung, Bremerhaven*, **3:** 30.

Ascomata: solitary, sometimes gregarious, globose to subglobose, superficial but sometimes immersed, subiculum present on materials such as sand grains and shells, with or without ostioles, papillate, periphyses absent, carbonaceous, dark brown to black. *Peridium*: variable arrangement of cells, usually comprised of dark, thick-walled cells, cells sphericsl to subspherical with inconspicuous pit connections. *Catenophyses*: thin-walled, and deliquescing at maturity. *Asci*: 8-spored, fusiform, to subclavate, short- stalked, unitunicate, thin-walled and deliquescing early. *Ascospores*: 1-3-septate, ellipsoidal, constricted at the septa, hyaline and appendaged. The genus is represented by five species (Jones et al., 2009a; Pang et al., 2011).

Type species: *Arenarimyces trifurcatus* Hohnk
Habitat: Saprobic on intertidal wood, bark, sand grains and shells, ascospores accumulate in foam along shore.
Description: Based on Jones et al. (1983a, 2009a).

Key to *Arenariomyces* species from India:

1 Ascospores 3-septate .... *Arenariomyces triseptatteus*

1' Ascospores 2-septate 2

2 Ascospores narrower than 6 μm wide, 16-18 x 3-4 μm .... *Arenariomyces parvulus*

2' Ascospores more than 7 μm wide 3

3 (2') Ascomata smooth, ascospores consistently with 3 appendages at each apex .... *Arenariomyces trifurcatus*

3'(2') Ascomata with spines on the peridium, ascospores with 3 or 4 appendages at each apex .... *Arenariomyces majusculus*

21.1. ***Arenariomyces majusculus*** Kohlm. & Volkm.-Kohlm., 1989b. *Mycol. Res.*, **92**: 411. [Fig. 21.1]

Ascomata: 200-350 μm diam., partley immersed in the substratum, ostiolate, papillate, coriaceous, dark brown, covered with spine-like short hyphae aboce, with long, brown, interwined hyphae below, single. *Necks*: 45-65 μm high, 40-60 μm diam., conical, without periphyses, ostiolar canal filled with small, light brown pseudoparenchymatous cells; this tissue is separated from the pseudoparenchyma of the centrum by an inconspicuous hyaline wall. *Peridium*: 6.5-10 μm thick, composed of 3-4 layers of flat cells with narrow lumina forming a *textura angularis*, dark brown. *Catenophyses*: thin-walled, and deliquescing at maturity. *Asci*: 8-

spored, clavate, pedunculate, unitunicate, thin-walled, and deliquescing early. *Ascospores*: 27-39 x 9-15 µm ellipsoidal, 1-septate, slightly constricted at the septum, hyaline, appendaged; at each end with 3 or 4 subtreminal appendages. *Appendages*: 18-24 µm long, 1-1.5 µm diam. at the slightly swollen base, slender, attenuate, rigid, curved, attached to the side of immature spore.

Habitat: Saprobic on intertidal wood, ascospores accumulate in foam along shore.

Description: Based on Kohlmeyer and Volkmann-Kohlmeyer (1989b).

Distribution: West Coast:-Maharashtra: On intertidal wood and *Avicennia marina* (Borse, 2000b; Borse et al., 2005a); **Gujarat:** On mangrove wood (Kamble et al., 2011); **Karnataka:** On intertidal wood (Prasannarai et al., 1999), on intertidal wood ((Prasannarai and Sridhar, 2001); **Pondicherry (Mahe):** Ascospores in foam samples (Borse and Pawar, 2005); **Kerala:** On intertidal wood (Raveendran and Manimohan, 2007; Nambiar and Raveendran, 2007, 2008b, 2010a, c); **Lakshadweep Islands:** On intertidal wood of *Avicennia marina* (Chinnaraj, 1992).

East Coast:-**Tamil Nadu:** on intertidal wood (Nambiar and Raveendran, 2009l); on intertidal woody debris (Ravikumar et al., 2009); **Pondicherry (Karaikkal):** on intertidal woody debris (Ravikumar et al., 2009); **West Bengal:** Ascospores in foam samples (Borse et al., 2001a), on intertidal wood (Borse et al., 2002a).

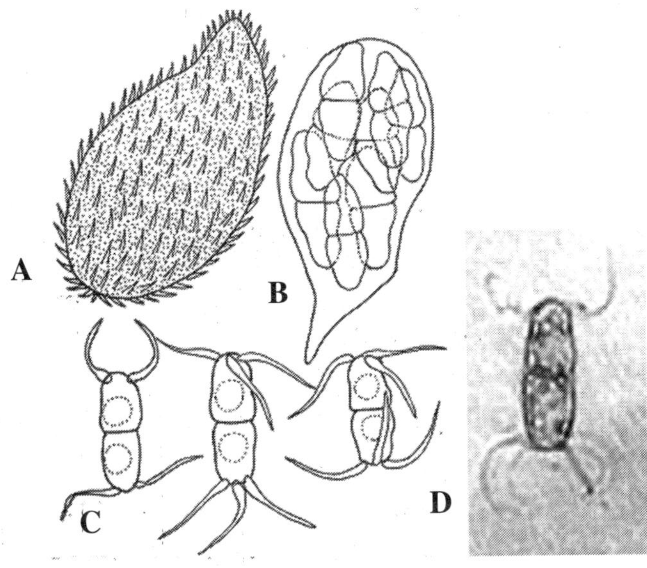

Fig. 21.1. ***Arenariomyces majusculus,*** A-D; A- ascocarp (100x), B- ascus (450x), C- ascospores (1000x), D- ascospore (1000x).

21.2. *Arenariomyces parvulus* Jorgen Koch, 1986. *Nordic J. Bot.*, **6**: 497. [Fig. 21.2]

Ascomata: 44-60 μm diam., globose or subglobose, brown-black, dull, densely covered by short brown hyphae 4-6 μm long, with subicula on grains of sand, gregarious or solitary, coriaceous, collapsing in drying, ostiole not seen. *Peridium*: membraneous, outer dark layer 1-2 μm thick composed of flat angular cells 4-5 μm in diam. *Asci*: early deliquescing, 8-spored, as seen from spore clusters. *Ascospores*: 16-18 x 3-4 μm, cylindrical, curved, 1-septate, slightly constricted at the septum, hyaline with 3 terminal appendages at both ends. *Appendages*: 10-12 μm long, 0.75 μm wide at the base, tapering towards the end, rigid, curved, mostly oriented towards septum.

Habitat: Saprobic on intertidal wood and sand grains, ascospores accumulate in foam along shore.

Description: Based on Koch (1986).

Distribution: West Coast:-Gujarat: On intertidal wood (Kamble et al., 2008); **Maharashtra:** On intertidal wood, *Rhizophora mucronata* and ascospores in foam samples (Borse, 2000b; Borse et al., 2005a); **Goa:** On intertidal animal substrates (Ananda et al., 1998), Ascospores in foam samples (Borse and Tuwar, 2006); **Karnataka:** On intertidal animal substrates (Ananda et al., 1998; Ananda and Sridhar, 2001, 2004), on intertidal wood (Prasannarai and Sridhar, 2001, 2000-2001), on intertidal mangrove wood (Ananda and Sridhar, 2003); **Pondicherry (Mahe):** Ascospores in foam samples (Borse and Pawar, 2005); **Kerala:** On intertidal wood (Raveendran and Manimohan, 2007), on intertidal mangrove wood (Nambiar and Raveendran, 2008b, 2010a, c); **Lakshadweep Islands:** On intertidal mangrove wood (Ananda and Sridhar, 2003).

East Coast:-Tamil Nadu: On intertidal wood (Nambiar and Raveendran, 2009k, l); on intertidal woody debris (Ravikumar et al., 2009); **Pondicherry (Karaikkal):** on intertidal woody debris (Ravikumar et al., 2009); **Orissa:** Ascospores in foam samples (Borse et al., 2001b); **West Bengal:** Ascospores in foam samples (Borse et al., 2001), on intertidal wood (Borse et al., 2002a).

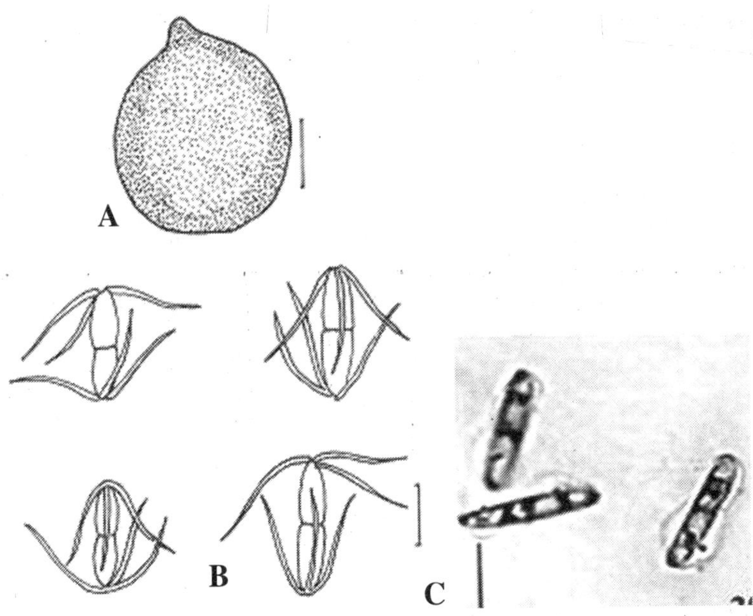

Fig. 21.2. ***Arenariomyces parvulus,*** A-C; A- ascocarp (bar= 20μm), B- ascospores (bar=10 μm) , C- ascospores (bar=10 μm).

21.3. ***Arenariomyces trifurcatus*** Hohnk, 1954. *Veroff. Inst. Meereforsch. Bremerhaven*, **3**: 30. [Fig. 21.3]

Basionym: *Halosphaeria trifurcata* (Hohnk) Cribb & J.W. Cribb, *Uni. Queen. Pap. Bot. Dept.*, **3**: 99 (1956).

≡ *Peritrichospora trifurcata* (Hohnk) Kohlm., *Nova Hedwigia,* **3**: 89 (1961).

≡ *Corollospora trifurcata* (Hohnk) Kohlm., *Ber. Deuts. Bot. Ges.*, **75**: 126 (1962).

Ascomata: 100-325 μm in diam., globose or subglobose, immersed or superficial, often flattened at the base and seated with subicula on grains of sand or calcareous animals shells, with or without ostioles, papillate or epapillate, carbonaceous or rarely coriaceous, black or dark brown, sometimes covered by short, brown hyphae, solitary or rarely gregarious. *Peridium*: 5-18 μm thick, composed of dark, thick-walled cells, roundish at the outside, oblong toward the center, forming a *textura angularis*; subicula prosenchymatous. *Necks*: up to 30 μm long, conical or subcylindrical. Catenophyses: thin-walled, polygonal or rounded cells, filling venter of young ascomata, deliquescing at spore maturity; cell walls with pit like thickenings, connecting plasmatic strands of neighboring cells. *Asci*: 60-80 x 19-23 μm, 8-spored, fusiform or subclavate, short pedunculate, unitunicate, thin-walled, aphysoclastic, early deliquescing; originating from an ascogenous tissue in the base of the ascomata venter. *Ascospores*: 24-39 x 7-17 μm (excluding appendages), fusiform, ellipsoidal or oblong, 1-septate, slightly constricted at the septum, hyaline, appendaged, at both ends with 3 or 4 terminal or subterminal appendages. *Appendages*: 15-39 x 1.5-2 μm, slender, attenuate, rigid, curved, sometimes with an apical thickening.

Habitat: Saprobic on intertidal wood and sand grains, ascospores accumulate in foam along shore.

Description: Based on Kohlmeyer and Kohlmeyer (1979).

Distribution: West Coast:-Gujarat: On intertidal wood (Patil & Borse, 2001), on intertidal wood (Kamble et al., 2008); **Maharashtra:** On intertidal wood and *Rhizophora mucronata* (Borse, 2000b, Borse et al., 2005a), Ascospores in foam samples (Patil and Borse, 1983b; Borse, 1988), on intertidal wood of *Avicennia alba* (Shrivastava, 1995); **Goa:** Ascospores in foam samples (Borse and Tuwar, 2006); **Karnataka:** Ascospores in sand and foam samples (Sridhar and Kavariappa, 1991), on intertidal wood (Prasannarai et al., 1999), on intertidal leaf and woody litter of mangroves (Ananda and Sridhar, 2004); **Pondicherry (Mahe):** Ascospores in foam samples (Borse and Pawar, 2005); **Kerala:** On intertidal wood (Prasannarai and Sridhar, 2001), on intertidal wood (Raveendran and Manimohan, 2007; Nambiar and Raveendran, 2007, 2008b, 2010a, c).

East Coast:-Tamil Nadu: On drift wood (Raghukumar, 1973), on intertidal wood (Nambiar and Raveendran, 20091); **Orissa:** Ascospores in foam samples (Borse et al., 2001b). **West Bengal:** Ascospores in foam samples (Borse et al., 2001a), on intertidal wood ((Borse et al., 2002a).

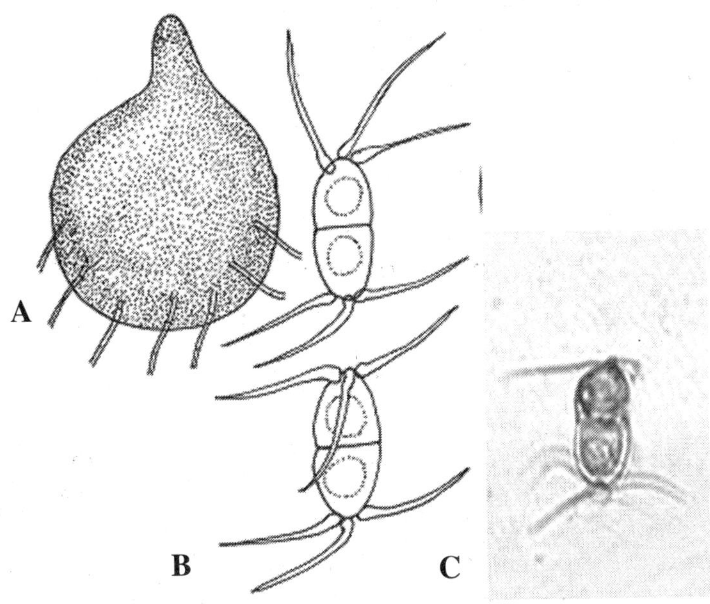

Fig. 21.3. ***Arenariomyces trifurcatus,*** A-C; A- ascocarp (100x), B- ascospore (1000x) , C- ascospores (1000x).

21.4. ***Arenariomyces triseptatus*** Kohlm., 1984. *P. S. Z. N. I. Marine Ecology*, **5:** 333. [For Fig. refer original paper]

Ascomata: 50-120 µm in diameter, globose or subglobose, or irregular, superficial or immersed, often flattened at the base and seated with subicula on grains of sand or calcareous animal shells, or inside foraminiferous shells, ostiolate, papillate, coriaceous, dark to light brown, sometimes covered by short hyphae, solitary or gregarious. *Peridium*: 14-16 µm thick, two-layered; wider outer layer light brown, composed of 2 or 3 layers of thin-walled, polygonal cells with wide lumina, forming a *textura angularis*; narrow inner layer dark brown, composed of 1 or 2 layers of thick-walled, compressed cells with narrow lumina; subicula prosenchymatous. *Necks*: up to 20-24 µm high, 16-25 µm in diameter, conical; ostiolar canal filled with small-celled pseudoparenchyma. Pseudoparenchyma: of thin-walled polygonal cells filling the centrum of young ascomata, deliquescing as ascospores maturity. *Asci*: 8-spored, ellipsoidal, unitunicate, thin-walled, aphysoclastic, deliquescing before ascospore maturation. *Ascospores*: 27-34 x 6-8.5 µm (excluding appendages), cylindrical, 3-septate, constricted at the septa, hyaline, appendaged:

at both ends with 3 (rarely 4 or 5) subterminal appendages; appendages 18-20 µm long, 2 µm in diameter at the slightly swollen base, 0.5 µm at the tip, slender, attenuate, rigid, curved, attached to the side of immature spores.

Habitat: Saprobic on wet rotting algae, seagrass leaves, or inside chambers of foraminifera; ascospores accumulate in foam along shore.

Description: Based on Kohlmeyer (1984).

Distribution: West Coast:-Kerala On intertidal wood (Nambiar and Raveendran, 2011a).

22. *Ascocratera* Kohlm., 1986. *Can. J. Bot.*, **64**: 3036.

Ascomata: conical, craterlike, erumpent and superficial when mature, seated on a black stroma, ostiolate, periphysate, epapillate, carbonaceous, black, with irregular bulges or ridges around the ostiole, gregarious. *Ostiole*: depressed, canal periphysate. *Peridium*: thick, composed of small irregular cells, encrusted with melanin particles, forming a *textura angularis*, enclosing cells of the host. *Pseudoparaphyses*: trabeculate, unbranched at the base and embedded in a gelatinous matrix, free at the tips, branching and anastomosing. *Asci*: 8-spored, cylindrical, pedunculate, thick-walled, fissitunicate, with a refractive apical ring in the endoascus, not bluing in IKI, arising from a basal ascogenous tissue. *Ascospores*: overlapping monostichously in the ascus, ellipsoidal, at first one-, later 3-septate, constricted at the central primary septum, hyaline, surrounded by a gelatinous sheath. The genus is monotypic (Jones *et al.*, 2009a).

Type species: *Ascocratera manglicola* Kohlmeyer

Habitat: Saprobic on intertidal wood of mangrove trees.

Description: Based on Kohlmeyer (1986).

22.1. *Ascocratera manglicola* Kohlm., 1986. *Can. J. Bot.*, **64**: 3037.

[Fig. 22.1]

Ascomata: 775-895 µm high, 1072-1370 µm in diam., conical with a flat base, craterlike, erumpent from the substrate and superficial when mature, seated on a thin black stroma, ostiolate, periphysate, epapillate, carbonaceous, black, with irregular bulges or ridges around the ostiole, gregarious, single or sometimes 2 or more attached to each other. *Ostioles*: circular, depressed, 55-90 µm in diam.,; ostiolar canal filled with short hyaline branching periphyses, 2.2-3 µm thick, embedded in a gelatinous matrix. *Peridium*: 92-193 µm on the sides, 45-75 µm at the base, composed of small irregular cells, encrusted with melanin particles, forming a *textura angularis*; cells of the host are incorporated in the peridium, especially in the basis of the lateral part; a *textura intricata*, composed of loose, yhick hyaline hyphae, foems a wedge between the lateral and basal part of the peridium. *Pseudoparaphyses*: 1.5-3 µm in diam., trabeculate, septate, simple at the base and embedded in a gelatinous matrix, free at the tips, branched and anastomosing above the asci. *Asci*: 323-450 x 21-27 µm, 8-spored, cylindrical, pedunculate, thick-walled, fissitunicate, with a refractive apical cap or ring, arising from a basal ascogenous tissue, composed of small cells, 1-2 µm in diam., 4-5 µm high, in the endoascus, surrounding the ocular chamber, nonamyloid, including the ring. *Ascospores*: 46-66 um x 14-22 µm, overlapping monostichously in the ascus, ellipsoidal, at first one-, later 3-septate, constricted at the central primary septum, hyaline, at maturity becoming light brown, surrounded by a gelatinous evanescent sheath.

Habitat: Saprobic on intertidal wood of mangrove trees.

Description: Based on Kohlmeyer (1986).

Distribution: West Coast:-Maharashtra: On intertidal wood of *Avicennia alba* (Borse and Shrivastava, 1994, Borse et al., 2005a); **Karnataka:** On intertidal mangrove wood (Maria and Sridhar, 2002a); **Kerala:** On intertidal mangrove wood (Maria and Sridhar, 2002a), on intertidal wood (Raveendran and Manimohan, 2007), on intertidal mangrove wood of wood of *Bruguiera* sp. (Nambiar and Raveendran, 2006, 2009d, g); **Lakshadweep Islands:** On intertidal wood (Chinnaraj, 1992).

East Coast:-Tamil Nadu: On wood of *Rhizophora mucronata* and *R. apiculata* (Ravikumar and Vittal, 1987, 1996), on intertidal wood (Nambiar and Raveendran, 2010c); **Andhara Pradesh:** On intertidal wood of *Avicennia marina* and *Rhizophora apiculata* (Sarma and Vittal, 1998-99; Sarma and Vittal, 2000; Sarma and Vittal, 2001; Sarma et al., 2001; Sarma and Vittal, 2004; Vittal and Sarma, 2005); **West Bengal:** On intertidal wood of *Rhizophora mucronata* (Pawar et al., 2009); **Andaman-Nicobar Islands:** On intertidal wood of *Avicennia marina, A. officinalis, Bruguiera gymnorhiza, Rhizophora apiculata, R. mucronata* and *Sonneratia alba* (Chinnaraj, 1993a).

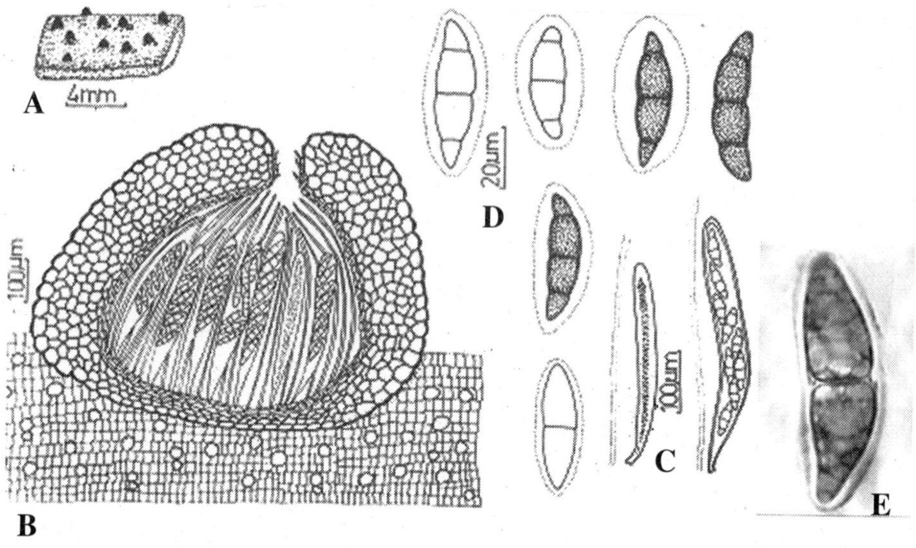

Fig. 22.1. ***Ascocratera manglicola,*** A-E; A- Habit, B- v.s.of ascocarp, C- ascus & pseudoparaphyses, D- ascospores, E- ascospore 1000x).

23. *Astrosphaeriella* Sydow & P. Sydow, 1913. *Ann.Mycol.,* 11: 260.

Ascomata: perithecioid, immersed and subepidermal, at maturity covered by epidermis except at the ostiole, or in some becoming superficial with host tissue rupturing as scales around the base to give them a stellate appearance, arising singly, or rarely 2-3 joined together at the base, conical or hemispherical, base flattened, dark purplish-brown to almost black, solitary or clustered. *Peridium*: relatively thick and strongly carbonized, composed of thick, dark angular pseudoparenchymatous cells, lumens of cells often occluded, base poorely developed. *Pseudoparaphyses*: trabeculate up to 1 um wide, hypha-like, numerous, embedded in a gelatinous matrix. *Asci*: 8-spored, cylindrical, cylindrical-clavate, or narrowly obclavate, pedunculate, bitunicate, with an ocular chamber, some with a faint ring, mostly persistent. *Ascospores*: 2-3-seriate, elongate-fusiform, mostly straight, often tapering towards the apices, 1-mult-septate, slightly constricted at the septum, hyaline, brown or reddish brown, concolourous, some with lighter end cells, wall smooth, verrucose or with striations, often with mucilaginous sheaths or appendages. There are seven species from freshwater (Cai *et al.*, 2002a) and four species recorded from marine water (Jones et al., 2009a).

Type species: *Astrosphaeriella fusispora* Sydow & P. Sydow
Habitat: On monocotyledon and submerged wood in aquatic habitats.
Description: Based on Cai et al. (2006).

23.1. *Astrosphaeriella striatapora* (K.D. Hyde) K.D. Hyde, *Bot. J. Linn. Soc.*, **110**: 97. [Fig. 23.1]

Basionym: *Trematosphaeria striatispora* K.D. Hyde, *Bot. J. Linn. Soc.*, **98**: 142 (1988).

Fruiting Bodies: black, slightly raised spots, with a central neck; in section lenticular, developing amongst the host corticle cells beneath the host epidermis, composed of a lenticular ascomata, with overlying clypeus; clypeus 390-520 µm in diam., black, composed of host epidermal cells, packed with fungal hyphae, encrusted with melanin paricles, 85-170 µm thick. *Ascomat*: 352-528 µm in diam., 176-355 µm high, subglobose-broadly conical, ostiolate, papilate, solitary or gregarious. *Peridium*: indistinguishable from clypeus, the side and lower layer, 42-57 µm thick, consisting of thin-walled elongate cells growing amongst host tissue, Necks: 193-484 µm long, 53-82 µm in diam., black, periphysate. Pseudoparaphyses: 0.8-2.1 µm in diam., filiform, trabeculate. *Asci*: 100-173 µm long, 11-23 µm in diam., 8-spored, cylindrical, pedunculate, bitunicate, thick-walled, with ocular chamber, developing from the base and sides of the ascomata. *Ascospores*: 31-38 x 6-9 µm, 2-seriate, fusiform, 3-septate, constricted at the septa, central cells larger, brown, end cells smaller, conical and light brown-yellow, wall covered in distinct longitudinal striations and surrounded by an indistinct gelatinous sheath.

Habitat: On intertidal mangrove wood.
Description: Based on Hyde (1988b).
Distribution: West Coast:-Maharashtra: On intertidal mangrove (seedlings of *Rhizophora mucronata*) substrate (Borse, 2000, Borse et al., 2005a); **Karnataka:** On intertidal wood (Prasannarai and Sridhar, 2001, 2000-2001).
East Coast:- Tamil Nadu: on intertidal wood (Prasannarai and Sridhar, 2001); **West Bengal:** On intertidal wood of *Phoenix poludosa* (Pawar et al., 2005); **Andaman & Nicobar Islands:** On intertidal frond of *Nypa fruticans* (Chinnaraj and Untawale, 1992), on intertidal wood of *Rhizophora mucronata* (Chinnaraj, 1993a).

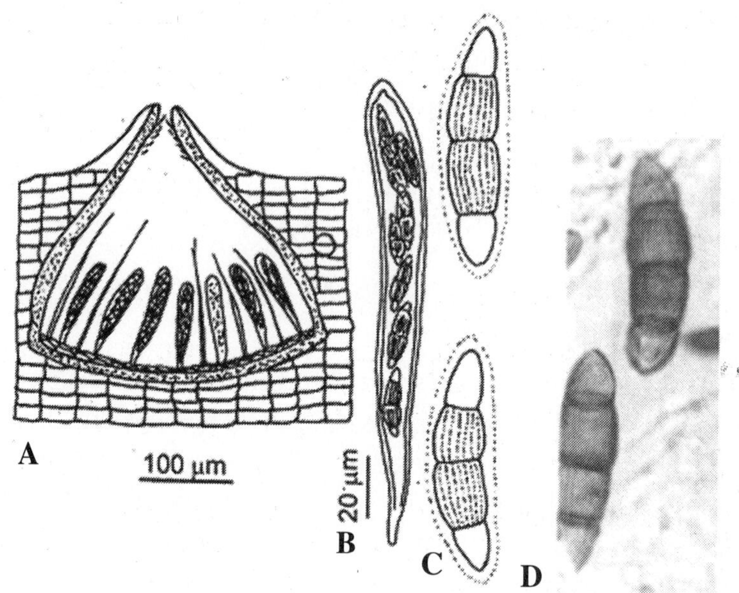

Fig. 23.1. ***Astrosphaeriella striatapora,*** A-D; A- v.s. of ascocarp, B- ascus, C- ascospores, D- ascospores (1000x).

24. *Bathyascus* Kohlm., 1977. *Rev. Mycol.*, **41**: 191.

Ascomata: solitary, subglobose or ellipsoidal, partly or completely immersed, ostiolate, papillate, coriaceous, brownish. *Peridium*: thin, forming a textura angularis. *Catenophyses*: absent, center of immature ascomata filled with thin-walled pseudoparenchymatous cells . *Asci*: 8-spored, fusiform to clavate, unitunicate, thin-walled, early deliquescing, maturing simultaneously on a small-celled ascogenous tissue at the base of the ascomata venter. *Ascospores*: filiform, 1-celled, hyaline, without appendages.

Type Species: *Bathyascus vermisporous* Kohlm.
Habitat: On intertidal wood.
Discription: Based on Kohlmeyer (1977).

Key for *Bathyascus* species from India:

1. Ascospores 1-septate, 70-100 x 8-9 µm .... *Bathyascus tropicalis*

1'. Ascospores non-septate 2

2. A deep sea species, 50-72 x 4-6 µm .... *Bathyascus vermisporus*

2'. A intertidal species 3

3. Ascospores 90-145 x 2.5-4 µm .... *Bathyascus avicenniae*

3'. Ascospores 205-300 x 9-15 µm .... *Bathyascus grandisporus*

3". Ascospores 125-160 x 2.5-5 µm .... *Bathyascus mangrovei*

24.1. *Bathyascus avicenniae* Kohlm., 1980. *Bot. Mar.*, **23**: 530.

[Fig. 24.1]

Ascomata: 200-250 µm long, 140-220 µm in diam., subglobose to ampulliform, immersed in rotten bark, ostiolate, papillate, coriaceous, thin-walled, black, base sometimes light-coloured, solitary. *Peridium*: 20-24 µm thick, composed of two strata; a dark outer stratum of 3-4 layers of irregular polygonal to 2-4 layers of flattened cells. *Necks*: 300-800 µm long, 40-70 µm in diam. at the base, 22-50 µm in diam. at the tip, cylindrical, extending above the surface of the substrate; ostiolar canal periphysate; periphyses 4 um long, 2.5 µm in diam., thin-walled, hyaline, possibly dissolving at ascospore maturity. *Paraphyse*: absent, deliquescent pseudoparenchymatous cells probably filling the venter of immature ascocarps; large thin-walled hyaline cells occur in the apex of the cavity of mature ascocarps, probably closing the base of the ostiolar canal before the release of ascospores. *Asci*: 140 x 12 µm, 8-spored, cylindrical to fusiform, unitunicate, thin-walled, deliquescing before ascospore maturity, without apical apparatuses, originating in a fascicle at the base of the ascocarp and releasing the spores into the venter. *Ascospores*: 90-145 x 2.5-4 µm, parallel in the ascus, filiform, curved, non-septate, without appendages or apical chambers, hyaline.

Habitat: Saprobic on bark of mangroves in the intertidal zone.

Discription: Based on Kohlmeyer (1980) and Hyde and Jones (1987).

Distribution: East Coast:-Tamil Nadu: On pneumatophores of *Avicennia marina* and *A. officinals* (Ravikumar and Vittal, 1987); **Andhara Pradesh:** On intertidal wood of *Avicennia marina* and *A. officinalis* (Sarma and Vittal, 1998-

99; Sarma and Vittal, 2000; Sarma and Vittal, 2001; Sarma et al., 2001; Sarma and Vittal, 2004; Vittal and Sarma, 2005); **Andaman & Nicobar Islands:** On intertidal wood of *Avicennia marina* (Chinnaraj, 1993a).

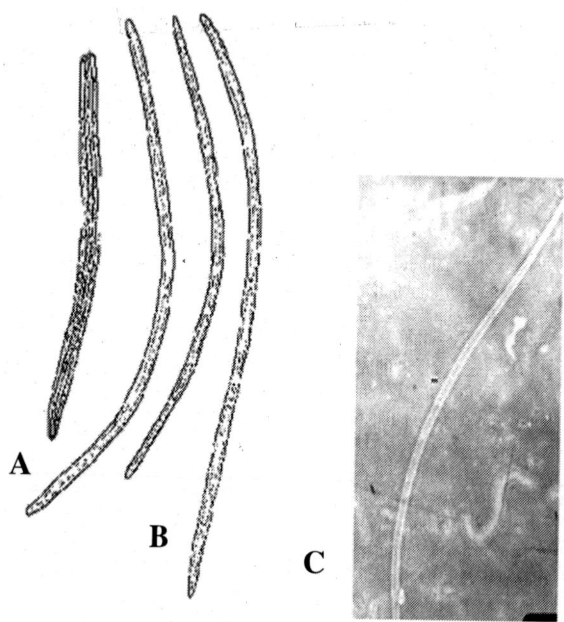

Fig. 24.1. ***Bathyascus avicenniae,*** A-C; A- ascus (450x), B-ascospores (1000x), C- ascospore (1000x).

24.2. ***Bathyascus grandisporus*** K.D. Hyde & E.B.G. Jones, 1987. *Bot. Mar.*, **30**: 413. [For Fig. refer original paper]

Ascomata: 360-480 µm long, 170-330 µm wide, 120-190 µm thick, pyriform-subglobose in view, ellipsoidal in end view, laterally depressed, lying horizontally immersed below the substrate surface, ostiolate, carbonaceous, black, solitary or gregarious. *Peridium*: 36-72 µm wide, composed of an outer layer of black thick-walled angular cells with large lumina and an inner thin layer of hyaline thin-walled irregularly elongate cells. *Necks*: 430-720 µm long, 36-48 µm in diam., often branching at the base into 2-3 necks, lower part black, upper third hyaline, composed of thick-walled oblong cells with large lumina and filled with periphyses. *Catenophyses*: absent. *Asci*: 300-365 x 37-65 µm, 8-spored, fusiform, short peduculate, thin-walled, unitunicate, early deliquescing, withoutan apical apparatus, developing at the base of the ascomata venter. *Ascospores*: 205-300 x 37-56 µm, elongate fusiform to cylindrical, straight to slightly curved, non-septate, hyaline, guttulate, thin-walled with rounded apices, in a bundle, arranged longitudinally in a asci, often spirally wound around other spores, without appendages.

Habitat: On intertidal mangrove wood.

Discription: Based on Hyde and Jones (1987).

Distribution: East Coast:-Andaman & Nicobar Islands: On intertidal wood of *Rhizophora mucronata* (Chinnaraj, 1993a).

24.3. *Bathyascus mangrovei* D.R. Ravikumar & B.P.R. Vittal, 1991a. *Mycol. Res.*, **95**: 370. [For Fig. refer original paper]

Ascomata: 120-240 µm high, 250-350 µm diam., subglobose to amulliform, immersed or superficial, ostiolate, papillate, coriaceous, black, solitary or gregarious. *Peridium*: 30-35 µm thick, composed of outer dark layer and inner hyaline layer. *Necks*: cylindrical, ostiolar canal periphysate. Catenophyses: absent. *Asci*: 132-153 x 7-10 µm, 8-spored, cylindrical to fusiform, thin-walled, unitunicate, early deliquescing, developing at the base of ascomata venter. *Ascospores*: 125-160 x 2.5-5 µm, parallel in the ascus, filiform, straight or slightly curved, aseptate, hyaline, multiguttulate.

Habitat: On intertidal mangrove wood.

Discription: Based on Ravikumar and Vittal (1991a).

Distribution: East Coast:-Tamil Nadu: On intertidal wood of *Rhizophora apiculata* and *mucronata* (Ravikumar and Vittal, 1991a, 1996).

24.4. *Bathyascus tropicalis* Kohlm., 1980. *Bot. Mar.*, **23**: 532.

[Fig. 24.4]

Ascomata: 140-230 µm high, 110-150 µm in diam., ellipsoidal, completely immersed in the softened wood, ostiolate, papillate, coriaceous, thin-walled, light brown, solitary. *Peridium*: 6-8 µm thick, composed of 2-3-layered of flat, thin-walled, light brown cells; mature ascospores are visible through the wall. *Necks*: 200-360 µm long, 26-28 µm in diam. at the base, 18-20 µm in diam. in the upper part, cylindrical, slightly inflatened at the tip, eccentric, extending above the wood surface; ostiolar canal without periphyses. *Paraphyses*: absent, deliquescing pseudoparenchymatous cells probably filling the venter of immature ascocarps; some thin-walled utricular cells, 13-16 x 8 µm, similar to catenophyses have been observed in one ascocarp. *Asci*: 8-spored, ellipsoidal, unitunicate, thin-walled, deliquescing before ascospore maturity, originating in a fascicle at the base of the ascocarp and releasing the spore into the venter. *Ascospores*: 70-100 x 8-9 (-10) µm, elongate-fusiform, rounded at the upper end, tapering towards the base, straight or slightly curved, 1-septate below the centre, not or slightly constricted at the septum, without appendages, hyaline.

Habitat: Saprobic on intertidal wood.

Description: Based on Kohlmeyer (1980) and Hyde and Jones (1987).

Distribution: West Coast:-Maharashtra: On intretidal wood and *Avicennia marina* (Borse, 1987a, 1988; Borse et al., 2005a); **Pondicherry (Mahe):** On intertidal mangrove wood (Nambiar and Raveendran, 2008); **Kerala:** On intertidal wood and *Avicennia* sp. (Raveendran and Manimohan, 2007, 2008b; Nambiar and Raveendran, 2009d, 2010c); **Lakshadweep Islands:** On intertidal wood (Raghukumar, 1988). **East Coast:-Tamil Nadu:** On intertidal wood (Nambiar et al., 2008b), on mangrove wood (Nambiar and Raveendran, 2008e, 2009k, l, 2010c); **Andhra Pradesh:** On intertidal wood (Nambiar and Raveendran, 2009j); **West Bengal:** On intertidal wood of *Avicennia marina* and *Rhizophora mucronata* (Pawar and Borse, 2005).

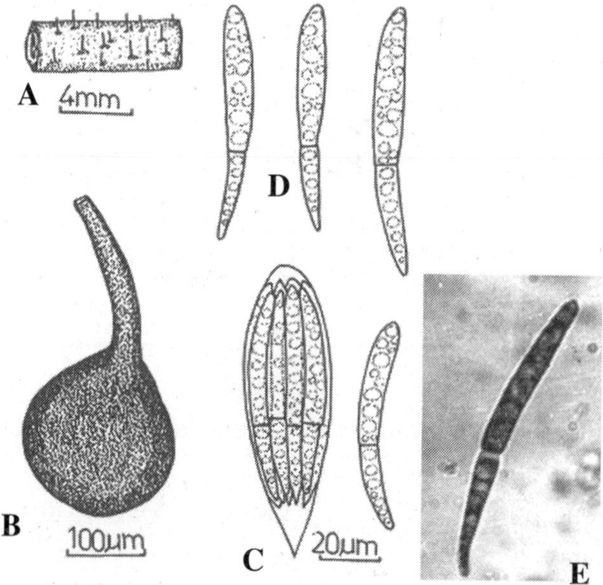

Fig. 24.4. ***Bathyascus tropicalis,*** A-E; A- Habit, B- ascocarp, C- ascus, D- ascospores, E- ascospore 1000x).

24.5. *Bathyascus vermisporus* Kohlm., 1977. *Rev. Mycol.*, **41**: 191.

[For Fig. refer original paper]

Ascomata; 150-260 µm high (including necks), 215-480 µm in diam., solitary, subglobose or ellipsoidal, partly or completely immersed, ostiolate, papillate, coriaceous, dark brown above, light brown to almost hyaline at the base. *Peridium*: 40-45 µm thick around the ostiole, 20-30 um at the base and sides, composed of 5 or 6 layers of thin-walled, polygonal, elongated cells with large lumina, forming a *textura angularis*, mearging into the pseudoparenchyma of the center. *Necks*: 60-380 µm long, 60-70 µm in diam., cylindrical or obtusely conical, dark brown; ostiolar canal about 15 µm in diam., periphyses absent; walls of some necks extend into lumina of wood vessels and, thus, appear warty. *Catenophyses*: absent, center of immature ascomata filled with hyaline, thin-walled pseudoparenchymatous cells that become compressed by the developing asci. *Asci*: 70-80 x 12-14 µm, 8-spored, fusiform to clavate, unitunicate, thin-walled, early deliquescing, maturing simultaneously on a small-celled ascogenous tissue at the base of the ascomata venter. *Ascospores*: 50-72 x 4-5.5 µm, filiform, or rarely elongate fusoid, straight or curved, attenuate, thick-walled, non-septate, hyaline, without appendages.

Habitat: On intertidal wood (deep sea).
Discription: Based on Kohlmeyer (1977) and Hyde and Jones (1987).
Distribution: West Coast: Deep sea sediments in the Arabian sea (Raghukumar and Raghukmar, 1998).

25. *Belizeana* Kohlm. & Volkm.-Kohlm., 1987a. *Bot. Mar.*, **30**: 196.

Ascomata: subglobose to ampulliform, immersed, later erumpent, ostiolate, epapillate or papillate, black above, light coloured below, single or gregarious. *Peridium*: thick. *Pseudoparaphyses*: simple or branched, embedded in a gelatinous matrix, growing down from the roof of the locule. *Asci*: 8-spored, cylindrical, pedunculate, thick-walled, fissitunicate, parallel, matuaring successively on an ascogenous tissue at the bottom of the locule. *Ascospores*: ellipsoidal, 1-septate, hyaline, thick-walled, tuberculate.

Type Species: *Belizeana tuberculata* Kohlm. & Volkm.-Kohlm.
Habitat: On intertidal wood of a mangrove.
Discription: Based on Kohlmeyer and Volkmann-Kohlmeyer (1987a).

25.1. *Belizeana tuberculata* Kohlm. & Volkm.-Kohlm., 1987a. *Bot. Mar.*, **30**: 196. [For Fig. refer original paper]

Ascomata: 210-260 µm high, 220-230 µm in diam., subglobose to broadly ampulliform, at first immersed, later erumpent, ostiolate, epapillate or shortly papillate, black and carbonaceous above, light brown and leathery on the sides and base, single or gregarious. *Ostioles*: up to 20 µm in diam., filled with a tissue of hyaline cells, forming a *textura angularis*. *Peridium*: 20-31 µm thick on the sides and base, 37-50 µm around the ostiole, enclosing some deteriorating wood particles, with melanin incrustations in the upper half, hyaline in the lower half, composed of about 9-12 layers of polygonal cells with small lumina. *Pseudoparaphyses*: 2 µm in diam., simple, rarely branched or anastomosing, embedded in a gelatinous matrix, growing down from the roof of the locule, merging with thinner interascicullar filaments. *Asci*: 140-155 x 25-29 µm, 8-spored, cylindrical, short pedunculate, thick-walled, fissitunicate, without apical apparatus, with a small ocular chamber, parallel, maturing successively on an ascogenous tissue at the bottom of the locule. *Ascospores*: 17-24 x 10-15 µm, monostichously in the ascus, broad ellipsoidal, 1-septate, constricted at the septum, hyaline, wall thick, two layered, mature spores with tuberculate ornamentations between the two layers; tubercles 0.7-1.5 µm in diam.; young spores smooth, becoming delicately verruculose during maturation; outer layer eventually gelatinizing and forming a sheath.

Habitat: On intertidal wood of a mangrove.

Discription: Based on Kohlmeyer and Volkmann-Kohlmeyer (1987a).

Distribution: East Coast:-Tamil Nadu: On submerged parts of *Rhizophora* spp. (Ravikumar and Vittal, 1996); **Andaman & Nicobar Islands:** On intertidal wood of *Rhizophora mucronata* (Chinnaraj and Untawale, 1992), on intertidal wood of *Avicennia marina* and *Rhizophora apiculata* (Chinnaraj, 1993a).

26. *Biatriospora* K.D. Hyde & Borse, 1986a. *Mycotaxon,* **26**: 263.

Ascomata: solitary or gregarious, large, subglobose to pyriform, immersed in the substratum, ostiolate, periphysaate, papillate, carbonaceous, black. *Pseudoparaphyses*: numerous, simple, thin, branched, hyaline embedded in in a gelatinous matrix. *Asci:* 8-spored, cylindrical, pedunculate, bitunicate, thick-walled, with an apical apparatus, developing in the hymenium at the base of the ascomata. *Ascospores*: fusiform, dark brown, septate towards each end, non-septate in the centre. The genus is monotypic (Jones *et al.*, 2009a).

Type Species: *Biatriospora marina* Hyde & Borse
Habitat: On intertidal mangrove wood.
Discription: Based on Hyde and Borse (1986a).

26.1. *Biatriospora marina* K.D. Hyde & Borse, 1986a. *Mycotaxon,* **26**: 263.

[Fig. 26.1]

Ascomata: 650-860 μm long, 350-510 μm in diam., lying horizontally, immersed in the substratum, subglobose to pyriform, solitary or gregarious, black, carbonaceous, ostiolate and papillate. *Necks*: 150-225 μm long, 60-75 μm in diam., canal filled with periphyses. Pseudoparaphyses: 0.6-1.7 μm in diam., numerous, simple, branched, filamentous, hyaline, embedded in a gelatinous matrix. *Asci*: 175-400 μm long, 22-40 μm in diam., 8-spored, cylindrical, long pedunculate, bitunicate, thick-walled, with an apical apparatus, developing in the hymenium at the base of the ascomata. *Ascospores*: 55-82 x 16-25 μm, uniseriate, fusiform, 2-4-septate towards each end, non-septate in the centre, not constricted at the septa, hyaline when immature, becoming brown to dark brown at maturity, and provided at each end with a hyaline, globose refractive chamber, 6-8 μm long, 4-6 μm in diam.

Habitat: On intertidal mangrove wood.
Discription: Based on Hyde and Borse (1986a).
Distribution: West Coast:-Maharashtra: On intertidal wood of *Avicennia alba* and *A. officinalis* (Hyde and Borse, 1986a; Srivastava, 1989), on mangrove wood (Borse, 1988); **Kerala:** On intertidal wood (Raveendran and Manimohan, 2007), on intertidal mangrove wood of *Avicennia* sp. and *Rhizophora* sp. (Nambiar and Raveendran, 2008a, b; 2009d, f, h, 2010c); **Lakshadweep Islands:** On intertidal wood of *Avicennia marina* (Chinnaraj, 1992).
East Coast:-West Bengal: On intertidal wood of *Avicennia alba, A. officinalis* and *Sonneratia apetala* (Pawar et al., 2005); **Andaman & Nicobar Islands:** On intertidal wood of *Avicennia marina, A. officinalis, Bruguiera gymnorhiza, Rhizophora apiculata, R. mucronata* and *Sonneratia alba* (Chinnaraj, 1993a).

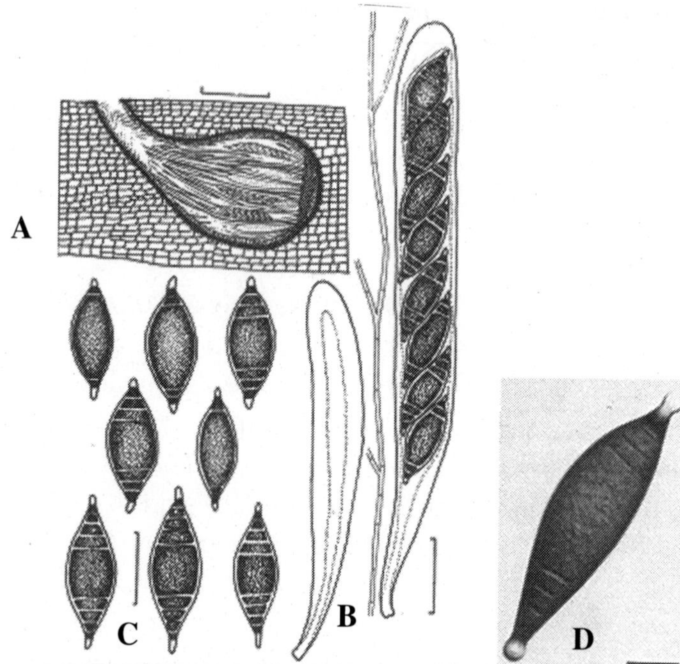

Fig. 26.1. ***Biatriospora marina,*** A-D; A- v.s. of ascocarp (bar=200 μm), B- asci and paraphysis (bar=40 μm) , C- ascospores (bar=30 μm), D- ascospore (1000x).

27. ***Biconiosporealla*** Schaumann, 1972. Veroff. Inst. Meereforsch. Bremerhaven, 14: 24.

Ascomata: ovoid, pyriform, flask-shaped, superficial or immersed, ostolate, papillate, coriaceous, brown to black, solitary or gregarious, periphysate, catenophyses present. *Asci*: 8-spored, cylindrical, short pedunculate, thin-walled, unitunicate, persistent, with an apical plate. *Ascospores*: biconical, unequally 3-septate, thick-walled, slightly constricted at the septa, dark brown, no sheath or appendages. The genus is monotypic (Jones *et al.,* 2009a).

Type Species: *Biconiosporella corniculata* Schaumann

Habitat: On intertidal wood.

Discription: Based on Kohlmeyer and Kohlmeyer (1979) and Jones et al. (2009a).

27.1. *Biconiosporella corniculata* Schaumann, 1972. *Veroff. Inst. Meereforsch. Bremerhaven*, **14:** 24. [For Fig. refer original paper]

Ascomata: 350-532 µm high (including necks), 155-297 µm in diameter, ovoid, pyriform, flask-shaped, superficial or immersed, ostolate, papillate, coriaceous, brown to black, solitary or gregarious, periphysate, catenophyses present. *Asci*: (155-) 173-205 x (18.5-) 20.5-22 µm, 8-spored, cylindrical, short pedunculate, thin-walled, unitunicate, persistent, with an apical plate. *Ascospores*: 26.5-43 x 16-28.5 µm, biconical, unequally 3-septate, thick-walled, slightly constricted at the septa, dark brown, no sheath or appendages.

Habitat: On intertidal wood.
Discription: Based on Kohlmeyer and Kohlmeyer (1979).
Distribution: West Coast:-Kerala: On intertidal wood (Nambiar and Raveendran, 2011b).

28. *Bicrouania* Kohlm. & Volkm.-Kohlm., 1990b. *Mycol. Res.,* **94**: 685.

Ascomata: subglobose to ellipsoidal, superficial, erumpent or rarely immersed, ostiolate, periphysate, short papillate or epapillate, subcarbonaceous, blakish-brown, rough, gregarious. *Peridium*: dark brown at the outside, hyaline near the venter and merging into pseudopapraphyses, cells irregular, occluded by dark deposits, enclosing some cells of the host, forming a *textura epidermoidea. Hamathecium*: pseudoparaphysoid, trabeculate. *Asci*: eight-spored, cylindrical, short pedunculate, non-amloid, without ocular chamber or apical apparatus, thick-walled, fissitunicate, maturing successively at the bottom of the locule; with hymenial one-celled photobionts in the dome of the locule. *Ascospores*: monostichous, ellipsoidal, one-septate in the middle, constricted at the septum, reddish-brown, darkest around the septum, thick-walled, smooth.

Type Species: *Bicrouania maritima* (Crouan & Crouan) Kohlm. & Volkm.-Kohlm.
Habitat: On dead stems of salt marsh plant *Halimione portulacoides*.
Discription: Based on Kohlmeyer and Volkmann-Kohlmeyer (1990b).

28.1. ***Bicrouania maritima*** (H. Crouan & P. Crouan) Kohlm. & Volkm.-Kohlm., 1990. *Mycol. Res.,* **94**: 685. [For Fig. refer original paper]

Basionym: *Sphaeria maritima* H. Crouan & P. Crouan, *Florule du Finistere*, Paris, 27: (1867).

≡ *Didymosphaeria maritima* (H. Crouan & P. Crouan) Sacc., *Syll. Fung.* 1: 703 (1882).

Ascomata: 270-420 µm high (including papilla), 300-450 µm in diam., subglobose to ellipsoidal, superficial, erumpent or rarely immersed, ostiolate, periphysate, short papillate or epapillate, subcarbonaceous, blakish-brown, rough, gregarious. *Peridium*: 30-70 µm thick, dark brown at the outside, hyaline near the venter and merging into pseudopapraphyses, cells irregular, occluded by dark deposits, enclosing some cells of the host, forming a *textura epidermoidea*. *Hamathecium*: pseudoparaphysoid, trabeculate. *Asci*: 130-145 x 20-24 µm, eight-spored, cylindrical, short pedunculate, non-amloid, without ocular chamber or apical apparatus, thick-walled, fissitunicate, maturing successively at the bottom of the locule; with hymenial one-celled photobionts in the dome of the locule. *Ascospores*: 24-32 x 12-15 µm, monostichous, ellipsoidal, one-septate in the middle, constricted at the septum, reddish-brown, darkest around the septum, thick-walled, smooth.

Habitat: On dead stems of salt marsh plant *Halimione portulacoides*.
Discription: Based on Kohlmeyer and Kohlmeyer (1979).
Distribution: West Coast:-Kerala: On woody litter of mangroves (Shini et al., 2009-10).
Note: This is a temperate water ascomycete collected on dead stems of salt marsh plant *Halimione portulacoides* from France (Crouan & Crouan, 1867).

29. ***Biflua*** Jorgen Koch & E.B.G. Jones, 1989. *Can. J. Bot.*, **67**: 1187.

Ascomata: globose, coriaceous, light coloured, nonostiolate, with appendages covering the free surfsce. *Peridium*: membranous, uniform, and composed of 1-(2-3) layers of flat angular cells. *Asci*: in a hymenium arising from a flattened cushion of pseudoparenchymatous cells at the base of the ascomata, claviforme, pedunculate, and the wall elastic in nature. *Paraphyses*: absent. *Ascospores*: unequally 2- (-4) celled, constricted at the septum, hyaline, with a uniform mucilaginous sheath. The genus is monotypic (Jones *et al.*, 2009a).

Type Species: *Biflua physasca* Jorgen Koch & E.B.G. Jones

Habitat: On intertidal wood (driftwood).
Discription: Based on Koch and Jones (1989).

29.1. *Biflua physasca* Jorgen Koch & E.B.G. Jones, 1989. *Can. J. Bot.*, **67**: 1187. [For Fig. refer original paper]

Ascomata: 75-112 µm in diam., globose or slightly flattened, coriaceous, white-light buff, gregarious, light coloured, free (attached to the wood by a few or several thick-walled sinuous hyphae 4 um in width), or embedded in the outer few layers of the substratum. Ascomata with numerous appendages and lacking an ostiole. Appendages 4-8 µm high, 2-3 µm broad at the base, the stem dividing into 2-3 (-6) tapering blunt branches. *Peridium*: membranous, with one wall layer, 1-2 cells thick, cells flat and angular up to 12 µm in diam. *Asci*: 8-spored, in a hymenium arising from a small, flat cushion of pseudoparenchymatous cells at the base of the ascomata, claviforme to cylindrical, pedunculate, persistent, and laxkine an apical apparatus. On release from the ascocarp, asci expand becoming broadly claviform to oblong spherical, 68-147 x 28-57 µm with the cytoplasm or spore cluster occupying only a minor part of the ascus. Ascus tip appears granulated or viscous. Ascospores released as a cluster through a broad circular opening of the ascus tip. Ascus wall elastic and sometimes bulging from the side or tip. *Paraphyses*: absent. *Ascospores*: unequally 2- (-4) celled, 10-16 x 6-9 µm, strongly constricted at the septum, hyaline, with a uniform mucilaginous sheath.

Habitat: On intertidal wood (driftwood).
Discription: Based on Koch & Jones (1989).
Distribution: West Coast:-Karanataka: On intertidal leaf and woody litter of mangroves (Ananda and Sridhar, 2004).
Note: This is a temperate water ascomycete collected on drift wood (Koch and Jones, 1989).

30. *Byssothecium* Fuckel, 1861. *Bot. Ztg.*, **19**: 251.

Ascomata: subglobose or ellipsoidal, immersed to erumpent, ostiolate, subcabonaceous, dark brown to black, gregarious. Pseudoparaphyses: septate, ramose. *Asci*: clavate to subcylindrical, short pedunculate, thick-walled, without an apical apparatuses. *Ascospores*: versicoloured, end cells hyaline, central cells brown, 3-septate, and constricted at the septa.

Type Species: *Byssothecium circinans* Fuckel
Habitat: Saprobic on wood.
Discription: Based on Jones et al. (2009a).

30.1. ***Byssothecium obiones*** (P. Crouan & H. Crouan) M.E. Barr, 2002. *Mycotaxon.*, **82**: 378. [For Fig. refer original paper]

Basionym: *Pleospora obioens* P. Crouan & H. Crouan, Florule Finistere (Paris): 22 (1867).

≡ *Passeriniella obiones* (P. Crouan & H. Crouan) Hyde & Mouzouras, Trans. Br. Mycol. Soc., 91: 183 (1988).

= *Leptosphaeria discors* Sacc. & Ellis, Michelia, 2: 567 (1882).

≡ *Leptosphaeria obiones* (P. Crouan & H. Crouan) Sacc., Syll. Fung. (Abellini), 2: 24 (1883).

= *Leptosphaeria obiones* f. *evolitior* Grove, J. Bot. (Lond.), 71: 289 (1933).

= *Didumosphaeria spartinae* Grove, J. Bot. (Lond.), 71: 259 (1933).

= *Metasphaeria discors* (Sacc. & Ellis) Sacc., Syll. Fung., 2: 173 (1883).

= *Passeriniella discors* (Sacc. & Ellis) Apinis & Chesters, Trans. Br. Mycol. Soc., 47: 432 (1964).

= *Passeriniella incarcerate* Berl., Icon. Fung., 1: 51 (1892).

Ascomata: 105-295 µm high, 200-440 µm wide, subglobose or ellipsoidal, immersed or erumpent, ostiolate, epapillate or perforatin the epidermis with a short papilla, carbonaceous or subcabonaceous, black above, brown or light brown below, sometimes clypeate, usually covered by brown hyphae, gregarious. *Peridium*: 36-47 µm thick in upper part, 15-25 µm thick in the sides and base, composed of irregular or compressed, thick-walled cells, forming a textura angularis, enclosing cells of the host. *Pseudoparaphyses*: septate, ramose, filling the ostiolar canal. *Asci*: 141-291 x 15-24 µm, 8-spored, clavate or subcylindrical, short pedunculate, bi-tunicate, thick-walled, physoclastic, without apical apparatuses; developing at the base of the ascomata venter. *Ascospores*: 25-38 x 9-14 µm, 1-seriate at the base, 2-seriate in the top, ellipsoidal or broadly fusoid, slightly constricted at the septa; central cells brown or yellowish brown, large; apical cells subhyaline, small. Spemagonial stage present.

Habitat: Saprobic on salt marsh plants and intertidal wood.
Description: Based on Kohlmeyer and Kohlmeyer (1979).
Distribution: West Coast:-Maharashtra: On intertidal mangrove wood (Borse, 1988).
East Coast:-Tamil Nadu: On wood of *Avicennia marina* (Ravikumar and Vittal, 1987), **Andhara Pradesh:** On intertidal wood of *Avicennia marina, Excoecaria agallocha* and *Rhizophora apiculata* (Sarma and Vittal, 1998-99; Sarma and Vittal, 2000; Sarma and Vittal, 2001; Sarma et al., 2001; Sarma and Vittal, 2004; Vittal and Sarma, 2005).
Note: This is a common ascomycete found on decaying *Spartina* culms (Jones et al., 2009a).

31. *Carbosphaerella* I. Schmidt, 1969. *Feddes Repert.*, **80**: 108.

Ascomata: solitary, globose or subglobose, subiculate, ostiolate or nonostiolate, papillate or epapillate, carbonaceous, black. *Paraphyses*: absent. *Asci*: 4-8 spored, obpyriform to subglobose, short pedunculate, unitunicate, thin-walled, early deliquescing. *Ascospores*: ellipsoidal or ovoid, 3-septate or muriform, apical cells lighter coloured than the others, surrounded with a gelatinous, striate sheath. The genus is represented by 2 species (Jones *et al.*, 2009a).

Type Species: *Carbosphaerella pleosporoides* I. Schmidt
Habitat: On intertidal wood.
Discription: Based on Kohlmeyer and Kohlmeyer (1979).

31.1. *Carbosphaerella leptosphaerioides* I. Schmidt, 1969. *Nat. Naturschutz Meeklenburg*, **7**: 9. [Fig. 31.1]

Ascomata: 90-210 µm in diam., globose to subglobose, superficial, sometimes seated with subicula on grains of sand, with or without ostiolates, papillate or epapillate, carbonaceous, black, solitary or gregarious. Peridium: 5-12 µm thick, composed of 2-3 layers of cells, subglobose at the outside, flattened toward the inside. Necks: when present, short, conical, near the basal subiculam and pointing downward. *Paraphyses*: absent. *Asci*: 65-90 x 45-60 µm (excluding peduncle), 4-8 spored, ovoidal or obpyriform , short pedunculate, unitunicate, thin-walled, aphysoclastc, without apical apparatus, early deliquescing. *Ascospores*: 27-42 x 16-24 µm (excluding sheath), ellipsoidal, 3-septate, not or slightly constricted at the septa, central large cells dark brown, apical small cells hyaline or light brown; septa with a central pore, ca. 1 μm in diam.; except for the apical cells, surrounded by a gelatinous, persistent, irregular sheath, 4-12.5 μm thick, hyaline or faintly brown, appearing striated or dotted with parallel fibers embedded in a matrix; germination from the apical cells only.

Habitat: On intertidal wood.

Discription: Based on Kohlmeyer and Kohlmeyer (1979).

Distribution: West Coast:-Kerala: On mangrove woody litter (Shini et al., 2009-10). **East Coast:-Tamil Nadu:** On intertidal wood ((Prasannarai and Sridhar, 2001); **West Bengal:** On intertidal wood (Borse and Pawar, 2001).

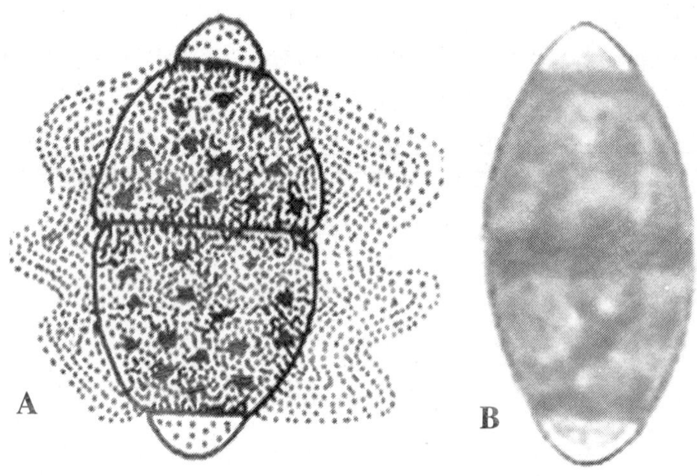

Fig. 31.1. ***Carbosphaerella leptosphaerioides,*** A-B; A- ascospore (1000x) , B- ascospore (1000x).

32. ***Caryosporarella*** Kohlm., 1985a. *Proc. Ind. Acad. Sci. (Plant Sci.),* **94**: 356.

Ascomata: gregarious, large, subglobose, superficial seated on a thin black stroma, ostiolate, periphysate, papillate, carbonaceous, black. *Paridium*: thick, 3-layered, yellow on the outside, brown in the middle, hyaline near the centrum, the outer two layers forming *textura epidermiodia*, the inner one a *textura prismatica*. *Pseudoparaphyses*: trabeculate, thin, simple between the asci, branched and anastomosing above them, embedded in the gelatinous matrix. *Asci*: 8-spored, cylindrical, pedunculate, bitunicate, thick-walled, with an apical cup-like apparatus, maturing successively on an ascogenous tissue at the bottom of the locule. *Ascospores*: ellipsoidal, 1-septate, dark brown with light-coloured tips, indistinctly verruculose.

Type Species: *Caryosporarella rhizophorae* Kohlm.
Habitat: On intertidal mangrove wood.
Description: Based on Kohlmeyer (1985a).

32.1. *Caryosporella rhizophorae* Kohlm., 1985a. *Proc. Ind. Acad. Sci. (Plant Sci.)*, **94**: 356. [For Fig. refer original paper]

Ascomata: 750-900 umhigh, 830-1050 µm in diam., gregarious, often growing close together, subglobose, superficial seated on a thin black stroma that covers the wood surface, ostiolate, periphysate, papillate, carbonaceous, thick-walled,appearing black, with a rough surface. *Necks*: 80-200 um high, 260-280 µm in diam., ostiolar canal 80-100 µm in diam. *Paridium*: 90-160 µm thick, 3-layered, outer layer 30-50 µm thick, golden yellow, middle layer 50-80 µm thick, deep brown, inner layer 15-25 µm thick, light brown to hyaline, the colour is caused by pigments encrusted on the walls; the outer two layers forming *textura epidermiodia*, the inner layer with elongate cells and large lumina, forming *textura prismatica*, giving rise to pseudoparaphyses; cells of the host are occasionally incorporated in the paridium. *Pseudoparaphyses*: 2-3 µm in diam., trabeculate, simple between the asci, branched and anastomosing above them, inconspicuously sepatate, embedded in a gelatinous matrix. *Asci*: 200-250 x 16-18 µm, 8-spored, cylindrical, long pedunculate, bitunicate, thick-walled, with an apical cup-like apparatus, IKI-negative apparatus, maturing successively on the ascogenous tissue at the bottom of the locule. *Ascospores*: 24-30 x 10-12 µm, ellipsoidal, 1-septate, slightly constricted at the septum, dark brown, lighter coloured at the thickened apices, indistinctly verruculose, thick-walled, overlapping monostichously arranged in the ascus.

Habitat: On intertidal mangrove wood.
Description: Based on Kohlmeyer (1985a).
Distribution: West Coast:-Goa: On intertidal aniiamal substrates (Ananda et al., 1998); **Karnataka:** On intertidal wood (Prasannarai and Sridhar, 1997, 2001, 2000-2001), on intertidal animal substrates (Ananda et al., 1998), on intertidal

wood of *Avicennia officinalis* and *Brugueiria gymnorrhiza, Rhizophora mucronata* (Maria and Sridhar, 2003a); **Kerala:** On intertidal wood (Prasannarai and Sridhar, 2001), on mangrove woody litter (Shini et al., 2009-10). **Lakshadweep Islands:** On intertidal mangrove wood (Ananda and Sridhar, 2003). **East Coast:-Andaman & Nicobar Islands:** On intertidal wood of *Rhizophora mucronata* (Chinnaraj and Untawale, 1992; Chinnaraj, 1993a).

33. *Ceriosporopsis* Linder, 1944. *Farlowia*, **1**: 408.

= *Ceriosporella* Cavaliere, *Nava Hedwigia,* **10**: 393 (1966).

= *Marinospora* Cavaliere, *Nava Hedwigia*, **11**: 548 (1966).

Ascomata: solitary or gregarious, subglobose to cylindrical, immersed or becoming exposed ostiolate, papillate, coriaceous or carbonaceous, light brown to black. *Catenophyses*: of large, thin-walled cells filling venter of young ascomat; deliquescing, or central core developing into catenophyses. *Asci*: 8-spored, clavate, pedunculate, unitunicate, thin-walled, early deliquescing. *Ascospores*: ellipsoidal, 1-septate, hyaline, surrounded by an exosporic sheath that forms apical or also lateral mucilaginous appendages. The genus is represented by six species (Sakayaroj et al., 2010a).

Type Species: *Ceriosporopsis halima* Linder
Habitat: On intertidal wood.
Description: Based on Kohlmeyer and Kohlmeyer (1979).

Key to *Ceriosporopsis* species from India:

1 Ascospores with small equatorial and long polar appendages,
ascospores (10-) 21-34 x 10-17 µm .... *C. caduca*

1' Ascospores lacking equatorial appendages 2

2 (1') Ascospore appendages more than 750 µm long, ascospores (26-) 30-38 (-42) x 10-12 µm .... *C. capillaceae*

2'(1') Ascospore appendages less than 750 µm long, 22.5-26 x 8-10 µm..........
.... *C. halima*

33.1. *Ceriosporopsis caduca* E.B.G. Jones & Zainal, 1988. *Mycotaxon,* **32**: 327. [Fig. 33.1]

Ascomata: 800-900 µm high, 900-1100 µm in diam., eillipsoidal to subglobose, immersed, ostiolate, papillate, cariaceous, light brown to brown. *Peridium*: 35 mm thick, two layerd. *Necks*: 150-1000 µm long, 85-95 µm in diam, cylindrical, centrally inserted, periphysate. Pseudoparenchyma thin-walled. *Asci*: 97-135 x 20-25 µm, 8-spored, clavate, pendunnculate, unitunicate, thin-walled, no apical apparatus, deliquescing at maturity, born at the base of the ascocarp. *Ascospores*: 21-34 x 10-17 µm (excluding appendages), ellipsoidal, 1-septate, hyaline with polar appendages at each end and an annulus-like appendage around the central septum, 5-7 µm thick. *Appendages*: Polar appendages arising as outgrowths of the spore wall, 10-54 x 3-5 µm and surrounded by a sac-like structure which is shed when mounted in water.

Habitat: On intertidal wood.

Description: Based on Jones and Zainal (1988).

Distribution: West Coast:-Maharashtra: On intertidal wood (Borse, 1987e, 1988); on intertidal wood of *Avicennia marina* (Borse et al., 2005a). **East Coast:-West Bengal:** On intertidal mangrove wood (Pawar & Borse, 2005).

***Note*:** This fungus was earlier reported as *C. cambriensis* Wilson but it is actually *C. caduca*. The records included as *C. cambriensis* by Borse (1987e, 1988) may be considered as *C. caduca* Jones and Zainal.

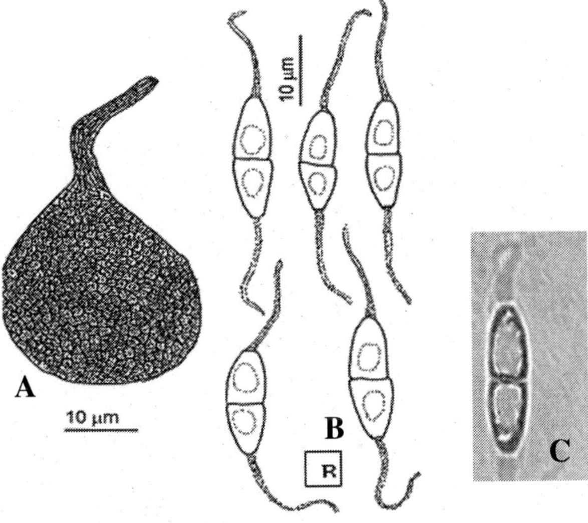

Fig. 33.1. *Ceriosporopsis caduca,* A-C; A- ascocarp, B- ascospores, C- ascospore (1000x).

33.2. *Ceriosporopsis capillacea* Kohlm., 1981. *Can. J. Bot.*, **59:** 1314.

[Fig. 33.2]

Ascomata: 120-150 µm high, 150-170 µm in diam., subglobose, eillipsoidal or cylindrical, superficial, ostiolate, papillate, coriaceous, thin-walled, reddish brown, solitary. *Peridium*: 5-6 µm thick, composed of 3 to 4 layers of flat thin–walled brown cells, merging into the pseudoparenchyma of the center. *Necks*: 50-180 µm long, 25-30 µm in diam., cylindrical; ostiolar canal without periphyses, at first field with a deliquescing, thin-walled pseudoparenchyma. *Catenophyses*: of thin-walled, polygonal cells filling venter of young ascomatas. *Asci*: unitunicate, thin-walled, deliquescing before ascospore maturity. *Ascospores*: 30-38 x 10-12.5 µm (excluding appendages), ellipsoidal, 1-septate, slightly or not constricted at the septum, hyaline, surrounded by an exosporic sheath that is pierced at each apex by a tube-like slender appendage. *Appendages*: 100-165 µm long, 2-2.5 µm in diam. at the base, about 0.3 µm in diam. at the tip, subcylindrical, tapering, capillary, non-gelatinous, persistent.

Habitat: On intertidal wood.

Desciption: Based on Kohlmeyer (1981).

Distribution: West Coast:-Maharashtra: On intertidal wood (Borse, 1987e); **Goa:** On intertidal wood (Raghukumar, 1988); **Kerala:** On intertidal wood (Nambiar and Raveendran, 2007, 2008b, 2010c).

East Coast:-Tamil Nadu: On intertidal wood (Nambiar and Raveendran, 2010a).

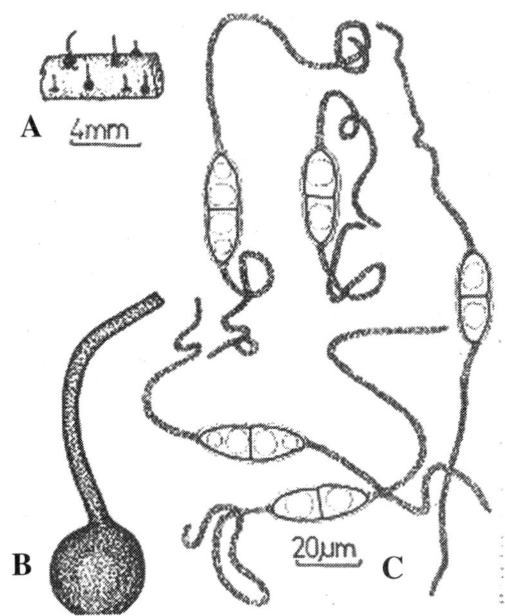

Fig. 33.2. ***Ceriosporopsis capillacea,*** A-C; A-Habit, B- ascocarp (100x), C-ascospore.

33.3. ***Ceriosporopsis halima*** Linder, 1944. *Farlowia,* **1**: 409. [Fig. 33.3]

= *Ceriosporopsis barbata* Hohnk, *Veroeff. Inst. Meeresf. Bremerhaven,* **3:** 210 (1955).

Ascomata: 80-481 µm high, 130-503 µm in diam., subglobose, eillipsoidal or cylindrical, immersed or superficial, ostiolate, papillate, cariaceous, dark brown to black reddish, brown when empty, solitary or gregarious. *Peridium*: 8-26 µm thick, composed of 3 to 4 layers of thin–walled, elongate cells with large lumina, merging into the pseudoparenchyma of the venter. *Necks*: up to 706 µm long, 16-30 µm in diam., centric or eccentric, cylindrical, ostiolar canal at first field with a small-celled pseudoparenchyma. *Catenophyses*: of thin-walled, polygonal cells filling venter of yonger ascomatas. *Asci*: 56-89 x 14-22 µm, 8-spored, ellipsoidal, subclavate or subfusiform, short pendunnculate, unitunicate, thin-walled, aphysoclastic, deliquescing before ascospore maturity, without apical apparatuses, developing at the base of ascomata venter. *Ascospores*: 18-27 x 6-12 µm (excluding appendages), ellipsoidal to fusiform, 1-septate, slightly or strongly constricted at the septum, hyaline, surrounded by a gelatinous exosporic sheath, that is pierced at each apex by an outward growing appendage. *Appendages*: 5-

8 μm in diam., of variable length, terminal, simple, subcylindrical, tapering, finally becoming viscous and filamentous.

Habitat: On intertidal wood.

Desciption: Based on Kohlmeyer and Kohlmeyer (1979).

Distribution: West Coast:-Daman: On intertidal wood and *Avicennia marina* (Borse et al., 2000b); **Gujarat:** On intertidal wood (Patil and Borse, 2001); **Maharashtra:** On intertidal wood and *Avicennia marina* (Borse, 1988, Borse et al., 2005a); **Goa:** On intertidal wood (Borse et al., 1999); **Karnataka:** On intertidal wood (Prasannarai et al., 1999), on intertidal wood (Prasannarai and Sridhar, 2001); **Kerala:** On intertidal wood (Prasannarai and Sridhar, 2001), on intertidal wood (Raveendran and Manimohan, 2007; Nambiar and Raveendran, 2007, 2008b, 2010a, c).

East Coast:-Tamil Nadu: On drift wood (Raghukumar, 1973); **West Bengal:** On intertidal wood of *Avicennia marina* (Pawar and Borse, 2004).

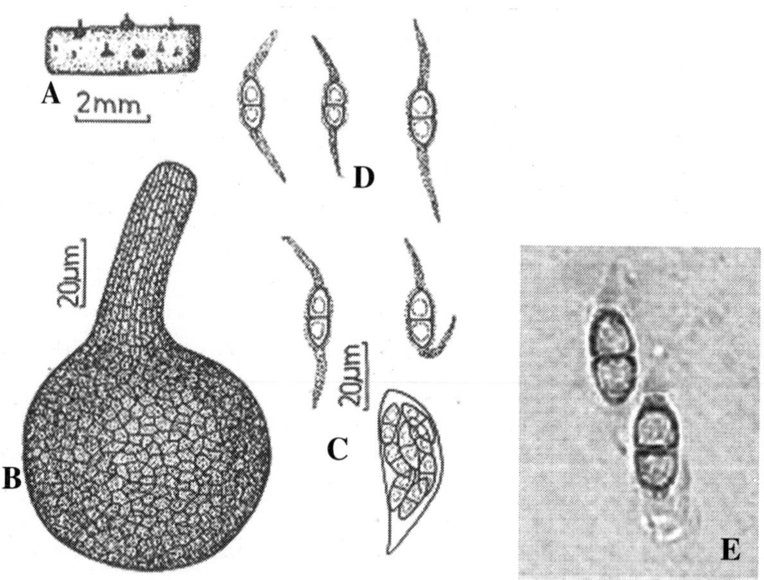

Fig. 33.3. ***Ceriosporopsis halima,*** A-E; A- Habit, B- ascocarp, C- ascus, D- ascospores E- ascospores (1000x).

34. *Corollospora* Werderm., 1922. *Notizbl. Bot. Gart. Berlin-Dahlem*, **8:** 248.

= *Peritrichospora* Linder, *Farlowia*, 1: 414 (1944).

Ascomata: globose or subglobose, superficial, erumpent, subiculate, oasiolate or lacking ostiole, papillate or epapillate, carbonaceous, pale brown to black, *Catenophyses*: deliquescing. *Asci*: fusiform or subclavate, apiculate, short pedunculate, unitunicate, thin-walled, lacking an apical apparatus, deliquescing. *Ascospores*: various fusiform, subellipsoidal, constricted at the septa, 1-14-septate, muriform with longitudinal and trens-septa, hyaline to pale brown to dark brown, with appendages. *Appendages*: of two kinds : (i) primary appendages as polar spine-like appendages arising as outgrowths of the mesosporium and episporium; (ii) secondary appendages as fine hair-like appendages formed by fragmentation of the exosporium. The genus is represented by 22 obligate marine species (Jones et al., 2009a; Al-Saadoon, 2006).

Type Species: *Corollospora maritima* Werderm.
Habitat: On intertidal wood and grains of sand, ascospores accumulate in foam along shore.
Description: Based on Jones et al. (2009a).

Key to *Corollospora* species from India:

1 Ascospores light to dark brown .... *Corollospora cinnamomea*

1' Ascospores hyaline 2

2 (1') Ascospores 1-septate 3

2'(1') Ascospores with more than 1-septate 4

3 (2) Ascospores wider than 8 µm, 22-33 x 8-10 µm .... *Corollospora maritima*

3'(2) Ascospores narrower than 8 µm, 26-45 x 3-7 µm, polar appendages 6.5-12 µm .... *Corollospora gracilis*

4 (2') Ascospores with 3-septate 5

4'(2') Ascospores with more than 3-septate 6

5 (4) Ascospores wider than 7 µm, 25-34 x 7-12 µm, polar appendages 10-14 µm .... *Corollospora intermedia*

5'(4) Ascospores narrower than 7 µm, 35-57 x 3-7.5 µm, polar appendages 3-8 µm .... *Corollospora angusta*

6 (4') Ascospores with polar appendages 11

6'(4') Ascospores without polar appendages 7

7 (6') Ascospores 5-septate, 50-85 x 5-8 µm .... *Corollospora luteola*

7' (6') Ascospores with more than 7-septate 8

8 (7') Ascospores predominately 7-septate 9

8' (7') Ascospores with more than 7-septate 10

9 (8) Ascospores 60-108 x 13-26 µm .... *Corollospora colossa*

9'(8) Ascospores 52-122 x 7-16 µm .... *Corollospora pulchella*

10 (8') Ascomata wall tuberculate, ascospores 7-11-septate, 65-98 x 8-12 µm .... *Corollospora pseudopulchella*

10'(8') Ascomata wall smooth, ascospores 13-septate, 73-120 x 5-8 µm .... *Corollospora filiformis*

11 (6) Ascospores 3-5-septate 12

11'(6) Ascospores more than 5-septate .... *Corollospora indica*

12 (11) Ascospores 39-63 x 10-19 µm, polar appendages 10-14 µm long .... *Corollospora lacera*

12'(11) Ascospores 38-59 x 8-10 µm, polar appendages 5-12 µm long .... *Corollospora quinqueseptata*

34.1. *Corollospora angusta* Nakagiri & Tokura, 1987. *Trans. Mycol. Soc. Jpn.*, **28**: 417. [Fig. 34.1]

Ascomata: 60-245 µm in diameter, globose to subglobose, superficial, black, carbonaceous, solitary or gregarious, seated with subicula. *Peridium*: 16–30 µm thick, composed of dark, thick–walled cells: polygonal or roundish on the outside, flattened on the inside. *Necks*: 32–57 µm long, 47-53 µm in diameter, conical to subcylindrical, near the basal subiculum, pointing downward. *Centrum*: of immature ascomata is filled with thin-walled, polygonal to rounded, deliquescing pseudoparenchyma cells with pit-connections in their walls. *Asci*: 68-90 x 17-25 µm, 8-spored, fusiform to subcylindrical, unitunicate, early deliquescing.

Ascospores: 35-57 x 3-7.5 µm (excluding polar appendages), fusiform, slender, 3(-5)-septate, hyaline. *Appendages*: of two kinds : (i) a single terminal appendage at each end of the spore, 3-8 µm long, spine- or thorn-like, attenuate; (ii) fibrous and peritrichous appendages at the terminal appendages (6-13 µm long) and around the central septum (18-25 µm long), developed by fragmentation and peeling off of the exosporium.

Habitat: Wood or other organic material between grains of beach sand, ascospores accumulate in foam along shore.

Description: Based on Nakagiri and Tokura (1987).

Distribution: West Coast:-Goa: On intertidal animal substrates (Ananda et al., 1998); **Karnataka:** On intertidal wood (Prasannarai and Sridhar, 1997, 2001, 2000-2001), on intertidal animal substrates (Ananda et al., 1998), on intertidal leaf and woody litter of mangroves (Ananda and Sridhar, 2003, 2004); **Pondicherry (Mahe):** Ascospores in foam samples (Borse and Pawar, 2005), **Kerala:** On intertidal wood (Prasannarai and Sridhar, 2001), on intertidal wood (Nambiar and Raveendran, 2007, 2008b, 2009b, 2010a, c).

East Coast:-Tamil Nadu: On intertidal wood (Nambiar et al., 2008b), on intertidal wood (Nambiar and Raveendran, 2009k, l, 2010a, c); on intertidal woody debris (Ravikumar et al., 2009); **Pondicherry:** On intertidal wood (Nambiar and Raveendran, 2008c, 2010a); **Orissa:** Ascospores in foam samples (Borse et al., 2001b), on intertidal wood (Borse et al., 2002b).

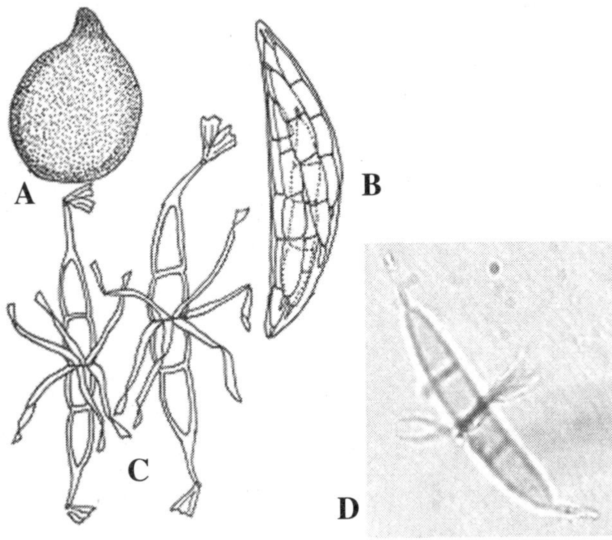

Fig. 34.1. ***Corollospora angusta,*** A-D; A- ascocarp (100x), B- ascus (1000x), C- ascospores (1000x), D- ascospore (1000x).

34.2. ***Corollospora cinnamomea*** Jorgen Koch, 1986. *Nordic J. Bot.*, **6**: 498.

[Fig. 34.2]

Ascomata: 67-105 µm in diam., globose or subglobose, black, smooth, brilliant, with subicula on grains of sand, solitary or gregarious, carbonaceous, ostiole not seen. *Peridium*: outer cell layer composed of flat, thick-walled angular cells with longest diam. 12 µm. *Asci*: 72-107 x 16-20 µm, 8-spored, fusiform or subclavate, thin-walled, without apical apparatus, deliquescing early. *Ascospores*: 18-25 x 6-9 µm (without polar spines), fusiform, 1-septate, not or slightly constricted at septum, asymmetric, brown, appendaged. Appendages: of two kinds a) polar spines 12-16 µm long, slender , 1 µm thick, slightly curved , hyaline, b) sheet like, soft, polar appendages up to 12 µm long from apical part of polar spine and equatorial appendages forming a double frill of thread-like spines 7-12 µm long.

Habitat: On intertidal wood and ascomata on the grains of sand, ascospores accumulate in foam along shore.

Description: Based on Koch (1986).

Distribution: West Coast:-Gujarat: On intertidal wood (Patil and Borse, 2001); **Maharashtra:** On intertidal wood and *Rhizophora mucronata* (Borse, 2000b; Borse et al., 2005a); **Goa:** On intertidal animal substrates (Ananda et al., 1998); **Karnataka:** On intertidal animal substrates (Ananda et al., 1998), on intertidal wood (Prasannarai and Sridhar, 2001), on intertidal leaf and woody litter of mangroves (Ananda and Sridhar, 2004); **Pondicherry (Mahe):** Ascospores in foam samples (Borse and Pawar, 2005).

East Coast:- Orissa: Ascospores in foam samples (Borse et al., 2001b), on intertidal wood (Borse et al., 2002b).

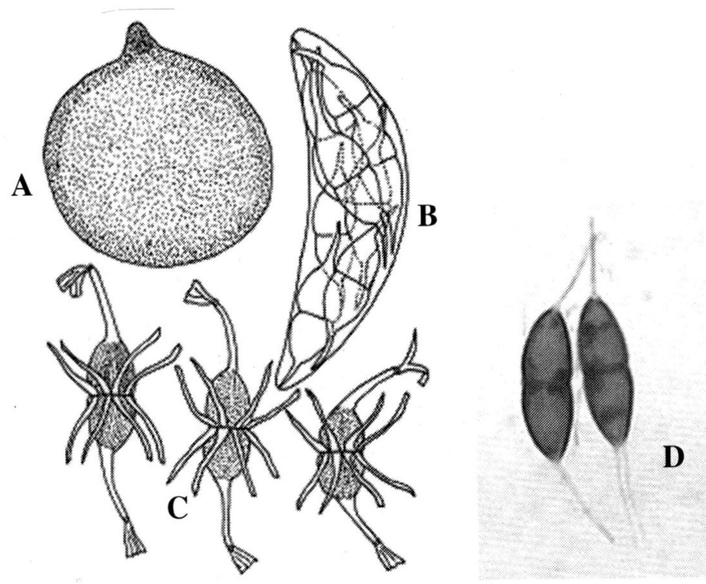

Fig. 34.2. *Corollospora cinnamomea*., A-D; A- ascocarp (100x), B- ascus (1000x), C- ascospores (1000x), D- ascospores (1000x).

34.3. *Corollospora colossa* Nakagiri & Tokura, 1987. *Trans. Mycol. Soc. Jpn.*, **28**: 418. [Fig. 34.3]

Ascomata: 318-515 μm in diam, globose to subglobose, superficial, seated with subicula on the sand grain, black, carbonaceous, solitary or gregarious. *Peridium*: 26-39 μm thick, composed of three layers of cells; large, roundish, paliform on the outside, polygonal or roundish in the middle, flattened on the inside. *Necks*: 63-130 μm long, 77-87 μm in diam, conical to subcylindrical, near the basal subiculum, pointing downward. *Centrum*: of immature ascomata is filled with thin-walled, polygonal to rounded, deliquescing pseudoparenchyma cells with pit-connections in their walls. *Asci*: 95-130 x 47-53 μm, 8-spored, fusiform, unitunicate, early deliquescing. *Ascospores*: 60-108 x 13-26 μm, fusiform to ellipsoidal, (6-) 7 (-8)-septate, hyaline. *Appendages*: fibrous, peritrichous at both ends of the spore (20-27 μm long) and around the central septum (20-28 μm long), developed by fragmentation and peeling off of the exosporium.

Habitat: Wood or other organic material between grains of beach sand, ascospores accumulate in foam along shore.

Description: Based on Nakagiri and Tokura (1987).

Distribution: West Coast:- Maharashtra: On intertidal wood and *Rhizophora mucronata* (Borse, 2000b; Borse et al., 2005a); **Gujarat:** On intertidal wood (Kamble, et al., 2011); **Goa:** Ascospores in foam samples (Borse and Tuwar, 2006); **Karnataka:** On intertidal wood (Prasannarai and Sridhar, 1997, 2001, 2000-2001; Prasannarai et al., 1999); **Pondicherry (Mahe):** Ascospores in foam samples (Borse and Pawar, 2005), on intertidal mangrove wood (Nambiar and Raveendran, 2007); **Kerala:** On intertidal wood (Prasannarai and Sridhar, 2001), on intertidal wood (Nambiar and Raveendran, 2007, 2008b, 2010a).**East Coast:- Tamil Nadu:** On intertidal wood (Prasannarai and Sridhar, 2001), on intertidal wood (Nambiar and Raveendran, 2009k, l, 2010c); **Pondicherry:** On intertidal wood (Nambiar and Raveendran, 2008c); **Orissa:** On intertidal wood (Borse et al., 2002b); **West Bengal:** Ascospores in foam samples (Borse et al., 2001a).

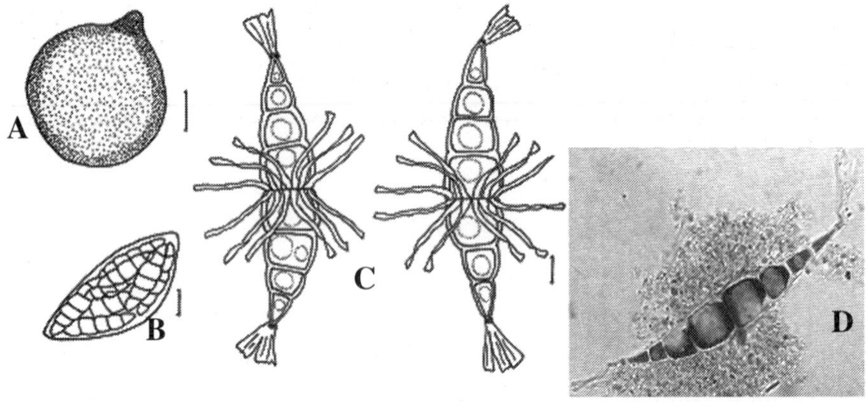

Fig. 34.3. ***Corollospora colossa*.,** A-D; A- ascocarp (100x), B- ascus (450x) , C- ascospores (1000x), D- ascospore (1000x).

34.4. ***Corollospora filiformis*** Nakagiri, 1987. *Trans. Mycol. Soc. Jpn.*, **28**: 422. [Fig. 34.4]

Ascomata: 243-418 µm in diam, globose to subglobose, superficial, black, carbonaceous, solitary, seated with subicula on sand grains. *Peridium*: 15-28 µm thick, composed of dark, thick-walled cells: large, polygonal, paliform on the outside, flattened on the inside. *Necks*: 34-133 µm long, 57-73 µm in diam, conical to cylindrical, near the basal subiculum, pointing downward. *Centrum*: of immature ascomata is filled with thin-walled, polygonal to rounded, deliquescing pseudoparenchyma cells with pit-connections in their walls. *Asci*: 130-158 x 20-28 μm, 8-spored, fusiform, unitunicate, early deliquescing. *Ascospores*: 87-120 x 5-8 μm, filiform, (9-) 13 (-17)-septate, hyaline. *Appendages*: fibrous, peritrichous, at both ends of the spore (18-25 μm long) and around the central septum (13-22 μm long), developed by fragmentation and peeling off of the exsoporium.

Habitat: Wood or other organic material between grains of beach sand (arenicolous), ascospores accumulate in foam along shore.

Description: Based on Nakagiri and Tokura (1987).

Distribution: West Coast:-Maharashtra: On intertidal wood (Borse et al., 2005a); **Karnataka:** On sand grains attached to intertidal wood (Sridhar and Prasannarai, 1993), on intertidal wood and teak panels (Prasannarai and Sridhar, 1997, 2001, 2000-2001; Prasannarai et al., 1999); **Kerala:** On intertidal wood (Prasannarai and Sridhar, 2001), on intertidal wood (Raveendran and Manimohan, 2007), on intertidal wood (Nambiar and Raveendran, 2007, 2008b, 2010a, c). **East Coast:-Tamil Nadu:** On intertidal wood (Nambiar et al., 2008b), on intertidal wood (Nambiar and Raveendran, 2009k, l, 2010a, c); on intertidal woody debris (Ravikumar et al., 2009); **West Bengal:** On drift wood of mangroves (Pawar et al., 2009); **Orissa:** On intertidal wood (Borse et al., 2002b).

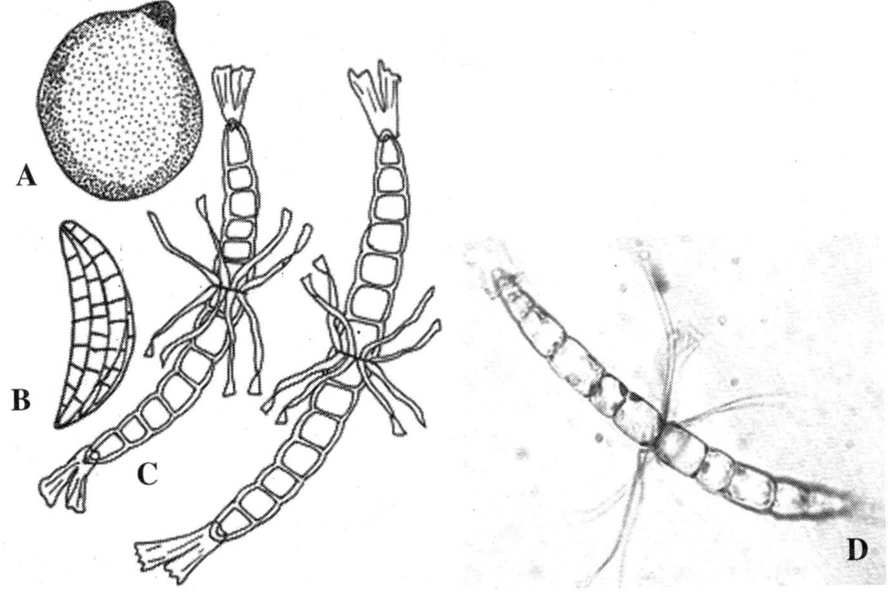

Fig. 34.4. *Corollospora filiformis*., A-D; A- ascocarp (100x), B- ascus (450x), C- ascospores (1000x), D- ascospore (1000x).

34.5. *Corollospora gracilis* Nakagiri & Tokura, 1987. *Trans. Mycol. Soc. Jpn.*, **28**: 426. [Fig. 34.5]

Ascomata: 88-195 µm in diam, globose to subglobose, superficial, black, carbonaceous, solitary or gregarious, seated with subicula. *Peridium*: 8-19 µm thick, composed of dark, thick-walled cells: polygonal or roundish on the outside, flattened on the inside. *Necks*: 18-62 µm long, 27-57 µm in diam, conical to subcylindrical, near the basal subiculum, pointing downward. *Centrum*: of immature ascomata is filled with thin-walled, polygonal to rounded, deliquescing pseudoparenchyma cells with pit-connections in their walls. *Asci*: 56-75 x 12-16 µm, 8-spored, fusiform to ellipsoidal, unitunicate, early deliquescing. *Ascospores*: 26-45 x 3-6 µm (excluding polar appendages), fusiform, slender, one-septate, hyaline. *Appendages*: of two kinds : (i) a single terminal appendage at each end of the spore, 6-12 µm long, spine- or thorn-like, attenuate; (ii) fibrous and peritrichous appendages at the terminal appendages (4-8 µm long) and around the central septum (12-20 µm long), developed by fragmentation and peeling off of the exosporium.

Habitat: Wood or other organic material between grains of beach sand, ascospores accumulate in foam along shore.

Description: Based on Nakagiri and Tokura (1987).

Distribution: West Coast:-Gujarat: On intertidal wood (Kamble et al., 2011); **Goa:** Ascospores in foam samples (Borse and Tuwar, 2006); **Pondicherry (Mahe):** Ascospores in foam samples (Borse and Pawar, 2005); **Kerala:** On intertidal wood (Nambiar and Raveendran, 2010c).

East Coast:- Tamil Nadu: On intertidal woody debris (Ravikumar et al., 2009); **Andhra Pradesh:** On intertidal wood (Nambiar and Raveendran, 2009j); **Orissa:** On intertidal wood (Borse et al., 2002b); **Pondicherry:** On intertidal wood (Nambiar and Raveendran, 2010a; Ravikumar, et al., 2009); **West Bengal:** Ascospores in foam samples (Borse et al., 2001a).

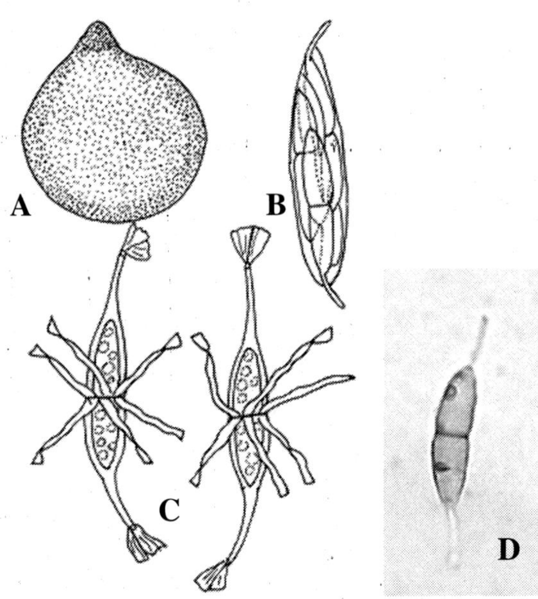

Fig. 34.5. *Corollospora gracilis*., A-D; A- ascocarp (100x), B- ascus (450x), C- ascospores (1000x), D- ascospore (1000x).

34.6. *Corollospora indica* Prasannarai, Ananda & K.R. Sridhar, 2000. *J. Envirn. Biol.*, **21**: 335. [Fig. 34.6]

Ascomata: 110–120 µm in diam., globose to subglobose, superficial, black, carbonaceous, solitary or gregarious, seated with subicula. *Necks*: conical or subcylindrical, near the basal subiculum, pointing downward. *Asci*: 75 -113 x 17-30 µm, 8-spored, fusiform, unitunicate, early deliquescing. *Ascospores*: 45-103 x 7-18 µm, fusiform, (3) 7-9 (-12)- septate, hyaline, cells bulged, septations often unequal, appendaged. *Appendages*: primarary, terminal, spine like appendages at each end of the spore, 5-15 µm long; fibrous, peritrichous, secondary appendages at each tip of the polar spine, 12-25 µm long and around the central septum,18-27 µm long.

Habitat: Wood or other organic material between grains of beach sand (arenicolous), ascospores accumulate in foam along shore.

Description: Based on Prassanarai et al., (2000).

Distribution: West Coast:-Gujarat: On intertidal wood (Kamble et al., 2008); **Karnataka:** On intertidal animal substrates (Prasannaria et al., 2000), on intertidal wood (Prasannarai and Sridhar, 2001, 2000-2001). **East Coast:-Tamil Nadu:** On intertidal wood (Nambiar and Raveendran, 2010a, c); on intertidal woody debris (Ravikumar et al., 2009); **Pondicherry (Karaikkal):** on intertidal woody debris

(Ravikumar et al., 2009); **Pondicherry:** On intertidal wood (Nambiar and Raveendran, 2008c, 2010a); **Andhra Pradesh:** On intertidal wood (Nambiar and Raveendran, 2009j); **Orissa:** On intertidal wood (Borse et al., 2002b).

Fig. 34.6. ***Corollospora indica.,*** A-D; A- ascocarp (100x), B- ascus (450x) , C- ascospores (1000x), D- ascospore (1000x).

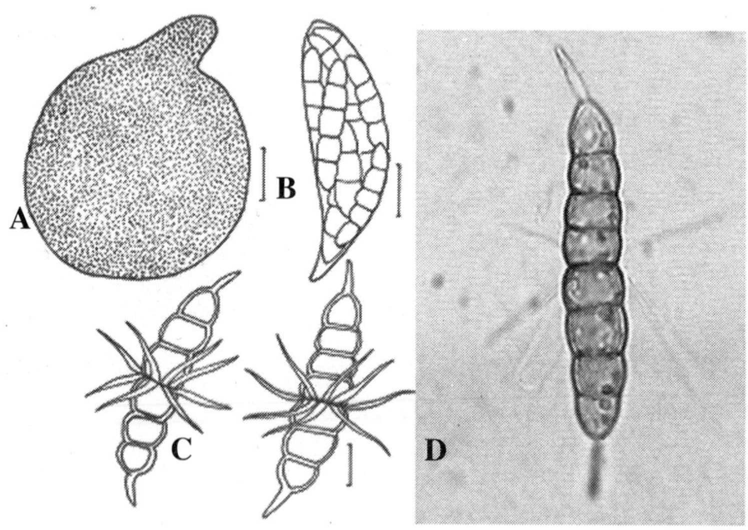

34.7. *Corollospora intermedia* I. Schmidt, 1969. *Nat. Naturschutz Meeklenburg*, **7:** 6. [Fig. 34.7]

Ascomata: 110-310 µm in diam, subglobose or ovoid, immerged or superficial, some times seated with subicula on grains of sand or cases of bryozoa, ostiolate, papillate, black, carbonaceous, solitary. *Peridium*: 10-20 µm thick, two-layered; outer layer composed of roundish cells with large, inner layer consisting of elongate cells with narrow lumina. *Necks*: 40-90 µm long, 36-54 µm in diam, conical near the basal subiculum and pointing downward; ostiolar canal at first filled with pseudoparenchymatous small cells. *Centrum*: of immature ascomata is filled with thin-walled, polygonal to rounded, deliquescing pseudoparenchyma cells with pit-connections in their walls. *Asci*: early deliquescing. *Ascospores*: 25-34 x 7-11.5 µm (excluding appendages), fusiform, straight or slightly curved, 3-septate, constricted at the septa, hyaline, appendaged. *Appendages*: at both ends with a single, terminal appendage, 9.5-14 x 1.5-2 µm, thorn-like, slender, attenuate, rigid, straight or some what curved, at the tip with a refractive body and bearing a small cap or fibers that develop by peeling off of the exospore, peritrichous around the central septum with 10-18 flexible setae, 10-16 µm long, which develop fragmentation of exosporium; setae attached to a narrow equatorial, belt-like thickening of the wall.

Habitat: Wood or other organic material between grains of beach sand, ascospores accumulate in foam along shore.

Description: Based on Kohlmeyer and Kohlmeyer (1979).

Distribution: West Coast:-Gujarat: On drift wood (Kamble et al., 2011); **Goa:** On intertidal animal substrates (Ananda et al., 1998), on intertidal wood (Prasannarai and Sridhar, 2001), Ascospores in foam samples (Borse and Tuwar, 2006); **Karnataka:** Ascospores in foam and sand samples (Sridhar and Kaveriappa, 1991, 2001), on intertidal animal substrates (Ananda et al., 1998, Ananda and Sridhar, 2001), on intertidal wood (Prasannarai et al., 1999), on intertidal mangrove wood (Prasannarai and Sridhar, 2000-2001); **Pondicherry (Mahe):** Ascospores in foam samples (Borse and Pawar, 2005); **Kerala:** On intertidal wood (Prasannarai and Sridhar, 2001). **East Coast:-Tamil Nadu:** On intertidal wood of *Prosopis juliflora* (Ravikumar and Purushothaman, 1988).

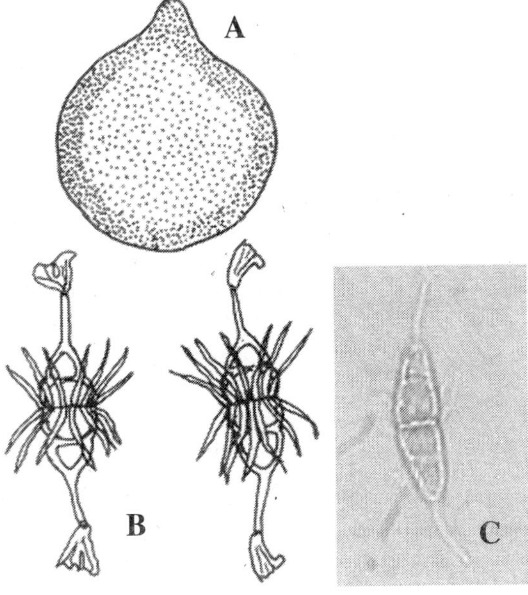

Fig. 34.7. ***Corollospora intermedia.,*** A-C; A- ascocarp (100x), B- ascospores (1000x), C- ascospore (1000x).

34.8. *Corollospora lacera* (Linder) Kohlm., 1962. *Ber. Dtsch. Bot. Ges.*, **75**: 126. [For Fig. refer original paper]

Basionym: *Peritrichospora lacera* Linder, In: Barghoorn & Linder, *Farlowia*, **1**: 415 (1944).

Ascomata: 307-468 µm in diam, subglobose or ovoid, immerged or superficial, sometimes seated with subicula on grains of sand, ostiolate, papillate or epapillate, black, carbonaceous, solitary. *Peridium*: 24-35 µm thick, composed of dark, thick-walled cells, roundish at the outside, oblong toward the center, forming a *textura angularis*; subicula prosenchymatous. *Necks*: 40 µm high, 70 µm in diam, conical near the basal subiculum and pointing downward; ostiolar canal 40 µm in diam, at first filled with pseudoparenchymatous hyaline cells, 6 µm in diam. *Centrum*: of immature ascomata is filled with thin-walled, polygonal to rounded, deliquescing pseudoparenchyma cells with pit-connections in their walls. *Asci*: 112-150 x 18-28 µm, 8-spored, fusiform to clavate, unitunicate, thin-walled, deliquescing at ascospore maturity. *Ascospores* : 39-60 x10-16 µm (excluding terminal appendage and fibres), fusiform, straight or curved, 5-septate, constricted at the septa, hyaline, appendaged. *Appendages*: at both ends with a single, terminal appendage, 9.5-24 µm long, 2.2-4.5 µm in diam, thorn-like, attenuate, rigid, straight or some what curved, at the tip with a refractive body and bearing a tube or fibers that develop by peeling off of the exospore, 19-33 x 2-6 µm; peritrichous around the central septum with flexible setae, 12-17 µm long, which develop by fragmentation of exosporium.

Habitat: Wood or other organic material between grains of beach sand (arenicolous), ascospores accumulate in foam along shore.

Description: Based on Kohlmeyer and Kohlmeyer (1979).

Distribution: West Coast:-Maharashtra: Ascospores in foam samples (Patil and Borse, 1983b; Borse, 1988), **Karnataka:** Ascospores in foam and sand samples (Sridhar and Kaveriappa, 1991), on intertidal wood ((Prasannarai and Sridhar, 2001, 2000-2001).

Note: This species is similar in ascospores morphology to other species of the genus with 5- more-septate ascospores (eg. *C. indica, C. quinqueseptata*). Reports based on ascospores encountered in foam samples may have been based on misidentification.

34.9. ***Corollospora luteola*** Nakagiri & Tubaki, 1982. *Trans. Mycol. Soc. Japan*, **23:** 102. [For Fig. refer original paper]

Ascomata: 105-260 µm in diam., globose or subglobose, superficial seated with subcula which are ostiolate, papillate, carbonaceous, black, and solitary or gregarious. *Peridium*: 12.5-37.0 µm thick, dark brown to black, composed of dark, thick–walled cells, round or polygonal at the outside, oblong towards the center, and of a *textura angularis*; the subicula prosenchymatous. *Necks*: 30-60(-200) µm high, 30-45 µm in diam., conical or subcylindrical, located laterally, or near the basal subiculum. Pseudoparenchyma composed of thin-walled, polygonal cells filling the venter of young ascocarps, deliquescing when the ascospores have matured. *Asci*: 56 -130 x 20-28 µm, 8-spored, fusiform or elliptical, unitunicate, thin-walled, deliquescing early. *Ascospores*: 50-85 x 4.8-7.5 µm (excluding the appendages), fusiform, slightly curved, 5 (4 to 6)-septate, hyaline, appendaged; the appendages developing by exospore fragmentation, seta-like, flexible, attached in a tuft at each apex and around the central septum.

Anamorph: *Sigmoidea luteola* Nakagiri & Tubaki

Habitat: Wood or other organic material between grains of beach sand (arenicolous), ascospores accumulate in foam along shore.

Description: Based on Nakagiri and Tubaki (1982).

Distribution: West Coast:-Kerala: On sand buried wood (Nambiar and Raveendran, 2010a).

East Coast:-Pondicherry: on intertidal mangrove wood (Nambiar and Raveendran, 2010a).

34.10. ***Corollospora maritima*** Werderm., 1922. *Notizbl. Bot. Gart.Mus. Berlin-Dahlem*, **8:** 248. [Fig. 34.10]

= *Arenariomyces cinctus* Hohnk, *Ver. Inst. Meeres.Bremerhaven,* **3:** 28 (1954).

= *Peritrichospora integra* Linder, *Farlowia,* **1:** 414 (1944).

Ascomata: 122–176 µm in diam., globose or subglobose, superficial or partly immersed, often seated with subcula on grains of sand or other hard surfaces with or without ostioles, papillate or epapillate, carbonaceous, metallic, thick, solitary or gregarious. *Peridium*: composed of dark, thick–walled cells, roundish or polygonal at the outside, oblong towards the center, forming a *textura angularis*, subicula prosenchymatous. *Necks*: 11-16 µm long, 18-24 µm in diameter, conical or subcylindrical, apically, laterally or near the basal subiculum and pointing downward. *Centrum*: of immature ascomata is filled with thin-walled, polygonal to rounded, deliquescing pseudoparenchyma cells with pit-connections in their walls. *Asci*: 68 -124 x 18-36 µm, 8-spored, fusiform or subclavate, some times apiculate, short stipitate, unitunicate, thin-walled, aphysoclastic, without apical apparatuses, early deliquescing. *Ascospores*: 18-36 x 6–9 µm (excluding apical thorns and setae), fusiform or subellipsoidal, one-septate , constricted at the septum, hyaline, appendaged, at both ends with a single terminal appendage 7.5-15 x 1-2 µm , spine or thorn-like, slender, attenuate, rigid, straight or some what curved, at the tip with a refractive body and bearing a small cap or fibers, 7-9 µm long, peritrichous around the septum with 8 or more flexible ribbon-shaped setae, 6-16 x 1 µm.

Habitat: Wood or other organic material between grains of beach sand (arenicolous), ascospores accumulate in foam along shore.

Description: Based on Kohlmeyer and Kohlmeyer (1979).

Distribution: West Coast:-Diu Island: On intertidal wood (Borse et al., 1999b); **Gujarat:** On intertidal wood and mangrove wood (Patil and Borse, 2001), on intertidal wood (Kamble et al., 2008); **Maharashtra:** Ascospores in foam samples, on wood of *Avicennia marina* and *Rhizophora mucronata* (Borse, 2000b; Borse et al., 2005a); (Patil and Borse, 1983a; Borse, 1988), on intertidal wood (Shrivastava, 1995, Borse, 2000b); **Goa:** On intertidal animal substrates (Ananda et al., 1998), on intertidal wood (Borse et al., 1999b), on intertidal wood (Prasannarai and Sridhar, 2001), Ascospores in foam samples (Borse and Tuwar, 2006); **Karnataka:** Ascospores in sand and foam samples (Sridhar and Kaveriappa, 1991), on intertidal animal substrates (Ananda et al., 1998), on intertidal wood (Prasannarai et al., 1999), on intertidal mangrove (*Avicennia officinalis*) wood (Maria and Sridhar, 2002a, 2004), on intertidal mangrove wood (Prasannarai and Sridhar, 2000-2001), on intertidal leaf and woody litter of mangroves (Ananda and Sridhar, 2003, 2004); **Pondicherry (Mahe):** Ascospores in foam samples (Borse and Pawar, 2005); **Kerala:** On intertidal wood (Prasannarai and Sridhar, 2001), on intertidal mangrove wood (Maria and Sridhar, 2002a), on intertidal mangrove wood (Nambiar and Raveendran, 2008b, 2009b, 2010a, c); **Lakshadweep Islands:** On intertidal mangrove wood (Ananda and Sridhar, 2003).

East Coast:-Tamil Nadu: On drift wood (Raghukumar, 1973; Prasannarai and Sridhar, 2001; Nambiar et al., 2008b), on intertidal wood (Nambiar and

Raveendran, 2009k, l, 2010c); **Pondicherry:** on intertidal mangrove wood (Nambiar and Raveendran, 2010a); **Andhra Pradesh:** On intertidal wood (Nambiar and Raveendran, 2009j); **Orissa:** Ascospores in foam samples (Borse et al., 2001b), on intertidal wood (Borse et al., 2002); **West Bengal:** Ascospores in foam samples (Borse et al., 2001a).

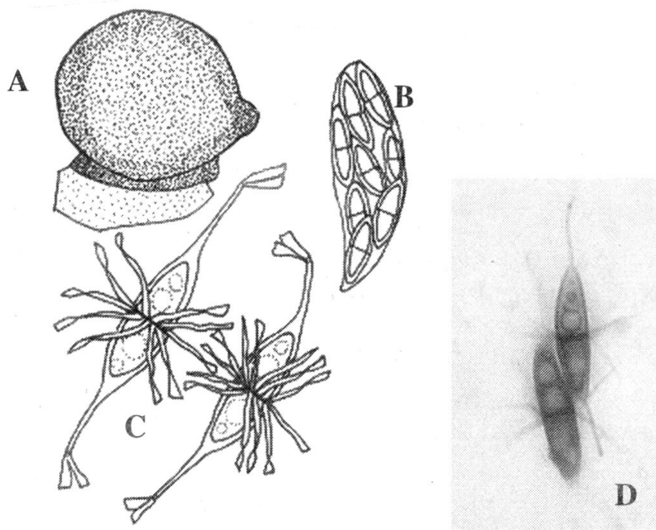

Fig. 34.10. ***Corollospora maritima,*** A-D; A- ascocarp (100x), B- ascus (450x), C- ascospores (1000x), D- ascospore (1000x).

34.11. ***Corollospora pseudopulchella*** Nakagiri & Tokura, 1987. *Trans. Mycol. Soc. Jpn.*, **28**: 428. [For Fig. refer original paper]

Ascomata: 180-245 µm in diam, globose to subglobose, superficial, seated with subicula on grains of sand or shells of marine mollusks, ostiolate, papillate, carbonaceous, black, tuberculate, solitary. *Peridium*: 8-13 µm thick, composed of dark, thick-walled cells: polygonal or roundish on the outside, flatteded on the inside. *Necks*: 20–55 µm long, 35–55 µm in diam, subcylindrical, near the basal subiculum, pointing downward. *Centrum*: of immature ascomata is filled with thin-walled, polygonal to rounded, deliquescing pseudoparenchyma cells with pit-connections in their walls. *Asci*: 82-113 x 20-25 µm, 8-spored, fusiform to ellipsoidal, early deliquescing. *Ascospores*: 65-98 x 8-12 µm (excluding appendages), fusiform, slender, 7-11-septate, hyaline, appendaged. *Appendages*: fibrous, peritrichous, at both ends of the spore (8-13 µm long) and around the central septum (18-31 μm long), developed by fragmentation and peeling off of the exosporium.

Habitat: Wood or other organic material between grains of beach sand, ascospores accumulate in foam along shore.

Description: Based on Nakagiri and Tokura (1987).

Distribution: West Coast:-Kerala: On intertidal wood (Raveendran and Manimohan, 2007), on intertidal wood (Nambiar and Raveendran, 2007, 2008b, 2010a, c).

East Coast:-Tamil Nadu: on intertidal wood (Nambiar and Raveendran, 2009l, 2010a).

34.12. *Corollospora pulchella* Kohlm., I. Schmidt & N.B. Nair, 1967. *Ber. Dtsch. Bot. Ges.*, **80**: 98. [Fig. 34.12]

Conidial State: *Clavatospora bulbosa* (Anastasiou) Nakagiri & Tubaki *Bot. Mar.*, **28:** 489, 1985.

≡ *Clavariopsis bulbosa* Anastasiou, *Mycologia,* **53:** 11, 1991.

Ascomata: 220-298 μm in diam., globose to subglobose, rarely subovoid, superficial on hard surfaces or immersed in soft substrates, sometimes seated with subicula on grains of sand or shells of marine mollusks, ostiolate, papillate, carbonaceous, black, tuberculate, solitary or gregarious. *Peridium*: dark, two–layered, outer layer composed of roundish cells with large lumens; inner layer consisting of elongate cells with narrow lumina. *Necks*: 12–18 μm long, 23–56 μm in diameter, cylindrical or conical, black, tuberculate, predominantly near the basal subiculum and pointing downward, ostiolar canal at first filled with pseudoparenchymatous small cells. *Centrum*: of immature ascomata is filled with thin-walled, polygonal to rounded, deliquescing pseudoparenchyma cells with pit-connections in their walls. *Asci*: 74-144 x 18-46 μm, 8-spored, cylindrical to fusiform, short pedunculate, unitunicate, thin-walled, aphysoclastic, without apical apparatuses, early deliquescing; originating from an ascogenous tissue of the ascomata venter; directed downword towards the ostiole. *Ascospores*: 24-36 x 8-11 μm (excluding appendages), fusiform, slightly curved, 7-septate (rarely 9-11-septate), constricted at the septa, hyaline, appendaged, setalike, flexible, attached in a tuft to a conical papilla at each apex and peritrichous around the central septum, about 7-apical appendages, 13-20x 1μm, about 15 lateral appendages, 14-24 x 1 μm.

Habitat: Wood or other organic material between grains of beach sand, ascospores accumulate in foam along shore.

Description: Based on Kohlmeyer and Kohlmeyer (1979).

Distribution: West Coast:-Daman: Ascospores in foam samples (Borse et al., 2000b); **Gujarat:** On intertidal wood (Patil and Borse, 2001), on intertidal wood (Kamble et al., 2008); **Maharashtra:** On intertidal wood and *Rhizophora mucronata* (Borse and Shrivastava, 1988,; Borse, 2000; Borse et al., 2005), on intertidal mangrove wood (Maria and Sridhar, 2002); **Goa:** On intertidal wood (Borse et al., 1999); **Karnataka:** Ascospores in sand and foam samples (Sridhar and Kaveriappa, 1991), on intertidal wood (Prasannarai et al., 1999), on intertidal wood (Prasannarai and Sridhar, 2001, 2000-2001), on intertidal mangrove wood (Maria and Sridhar, 2002), on intertidal leaf and woody litter of mangroves (Ananda and Sridhar, 2003, 2004); **Pondicherry (Mahe):** Ascospores in foam samples (Borse and Pawar, 2005); **Kerala:** On intertidal wood panels of *Mangifera indica* (Kohlmeyer *et al.,* 1967; Nair, 1970), on intertidal mangrove wood (Maria and Sridhar, 2002), on intertidal wood (Nambiar and Raveendran, 2007, 2008b, 2010a, c), on mangrove woody litter (Shini et al., 2009-10).

East Coast:-Tamil Nadu: On drift wood (Raghukumar, 1973; Nambiar et al., 2008b; Nambiar and Raveendran, 2010c); on intertidal woody debris (Ravikumar et al., 2009); **Pondicherry (Karaikkal):** on intertidal woody debris (Ravikumar et al., 2009); **Andhara Pradesh:** On intertidal wood of *Rhizophora apiculata* (Sarma and Vittal, 1998-99; Sarma and Vittal, 2000; Sarma and Vittal, 2001; Sarma et al., 2001; Sarma and Vittal, 2004; Vittal and Sarma, 2005), on intertidal wood (Nambiar and Raveendran, 2009j); **Orissa:** Ascospores in foam samples (Borse et al., 2001), On intertidal wood (Borse et al., 2002); **West Bengal:** Ascospores in foam samples (Borse et al., 2001).

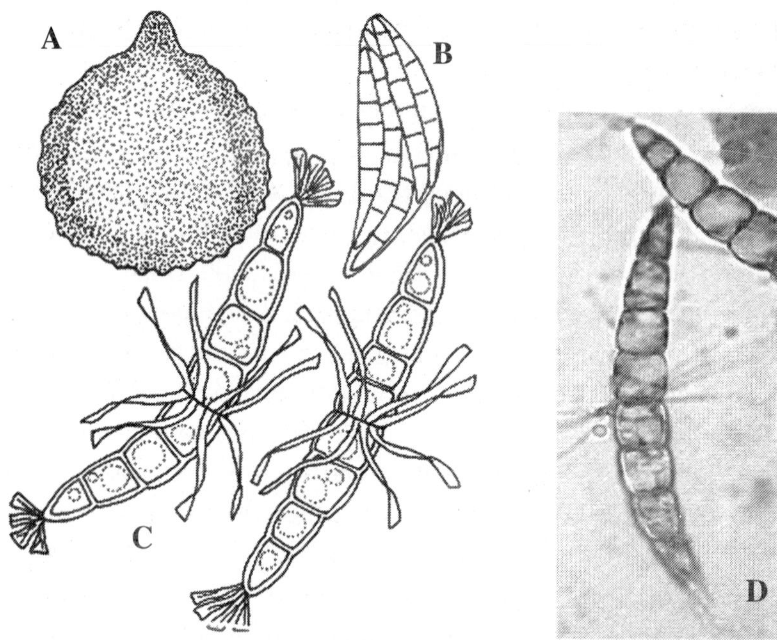

Fig. 34.12. ***Corollospora pulchella,*** A-D; A- ascocarp (100x), B- ascus (450x), C- ascospores (1000x), D- ascosporeS (1000x).

34.13. ***Corollospora quinqueseptata*** Nakagiri, 1987. *Trans. Mycol. Soc. Jpn.*, **28**: 430. [Fig. 34.13]

Ascomata: 170-265 µm in diam, globose to subglobose, superficial, black, carbonaceous, solitary or gregarious, seated with subicula on hard surfaces or partly immersed in decomposed algal thallus in nature. *Peridium*: 18-28 µm thick, composed of dark, thick-walled cells: polygonal or roundish on the outside, flattened on the inside. *Necks*: 40-63 µm long, 45-60 µm in diam, subcylindrical to cylindrical, near the basal subiculum, pointing downward. *Centrum*: of immature ascomata is filled with thin-walled, polygonal to rounded, deliquescing pseudoparenchyma cells with pit-connections in their walls. *Asci*: 60-113 x 18-25 µm, 8-spored, fusiform to subcylindrical, unitunicate, early deliquescing.. *Ascospores*: 42-58 x 7.5-10 µm (excluding polar appendages), fusiform, (3-) 5 (-8)-septate, hyaline. *Appendages*: of two kinds: (i) a single terminal appendage at each end of the spore, 5-12 µm long, spine-or thorn-like, attenuate; (ii) fibrous and peritrichous appendages at the terminal appendages (8-12 µm long) and around the central septum (17-25 µm long), developed by fragmentation and peeling of the exosporium.

Habitat: Wood or other organic material between grains of beach sand, ascospores accumulate in foam along shore.

Description: Based on Nakagiri and Tokura (1987).

Distribution: West Coast:-Karnataka: On intertidal wood (Prasannarai and Sridhar, 2001, 2000-2001); **Pondicherry (Mahe):** Ascospores in foam samples (Borse and Pawar, 2005); **Kerala:** On intertidal wood (Prasannarai and Sridhar, 2001).

East Coast:-Orissa: On intertidal wood (Borse et al., 2002b).

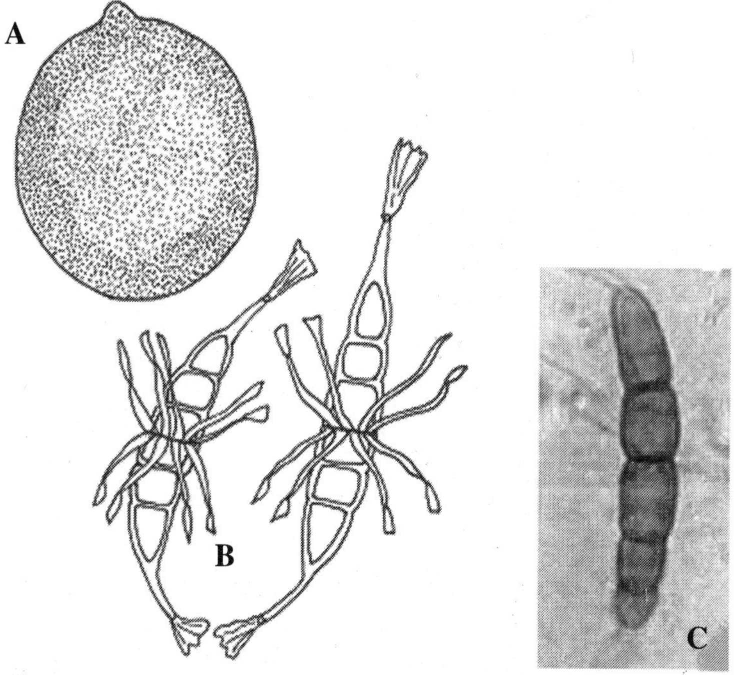

Fig. 34.13. ***Corollospora quinqueseptata,*** A-C; A- ascocarp (100x), B- ascospores (1000x),C- ascospore (1000x).

35. *Coronopapilla* Kohlm. & Volkm.-Kohlm., 1990b. *Mycol. Res.,* 94: 686.

Ascomata: subglobose, immersed, ostiolata, periphysate, papillate, clypeate, coriaceous, black above, light-coloured below, single or gregarious. *Necks*: conical, crown-like, tearing at the top in cabonaceous sections, emersed. *Peridium*: hyaline, uniform, forming a textura angularis. *Hamathecium*: pseudoparaphysoid, trabeculate, with gelatinous matrix. *Asci*: 8-spored, cylindrical, long pedunculate, non-amyloid, endotunica with oculus, thick-walled, fissitunicate, maturing successively at the bottom of the locule. *Ascospores*: monostichous, ellipsoidal, 1-3-septate, brown, thick-walled, wall 2-layered. The genus is monotypic (Jones *et al.*2009a).

Type Species: *Coronopapilla mangrovei* (Hyde) Kohlm. & Volkm.-Kohlm.
Habitat: On intertidal wood and grains of sand.
Description: Based on Kohlmeyer and Volkmann-Kohlmeyer (1990b).

35.1. *Coronopapilla mangrovei* (K.D. Hyde) Kohlm. & Volkm.-Kohlm., 1991b. *Bot. Mar.*, **34**: 19. [For Fig. refer original paper]

Basionym: *Caryospora mangrovei* K. D. Hyde, *Trans. Mycol. Soc. Japan.*, **30**: 336 (1989).

= *Coronopapilla avellina* Kohlm. & Volkm.-Kohlm., *Mycol. Res.,* **94**: 687 (1990).

Ascomata: 600-850 µm high, 450-650 µm in diam, subglobose, immersed in wood, ostiolate, papillate, clypeat, coriaceous, light-coloured, single or in groups. *Necks*: 80-160 µm high, 245-430 µm in diam, conical, black, crown-like, tearing at the top into 3-5 cabonaceous sections surrounded by a shield-like, black clypeus; ostiolar canal 95-180 µm long, 70-95 µm in diam, periphysate. *Peridium*: 25-45 µm thick, composed of 8-15 layers of polygonal hyaline cells, forming a textura angularis. *Hamathecium*: pseudo-paraphysoid, trabeculate, forming a network of anastomosing filaments, embedded in a gelatinous matrix. *Asci*: 350-430 x 28-37 µm, 8-spored, cylindrical, long pedunculate, non-amyloid, endotunica with oculus, thick-walled, fissitunicate, with a faint apical ring, maturing successively at the bottom of the locule. *Ascospores*: 40-54 x 18-22.5 µm, monostichous, ellipsoidal, 1-3-septate, constricted at the centre septa, brown, thick-walled, wall 2-layered.

Habitat: On intertidal mangrove wood.
Description: Based on Kohlmeyer and Volkmann-Kohlmeyer (1990b).
Distribution: West Coast:-Karnataka: On intertidal wood ((Prasannarai and Sridhar, 2001, 2000-2001), on intertidal wood of *Rhizophora mucronata* and *Sonneratia caseolaris* (Maria and Sridhar, 2003a).
East Coast:-Andaman & Nicobar Islands: On intertidal wood of *Bruguiera gymnorhiza* and *Rhizophora mucronata* (Chinnaraj, 1993a).

36. *Crinigera* I. Schmidt, 1969. *Nat. Natursch. Meckl.*, **7**: 11.

Ascomata: globose, slightly falttened, coriaceous, light brown, solitary, attached to surface with a subiculum. Substratum forming a bowl around the upper part of the ascomata, coposed of stromatic tissue of interlacing hyphae, thin-walled, *Peridium*: stratified, with an outer layer of thin-walled cells, and an inner layer of flat, prismatic cells. No appendages present on the peridium. *Asci*: 8-spored, early deliquescing. *Ascospores*: numerous, 1-septate, hyaline, appendaged. *Appendages*: forming two subpolar whorls of soft threads or spines. The genus is monotypic (Jones et al., 2009a).

Type Species: *Crinigera maritima* I. Schmidt

Habitat: On brown algae *Fucus vesiculosus* L.

Description: Based on Kohlmeyer and Kohlmeyer (1979, 1991) and Koch and Jones (1989).

36.1. *Crinigera maritima* I. Schmidt, 1969. *Nat. Naturs. Meekl.,* **7:** 11.

[Fig. 36.1]

Ascomata: 310-324 µm (exclusive of subiculum), globose, slightly falttened, coriaceous, light brown, solitary, attached to surface with a subiculum 30-36 µm thick. Substratum forming a bowl around the upper part of the ascomata, coposed of stromatic tissue of interlacing hyphae, thin-walled, 3-4 µm in diam. *Peridium*: stratified, 24-39 µm thick, with an outer layer of thin-walled cells, 10-16 µm in diam, and an inner layer of flat, prismatic cells 3 um thick and 6 µm long. No appendages present on the peridium. *Asci*: 8-spored, early deliquescing. *Ascospores*: 10-14 x 4-5 µm, numerous, 1-septate, hyaline, appendaged. *Appendages*: forming two subpolar whorls of soft threads or spines 8-10 µm long.

Habitat: Perthophytic in scenescent brown algae *Fucus vesiculosus* L.

Description: Based on Kohlmeyer and Kohlmeyer (1979, 1991) and Koch and Jones (1989).

Distributiin: West Coast:-Goa: On intertidal wood ((Prasannarai and Sridhar, 2001); **Karnataka:** On intertidal wood, and wood of *Bambusa* sp. (Prasannarai and Sridhar, 1997, 2001, 2000-2001), on intertidal wood and teak panels (Prasannarai et al., 1999); **Kerala:** On intertidal wood ((Prasannarai and Sridhar, 2001).

East Coast:- Tamil Nadu: on intertidal woody debris (Ravikumar et al., 2009); **Pondicherry (Karaikkal):** on intertidal woody debris (Ravikumar et al., 2009); **West Bengal:** On intertidal drift wood mangroves (Pawar et al., 2009); **Orissa:** On intertidal mangrove wood (Borse et al., 2005b).

Note: A genus of doubtful validity collected on the seaweed (Jones et al., 2009a). Reports from India on intertidal wood and mangrove wood may be misidentification. According to Kohlmeyer & Volkmann-Kolhmeyer (1991) tropical ascomycetes, resembling *Crinigera maritima* is most likely different species.

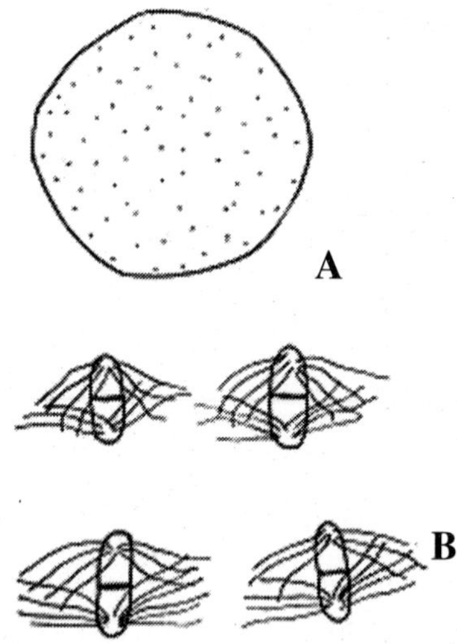

Fig. 36.1. *Crinigera maritima,* A-B; A- ascocarp (100x), B- ascospores (1000x).

37. *Cryptosphaeria* Ces. & De Not., 1863. *Comm. Soc. Critt. Ital.*, **1:** 231.

Stromata: widely effuse within bark tissue that are slightly elevated, usually unchanged, but sometimes discoloured or darkened, with separate, rounded or stellate ostioles at the surface, formed by fungus tissue enclosing host elements (pseudostroma), reduced to black clypeal zones or may be absent, usually covered by a periderm and a ventral blackened line is often present. *Ascomata*: scattered in a single layer, or in small groups, globose to subglobose, immersed, coriaceous, ostiolate, papillate or epapillate, brown to black, neck periphysate, thin, entostroma under the peridium. *Paraphyses*: wide, branched, hypha-like. *Asci*: 8-spored, clavate or cylindrical, unitunicate, thin-walled, usually with long pedunculate, with an apical apparatus, J- ring. *Ascospores*: sausage-shaped or allantoid, hyaline or brownish, up to 4-septate, no sheath or appendages. A large terrestrial genus, one species is known from mangroves (Jones et al., 2009a).

Type Species: *Cryptosphaeria eunomia* (Fr.) Fuckel
Habitat: Saprophytic on wood.
Description: Based on Vasilyeva and Stephenson (2006), Pande (2008) and Jones et al. (2009a).

37.1. ***Cryptosphaeria mangrovei*** K.D. Hyde, 1993a. *Trans. Mycol. Soc. Jpn.*, **34**: 311. [For Fig. refer original paper]

Ascomata: 325-520 µm high, 650-910 um wide, immersed, globose to subglobose, coriaceous, brown to black, ostiolate, papillate, developing singly. entostroma prosenchymatous, limited by thin black line, running under the peridium and 1.5-2 mm deep in the wood. *Necks*: 260-390 µm high, 130-190 µm diam, mostly immersed, seen as a blackened projecting shiny nipple on the surface, with internal periphyses. *Paraphyses*: uo to 75 µm thick, an inner stratum of compressed hyaline cells, an outer stratum of melanized *textura intricata*, fusing at the outside with the host tissue. *Paraphyses*: irregular, wide, branched, hyphae-like, with thin walls, up to 9 um thick. *Asci*: 62-80 x 9-13 µm, spore bearing part 44-58 µm long, stalk 14-40 µm long, 8-spored, cylindrical, thin-walled, long, pedunculate, peduncle long and tapering, asci apically truncate with a ring-like subapical apparatus, not staining blue in Melzer's reagent, developing from the base and sides of the ascomata. *Ascospores*: 10-16 x 5-7 µm, light-brown, allantoid, unicellular or bicellular. hyaline or light yellowish, unicellular or 1-3-septate, no sheath or appendages.

Habitat: On mangrove wood.
Description: Based on Hyde (1993a).
Distribution: East Coast:-Andhara Pradesh: On intertidal wood of *Rhizophora apiculata* (Sarma and Vittal, 1998-99; Sarma and Vittal, 2000; Sarma and Vittal, 2001; Sarma et al., 2001; Sarma and Vittal, 2002; Sarma and Vittal, 2004; Vittal and Sarma, 2005).

38. ***Cryptovalsa*** Ces. & De Not. ex Fuckel, 1870. *Jahrb. Nassauischen Vereins Naturk.*, **23-24**: 112.

Stromata: rather variable, effuse, eutypoid, but when erumpent separately diatrypelloid, offen immersed in the wood but sometimes invading bark tissue, mostly on twigs, rarely on wood, internally dirty brown. *Ascomata*: in groups in a stroma, immersed becoming not much erumpent, ectostroma effuse, subglobose to broadly ellipsoidal, ostiolate, papillate. *Necks*: periphysate, ostiolate. *Paraphyses*: septate, hyaline, deliquescent. *Asci*: polysporous, cylindrical or clavate, long, short or long pedunculate, apical refractive non-amyloid ring. *Ascospores*: crowded, allantoid, unicellular, yellowish, no appendages or sheath. A large terrestrial genus, 3 species are known from mangroves and salt marsh plants (Inderbitzin *et al.*, 1999).

Type Species: *Cryptovalsa protracta* (Pers. & Fr.) De Not.
Habitat: Saprobic on plant material.
Description: Based on Vasilyeva and Stephenson (2005), Jones et al. (2009a) and Pande (2008).

Key to *Cryptovalsa* species from India:

1 Asci 64-spored, ascospores pale yellow- pale brown, verrucose, 5.6-11(-18) x 1.6-3.6 µm .... *Cryptovalsa mangrovei*

1' Asci multi-spored, hyaline, 5-7 x 1-1.2 µm .... *Cryptovalsa suaedicola*

38.1. ***Cryptovalsa mangrovei*** Abdel-Wahab & Inderbitzin, 1999. *Mycol. Res.*, **103**: 1628. [For Fig. refer original paper]

Ascomata: 240-270 µm high, 260-350 µm wide, subglobose to broadly ellipsoidal, with a flattened base, in groups of 3-10, sometimes confluent, in 1-3 rows, immersed, raising the substratum, or erumpent. *Entostroma*: effuse, a faint black line between the groups sometimes present, wood softened; a thin layer of white pulvinate fungal material sometimes present around the ascomata venter. *Necks*: 180-230 µm long, 190-210 µm wide, ostolate, periphysate (36 x 1 µm), comprising an outer layer of completely melanized cells 32-40 µm wide, and an inner layer, up to 32 µm wide, of less pigmented, elongate cells; singly erumpent, protruding up to 80 µm above the level of the raised substratum, straight; the outermost ascomata in an aggregate often bent toward the centre of the group, with four narrow furrows on the dome-shaped apex. *Peridium*: in transverse section with an outer, layer of small, thick-walled, melanized, rounded cells 8-20 um wide, and an inner layer, up to 12 µm wide, of hyaline, elongate cells in *textura angularis*. *Paraphyses*: hyaline, septate, deliquescent, up to 160 x 11 µm. *Asci*: 88-180 x 9.5-15 µm, spore bearing part 45-96 µm long, multispored, apex up to 4 µm thick, clavate, truncate, with a stalk, with a refractive subapical, non-amyloid ring at the base of an invagination of the ascus apex. *Ascospores*: 8-15.5 x 1.5-3.5 µm, ca 128 per ascus, allantoid, pale yellow to pale brown.

Habitat: On intertidal mangrove wood.

Description: Based on Inderbitzin et al. (1999).

Distribution: West Coast:-Diu Island: On intertidal wood of *Avicennia marina*, (Borse et al., 1999b); **Maharashtra:** On intertidal mangrove wood (Borse, 2000b).

Note: This fungus was earlier misidentified as *C. halosarceicola* Hyde (Borse et al., 1999a; Borse, 2000b). All the records included as *C. halosarceicola* be considered as *C. mangrovei* Hyde

38.2. *Cryptovalsa suaedicola* Spooner, 1981. *Trans. Br. Mycol. Soc.*, **76**: 265-301. [Fig. 38.2]

Ascomata: 285-330 µm high, 300-350 µm diam, flask-shaped, black, immersed in a wide spreading entostroma in the surface layers of the host tissue, with protruding necks, blackened zone around ascomata present, ostioles with 4 vertical furrows. *Necks*: 180-280 µm long, 65-110 µm in diam, of dark angular cells, with periphyse. *Peridium*: 2-layered, inner layer of hyaline elongate cells and outer layer of dark, fungal hyphae fusing with the host tissue at the apex. Paraphyses: filiform, filamentous and hyphae-like, persistent. *Asci*: 95-135 x 7-9 µm, multispored, cylindric-fuscoid with tuncate base at the apex having a subapical ring, faint in IKI and tapering towards the peduncle. *Ascospores*: 5-7 x 1-1.2 µm, allantoid, aseptate, hyaline.

Habitat: On wood of *Suaeda fruticosa* Euras.

Description: Based on Spooner (1981) and Hyde (1993b).

Distribution: East Coast:-Andhara Pradesh: On intertidal wood of *Avicennia marina,* and *A. officinalis* (Sarma and Vittal, 2004).

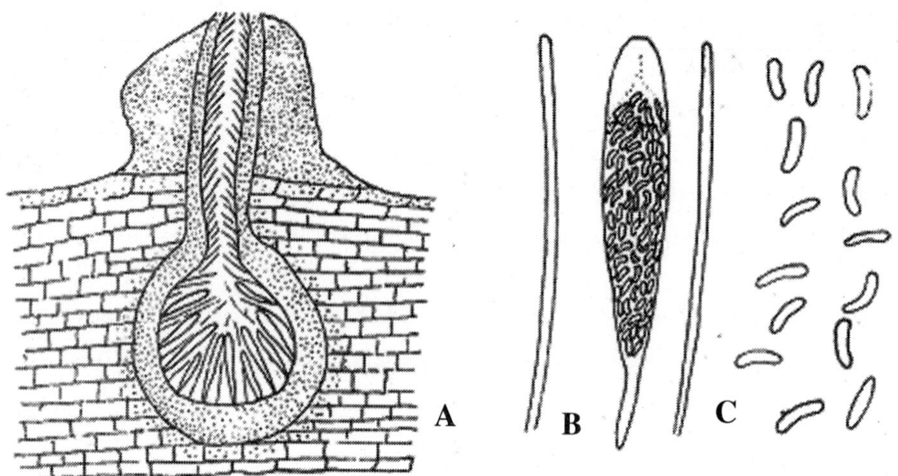

Fig. 38.2. ***Cryptovalsa mangrovei,*** A-D; A- v.s. of ascocarp (100x), B- ascus (1000x) , C- ascospores (1000x).

39. ***Dactylospora*** Korb., 1855. *Syst. Lich.Germ.,* pp. 271.

= *Kamadiscus* Kohlm. & Kohlm., *Mycol. Res.*, **63:** 837 (1971).

Ascomata: initially sub-globose, becoming subglobose or discoid, flat or convex, superficial, sessile, leathery, dark reddish–brown, becoming black, solitary or gregarious. *Paraphyses*: swollen at the spex, septate, hyaline, branching at the upper part. *Asci*: 8-spored, clavate, short pedunculate, unitunicate, apically thick-walled, without an apical appartus. *Ascospores*: ellipsoidal or obovoid, 1-septate, constricted at the septum, with longitudinal or verrucose ornamentations, light brown. A large terrestrial genus, 3 species are known from intertidal wood and mangroves (Jones *et al.,* 2009a).

Type species: *Dactylospora floerkei* Korb.

Habitat: Saprophytic on wood.

Description: Based on Jones et al., (2009a).

39.1. ***Dactylospora heliotrepha*** (Kohlm. & E. Kohlm.) Hafellner, 1979.

Beih.Nova Hedwigia, **62:** 111. [Fig. 39.1]

Basionum: *Buellia heliotrepha* Kohlm. & E. Kohlm., *Nova Hedwigia,* **9:** 90 (1965).

≡ *Kymadiscus haliotrephus* (Kohlm. & E. Kohlm.) Kohlm. & Kohlm., *Mycologia*, **63**: 837 (1971).

Ascomata: 200-320 µm high, up 360-1000 µm in diam, at first subglobose, becoming semiglobose or discoid, flat or convex, apothesia-like superficial, sessile, rooted in the substratum with an obconical foot, fleshy-leathery, dark reddish-brown, appearing almost black when dry, solitary or gregarious. *Excipula*: composed of thick-walled, subglobose to polygonal cells with large lumina, developing from the center in parallel rows, mearging into the sclerenchymatous foot, forming a *textura angularis*; barely arching over the hymenium. *Hypothecium*: composed of thin-walled cells, forming *textura angularis*, claret-coloured, mearging basally into the hyaline foot. *Epithecia*: brownish, crumbling, finally exposing the hymenium. *Pseudoparaphyses*: 85-125 x 1-2 µm, at the swollen apex 2.5-4 µm in diam., clavate, simple, rarely branching, septate, hyaline, surpassing the asci. *Asci*: 70-100 x 18-20 µm, 8-spored, clavate, short pedunculate, bitunicate, apically thick-walled, thinner towards the base, without apical appartusses. *Ascospores*: 18-28 x 8-12 µm, obliquely 1 or 2-seriate, ellipsoidal or obovoid, 1-septate in the lower third, constricted at the septum, at first grayish-green, becoming brownish; epispore with delicate, forked longitudinal striations.

Habitat: Saprophytic on intertidal or washed-up wood.

Description: Based on Kohlmeyer and Kohlmeyer (1979).

Distribution: West Coast:-Diu Island: On intertidal wood (Borse et al., 1999b); **Daman:** On intertidal wood and *Avicennia marina* (Borse et al., 2000b); **Gujarat:** On intertidal wood of *Avicennia marina* and *Ceriops tagal* (Borse et al., 2000a), on intertidal mangrove wood (Patil and Borse, 2001); **Maharashtra:** On dead roots of *Avicennia alba* (Patil and Borse, 1983b, 1986; Borse et al., 2005a), intartidal wood of *Rhizophora mucronata* (Patil and Borse, 1986), on intertidal wood (Borse, 1988), on intertidal wood of *Sonneratia acida* (Srivastava, 1989), on intertidal mangrove wood (Maria and Sridhar, 2002a); **Goa:** On intertidal wood of *Rhizophora mucronata* (Nandan et al., 1993), on intertidal wood (Prasannarai and Sridhar, 2001), on intertidal mangrove wood of *Avicennia* *officinalis* (Maria and Sridhar, 2003); (Maria and Sridhar, 2002a); **Karnataka:** On intertidal wood and teak panels (Prasannarai and Sridhar, 1997, 2001, 2000-2001), on intertidal mangrove wood (Maria and Sridhar, 2002a), on intertidal leaf and woody litter of mangroves (Ananda and Sridhar, 2003, 2004); **Pondicherry (Mahe):** On intertidal mangrove wood of *Avicennia* sp. and *Rhizophora* sp. (Nambiar and Raveendran, 2007; 2008a; 2009d); **Kerala:** On intertidal wood (Prasannarai and Sridhar, 2001), on intertidal mangrove wood (Maria and Sridhar, 2002a), on intertidal wood (Raveendran and Manimohan, 2007), on intertidal wood and mangrove wood of *Avicennia* sp. (Nambiar and Raveendran, 2006, 2008a, b, d, 2009a, b, e, f, g, h, 2010b, c; Nambiar et al., 2006, 2008a), on mangrove woody litter (Shini et al., 2009-10).

Lakshadweep Islands: On intertidal wood of *Ceriops tagal* (Chinnaraj, 1992); on intertidal mangrove wood (Ananda and Sridhar, 2003).

East Coast:-Tamil Nadu: On roots of *Avicennia* sp. (Raghukumar, 1973), on submerged parts of *Rhizophora* spp. (Ravikumar and Vittal, 1996), on intertidal wood (Nambiar et al., 2008b), on mangrove wood (Nambiar and Raveendran, 2008e, 2009k, 2010c); **Andhara Pradesh:** On intertidal wood of *Avicennia marina, A. officinalis* and *Rhizophora apiculata* (Sarma and Vittal, 1998-99; Sarma and Vittal, 2000; Sarma and Vittal, 2001; Sarma et al., 2001; Sarma and Vittal, 2002; Sarma and Vittal, 2004; Vittal and Sarma, 2005), on intertidal wood (Nambiar and Raveendran, 2009j); **West Bengal:** On intertidal wood of *Rhizophora mucronata* (Pawar et al., 2009); on mangrove woody litter (Shini et al., 2009-10); **Orissa:** On intertidal wood of *Avicennia marina, A. officinalis* and *Ceriops decandra* (Borse et al., 2009); **Andaman & Nicobar Islands:** On intertidal wood of *Avicennia marina, A. officinalis, Bruguiera gymnorhiza, Rhizophora apiculata, R. mucronata* and *Sonneratia alba* (Chinnaraj, 1993a).

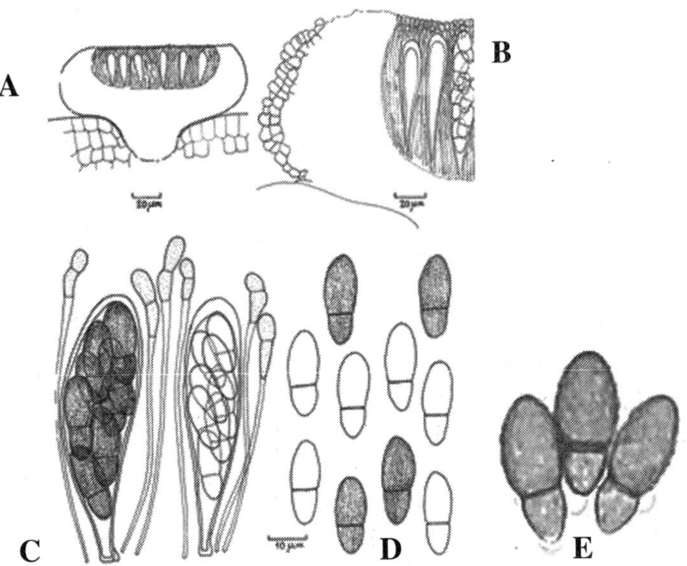

Fig. 39.1. ***Dactylospora heliotrepha,*** A-E; A- v. s. of ascocarp, B- v. s. of ascocarp enlarged, C- asci and paraphyses, D- ascospores, E- ascospores (1000x).

40. *Didymella* Sacc.

Michelia, **2**: 57 (1880).

Ascomata: solitary, globose, depressed or flattened, without well developed stroma, immersed or erumpent, ostiolate, papillate, dark. *Pseudopaphyses*: filiform, simple, often connected to each other by a gelatinous matrix. *Asci*: 8-spored, bitunicate, cylindrical, short pedunculate, rounded and thickened at the apex. *Ascospores*: ovoid to ellipsoidal, 1-septate in or below the center, rarely above the middle, mostly constricted at the septum, hyaline. A large terrestrial and freshwater genus, 4 species are known from larger marine algae and mangrove wood (Jones *et al.,* 2009a).

Type species: *Didymella exigua* (Niessl) Saccardo
Habitat: Parasitic on leaves and maturing on dead substrates.
Description: Based on Kohlmeyer and Kohlmeyer (1979).

40.1. *Didymella avicenniae* S.D. Patil & Borse, 1985. *Trans. Mycol. Soc. Japan.*, **26**: 271. [Fig. 40.1]

Ascomata: 275-350 µm high, 250-275 µm wide, immersed, subglobose, black. *Peridium*: composed of irregular and thick-walled cells, 15-20 µm thick. *Pseudoparaphyses*: forming an indistinct network in immature ascomata, absent in mature ascomata. *Asci*: 140-190 x 28-32 µm, 8-spored, bitunicate, ellipsoidal to clavate. *Ascospores*: 25-32 x 10-15 µm, ellipsoidal to ovoid, 1-septate, hyaline, rounded at the tip, constricted at the septum, narrowing to the base, surrounded by a gelatinous sheath, which slowly swells in water and eventually dissolves.

Habitat: Saprophytic on intertidal wood of mangroves.

Description: Based on Patil and Borse (1985).

Distribution: West Coast:-Maharashtra: On intertidal dead and decaying wood of *Avicennia alba* (Patil and Borse, 1985, 1986; Borse, 1988; Srivastava, 1989; Borse et al., 2005a), on intertidal mangrove wood (Maria and Sridhar, 2002a); **Karnataka:** On intertidal mangrove wood *Avicennia officinalis* and *Sonneratia caseolaris* (Maria and Sridhar, 2002a, 2003a).

East Coast:-Tamil Nadu: On submerged parts of *Rhizophora* spp. (Ravikumar and Vittal, 1996).

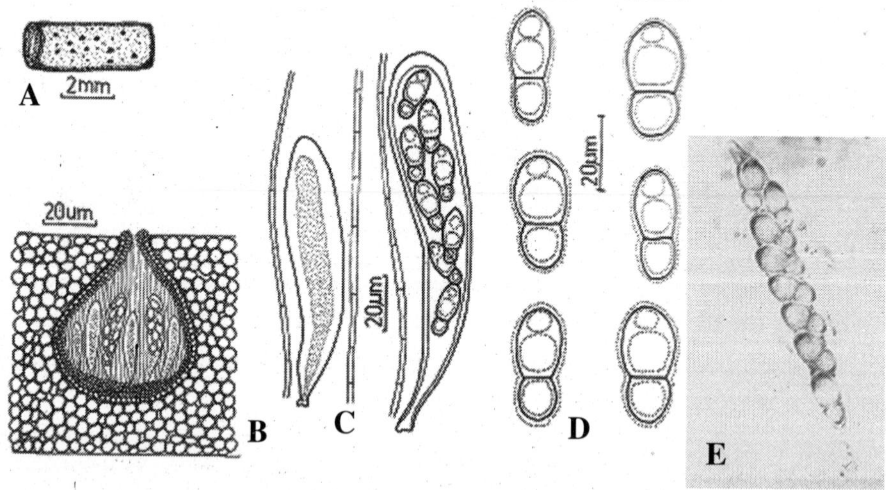

Fig. 40.1. *Didymella avicenniae,* A-E; A- Habit, B- v.s. of ascocarp, C- asci and paraphyses, D- ascospores, E- ascus (1000x).

40.2. *Didymella fucicola* (G.K. Sutherl.) Kohlm., 1968. Phytopath. Z., **63**: 342. [For Fig. refer original paper]

Basionym: *Didymosphaeria fucicola* G.K. Sutherl., New Phytol., **14**: 188 (1915).

= *Didymosphaeria pelvetiana* G.K. Sutherl., New Phytol., **14**: 185 (1915).

Habitat: Perthophytic (parasitic) or saprophytic on midrib and air vesicles of lower branches of live brown algae *Fucus spiralis* and *F. vesiculosus*, vegetative thalli of *Pelvetia canaliculata*, often associated with bases of *Elachista clandestine* and *E. ficicola*.

Distribution: East Coast:-West Bengal: On mangrove wood (Shini et al., 2009-10).

Note: This is a temperate water ascomycete collected on living or dead brown algae (Kohlmeyer and Kohlmeyer, 1979).

41. *Didymospharia* Fuckel, 1870. *Jahrb. Nassau. Ver. Naturkd.*, **23-24**: 140.

Ascomata: globose or flask-shaped, immersed or erumpent, often with a dark clyeus or hyphoid stroma, ostiolate, short papillate. *Peridium*: composed of several layers of roundish, isodimetric or slightly flattened mostly small, thick-walled cells. *Pseudoparaphyses*: thin, filiform, branched. *Asci*: 4 or 8-spored, cylindrical or claviform, short or long pedunculate, slightly constricted at the apex, bitunicate, developing at the base of the ascomata venter. *Ascospores*: uni- or biseriate, elongate, 1-septate near the middle, brown. A large terrestrial genus, one species is known from intertidal wood and mangroves (Jones *et al.,* 2009).

Type species: *Didymospharia futilis* (Berk. & Broome) Rehm.
Habitat: Saprobic on wood.
Description: Based on Kohlmeyer and Kohlmeyer (1979).

41.1. *Didymospharia lignomaris* Strongman & J.D. Miller, 1985. *Proc. N. S. Inst. Sci.*, **35**: 102. [For Fig. refer original paper]

Ascomata: 225-250 µm high, 220-313 µm in diam, subglobose, partly or tatally immersed, ostiolate, papillate, black. Necks: 35 µm high, 100-120 µm in diam, ostiolar canal 30-45 µm in diam. *Pseudoparaphyses*: 1-2 µm in diam, filiform, attached at both ends. *Asci*: 120-135 µm long, 8-spored, cylindrical, bitunicate, thick-walled, without apical apparatus, developing in a layer at the base of the ascomata venter. *Ascospores*: 25-26 x 10-12 µm, uniseriate, ellipsoidal, 1-septate in the middle, constricted at the septum, finaly echinulate, thick-walled, thickenings around the apicies, dark around the septum.

Habitat: Saprophytic on intertidal wood.
Description: Based on Strongman et al. (1985).
Distribution: West Coast:-Goa: On intertidal wood (Prasannarai and Sridhar (2001); **Karnataka:** On intertidal wood (Prasannarai et al., 1999), on intertidal wood (Prasannarai and Sridhar, 2001, 2000-2001), on intertidal leaf and woody litter of mangroves (Ananda and Sridhar, 2004).

42. ***Dryosphaera*** Jorgen Koch & E.B.G. Jones, 1989. *Can. J. Bot.,* **67:** 1184.

Ascomata: globose, coriaceous, whitish, gregarious, erumpent, becoming free without ostioles. When free attached to surface with thick-walled hyphae originating from basal part of ascomata. Free surface covered by numerous short appendages each comprised of a thick-walled stem bearing a terminal crown of repeatedly branched arms. *Peridium*: membranous, one-layered. *Asci*: numerous, in an irregular hymenium on the expanded upper part of a stem-like structure arising from the base of the ascomata. *Asci*: thick-walled, clavate, pedunculate, persistent. *Paraphyses*: absent. *Ascospores*: two-celled, hyaline, appendaged. *Appendages*: threadlike, spirally unfolding from side of spore. The genus is represented by 3 species in marine habitats (Jones *et al.,* 2009a).

Type species: *Dryosphaera navigans* Jorgen Koch & E.B.G. Jones
Habitat: Saprophytic on intertidal or washed-up wood.
Description: Based on Koch and Jones (1989).

42.1. ***Dryosphaera navigans*** Jorgen Koch & E.B.G. Jones, 1989. *Can. J. Bot..,* **67**: 1185. [For Fig. refer original paper]

Ascomata: 150-220 μm wide, slightly flattened, globose, coriaceous, white-light buff, non-ostiolate, gregarious, erumpent, free, when free attached to the surface with few or many long, rigid, thick-walled, unbranched, pale yellow hyphae 3-4 μm in diam. originating from the basal part of the ascomata and often terminating in a branched attachment disc. The entire free surface is covered by numerous short appendages each comprised of a thick-walled stalk 12-14 μm high, 5-6 μm broad, broader at the base, with a terminal flat crown 12-16 μm broad, 4-6 μm high, of short, repeatedly divided branches ending in small conical tips. *Peridium*:

membranous, up to 8 µm thick, one layered, compressed of 1-(2-3) cells that are flat, thick-walled, and angular with diam up to 10 ìm. *Asci*: 36-48 x 13-20 µm, 8-spored, numerous, in an irregular hymenium on the expanded and lobed upper part upper part of a stem-like structure arising from the base of the ascomata and stretching about one-third up into the centrum, thick-walled, clavate, pedunculate, persistent. *Paraphyses*: absent. *Ascospores*: 8-12 x 3-4 µm, two-celled, constricted at the septum, hyaline, appendaged. *Appendages*: thread-like, spirally unfolding from side of spore.

Habitat: Saprophytic on intertidal or washed-up wood.

Description: Based on Koch and Jones (1989).

Distribution: West Coast:-Karnataka: On intertidal wood ((Prasannarai and Sridhar, 2001, 2000-2001).

Note: According to Kohlmeyer and Volkmann-Kohlmeyer (1991) tropical ascomycetes, resembling *Dryosphaera navigans* is most likely different species.

42.2. ***Dryosphaera tropicalis*** Kohlm. & Volkm.-Kohlm., 1993. *Can. J. Bot.*, **71:** 992. [Fig. 42.2]

Ascomata: 140-245 µm in diam, subglobose, superficial, without ostiole, coriaceus, cream colored, covered by tree-like appendages, attached to the substrate with some thick-walled, simple, sometimes spiral-shaped hyphae, solitary or gregarious. *Ascomata appendages*: 14-36 µm long, thick-walled, refractive; stalk 3.5-7 µm in diameter, cylindrical to conical; head 10-17 µm in diam, 6-9 µm high, divided into numerous short, tapering branches. *Peridium*: 4-7 µm thick, composed of 1 or 2 layers of polygonal, thick-walled cells, staining blue in aniline blue, forming a *textura angularis,* merging with the central thin-walled pseudoparenchyma. *Paraphyses*: absent. *Pseudoparenchyma*: of thin-walled, polygonal cells filling the centrum of immature ascomata. *Asci*: 50 x 19-22 µm, 8-spored, clavate, pedunculate, thick-walled, nonamyloid, developing successively at the periphery of an irregular head-like ascogenous tissue. The centrally located hymenium originates on a stalk composed of large pseudoparenchymatous cells at the base of the locule. *Ascospores*: 12-15 x 4–6 µm, ellipsoidal, 1-septate, slightly constricted at the septum, hyaline, appendaged. *Appendages*: at first at each end with a cap that develops into numerous subterminal appendages, 14-21µm long, filamentous, sometimes with an inflated tip.

Habitat: Saprophytic on intertidal or washed up wood.

Description: Based on Kohlmeyer and Volkmann-Kohlmeyer (1993a).

Distribution: West Coast:-Maharashtra: On intertidal wood and *Rhizophora mucronata* (Borse, 2000b; Borse et al., 2005a); **Gujarat:** On intertidal wood (Kamble et al., 2011); **Karnataka:** On intertidal wood (Prasannarai and Sridhar, 2001, 2000-2001); **Pondicherry (Mahe):** On intertidal wood (Nambiar and Raveendran, 2007); **Kerala:** On intertidal mangrove wood (Nambiar and Raveendran, 2008b, 2010a, c), on mangrove woody litter (Shini et al., 2009-10). **East Coast:-Tamil Nadu:** On intertidal wood (Nambiar et al., 2008b), on intertidal wood (Nambiar and Raveendran, 2009k, l, 2010a, c); **Pondicherry:** on intertidal wood (Nambiar and Raveendran, 2010a); **West Bengal:** On intertidal wood (Borse and Pawar, 2001).

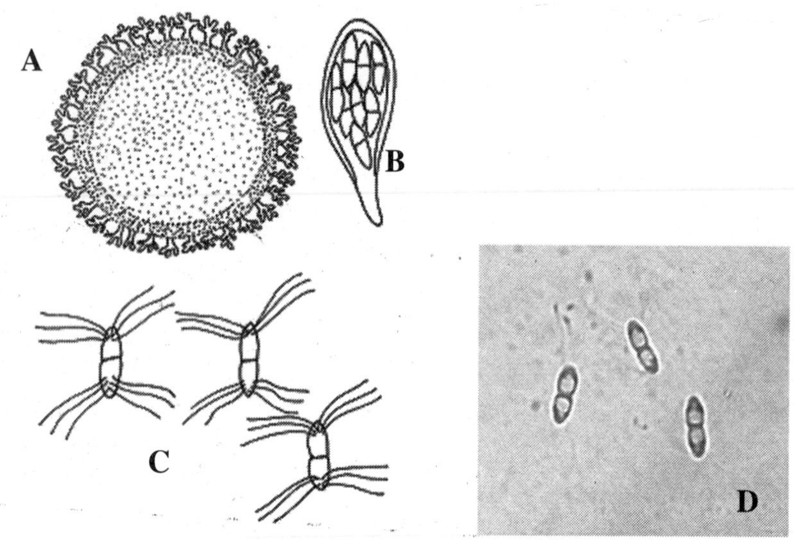

Fig. 42.2. ***Dryosphaera tropicalis,*** A-D; A- ascocarp (100x), B- ascus (1000x), C- ascospores (1000x), D- ascospores (1000x).

43. *Etheirophora* Kohlm. & Volkm.-Kohlm., 1989b. *Mycol. Res.*, **92**: 414.

Ascomata: depressed, immersed, ostiolate, periphysate, papillate, clypeate, coriaceous, light-coloured, paraphysate. *Peridium*: forming a *textura angularis. Asci*: 8-spored, cylindrical or oblong-ventricose, pedunculate, non-amyloid, thin-walled, sometimes thicker at the top, aphysoclastic, persistent, maturing successively on an ascogenous tissue at the bottom of the locule. *Ascospores*: ellipsoidal, 1-septate, hyaline, with filamentous appendagesat one or both ends.

Type species: *Etheirophora bijubata* Kohlm. & Volkm.-Kohlm.
Habitat: Saprophytic on intertidal wood and mangroves.
Description: Based on Kohlmeyer and Volkmann-Kohlmeyer (1989b).

Key to *Etheirophora* species from India:

1. Ascospores with appendages at both ends .... *Etheirophora bijubata*

1'. Ascospore appendages at one end only 2

2. Exclusively in bark of *Rhizophora* sp., ascospores up to 21 µm long, up to 7 appendages .... *Etheirophora blepharospora*

2'. In other substrates, ascospores up to 29 µm long, more than 7 appendages .... *Etheirophora unijubata*

43.1. *Etheirophora bijubata* Kohlm. & Volkm.-Kohlm., 1989. *Mycol. Res.*, **92**: 415. [For Fig. refer original paper]

Ascomata: 100-160 µm high, 200-400 µm in diam., depressed, immersed, ostiolate, papillate, clypeate, coriaceous, light-coloured, single. *Necks*: 100-200 µm high, 40-75 µm in diam., conical, brown, periphysate, projecting above the wood surface, surrounded by an irregular small, brown clypeus that penetrates into the lumina of wood vessels. *Peridium*: 8.5-15 µm thick, composed of 4-6 layers of polygonal cells, forming a *textura angularis,* light brown. *Paraphyses*: septate, sparingly branched, surrounded by a gelatinous matrix. *Asci*: 65-95 x 13-18 µm, 8-spored, cylindrical, pedunculate, non-amyloid, thin-walled, slightly thicker at the flat apex, aphysoclastic, persistent, developing on an ascogenous tissue at the bottom of the locule. *Ascospores*: 16.0-21.5 x 6.0-8.0 µm, ellipsoidal, 1-septate, not constricted at the septum, hyaline, appendaged. *Appendages*: at each end with 5-8 apical appendages; 11-17 µm long, bristle-like, rigid, slightly curved, in immature spores laterally attached.

Habitat: On intertidal wood.
Description: Based on Kohlmeyer and Volkmann-Kohlmeyer (1989b).
Distribution: West Coast:-Karala: On driftwood (Khan and Manimohan, 2011).

43.2. *Etheirophora blepharospora* (Kohlm. & E. Kohlm.) Kohlm. & Volkm.-Kohlm., 1989b. *Mycol. Res.*, **92**: 415. [For Fig. refer original paper]

Basionym: *Keissleriella blepharospora* Kohlm. & E. Kohlm., Nova Hedwigia, 9: 97 (1965).

Ascomata: 125-230 µm high, 225-425 µm in diam (including dark clypei), depressed ellipsoidal, subglobose or pyriform, immersed in the bark, ostiolate, papillate, clypeate, coriaceous, subhyaline, solitary or gregarious. *Peridium*: 10-15 µm thick, hyaline to light brown, composed of several layers of subglobose, ellipsoidal or elongate, thick-walled cells, forming a *textura prismatica*; at the edge intimately connected to decomposing bark cells. *Necks*: short, subconical, surrounded by a brownish-black, 30-37.5 µm thick pseudoclypeus that encloses bark cells of the host; ostiolar canal 12.5-27.5 µm in diam, periphysate. *Pseudoparaphyses*: 15-2.5 µm in diam, sparsely septate, rarely branched, sometimes connected by mucilage to one another, attached at the base and apex. *Asci*: 62-75 x 11-15 µm, 8-spored, cylindrical or oblong-ventricose, apically truncate, short pedunculate, non-amyloid, thin-walled, indistinctly bitunicate, thick-walled at the apex and provided with an apical plate, aphysoclastic, persistent, maturing successively on an ascogenous tissue at the bottom of the locule. *Ascospores*: 12-21 x 6-8 µm (excluding appendages), one or biseriate, ellipsoidal, 1-septate, not or slightly constricted at the septum, thin-walled, hyaline, appendaged. Appendages: at one ends, 4-7 radiating, terminal setae, deciduous, up to 13 µm long.

Habitat: On intertidal mangrove wood with bark.

Description: Based on Kohlmeyer and Kohlmeyer (1965).

Distribution: East Coast:-Andaman & Nicobar Islands: On intertidal wood of *Rhizophora mucronata* (Chinnaraj, 1993a).

43.3. *Etheirophora unijubata* Kohlm. & E. Kohlm., 1989b. *Mycol. Res.*, **92**: 415. [For Fig. refer original paper]

Ascomata: 150-280 µm high, 100-160 µm in diam., depressed, immersed in the bark, ostiolate, papillate, clypeate, coriaceous, light-coloured, single. *Necks*: 40-80 µm high, 35-45 µm in diam., conical, periphysate, surrounded by a shield-like, black clypeus. *Peridium*: 7.5-12 µm thick, composed of 3-5 layers of polygonal cells, forming a *textura angularis,* hyaline. *Paraphyses*: simple, attached at the flat apex. *Asci*: 85-105 x 11-15 µm, 8-spored, cylindrical, pedunculate, non-amyloid, slightly thicker at the flat apex, aphysoclastic, persistent, developing on an ascogenous tissue at the bottom of the locule. *Ascospores*: 18.5-24.0 x 5.5-7.0 µm, ellipsoidal, 1-septate, not constricted at the septum, hyaline, appendaged. *Appendages*: 12-18 µm long, forming a crown at one end.

Habitat: On intertidal mangrove wood with bark.
Description: Based on Kohlmeyer and Volkmann-Kohlmeyer (1989b).
Distribution: West Coast:-Karnataka: On intertidal leaf and woody litter of mangroves (Ananda and Sridhar, 2004).

44. *Eutypa* Tul. & C. Tul., 1863. *Sel. Fung. Carp.*, **2**: 52.

Stromata: widely effuse within wood tissue causing its blackening. *Ascomata*: scattered evenly in a single layer just beneath the surface, flask-shaped, the sulcate ostioles often prominent, some times distinctly beaked, black, sometime ornamented with radial furrows. *Asci*: small, clavate or cylindrical, 8-spored, usually with long stalks, paraphysate, with a well developed apical ring in some species. *Ascospores*: allantoid, hyaline or yellowish. A large terrestrial genus, one species are known from intertidal wood and mangroves (Jones *et al.,* 2009a).

Type species: *Eutypa maura* (Fr. & Fr.) Fuckel
Habitat: On dead wood.
Description: Based on Vasilyeva and Stephenson (2006) and Jones *et al.* (2009a).

44.1. *Eutypa bathurstensis* K.D. Hyde & Rappaz, 1993. *Mycol. Res.*, **97**: 861. [Fig. 44.1]

Stromata: effuse up to 4.5 cm long, blackening the wood surface, entostroma prosenchymatous, poorly developed, dorsally limited by a black zone binding the fruiting areas. *Ascomata*: spherical to flattened, 0.3-0.6 µm high, up to 0.8 mm wide, submerged in the wood, regularly spaced, and occasionally deeply buried. *Necks*: up to 0.5 µm long, not raising the uniformly blackened wood surface, or occasionally weakly raised, then wood surface blackened only under one or a few ascomata. *Peridium*: enclosing fragments of crystals in the external layer, 35-50 µm wide or wider near the ostiolar canal. *Ostioles*: poorly developed or conical, up to 300 µm high, 150-200 µm diam., not sulcate. *Hymenium*: lining the whole cavity. *Paraphyses*: numerous, persistent. *Asci*: spore part 30-45 x 8-10 µm, pedicel 10-30 µm long, total 50-65 µm long, clavate, 8-spored, apical annulus not blueing in iodine, ascal apex showing a fissure on the side of the manubrium, probably limitinh the exotunica from the endotunica. *Ascospores*: 7-10 x 2-2.8 µm, olive-brown, allantoid, containing oil droplets and limited by a thin episporium.

Habitat: On dead branches.

Description: Based on Hyde and Rappaz (1993).

Distribution: West Coast:-Gujarat: On intertidal mangrove wood (Patil and Borse, 2001); **Maharashtra:** On intertidal mangrove (*Avicennia marina*) wood (Borse, 2000b; Borse et al., 2005a); **Goa:** On intertidal mangrove wood (Maria and Sridhar, 2002a); **Karnataka:** On intertidal mangrove wood on intertidal wood of *Avicennia officinalis, Rhizophora mucronata* and *Sonneratia caseolaris* (Maria and Sridhar, 2002a, 2003a); **Kerala:** On intertidal mangrove wood (Maria and Sridhar, 2002a)

East Coast:-Tamil Nadu: On intertidal mangrove wood (Ravikumar, 1991); **Andhara Pradesh:** On intertidal wood of *Avicennia marina,* and *A. officinalis* (Sarma and Vittal, 1998-99; Sarma and Vittal, 2000; Sarma and Vittal, 2001; Sarma et al., 2001; Sarma and Vittal, 2004; Vittal and Sarma, 2005); **West Bengal:** On intertidal wood of *Avicennia marina* (Pawar et al., 2009).

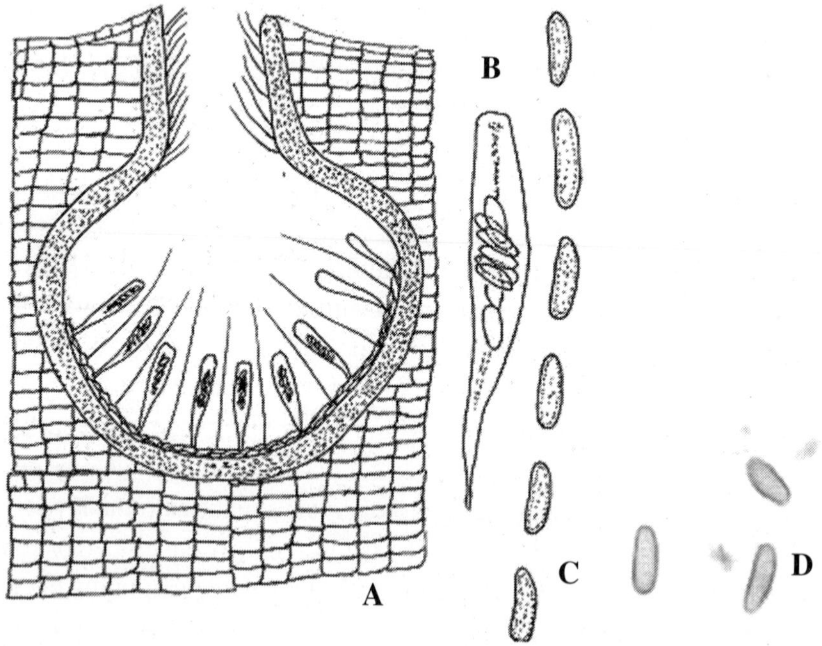

Fig. 44.1. ***Eutypa bathurstensis,*** A-D; A- v.s. of ascocarp (100x), B- ascus (450x), C- ascospores (1000x), D- ascospores (1000x).

45. ***Eutypella*** (Nitschke) Sacc., 1875. *Atti. Soc. Ven. Trent. Sci. Nat.*, **4**: 80.

Stromata: erumpent from bark as crowded perithecial beaks, mostly sulcate, sometimes rounded, usually without ectostromatic disc but united below by a common tissue that sometimes strongly developed and causing the appearance of pulvinate prickly globules. *Asci*: clavate, 8-spored, with long stalks. *Ascospores*: allantoid, hyaline or yellowish. A large terrestrial genus, one species is known from intertidal mangroves wood (Jones et al., 2009a).

Type species: *Eutypella cerviculata* (Fr.) Sacc.
Habitat: On dead branches.
Description: Based on Vasilyeva and Stephenson (2006) and Jones et al. (2009a).

45.1. ***Eutypella naqsii*** K.D. Hyde, 1995. *Mycol. Res.*, **99**: 1462.

[Fig. 45.1]

Stromata: up to 80 mm thick, forming a blackened crust on the host surface, with ectostromal cushions occurring around each ascomata, comporising host cells and intracellular blackened fungal hyphae. *Ascomata*: 520-650 µm high, 650-840 µm in diam, forming beneath, raised, blackened, crust-like stromata on the host surface, with 3-5 ascomata per stroma, globose or subglobose, with periphysate necks collectively erumpent through postulate discs. *Peridium*: up to 50 µm wide, comprising an inner layer of thin-walled, compressed, hyaline cells and an outer layer of pigmented, globose, fungal and host cells, fusing with the stroma at the outside. *Paraphyses*: up to 8 µm wide at the base, hyphae-like, sparse, septate, hyaline, and tapering distally. *Asci*: 28-50 x 4-6.4 µm, spore bearing part 22-30 µm long, 8-spored, cylindric-clavate, long pedunculate, thin-walled, unitunicate, apically truncate, with a J-, apical thickening, arising from the base and sides of the ascomata. *Ascospores*: 6.4-8 x 1.2-1.6 µm, allantoid, straight or mostly curved, hyaline to pale-yellow in mass and unicellular.

Habitat: On intertidal mangrove wood.
Description: Based on Hyde (1995).
Distribution: East Coast:-West Bengal: On intertidal wood of *Avicennia marina* and *Rhizophora mucronata* (Pawar and Borse, 2005a).

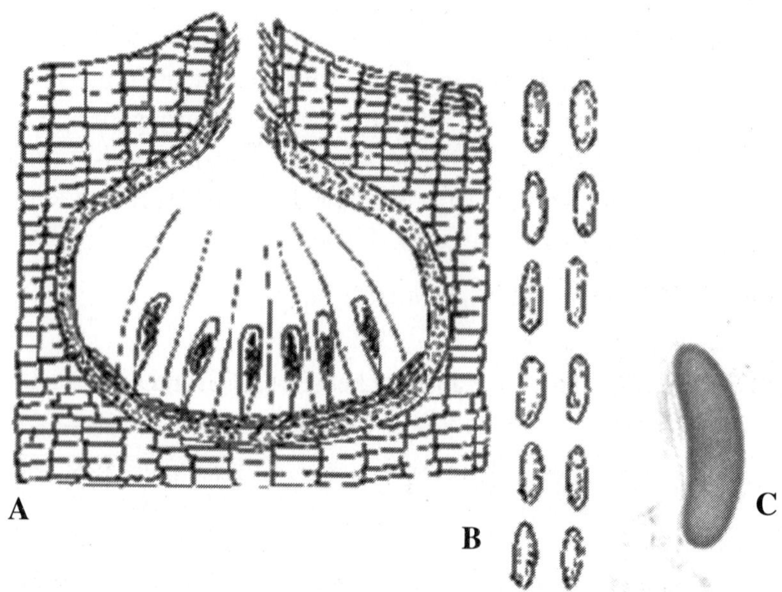

Fig. 45.1. ***Eutypella naqsii,*** A-C; A- v.s. of ascocarp (100x), B- ascospores (1000x), C- ascospore (1000x).

46. *Haiyanga* K.L. Pang & E.B.G. Jones, 2008. *The Raffl. Bull. Zool.*, **19**: 8.

Ascomata: globose to subcylindrical, immersed, ostiolate, papillate, membranous or coriaceous, dark brown to black, periphysate, catenophyses. *Asci*: clavate to ellipsoidal, short pedunculate, thin-walled, unitunicate, no apical apparatus, deliquescing early. *Ascospores*: ellipsoidal, 1-septate, slightly constricted at the septum, hyaline with appendages at both ends of the spore. *Appendages*: obclavate, curved, attenuate and release a drop of mucilage from their tips. The genus is monotypic (Jones *et al.*, 2009a).

Type species: *Haiyanga salina* (Meyers) Pang & Jones
Habitat: On interdidal wood.
Description: Based on Jones et al. (2009a).

46.1. ***Haiyanga salina*** (Meyers) K.L Pang & E.B.G. Jones, 2008. *The Raffles Bull. Zool.*, **19**: 8. [Fig. 46.1]

Basionym: *Arenariomyces salina* Meyers, *Mycologia*, **49**: 505 (1957).

≡ *Remispora salina* (Meyers) Kohlm., *Mycologia,* **60**: 262 (1968).

≡ *Halosphaeria salina* (Meyers) Kohlm., *Can. J. Bot.*, **50**: 1957 (1972).

≡ *Antennospora salina* (Meyers) Yusoff, E.B.G. Jones & S.T. Moss, *Mycol. Res.*, **98**: 1003 (1994).

Ascomata: 100-425 µm in diam, globose or subcylindrical, completely or partly immersed or becoming exposed, ostiolate, papillate, membranaceous or coriaceous, dark brown or black, surrounded by brown hyphae, solitary or gregarious. *Peridium*: 8-12 µm thick, composed of 3 to 5 layers of flat, thick-walled cells with large lumina forming a *textura angularis*, merging into the pseudoparenchyma of the venter. *Necks*: 68-165 µm long, 28-52 µm in diam, cylindrical or conical, ostiolar canal in the upper part filled with thin- walled, hyaline cells. Pesudoparenchyma of thin-walled large polygonal to ellipsoidal cells filling venter of young ascomata. *Asci*: 60-80 x 29-41µm, 8-spored, clavate or ellipsoidal, short peduculate, unitunicate, thin-walled, aphysoclastic, without apical appartuses, early deliquising; developing at the base of ascomata venter on a small-celled ascogenous tissue. *Ascospores*: 19-28 x 8-13.5 µm (excluding appendages), ellipsoidal, 1-septate, not or slightly constricted at the septum, hyaline, appendaged, Appendages: at each end 3 or 4 (rarely five) radiating appendages, 12-19 µm long, 1.5-2.5 µm in diam at the base, subterminal, obclavate, curved, attenuate, semirigid, indistinctly spoon shaped at the base and release a drop of mucilage from their tips.

Habitat: On intertidal wood.

Description: Based on Kohlmeyer and Kohlmeyer (1979).

Distribution: West Coast:-Diu Island: On intertidal wood (Borse et al., 1999b); **Gujarat:** On intertidal wood and mangrove wood (Patil and Borse, 2001); **Maharashtra:** On intertidal wood and *Avicennia marina* (Borse,1987a, 1988; Borse et al., 2005a); **Goa:** On intertidal wood (Borse et al., 1999a), on intertidal wood (Prasannarai and Sridhar, 2001); **Karnataka:** On intertidal wood and teak panels (Prasannarai et al., 1999), on intertidal wood (Prasannarai and Sridhar, 2001); **Kerala:** On intertidal wood (Prasannarai and Sridhar, 2001), on intertidal wood (Raveendran and Manimohan, 2007; Nambiar and Raveendran, 2007, 2010a, c).

East Coast:-Tamil Nadu: On intertidal wood ((Prasannarai and Sridhar, 2001), on intertidal wood (Nambiar and Raveendran, 2009k, l); **West Bengal:** On intertidal wood of *Avicennia marina* and *A. officinalis* (Pawar and Borse, 2004); **Andaman & Nicobar Islands:** On intertidal wood of *Sonneratia alba* (Chinnaraj, 1993a).

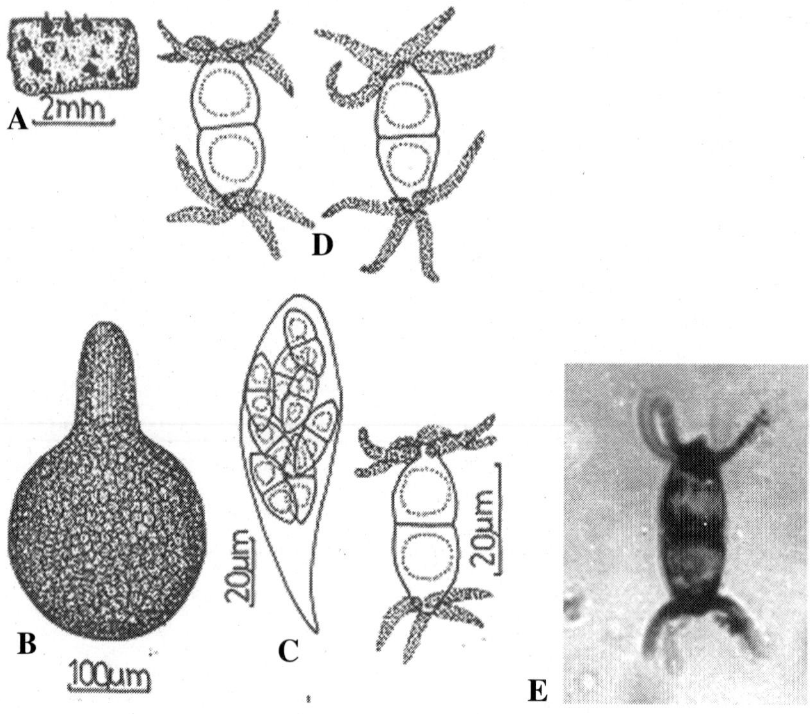

Fig. 46.1. ***Haiyanga salina,*** A-E; A- Habit, B- ascocarp, C- ascus, D- ascospores, E- ascospore 1000x).

47. ***Halographis*** Kohlm. & Volkm.-Kohlm., 1988. *Can. J. Bot..*, **66**: 1138.

Association: crustose, endolithic, inconsoicuous. Hyphae: forming a network in calcareous substrate, surrounding cyanobacterial cells without producing haustoria. *Ascomata*: lirelliform, immersed, opening with a longitudinal slit, light coloured with a dark brown margin. *Peridium*: composed of loosely interwined hyphae embedded in a gelatinous matrix, around the ostiolar slit forming a dark textura prismatica. *Hamathecium*: paraphysoid, branched and anastomosing, surrounded by gelatinous matrix. *Asci*: 8-spored, clavate, thick-walled, bitunicate, nonamyloid.

Ascospores: ellipsoidal, one-septate, hyaline. *Pycnidia*: (spermogonia?): subglobose or elongate, immersed with round or cleft-like ostiole, light brown. *Conidiogenous cells*: cylindrical, phialidic. *Conidia*: one-celled, hyaline. The genus is monotypic (Jones *et al*., 2009a).

Type species: *Halographis runica* Kohlm. & Volkm.-Kohlm.
Habitat: On shells of snails, worm tube, and coral rock.
Description: Based on Kohlmeyer and Volkmeyer-Kohlmeyer (1988a).

47.1. *Halographis runica* Kohlm. & Volkm.-Kohlm., 1988a. *Can. J. Bot*.., **66**: 1138. [For Fig. refer original paper]

Association: crustose, endolithic, inconsoicuous. Hyphae: 1.5-2.5 µm diam, forming a network in calcareous substrate, surrounding cyanobacterial cells without producing haustoria. *Ascomata*: 160-210 µm high, 140-200 µm wide, 1-1.5 µm long, lirelliform, simple or irregularly branched, completely immersed in calcareous substrates, opening with a longitudinal slit, light coloured with a dark brown margin, without periphyses, single or gregarious, surrounded by cyanobacteria. *Peridium*: 25-35 µm thick on the sides and base, 45-75 µm around the ostiole, along the sides and base composed of loosely interwined hyphae that are embedded in a gelatinous matrix, light brown, around the ostiolar slit forming a dark *textura prismatica*. *Hamathecium*: paraphysoid, 1-1.5 µm in diam, hyaline, branched and anastomosing, surrounded by gelatinous matrix, forming a network that tightly encloses the asci. *Asci*: 60-100 x 15-20 µm, 8-spored, clavate, thick-walled, bitunicate, pedunculate, nonamyloid, without apical apparatus, but with an ocular chamber, arise from an ascogenous tissue at the base of the locule. *Ascospores*: 17-29 x 6-9 µm, ellipsoidal to fusiform, one-septate, not constricted at the septum, hyaline. *Pycnidia*: (spermogonia?): 95-140 µm high, 100-180 µm in diam, subglobose or elongate, immersed with round or cleft-like ostiole, light brown, gregarious. *Peridium*: like that of ascomata. *Conidiogenous cells*: 8-11 x 1-2 µm, cylindrical, phialidic, lining inside of the peridium. *Conidia*: 2-3 x 1-1.5 µm, ellipsoidal, one-celled, hyaline.

Habitat: On shells of snails, worm tube, and coral rock.
Description: Based on Kohlmeyer and Volkmann-Kohlmeyer (1988a).
Distribution: West Coast:-Karanataka: On intertidal leaf and woody litter of mangroves (Ananda and Sridhar, 2004).

48. ***Halomassarina*** Suetrong, Sakayaroj, E.B.G. Jones, Kohlm., Volkm.- Kohlm. & C.L. Schoch, 2009. *Studies in Mycology*, **64**: 161.

Ascomata: subglobose to pyriform, immersed or erumpent, ostiolate, periphysate, papillate or apapillate, clypeate, coriaceous, brown, single. *Peridium:* of flattened, thick-walled cells, forming a *textura angularis. Hamathecium:* of simple, rarely anastomosing pseudoparaphyses. *Asci:* 8-spored, cylindrical to clavate, pedunculate, thick-walled, fissitunicate, with ocular chamber but without apical apparatus, I-negative. *Ascospores: distichous, ellipsoidal, 3-septate, hyaline, surrrounded by a gelatinous sheath.*

Type species: *Halomassarina thalassiae* (Kohlm. & Volkm.-Kohlm.) Suetrong, Sakayaroj, E.B.G. Jones, Kohlm., Volkm.-Kohlm. & C.L. Schoch
Habitat: On intertidal and subtidal mangrove wood or fishing crafts.
Description: Based on Suetrong et al. (2009).

48.1 ***Halomassarina thalassiae*** (Kohlm. & Volkm.-Kohlm.) Suetrong, Sakayaroj, E.B.G. Jones, Kohlm., Volkm.-Kohlm. & C.L. Schoch, 2009. *Studies in Mycology*, **64**: 161. [Fig. 48.1]

Basionym: *Massarina thalassiae* Kohlm. & Volkm.-Kohlm., 1987a, *Canad. J. Bot.*, **65**: 575.

Ascomata: 270-425 x 240-335 µm, subglobose to pyriform, immersed in the substrate, ostiolate, periphysate, epapillate or short papillate, clypeate, coriaceous, brown. *Necks*: 140-150 x 150-180 µm, ostiolar canal filled with short hyaline periphyses. *Peridium*: 12-16 µm thick, composed of flattened thick-walled cells. Pseudoparaphyses: 1.5-2 µm wide, simple, rarely anastomosing, septate. *Asci*: 100-160 x 28-34 µm, 8-spored, cylindrical to clavate, short pedunculate, thick-walled, fissitunicate, with a distinct ocular chamber, arising from a basal ascogenous tissue. *Ascospores*: 34-49 x 9-13 µm, distichously in the ascus, ellipsoidal, at frist 1-septate, later 3-septate, constricted at the central septum, surrounded by a gelatinous sheath, senescent ascospores become brownish and verruculose.

Habitat: On intertidal mangrove wood.
Description: Based on Kohlmeyer and Volkmann-Kohlmeyer (1987a).
Distribution: West Coast:-Gujarat: On intertidal wood (Patil and Borse, 2001); **Maharashtra:** On intertidal wood of *Sonneratia apetala* (Borse and Shrivastava, 1994; Borse et al., 2005a); **Lakshadweep Islands:** On intertidal wood of *Pemphis acidula*, wood and pneumatophores of *Ceriops tagal* (Chinnaraj, 1992).

East Coast:-Tamil Nadu: On mangrove wood (Kohlmeyer and Kohlmeyer, 1987), on submerged parts of *Rhizophora* spp. (Ravikumar and Vittal, 1996); **Andhara Pradesh:** On intertidal wood of *Aegiceras corniculatum, Excoecaria agallocha* and *Rhizophora apiculata* (Sarma and Vittal, 1998-99; Sarma and Vittal, 2000; Sarma and Vittal, 2001; Sarma et al., 2001; Sarma and Vittal, 2004; Vittal and Sarma, 2005); **West Bengal:** On intertidal wood of *Avicennia marina* (Pawar et al., 2005); **Andaman & Nicobar Islands:** On intertidal wood of *Bruguiera gymnorhiza* and *Rhizophora mucronata* Chinnaraj (1993a).

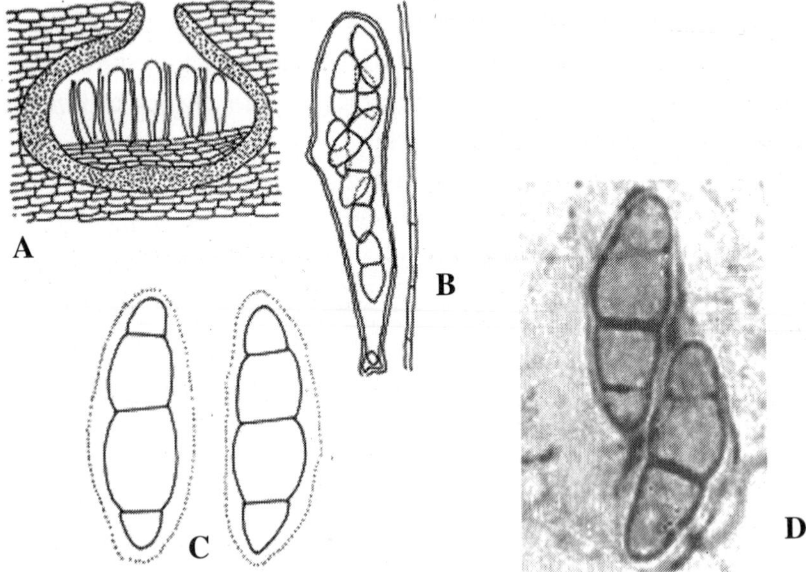

Fig. 48.1 ***Halomassarina thalassiae,*** A-D; A- v.s. of ascocarp (100x), B- ascus and pseudoparaphysis (450x) , C- ascospores (1000x), D- ascospores (1000x).

49. ***Halorosellinia*** Whalley, E.B.G. Jones, K.D. Hyde & Laessoe, 2000. *Mycol. Res.*, **104**: 308.

Pseudostromata: poorly developed and lacking extractable pigments. *Ascomata*: immersed in pseudostroma. subglobose or hemispherical, soft to leathery, black, ostiole papillate. *Asci*: 8-spored, cylindrical, unitunicate, with relatively large apical apparatus, rk blue in Melzer's reagent, datapering with a distinct apical rim. *Ascospores*: unicellular, dark brown, ellipsoidal, prominent straight germ slit on ventral side, smooth walled. The genus is monotypic (Jones *et al.*, 2009a).

Type species: *Halorosellinia oceanica* Whalley, E.B.G. Jones, K.D. Hyde & Laessoe

Habitat: On intertidal mangrove wood.

Description: Based on Whalley et al. (2000).

49.1. *Halorosellinia oceanica* (S. Schatz) Whalley, E.B.G. Jones, K.D. Hyde & Laessoe, 2000. *Mycol. Res.,* **104:** 368. [Fig. 49.1]

Basionym: *Hypoxylon oceanicum* S. Schatz, *Mycotaxon,* **33:** 413 (1988).

Pseudostromata: seated on decorticated wood, occasionally embedded at the base, pulviante to hemispherical, 0.4-0.8 µm diam., single, or in clusters, linear to suborbicular, surface leathery in fresh material, when young covered with a whitish hyphal layer bearing the anamorph, at maturity black, generally with conspicuous ascomatal projections. In section pseudostromata comprising host cells filled with light brown fungal cells in the form of *textura globulosa* or amorphous black fungal material. *Ascomata*: 0.4-0.6 µm in diam, immersed in pseudostroma, subglobose to hemispherical, soft to leathery, black, ostioles papillate. *Peridium*: 25-35 µm wide, comprising tissue of *textura porrecta*, fusing at the outside with the pseudostromata. *Paraphyses*: 2-2,5 µm wide at the base, abundant, persistent, remotely septate. *Asci*: 177-219 µm long, spore-bearing part 135-140 µm long, stipe 37-79 µm long, 8-spored, cylindrical, unitunicate, subapical apparatus dark blue in Melzer's reagent, tapering with a distinct apical rim. 5.6-6.6 x 4.2-4.7 µm. *Ascospores*: 19-26 x 7.5-13 µm, uniseriate to obliquely uniseriate or partially biseriate at the upper end of the ascus, dark grey-olive to opaque brown, more or less inequilaterally ellipsoid, ventral side varying in degree of convex curvature, upper end broadly rounded, lower end slightly pointed, one-celled, biguttulate, wall smooth and relatively thick, without appendages, germination slit usually clearly seen on the venral side, straight, conspicuous, ½- ¾ total length of the spore.

Habitat: On mangrove wood.

Description: Based on Schatz (1988) and Whalley et al. (2000).

Distribution: West Coast:-Gujarat: On intertidal wood of *Avicennia marinat* (Borse et al., 2000a), on intertidal wood and mangrove wood (Patil and Borse, 2001); **Maharashtra:** On intertidal wood and mangrove wood of *Avicennia marina, Rhizophora mucronata* and *Sonneratia alba* (Borse, 2000b), on intertidal mangrove wood (Maria and Sridhar, 2002a); **Goa:** On intertidal mangrove wood (Chinnaraj and Untawale, 1992), on intertidal wood (Prasannarai and Sridhar, 2001); **Karnataka:** On intertidal mangrove wood (Chinnaraj and Untawale, 1992), on intertidal wood (Prasannarai and Sridhar, 1997, 2001, 2000-

2001), on animal subsrate (Ananda and Sridhar, 2001), on intertidal mangrove wood on intertidal wood of *Brugueiria gymnorrhiza, Rhizophora mucronata* and *Sonneratia caseolaris* (Maria and Sridhar, 2002a, 2003a); **Pondicherry (Mahe):** On intertidal wood (Nambiar and Raveendran, 2007; 2008a); **Kerala:** On intertidal mangrove wood (Maria and Sridhar, 2002), on intertidal wood (Raveendran and Manimohan, 2007), on intertidal wood, mangrove wood of *Avicennia* sp., *Bruguiera* sp. and *Rhizophora* sp. (Nambiar and Raveendran, 2006, 2007, 2008a, b, d, 2009a, d, e, f, g, h; Nambiar et al., 2006). **East Coast:-** On intertidal mangrove wood (Ravikumar, 1991); on intertidal wood (Prasannarai and Sridhar, 2001; Nambiar et al., 2008b), on mangrove wood (Nambiar and Raveendran, 2008e, 2009k, l); **Andhara Pradesh:** On intertidal wood of *Avicennia marina, A. officinalis, Excoecaria agallocha, Rhizophora apiculata* and *Sonneratia apetala* (Sarma and Vittal, 1998-99; Sarma and Vittal, 2000; Sarma and Vittal, 2001; Sarma et al., 2001; Sarma and Vittal, 2004; Vittal and Sarma, 2005), on intertidal wood (Nambiar and Raveendran, 2009j); **West Bengal:** On intertidal wood of *Rhizophora mucronata* (Pawar et al., 2009); **Orissa:** On intertidal wood of *Aegiceras corniculatum, Avicennia marina, A. officinalis, Excoecaria agallocha, Heritiera fomes* and *Kandelia candel* (Borse et al., 2009); **Andaman & Nicobar Islands:** On intertidal wood of *Aegiceras corniculatum, Avicennia marina, Excoecaria agallocha* and *Rhizophora mucronata* (Chinnaraj and Untawale, 1992), on intertidal wood of *Avicennia marina, A. officinalis, Bruguiera gymnorhiza, Rhizophora apiculata, R. mucronata* and *Sonneratia alba* (Chinnaraj, 1993a).

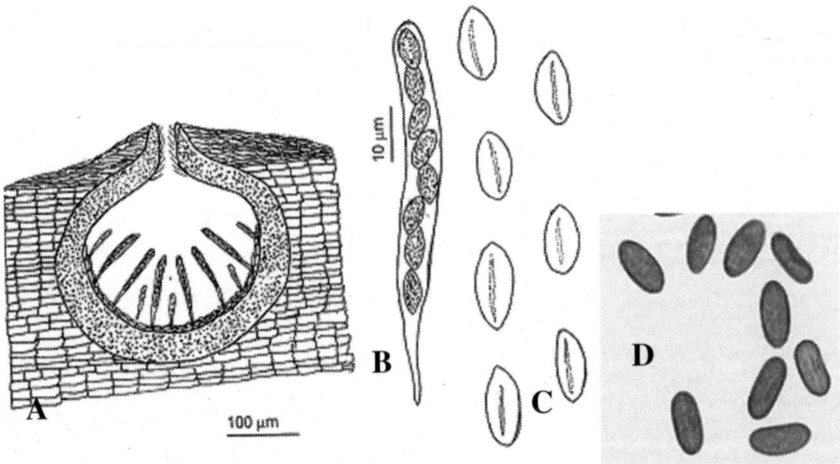

Fig. 49.1. *Halorosellinia oceanica,* A-D; A- v.s. of ascocarp, B- ascus, C- ascospores (1000x), D- ascospores (1000x).

50. *Halosarpheia* Kohlm. & E. Kohlm., 1977. *Trans. Br. Mycol. Soc.*, **68**: 208; Emend. Abdel-Wahab & Nagahama, Mycol. Progress, DOI 10.1007/s11557-010-0731-0

Ascomata: solitary or gregarious, obpyriform, conical, subglobose, immersed to erumpent, ostiolate, papillate, coriaceous, brown to black. Pseudoparenchyma of large, thin-walled cells filling the venter of young ascocarps, breaking up to form catenophyses. *Asci*: 8-spored, clavate, pedunculate, unitunicate, thin-walled with or without apical thickening, persistent or deliquesce early, developing at the base of the ascomatal venter. *Ascospores*: ellipsoidal, globose, subglobose, ovate, 0-1-septate, with one or two polar appendages. *Appendages*: coiled filaments enclosed in gelatinous membrane or amorphous material enclosed in cellular sheath that dissolves in water and the appendage swell to form huge tree-like appendage. The genus is represented by 9 species in marine habitats (Jones *et al.*, 2009). The genus *Halosarpheia sensu stricto* (Campbell *et al.* 2003; Jones *et al.*, 2009a) is represented by 4 species in marine habitats (Abdel-Wahab & Nagahama, 2011a)

Type species: *Halosarpheia fibrosa* Kohlm. & E. Kohlm.
Habitat: On mangrove wood.
Description: Based on Kohlmeyer and Kohlmeyer (1977) Abdel-Wahab and Nagahama (2011a).

Key to *Halosarpheia* species from India:

1. Ascospores less than 14 µm wide 3

1' Ascospores more than 14 um wide 2

2. Ascospores 32-44 x 18-24 µm, ... *Halosarpheia fibrosa*

2' Ascospores 23-32 x 14-18 µm, ...*Halosarpheia trullifera*

3 Ascospores less than 15 µm long (10-14 x 4-6 µm), ... *Halosarpheia minuta*

3' Ascospores more than 15 µm long (18-23 (-26) x 9-12 µm), ... *Halosarpheia marina*

50.1. ***Halosarpheia fibrosa*** Kohlm. & E. Kohlm., 1977. *Trans. Br.Mycol. Soc.*, **68**: 208. [Fig. 50.1]

Ascomata: 380-440 µm high, 340-450 µm diam, solitary or gregarious, obpyriform to subglobose, immersed part lighter coloured than the exposed neck and top, ostiolate, papillate, coriaceous, brown to black. *Peridium*: 40-60 µm thick, 2-layered, forming a *textura angularis*; outer stratum composed of 2-3 layers of polygonal or subglobose dark or light brown cells with large lumina; inner stratum composed of 7-10 layers of elongated, hyaline cells with narrow lumina, mearging with the pseudoparenchyma of the venter; brown hyphae are sometimes attached to the outside of ascomata. *Necks*: 140-530 µm long, 130-200 µm diam, at the base, 90-110 µm at the apex, subconical to cylindrical; ostiolar canal periphysate; periphyses 15-35 x 2-3 µm. *Pseudoparenchyma*: of large, thin-walled cells filling the venter of young ascomata, breaking up to form catenophyses. *Asci*: 160-220 x 34-46 µm, 8-spored, clavate, pedunculate, unitunicate, thick-walled below the apex, thin-walled in peduncle, persistent, without apical apparatuses, developing at the base of the ascomata venter on a small-celled acsogenous tissue which is separated from the peridium by thin-walled, polygonal cells of the sterile pseudoparenchyma; asci mature successively breaking off basally from the ascogenous tissue without releasing the ascospores. *Ascospores*: 32-44 x 18-24 µm (excluding appendages), broad ellipsoidal, 1-septate, not constricted at the septum, hyaline, with apical appendages. *Appendages*: exosporic remains, at first cap like, stiff and homogenous, attached to the ascospore apices, 3-5 µm thick, 6-8 µm diam, at maturity becoming soft and transforming into a coil of delicate fibres.

Habitat: On mangrove wood.

Description: Based on Kohlmeyer and Kohlmeyer (1977).

Distribution: West Coast:-Gujarat: On intertidal wood (Patil and Borse, 2001); **Maharashtra:** On intewrtidal wood, (Patil and Borse, 1982), on intertidal wood, *Avicennia marina* and prop roots of *Rhizophora mucronata* (Patil and Borse, 1986; Borse, 1987d, 1988; Borse et al., 2005a), on intertidal mangrove wood (Maria and Sridhar, 2002a); **Goa:** On intertidal wood (Prasannarai and Sridhar, 2001); **Karnataka:** On intertidal mangrove wood on intertidal wood of *Avicennia officinalis, Brugueiria gymnorrhiza,* and *Sonneratia caseolaris* (Maria and Sridhar, 2002a, 2003a); **Kerala:** On intertidal mangrove wood (Maria and Sridhar, 2002a).

East Coast:-Tamil Nadu: On intertidal mangrove wood (Ravikumar, 1991); **West Bengal:** On intertidal mangrove wood (Pawar et al., 2005).

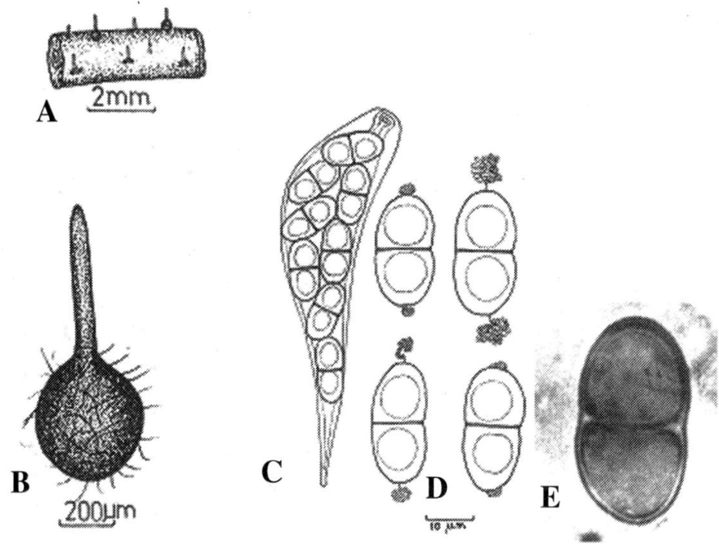

Fig. 50.1. ***Halosarpheia fibrosa,*** A-E; A- Habit, B- ascocarp, C- ascus (450x), D- ascospores (1000x), E- ascospore 1000x).

50.2. ***Halosarpheia marina*** (Cribb & J.W. Cribb) Kohlm., 1984. *Mar. Ecol. (P. S. Z. N. I.)*, **5**: 345. [Fig. 50.2]

Basionym: *Gnomonia marina* Cribb & J.W, Cribb, *Pap. Univ. Queensl. Dept. Bot.*, **3**: 100 (1965).

Ascomata: 140-300 µm in diam, bottle-shaped, mostly immersed, ostiolate, papillate, membranaceous, surrounded by brown hyphae 3-5 µm thick, sub-hyaline, light brown or fuscous, solitary or gregarious. *Necks*: 100-560 x 42-140 µm, lighter coloured than the ascomat, cylindrical, simple or rarely bifurcate, immersed or projecting over the surface of the substrate; ostiolar canal periphysate. *Paraphyses*: absent. *Asci*: 95-122 x 18-28 µm, 8-spored, clavate to subcylindrical, apically somewhat truncate, short pedunculate, unitunicate, thin-walled, but thickened at the apex with an apical pore, persistent, physoclastic. *Ascospores*: 18-23 x 9-12 µm (excluding appendages), cylindrical-ellipsoidal, apically rounded, 1-septate at the middle, not or slightly constricted at the septum, hyaline, with a

gelatinous apical appendage at each apex. *Appendages*: 1.5-3 µm long, apical or rarely subapical, cap like, round or apiculate, sometimes deciduous or deliquescing.

Habitat: On intertidal wood.

Description: Based on Kohlmeyer and Kohlmeyer (1979).

Distribution: West Coast:-Maharashtra: On intertidal wood and *Rhizophora mucronata* (Borse, 1987d, 1988; Borse et al., 2005a), intertidal wood of *Rhizophora mucronata, Sonneratia acida* (Borse, 1987d) **Karnataka:** On intertidal wood (Prasannarai and Sridhar, 2001, 2000-2001); **Pondicherry (Mahe):** On intertidal wood (Nambiar and Raveendran, 2007); **Kerala:** on intertidal wood (Prasannarai and Sridhar, 2001), on intertidal mangrove wood (Maria and Sridhar, 2002a), on intertidal wood (Raveendran and Manimohan, 2007), on intertidal wood and mangrove wood of *Avicennia* sp. and *Rhizophora* sp. (Nambiar and Raveendran, 2006, 2007, 2008b, d, 2009a, b, d, e, f, h, 2010b, c; Nambiar et al., 2006,); **Lakshadweep Islands:** On intertidal wood of *Ceriops tagal* (Chinnaraj, 1992).

East Coast:-Tamil Nadu: On wood of *Avicennia marina* and *Rhizophora mucronata* (Ravikumar and Vittal, 1987), On submerged parts of *Rhizophora* spp. (Ravikumar and Vittal, 1996), on mangrove wood (Nambiar and Raveendran, 2008e); **Andhara Pradesh:** On intertidal wood of *Avicennia officinalis* and *Rhizophora apiculata* (Sarma and Vittal, 1998-99; Sarma and Vittal, 2000; Sarma and Vittal, 2001; Sarma et al., 2001; Sarma and Vittal, 2002; Sarma and Vittal, 2004; Vittal and Sarma, 2005).

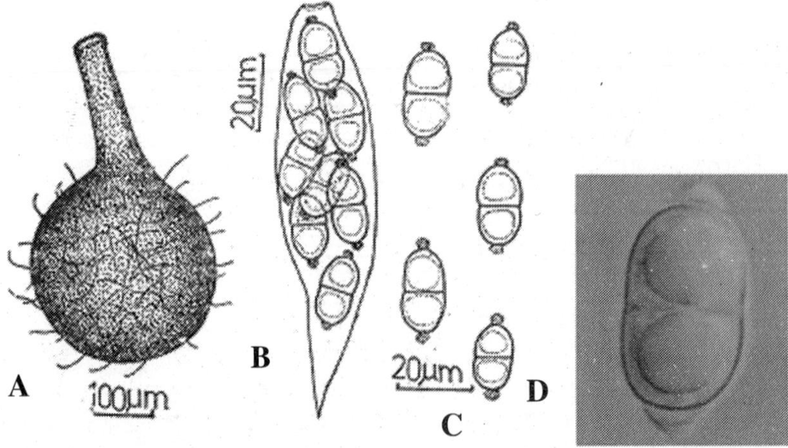

Fig. 50.2. ***Halosarpheia marina,*** A-D; A- ascocarp, B- ascus, C- ascospores, D- ascospore (1000x).

50.3. *Halosarpheia minuta* W.F. Leong, 1991. *Can. J. Bot..*, **69**: 883.

[Fig. 50.3]

Ascomata; 214-437 µm high, 175-318 µm in diam, globose to ellipsoidal, main axis horizontal with lateral, recurved neck, black, solitary, immersed to slightly superficialnon the intact bark or substratum, mambraneceous ostiolate, papillate. *Peridium*: thin, 6-10 µm thick, hyaline-brown, of elongate cells. In the upper region elongate cells are interdispersed amongst host cells. *Necks*: 143-954 µm long, 32-56 µm in diam, subhyaline, short or elongate, cylindrical, laterally inserted, with periphyses. *Asci*: 40-68 x 10-14 µm, 8-spored, cylindrical-clavate, pedunculate, unitunicate, persistent, apex flattened, slightly thickened, refractive, with an indistinct apical pore, developed in a hymenium at the base of the ascomata, spores irregularly arranged. *Ascospores*: 10-14 x 4-6 µm, ellipsoidal, hyaline, 2-celled, median septate, not or slightly constricted at the septum, thin-walled, smooth, with apical appendages. *Appendages*: terminal, caplike and stiff, 6.3-7.5 x 1 µm unferling in water into long, thin filament.

Habitat: On bark of intertidal submerged mangrove wood.

Description: Based on Leong et al. (1991).

Distribution: **West Coast:-Maharashtra:** On bark of intertidal mangrove (*Rhizophora mucronata*) wood (Borse, 2000b; Borse et al., 2005a); **Pondicherry (Mahe):** On intertidal wood (Nambiar and Raveendran, 2007); **Kerala:** On intertidal wood (Raveendran and Manimohan, 2007), on intertidal mangrove wood of *Avicennia* sp. (Nambiar and Raveendran, 2006, 2008b, d, 2009a, d, e, f, g, h, 2010c).

East Coast:-Tamil Nadu: On mangrove wood (Nambiar and Raveendran, 2008e); **Andhara Pradesh:** On intertidal wood of *Avicennia marina* and *A. officinalis* (Sarma and Vittal, 1998-99; Sarma and Vittal, 2000; Sarma and Vittal, 2001; Sarma et al., 2001; Sarma and Vittal, 2004; Vittal and Sarma, 2005); **Andaman & Nicobar Islands:** On intertidal wood of *Avicennia marina* and *Rhizophora mucronata* (Chinnaraj, 1993a).

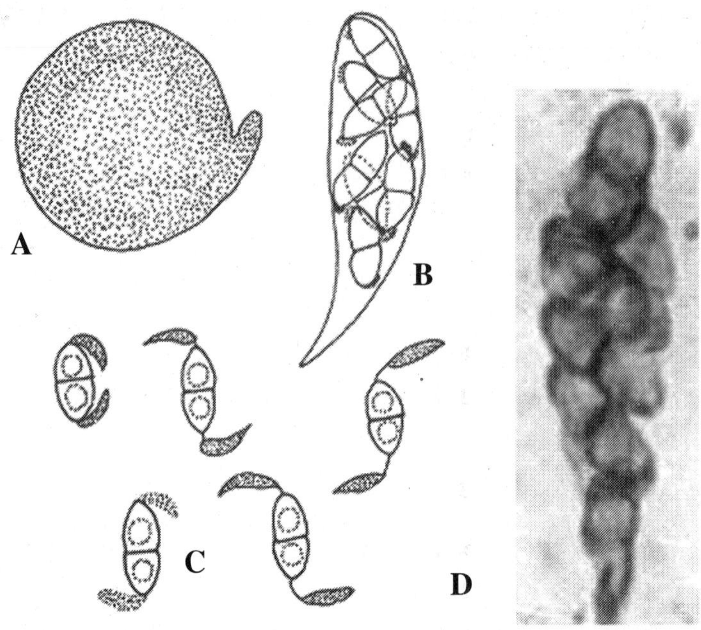

Fig. 50.3. *Halosarpheia minuta,* A-D; A- ascocarp (100x), B- ascus (1000x), C- ascospores (1000x), D- ascus (1000x).

50.4. *Halosarpheia trullifera* (Kohlm.) E.B.G. Jones, S.T. Moss & Cuomo, 1983. *Trans. Br. Mycol. Soc.*, **80:** 200. [For Fig. refer original paper]

Basionym: *Remispora trullifera* Kohlmeyer, *Nova Hedwigia*, **6:** 321 (1963).

≡ *Halosphaeria trullifera* (Kohlm.) Kohlm., *Can. J. Bot.*, **50:** 1957 (1972).

Ascomata; 200-745(-820) µm in diam, globose, immersed or becoming exposed, ostiolate, papillate, coriaceous, cream-coloured or brown, rarely dark brown or black, sometimes light coloured below, darker above, surrounded by brown hyphae. *Peridium*: 27.5-41 µm thick, consisting on the outside of 5 or 6 layers of yellowish, polygonal, thick-walled cells, forming a textura angularis, mearging into the thin-walled pseudoparenchyma of the venter. *Necks*: 240-680 µm long, 61-172 µm in diam, subconical to cylindrical, dark near the venter, lighter coloured at the apex; ostiolar canal periphysate. *Pseudoparenchyma*: of thin-walled, large, polygonal cells filling venter of young ascocarps; eventually breaking up into catenophyses. *Asci*: 118-182 x 29-40.5 µm, 8-spored, clavate, basely attenuate and pedunculate, unitunicate, thin-walled, aphysoclastic, without apical apparatuses,

early deliquescing, developing at the base of the ascocarp venter on a small celled ascogenous tissue. *Ascospores*: (23-)24-32.5(-35.5) x 14-17.5(-20) µm (excluding appendages), ellipsoidal or oblong ellipsoidal, one-septate, not or slightely constricted at the septum, hyaline (or pale grayish-brown?), appendaged. *Appendages*: at each end a scoop-shaped, terminal appendage, 3.5-4.5 µm high, 6-10 µm in diam., developing from the exospore, finally becoming viscous and forming long threads.

Habitat: On intertidal wood.

Description: Based on Kohlmeyer and Kohlmeyer (1979).

Distribution: **East Coast:-Tamil Nadu:** On mangrove wood (Ravikumar, 1991).

51. ***Halosphaeria*** Linder, 1944. *Farlowia*, **1:** 412.

Ascomat: globose or subglobose, ellipsoidal, immersed erumpent, ostiolate, papillate, coriaceous or membranous, pale to dark brown to black, solitary or gregarious, catenophyses deliquescing. *Asci*: 8-spored, clavate or subfusiform, pedunculate, unitunicate, thin-walled, deliquescing early. *Ascospores*: ellipsoidal, 1-septate, hyaline, with 3-4 spoon-shaped equatorial appendages and one at each end of the spore. Kohlmeyer (1972) referred 12 species to this genus but subsequent studies have assigned them to other genera. Pang et al. (2004) and Sakayaroj et al. (2005) accepted only one species in this genus.

Type species: *Halosphaeria appendiculata* Linder

Habitat: Saprobic on intertidal wood.

Description: Based on Kohlmeyer and Kohlmeyer (1979) and Jones et al. (2009a).

51.1. ***Halosphaeria appendiculata*** Linder, 1944. *Falowia*, **1:** 412.

[For Fig. refer original paper]

Ascomata: 140-330 µm high, 140-500 µm in diameter, globose or subglobose, ellipsoidal, immersed erumpent, ostiolate, papillate, coriaceous or membranous, pale to dark brown to black, solitary or gregarious, catenophyses deliquescing. *Asci*: 50-108 x 10-24 µm, 8-spored, clavate or subfusiform, pedunculate, unitunicate, thin-walled, deliquescing early. *Ascospores*: (16-) 18-29 x (6-) 8-12 µm, ellipsoidal, 1-septate, hyaline, with 3-4 spoon-shaped equatorial appendages and one at each end of the spore.

Habitat: On intertidal wood.
Description: Based on Kohlmeyer and Kohlmeyer (1979).
Distribution: West Coast:-Kerala: On intertidal wood (Nambiar and Raveendran, 2011a).

52. *Halotthia* Kohlm., 1963. *Nova Hedwigia*, **6**: 9.

Stromata: black, subepidermal. *Ascomata*: solitary, gregarious or confluent, conical or semiglobose, at first immersed, then erumpent, ostiolate, epapillate, carbonaceous, thick-walled, black. *Pseudoparaphyses*: numerous, persistent, ramose. *Asci*: 8-spored, cylindrical, pedunculate, bitunicate, thick-walled, persistent, developing in the angle between covering and basal wall of the ascomata. *Ascospores*: ellipsoidal, 1-septate, brown, thick-walled, septum with a central pore. *Spermagonia*: associated with ascomata, gregarious or confluent, scutellate, immersed, then erumpent, ostiolate, epapillate, carbonaceous, black, irregularly chambered. *Spermatiophores*: obclavate, simple. *Spermatia*: subglobose to ellipsoidal, 1-septate, hyaline. The genus is monotypic (Jones *et al.*, 2009a).

Type species: *Halotthia posidoniae* (Durieu & Montangne) Kohlm.
Habitat: On marine phanerograms (*Posidonia*).
Description: Based on Kohlmeyer (1963).

52.1. *Halotthia posidoniae* (Durieu & Mont.) Kohlm., 1963. *Nova Hedwigia*, **6**: 9. [For Fig. refer original paper]

Stromata: 55-110 µm thick, black, subepidermal in cortex of the foliated rhizomea of the host. *Ascomata*: 770-1075 µm high, 1.5-2.1 µm in diam, broadly conical or semiglobose, flattened at the base, enclosed in a stroma, at first immersed, then erumpent, ostiolate, epapillate, carbonaceous, thick-walled, black, solitary, gregarious or confluent. Peridium: 230-285 µm thick at the top. 165-275 µm thick in the sides, plectenchymatous, enclosing cells of the host. Necks: absent, ostiole 110-140 µm in diam toward the outside; canal obconical, 90-110 µm in diam near the cavity, filled with tips of entwined pseudoparaphyses. *Pseudoparaphyses*: 1.5-2 µm in diam, septate, numerous, persistent, ramose. *Asci*: 275-290 x 25-35 µm, 8-spored, cylindrical, attenuate at the base, short pedunculate, bitunicate, thick-walled, physoclastic, persistent, with a canal in the tip, developing in the angle between covering and basal wall of the ascomata. *Ascospores*: 37-70 x 16-26 µm, ellipsoidal, 1-septate, brown, thick-walled,

septum with a central pore. *Spermagonia*: associated with ascomata, gregarious or confluent, scutellate, immersed, then erumpent, ostiolate, epapillate, carbonaceous, black, irregularly chambered. *Spermatiophores*: obclavate, simple. *Spermatia*: subglobose to ellipsoidal, 1-septate, hyaline.

Habitat: On marine phanerograms (*Posidonia*).
Description: Based on Kohlmeyer and Kohlmeyer (1979).
Distribution: West Coast:-Karanataka: On intertidal wood (Prasannarai et al., 1999), on intertidal wood ((Prasannarai and Sridhar, 2001).
Note: the present fungus is described from the rhizomes of *Posidonia oceanica* (Kohlmeyer and Kohlmeyer, 1963), appears to be host specific as is *Ponoporia biturbinata* (Durieu and Montagne) Kohlmeyer (Jones et al., 2009a).

53. *Heleococcum* C.A. Jorgnesen, 1922. Bot. Tidsskr., 37: 417.

Ascomata: golbose, superficial, nonostiolate, membranaceous, white to orange. *Peridium*: thin, transperant, pseudoparenchymatous, smooth, dissolving irregularly at maturity. *Paraphyses*: absent. *Asci*: 8-spored, globose, sessile, unitunicate, thick-walled, scattered irregularlyin the venter of ascomata. *Ascospores*: ellipsoidal, 1-septate, smooth, thin-walled, hyaline to orange.

Type species: *Heleococcum aurantiacum* Jorgnesen
Habitat: On wood or soil.
Description: Based on Kohlmeyer and Kohlmeyer (1979).

53.1. *Heleococcum japonense* Tubaki, 1967. *Trans. Mycol. Soc. Jpn*, **8**: 5. [For Fig. refer original paper]

Ascomata: 250-400 µm in diam, golbose, superficial, nonostiolate, cleistothecial, membranaceous, white to orange, or pale brown, gregarious. *Peridium*: thin, transperant, pseudoparenchymatous, composed of several layers of angular or rounded cells, smooth, dissolving irregularly at maturity. *Paraphyses*: absent. *Asci*: 44-50 x 30-40 µm, 8-spored, globose or subglobose, sessile, unitunicate, thin-walled, aphysoclastic, without apical apparatuses, scattered irregularly in the venter of ascomata, deliquescing at spore maturity. *Ascospores*: 18-21 x 10-13 µm, ellipsoidal or ovoidal, 1-septate in the middle, not constricted at the septum, smooth or slightly rough, thick-walled, hyaline. *Conidiophores*: 20-28 x 3-4 µm, mostly unbranched, erect, developing from aerial hyphae. *Conidia*: 8-12 x 3-4 µm,

ellipsoidal, hyaline, originating as blown-out ends of the conidiophores and accumulating in the loose clusters; imperfect state of the *Trichothecium* type.

Habitat: On intertidal wood test panels of *Abies firma* and balsa.

Description: Based on Tubaki (1967).

Distribution: East Coast:-Tamil Nadu: On wood of *Avicennia marina* (Ravikumar and Vittal, 1987), on submerged parts of *Rhizophora* spp. (Ravikumar and Vittal, 1996); **Andhara Pradesh:** On intertidal wood of *Avicennia marina* and *Rhizophora apiculata* (Sarma and Vittal, 1998-99; Sarma and Vittal, 2000; Sarma and Vittal, 2001; Sarma et al., 2001; Sarma and Vittal, 2002; Sarma and Vittal, 2004; Vittal and Sarma, 2005).

54. *Helicascus* Kohlm., 1969b. *Can. J. Bot..*, **47**: 1471.

Stromata: with several locules, black, immersed in tha substratum. *Ascomata*: depressed ampuliform, horizontally arranged under a black paeudoclypeus, united radially into valsoid groups; ostioles connected into a common, central, periphysate pore. *Pseudoparaphyses*: numerous, persistent, delicate threads. *Asci*: 8-spored, subcylindrical to oblong clavate, pedunculate, bitunicate, thick-walled, physoclastic, with apical apparatuses; ectoascus forming a third layer at the base; endoascus swelling, stretching and coiling at maturity. *Ascospores*: obovoid, 1-septate, constricted at the septum, brown; walls two layered; at first covered by a gelatinous sheath.

Type species: *Helicascus kanaloanus* Kohlm.

Habitat: On intertidal mangrove wood.

Description: Based on Kohlmeyer (1969b).

54.1. *Helicascus kanaloanus* Kohlm., 1969b. *Can. J. Bot..*, **47**: 1471.

[For Fig. refer original paper]

Stromata: 0.6-0.78 µm high, 1.25-2.75 µm wide, lenticular, immersed, black, carbonaceous, enclosing 3-4(-5) loculi, covered by a pseudoclypeus. *Ascomata*: 235-270 µm high, 440-800 µm long, depressed ampulliform, united radially into valsoid groups, horizontally arranged under a black pseudoclypeus; ostioles 70-170 µm in diam., united into a common, central pore, stromatic tissue aroud the loculi composed of more or less isodimetric cells, at the periphery composed of vertically arranged, longitudinal cells; pseudoclypeus composed of host cells

enclosed in black fungal hyphae partitions between loculi formed of brown, isodimetric or elongated cells. *Necks*: 270-435 µm high, 255-300 µm in diam., canal filled with periphyses. *Hamathecium*: hyphae-like, numerous, persistent, filamentous, anastomosing. *Asci*: 250-335 x 25-30 µm, 8-spored, subcylindrical, finally oblong clavate, pedunculate, bitunicate, thick-walled, apically multilayered, with an ocular chamber; at maturity endoascus swelling in water and becoming coiled, finally stretching and pushing the ascus into the ostiolar canal. *Ascospores*: 30-55 x 17-25 µm, uniseriate, obovoidal, unequally two celled, highly constricted at the septum, dark brown, sometimes at one or both ends apiculate with a deliquescing sheath.

Habitat: Saprobic on intertidal mangrove wood.

Description: Based on Kohlmeyer & Kohlmeyer (1979) and Hyde (1991a).

Distribution: East Coast:-Tamil Nadu: On mangrove wood (Ravikumar, 1991); **Andaman & Nicobar islands:** On intertidal wood of *Rhizophora mucronata* and *Sonneratia alba* (Chinnaraj, 1993a).

Note: The genus *Helicascus* differ from *Salsuginea* Hyde mainly in the morphology of the fruiting body and ascospores. The collections named as *Helicascus kanaloanus* in the publications of Patil and Borse (1985), Borse et al. (1988), Ramesh and Borse (1989), Srivastava (1989) belongs to *Salsuginea ramicola* Hyde. Ravikumar (1991) and Chinnaraj (1993a) recorded the present fungus on mangrove wood without details of ascomata, description and illustrations. It appears to be misidentification.

55. *Julella* Fabre, 1879. *Ann. Sci. Nat. Bot.*, **9:** 113.

Ascomata: immersed, medium sized with a pseudoparenchymatous peridium that is thickened above by clypeal tissues. *Asci*: basal and interspersed by cellular pseudoparaphyses that branch above asci in a gelatinous matrix *Ascospores*: hyaline, with numerous transverse and longitudinal septa, surrounded by a gelatinous coating. The genus is represented by two terrestrial and two marine species (Aptroot and Boom, 1995; Jones et al., 2009a).

Type species: *Julella buxi* Fabre

Habitat: Saprophytic on bark, intertidal wood, mangrove wood, and senescent leaves of *Juncus* sp.

Description: Based on Barr (1985) and Aptroot and van den Boom (1995).

55.1. *Julella avicenniae* (Borse) K.D. Hyde, 1992a. *Mycol. Res.*, **96**: 939.

[Fig. 55.1]

Basionym: *Pleospora avicenniae* Borse, *Curr. Sc.,* **56:** 1109 (1987b).

Stromata: appearing on the host surface as irregular dome-shaped blackened spots, 460-650 μm in diam, with a central ostiole, solitary or in groups of 2-3 and with immersed ascomata. *Ascomata*: 180-380 μm high, 130–180 μm in diam, globose or subglobose, immersed beneath a clypeus, membranous, ostiolate. *Peridium*: up to 30 μm wide, inwardly hyaline, outwardly light-brown, a single stratum of elongated cells, fusing at the outside with the host tissue. Some cells at the sides are angular. *Hamathecium*: composing pseudoparaphyses which are filamentous, numerous, septate, hyphae-like, narrow, anastomosing and in a gel. *Asci*: 130–200 x 26-31 μm, 8-spored, clavate, pedunculate, thick-walled, bitunicate, fissitunicate, apically rounded, with an ocular chamber surrounded by a faint ring, and developing from the base of the ascoma. *Ascospores*: 28-36 x 12-16 μm, hyaline, with a central septum when young, brown, with (-6)7-transverse septa when mature, constricted particularly at the central septum, with up to 3 longitudinal septa, ellipsoidal and surrounded by a large spreading sheath.

Habitat: On intertidal wood and mangrove wood.

Description: Based on Borse (1987b) and Hyde (1992a).

Distribution: West Coast:-Diu Island: On intertidal wood of *Avicennia marina* (Borse et al., 1999b); **Gujarat:** On intertidal wood of *Avicennia marina* (Borse et al., 2000a), on intertidal mangrove wood (Patil and Borse, 2001); **Maharashtra:** On intertidal roots and pneumatophores of *Avicennia marina* (Borse, 1987b, 1988; Borse *et al.,* 2005a), on intertidal wood (Shrivastava, 1995); **Goa:** On intertidal wood (Prasannarai and Sridhar, 2001); **Karnataka:** On intertidal wood (Prasannarai and Sridhar, 2001), **Kerala:** On mangrove woody litter (Shini et al., 2009-10).

East Coast:-Tamil Nadu: On mangrove wood (Ravikumar, 1991); on intertidal woody debris (Ravikumar et al., 2009); **Andhara Pradesh:** On intertidal wood of *Avicennia marina,* and *A. officinalis* (Sarma and Vittal, 1998-99; Sarma and Vittal, 2000; Sarma and Vittal, 2001; Sarma et al., 2001; Sarma and Vittal, 2004; Vittal and Sarma, 2005); **West Bengal:** On intertidal wood of *Avicennia marina* (Pawar et al., 2009); **Orissa:** On intertidal wood of *Avicennia marina* (Borse et al., 2009); **Andaman & Nicobar Islands:** On intertidal wood of *Avicennia marina* (Chinnaraj, 1993a).

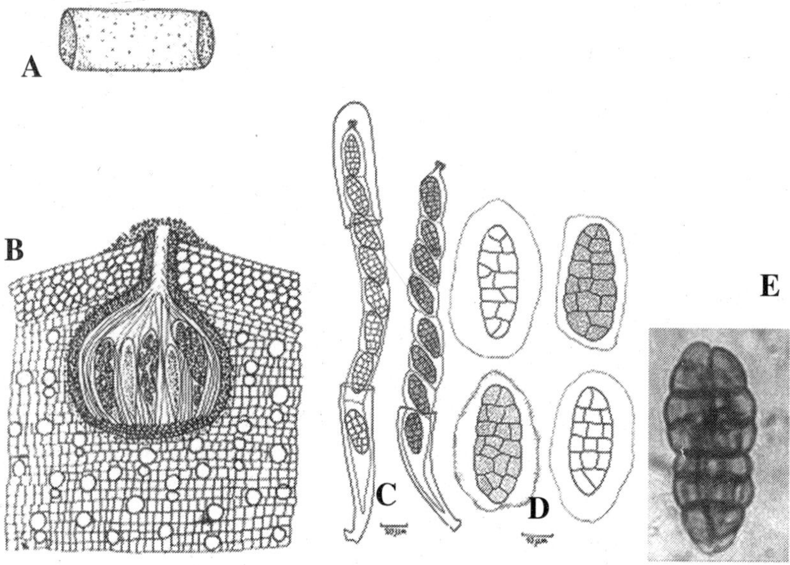

Fig. 55.1. ***Julella avicenniae,*** A-E; A- Habit, B- v.s. of ascocarp (100x), C- asci, D- ascospores, E- ascospore 1000x).

56. ***Kallichroma*** Kohlm. & Volkm.-Kohlm., 1993b. *Mycol. Res.*, **97**: 759.

Ascomata: depressed subglobose to ellipsoidal, at first immersed, erumpent, ostiolate, periphysate (periphyses surrounded by gelatinous matrix), epapillate, indistinctly clypeate, fleshy-leathery, orange-yellowish by carotenoid-containing lipid globules, gregarious or frequently confluent; the clypeoid tissue extends from the peridium around the ostiole and often connects adjoining ascomata. *Peridium*: hyaline, three–layered; on the outside composed of polygonal cells with large lumina, central part of thick-walled cells with small lumina, on the inside of elongate, flattened thick–walled cells forming a *textura angularis*. *Hamathecium*: composed of apical paraphyses, filamentous, septate, simple, attached to the dome of the venter, merging with periphyses. *Asci*: 8-spored, subcylindrical to clavate, short pedunculate, thin-walled at maturity except for the narrow, thick-walled apex, persistent, without apical apparatus, maturing successively on the ascogenous tissue at the bottom of the locule. *Ascospores*: biseriate, ellipsoidal, 1-septate in the middle, hyaline, longitudinally striate by thin ribs or smooth, with or without an early dissolving mucilaginous sheath. The genus is presented by two marine species (Jones et al., 2000a).

Type species: *Kallichroma tethys* (Kohlm.) Kohlm. & Volkm.-Kohlm.
Habitat: Saprobic on wood in marine habitata.
Description: Based on Kohlmeyer and Volkmann-Kohlmeyer (1993b).

Key to *Kallichroma* species from India:

1. Ascospore wall smooth .... *Kallichroma glabrum*

1. Ascospore wall with longitudinal ridges .... *Kallichroma tethys*

56.1. ***Kallichroma glabrum*** (Kohlm.) Kohlm. & Volkm.-Kohlm., 1993b.

Mycol. Res., **97:** 759. [Fig. 56.1]

Basionym: *Hydronectria tethys* var. *glabra* Kohlm. *Mar.Ecol. (P.S.Z.N. I),* **5:** 351 (1984).

Ascomata, Peridium, and *Asci* are similar to *K. tethys* (Kohlmeyer and Kohlmeyer) Kohlmeyer and Volkmann-Kohlmeyer. *Ascospores*: 17-24 (-27) x 8-10 µm, ovoid, 1-septate, somewhat below the center, slightly constricted at the septum, hyaline, smooth.

Habitat: Saprobic on wood in marine habitats.
Description: Based on Kohlmeyer (1984).
Distribution: West Coast:-Gujarat: On intertidal wood and mangrove wood (Patil and Borse, 2001); **Maharashtra:** On intertidal mangrove (*Avicennia marina*) wood (Borse, 2000b; Borse et al., 2005a).
East Coast:-Orissa: On intertidal wood of *Avicennia marina* (Borse et al., 2005b); **Andaman & Nicobar Islands:** On intertidal wood of *Avicennia marina* (Chinnaraj and Untawale, 1992).

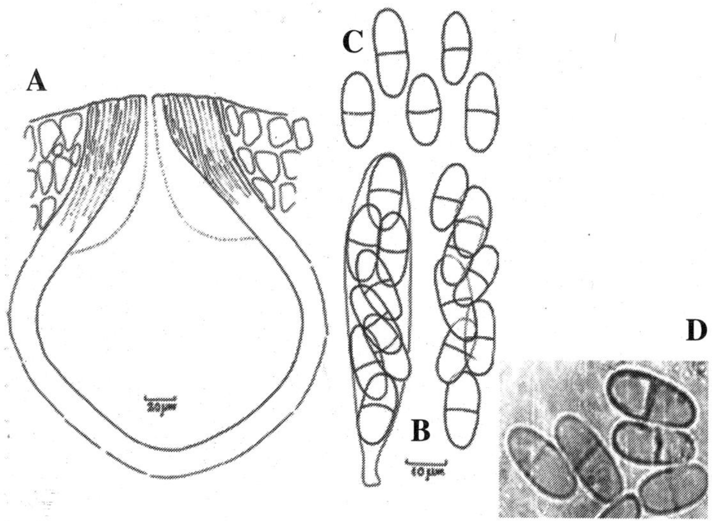

Fig. 56.1. ***Kallichroma glabrum,*** A-D; A- v.s. of ascocarp, B- ascus and ascospores, D- ascospores (1000x).

56.2. ***Kallichroma tethys*** (Kohlm. & E. Kohlm.) Kohlm. & Volkm.-Kohlm., 1993b. *Mycol. Res.*, **97**: 759. [Fig. 56.2]

Basionym: *Hydronectria tethys* Kohlm. & E. Koíhm., *Nova Hedwigia*, **9**: 95 (1965).

Ascomata: 275-380 μm high, 315-460 μm in diam, subglobose or depressed-ellipsoidal, partly or rarely completely immersed, ostiolate, epapillate or clypeoid, thickened around the ostiole, fleshy-leathery, orange-yellowish, gregarious or frequently confluent; the color results from orange-colored oil droplets concentrated in the apical part of the peridium and the paraphyses. *Peridium*: above and at the sides 70-75 μm thick, at the base 20-30 μm, two-layered; on the outside composed of irregularly polygonal thick-walled cells, on the inside of elongate flattened cells that merge into paraphyses. *Ostiolar canal*: 17.5-32.5 μm in diam, conical, lined with periphyses, 1-2 μm in diam. *Apical paraphyses*: 2 μm in diam, septate, in the dome of the venter attached to large isodiametric cells with large lumina, finally deliquescing. *Asci*: 90-105 x 15-23 μm, 8-spored, clavate or subcylindrical, short pedunculate, unitunicate, thin-walled at maturity, aphysoclastic, without apical apparatuses, developing at the base of the ascomata venter. *Ascospores*: 17-26 x 8.5-13 μm, ellipsoidal or ovoid, 1-septate, somewhat below the center, slightly

constricted at the septum, hyaline, with about 6 longitudinal ridges, 0.5-0.8 µm in diam, running around the spore from one pole to the other.

Habitat: Saprobic on wood in marine habitats.

Description: Based on Kohlmeyer and Kohlmeyer (1965).

Distribution: West Coast:-Gujarat: On intertidal wood of *Avicennia marina* and *Ceriops tagal* (Borse et al., 2000a); **Maharashtra:** On intertidal wood of *Avicennia marina* (2000b; Borse et al., 2005a); **Goa:** On intertidal wood ((Prasannarai and Sridhar, 2001), on intertidal mangrove wood (Maria and Sridhar, 2002a); **Karnataka:** On intertidal wood (Prasannarai et al., 1999), on intertidal wood ((Prasannarai and Sridhar, 2001, 2000-2001), on animal substrate (Ananda and Sridhar, 2001), on intertidal mangrove wood on intertidal mangrove wood of *Acanthus ilicifolius, Avicennia officinalis, Brugueiria gymnorrhiza, Rhizophora mucronata* and *Sonneratia caseolaris* (Maria and Sridhar, 2002a, 2003a), on intertidal leaf and woody litter of mangroves (Ananda and Sridhar, 2003, 2004); **Kerala:** On intertidal mangrove wood (Maria and Sridhar, 2002a), on mangrove woody litter (Shini et al., 2009-10); **Lakshadweeop Islands:** On intertidal mangrove wood (Ananda and Sridhar, 2003).

East Coast:-Tamil Nadu: On wood of *Avicennia marina* (Ravikumar and Vittal, 1987); on intertidal woody debris (Ravikumar et al., 2009); **Pondecherry (Karaikkal):** on intertidal woody debris (Ravikumar et al., 2009); **Andhara Pradesh:** On intertidal wood of *Avicennia marina* and *Rhizophora apiculata* (Sarma and Vittal, 1998-99; Sarma and Vittal, 2000; Sarma and Vittal, 2001; Sarma et al., 2001; Sarma and Vittal, 2002; Sarma and Vittal, 2004; Vittal and Sarma, 2005); **Andaman & Nicobar Islands:** On intertidal wood of *Avicennia marina, A. officinalis, Bruguiera gymnorhiza, Rhizophora apiculata, R. mucronata* and *Sonneratia alba* (Chinnaraj, 1993a); **West Bengal:** On intertidal wood of *Acanthus ilicifolius* (Pawar *et al.*, 2009); on mangrove woody litter (Shini et al., 2009-10).

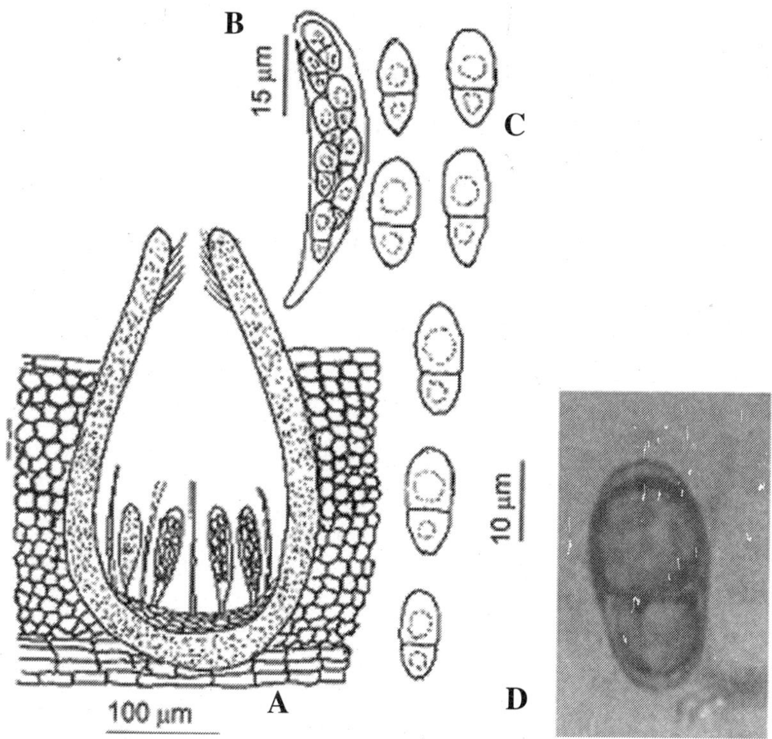

Fig. 56.2. ***Kallichroma tethys,*** A-D; A- v.s. of ascocarp, B- ascus, C- ascospores, D- ascospores (1000x).

57. *Kirschsteiniothelia* D. Hawksw., 1987a. *Bot J. Linn. Soc.*, **91:** 183.

Ascomata: perithecioid, immersed to superficial, when superficial base immersed in the substratum, mammiform, golbose to flattened globose, coriaceous, dark brown to black, ostiolate, papillate, solitary or clustered. *Ostiole*: mostly central, short, brown to black. *Peridium*: wide, comprising thin-walled pseudoparenchymatous cells. *Pseudoparaphyses*: cellular, hyphae-like, numerous, embedded in gelatinous matrix. *Asci*: 8-spored, clavate, pedunculate,apically rounded, with a cylindrical apical chamber, fissitunicate; sometimes with endoascus separating from ectoascus at time of ectoascus dehiscence, base of endoascus long, narrow and coiled within ectoascus, endoascus uncoiling to form a long tail. *Ascospores*: biseriate, ellipsoid, rounded at apices, 1-euseptate, constricted at the septum, apical cell usually longer than basal cell, brown, verruculose, thin-walled, guttulate. The genus is represented by one marine species (Jones *et al.*, 2009a).

Type species: *Kirschsteiniothelia aethiops* D. Hawksw.
Habitat: On driftwood and bark, or coniferous wood.
Description: Based on Hawksworth (1987a), Cai et al. (2006) and Jones et al. (2009a).

57.1. *Kirschsteiniothelia maritima* (Linder) D. Hawksw., 1985. *Bot. J. Linn. Soc.*, **91**: 193. [For Fig. refer original paper]

Basionym: *Amphisphaeria maritima* Linder, Farlowia, 1: 411 (1944).

Ascomata: 57-128 µm high, 104-268 µm in diam, subconical, semiglobose, rarely subglobose, with a flattened base, occasionally membranous at the edge, superficial or partly immersed, ostiolate, short papillate or epapillate, carbonaceous, black, gregarious. *Peridium*: 30-35 µm thick, 2-layered, outer layer brownish-black, inner layer hyaline; composed of subglobose, thick-walled cells with large lumina, frming textura angularis; base of peridium crustlike, 2.5-5 µm thick. *Ostiolar canal*: 20-30 µm in diam, periphysate. *Pseudoparaphyses*: ramose, attached at both ends. *Asci*: 37-60 x 8-14 µm 8-spored, clavate or elongate-ellipsoidal, pedunculate, bitunicate, thick-walled at the apex, physoclastic, without apical apparatuses; developing at the base of the ascomata venter. *Ascospores*: 13.5-21 x 5-7.5 µm, bi- or rarely triseriate, ellipsoidal to subfusiform, 1-septate somewhat below the center, constricted at the septum, brown.

Habitat: Saprobic on intertidal wood and drift wood,
Description: Based on Kohlmeyer and Kohlmeyer (1979).
Distribution: East Coast:-Andhara Pradesh: On intertidal wood of *Avicennia officinalis* (Sarma and Vittal, 1998-99; Sarma and Vittal, 2000; Sarma and Vittal, 2001; Sarma et al., 2001; Sarma and Vittal, 2004; Vittal and Sarma, 2005).
Note: Recently, Boonmee et al. (2011) established a new genus namely *Halokirschsteinionthelia* Boonmee and Hyde for the present species on the basis of molecular systematics.

58. *Kochiella* Sakayaroj, K.L. Pang & E.B.G. Jones, 2010. *Fungal Diversity*, **46:** 87-109.

Ascomata: subglobose to ellipsoidal, superficial to immersed, ostiolate, papillate, periphysate, coriaceous, thinwalled, hyaline to cream coloured; *Asci*: ellipsoidal to clavate, thin-walled, deliquescing at maturity; *Ascospores*: 1-septate, hyaline, appendages pleomorphic.

Type species: *Kochiella crispa* (Kohlm.) Sakayaroj, K.L. Pang & E.B.G. Jones
Habitat: Saprobic on intertidal wood.
Description: based on Sakayaroj et al. (2010a).

58.1. *Kochiella crispa* (Kohlm.) Sakayaroj, K.L. Pang & E.B.G. Jones, 2010. *Fungal Diversity*, **46:** 87-109.

Basionym: *Remispora crispa* Kohlm., *Can. J. Bot.*, **59:** 1317 (1981).

Ascomata: 200-350 µm in diam., subglobose to ellipsoidal, superficial or immersed, ostiolate, papillate, coriaceous, thin-walled, subhyaline, cream-coloured to grey, darker around the base of the neck, solitary. *Peridium*: 12-20 µm thick, composed of 4 to 5 layers of flat, thin-walled cells, merging into the pseudoparenchyma of the center; mature spores are visible through the peridium. *Necks*: 70-220 µm long, 44-50 µm in dium., cylindrical; ostiolar canal with delicate periphyses. *Paraphyses*: absent; center of immature ascomata filled with a hyaline, thin-walled deliquescing pseudoparenchyma. *Asci*: 55-90 x 26-28 µm, 8-spored, ellipsoidal to clavate, short stipitate, unitunicate, thin-walled, dissolving at spore maturity. *Ascospores*: 22-34 x 8-11(-12) µm, ellipsoidal, 1-septate, slightly or not constricted at the septum, hyaline, at first at both apices with a wide cap that extends along one side to the middle of each cell. *Appendages*: initially sub-gelatinous, upon release into the water becoming detached from thc side of the spore, but remaining attached with their bases to the spore apices; the lower part of the appendages swells and parallel fibers become apparent that emerge in a fountain-like fashion from the thickened tip of the spore wall; fibers are straight near the point of attachment, but become wavy above; the whole appendage is transformed into fibers, except for the part that was initially attached to the side of the spore; this portion which now becomes the triangular, spoon-shaped tip of the appendage remains smooth and uniform; developed appendages are at first yoke-like, later scoop-shaped when they become detached from the lateral spore wall and invert.

Habitat: On subtidal branches (2-3 m deep) with wood bores and algae.

Description: Based on Kohlmeyer (1980).
Distribution: East Coast:-Orissa: On intertidal wood (Borse and Borse, 2001).

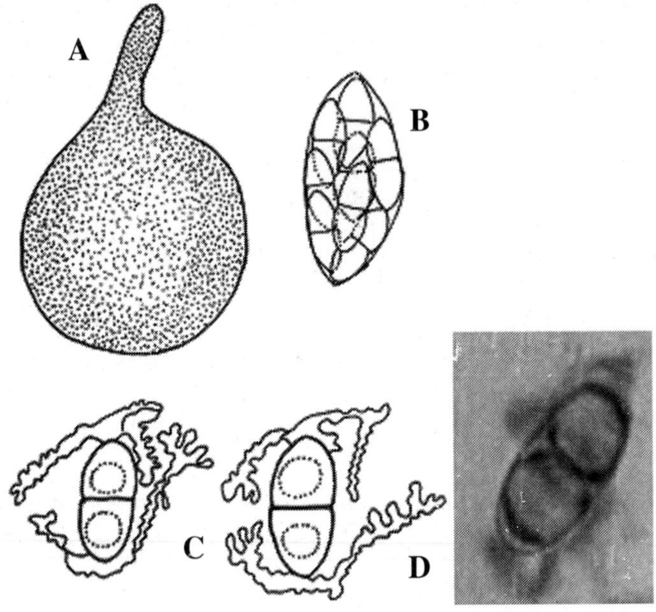

Fig. 58.1. ***Kochiella crispa,*** A-D; A- ascocarp (100x), B- ascus (450x) , C- ascospores (1000x), D- ascospore (1000x).

59. ***Koralionastes*** Kohlm. & Volkm.-Kohlm., 1987b. *Mycologia*, **79**: 765. [For Fig. refer original paper]

Characterised by large ascomata, simple septate paraphyses, clavate asci, short pedunculate, deliquescing at maturity, no apical apparatus, ascospores hyaline, thick-walled and septate near the apices, no appendages, with spermatia. The genus is represented by 5 species in marine habitata (Kohlmeyer and Volkm.-Kohlm., 1990a; Jones et al., 2009).

Type species: *Koralionastes ovalis* Kohlm. & Volkm.-Kohlm.
Habitat: On coralline-coated rocks and sponges.
Description: Based on Kohlmeyer and Volkmann-Kohlmeyer (1987b) and Jones et al. (2009a).

59.1. *Koralionastes* sp.

Ascomata: perithecioid, 470-600 x 360-400 µm, ellipsoidal, dark brown, partially immersed or superficial, ostiolate, papillate. *Necks*: 180-200 x 190-250 µm, straight or slightly curved, light brown or brown; ostiolar canal lined with periphyses. *Peridium*: 28-35 µm thick, multilayered with dark brown outer layers and light brown inner layers, composed of polygonal cells. *Paraphyses*: 130-165 x 2-4 µm, simple, septate, hyaline. *Asci*: 200-230 x 7-10 µm, 8-spored, clavate, short pedunculate, thin-walled, unitunicate. *Ascospores*: 27-34 x 6-9 µm, uniseriate, ellipsoidal, initially aseptate, becoming septate with up to 8-transverse septa, thick-walled, smooth, not constricted at the septa, hyaline.

Habitat: On intertidal wood.

Description: Based on Raveendran and Manimohan (2007).

Distribution: West Coast:-Kerala: On intertidal wood (Raveendran and Manimohan, 2007).

Note: The genus is represented by 5 species in marine habitats (Jones et al., 2009a). They occur on coralline-coated rocks and ascomata surrounded or covered by crustose sponges and thus appear to be substrate specific.

60. *Lautitia* S. Schatz, 1980b. *Can. J. Bot.*, **62:** 31.

Ascomata: ampulliform to subglobose, immersed in the cystocarps and tetracarps of *Chondrus crispus*, ostiolate, papillate, coriaceous, pale-coloured, clypeus black, gregarious. *Pseudoparaphyses*: thin, septate, simple or branched. *Asci*: subcalvate to subcylindrical, short pendunculate, thick-walled, bitunicate. *Ascospores*: elongate fusiform, 1-septate, slightly constricted at the septum, hyaline, no sheath or appendages. The genus is monotypic (Jones *et al*., 2009a).

Type species: *Lautitia danica* (Berl.) Schatz

Habitat: On cystocarps and tetracarps of red alga *Chondrus crispus*

Description: Based on Schatz (1980b) and Jones et al. (2009a).

60.1. *Lautitia danica* (Berl.) S. Schatz, 1984. *Can. J. Bot..*, **62**: 31.

[For Fig. refer original paper]

Basionym: *Leptosphaeria danica* Berl., *Icones Fung.*, **1**: 87 (1892).

= *Leptosphaeria chondri* Rosenv., *Bot. Tidsskr.*, **27**: 35 (1906).

= *Leptosphaeria marina* Rostr. *Bot. Tidsskr.*, **17**: 234 (1889).

≡ *Didymosphaeria danica* (Berl.) Wilsans & Knoyle, *Trans. Br. Mycol. Soc.*, **44**: 55 (1961).

= *Sphaerella chondri* H. L. Jones, *Oberlin Coll. Lab.*, **9**: 3 (1898).

= *Guiganardia chondri* (H. L. Jones) Estee, *Publ. Staz. Zool. Napoli*, **15**: 378 (1936).

Ascomata: 125-267 µm high, 110-300 µm in diam, ampulliform to subglobose, immersed, confined to cystocarps and tetrasporic pustules of *Chondrus crispus*, ostiolate, papillate, coriaceous, pale, except for the black pseudoclypeus, gregarious. *Peridium*: composed of interwoven hyphae, small-celled, indistinct above, particularly in the ostiolar region. *Necks*: elongate, conical or cylindrical. *Pseudoparaphyses*: thin, septate, simple or branched, attached at both ends. *Asci*: 70-94 x 10-15 µm, 8-spored, subcalvate to subcylindrical, short pendunculate, thick-walled, bitunicate, physoclastic; developing at the base and lower side of the ascomata venter. *Ascospores*: 33-40 x 5-7 µm, irregularly biseriate, elongate fusiform, 1-septate below the center; lower cell cylindrical, rrounded; upper cell broad in the lower half and pointed apically; slightly constricted at the septum, hyaline, but also becoming pale yellowish in age, no sheath or appendages. *Spermagonia*: 100-192 x 63-113 µm. *Spermatiophores*: 16-18 µm long. *Spermatia*: 4 x 1 µm, oblong to ellipsoidal, 1-celled, hyaline, originating in mucilaginous, basipetal strings at the apices of the spermatiophores. The genus is monotypic (Jones *et al.*, 2009a).

Habitat: Saprobic on intertidal and mangrove wood.

Description: Based on Kohlmeyer and Kolhmeyer (1979).

Distribution: West Coast:-Karanataka: On intertidal wood of *Acanthus ilicifolius* (Maria and Sridhar, 2003a).

Note: The occurrence of *L. danica* on intertidal wood of *Acanthus ilicifolius* as listed by Maria and Sridhar (2003) is unlikely, because the fungus is a parasite on submerged cystocarps and tetracarps of a red alga *Chondrus crispus* (Kohlmeyer

and Kohlmeyer, 1979, Kohlmeyer and Volkmann-Kohlmeyer, 1991, Jones et al., 2009a).

61. *Lautospora* K.D. Hyde & E.B.G. Jones, 1989a. *Bot. Mar.*, **32**: 482.

Ascomata: immersed, long axis parallel to the host surfsce, ostiolate, brown, solitart. *Peridium*: of thin-walled cells. *Hamathecium*: cellular, septate, and hyaline pseudoparaphyses. *Asci*: 4-spored, cylindrical, bitunicate with an ocular chamber and ring. *Ascospores*: muriform, fusiform, hyaline, very thick-walled. The genus is represented by two species in marine habitats (Jones et al., 2009a).

Type species: *Lautospora gigantea* Hyde & Jones
Habitat: Saprobic on intertidal mangrove wood.
Description: Based on Hyde and Jones (1989a).

61.1. *Lautospora gigantea* K.D. Hyde & E.B.G. Jones, 1989a. *Bot. Mar.*, **32**: 479. [For Fig. refer original paper]

Ascomata: 525-584 µm long, 270-363 µm wide, 153-197 µm high, subglobose, in frontal view, fusiform in side view, totally immersed in the substratum, lying horizontal to the wood surface, coriaceous, ostiolate, brown above, light brown to hyaline below, solitary. *Peridium*: 17-65 µm thick, composed of several layers of thin-walled cells. *Hamathecium*: (pseudoparaphyses) 5-9 µm thick, cellular, septate, and hyaline. *Asci*: 360-545 x 45-68 µm, 4-spored, cylindrical, thick-walled. bitunicate with an ocular chamber and ring. *Ascospores*: 140-195 x 36-45 µm, uniseriate, muriform, fusiform, 18-25 trans-septa, 4-7 longi-septa, hyaline, very thick-walled, wall appearing as a wing-like structure, 3-7.5 µm thick.

Habitat: Saprobic on intertidal and mangrove wood.
Description: Based on Hyde and Jones (1989a).
Distribution: West Coast:-Kerala: On intertidal mangrove wood (Maria and Sridhar, 2002a).
East Coast:-Andhara Pradesh: On intertidal wood of *Avicennia officinalis* (Sarma and Vittal, 1998-99; Sarma and Vittal, 2000; Sarma and Vittal, 2001; Sarma et al., 2001; Sarma and Vittal, 2004;Vittal and Sarma, 2005).

62. *Leptosphaeria* Ces. & De Not., 1863. *Comment. Soc. Critto. Ital.* **1:** 234.

Ascomata: perithecioid, conical to subglobose, obpyriform, erumpent, becoming superficial, ostiolate, papillate, coriaceous or carbonaceous, light brown to brown to black, solitary or gregarious *Ostiole*: central, papilla rounded or slot-like, brown, periphysate. *Peridium*: thick, comprising several layers of brown, thick-walledpolygonal scleroplectenchymatous cells, darker and thick-walled outwardly. *Pseudoparaphyses*: 2-3 µm wide, cellular, hyphae-like, numerous, septate, branched, anastomosing above the asci, embedded in a gelatinous matrix. *Asci*: 8–spored, numerous, cylindrical or cylandric-clavate, short pedunculate, thick-walled, fissitunicate, apically rounded, lacking or with an apical apparatus. *Ascospores*: overlapping uniseriate or biseriate, cylindric or clavate, fusiform or narrowly fusiform, 3 or more septate, sometimes constricted at the septa, often asymmetrical, yellow-brown to brown, with or without a sheath. The genus is represented by 5 marine species among several terrestrial species (Jones et al., 2000a).

Type species: *Leptosphaeria doliolum* (Pers. & Fr.) Ces. & De Not.
Habitat: Saprobic on wood.
Description: Based on Cai et al. (2006), Pande (2008) and Jones et al. (2009a).

Key to *Leptosphaeria* species from India:

1.Ascospores lacking a sheath, 19-27 x 6-9 µm, ... *L. austaliensis*

1'. Ascospores with a sheath, 18-25 x 6-8 µm, ...*L. avicenniae*

1". Ascospores with a sheath, 28-44 x 8-12 µm, ... *L. pelagica*

62.1. *Leptosphaeria avicenniae* Kohlm. & E. Kohlm., 1965. *Nava Hedwigia*, **9**: 98. [Fig. 62.1]

Ascomata: 340-420 µm high, 260-300 µm in diam, pyriform, half immersed, ostiolate, papillate, cabonaceous, black, gregarious, developing in light-coloured spots on pneumatophores. *Peridium*: 28-36 µm thick, composed of irregular, thick-walled cells, forming a textura angularis, merging externally into a stromatic tissue covered with brownish hyphae. *Necks*: conical, short, ostiolar canal 70 µm in diam., periphysate. *Pseudoparaphyses*: septate, simple or ramose. *Asci*: 112-158 x 7.5-10 µm, 8-spored, cylindrical, short pedunculate, bitunicate, thick-walled,

physoclastic, without apical apparatuses; developing at the base of the ascomata venter. *Ascospores*: 17.5-25 x 5.5-8 µm, uniseriate, ellipsoidal, 3-septate, slightely constricted at the septa, hyaline, covered by a gelatinous sheath, 2.5-5 µm thick.

Habitat: Parasitic or saprobic on pnematophores, bark and wood of submerged branches and tree trunks of *Avicennia* sp.

Description: Based on Kohlmeyer and Kohlmeyer (1965).

Distribution: East Coast:-West Bengal: On intertidal mangrove woody litter (Shini et al., 2009-10).

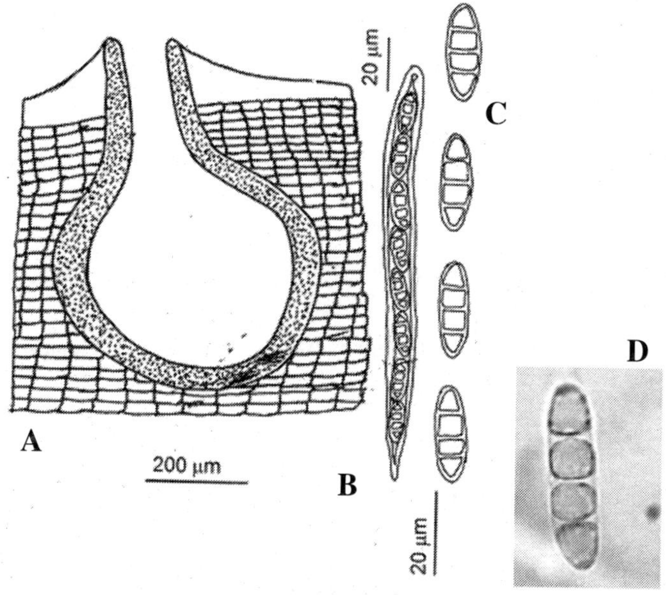

Fig. 62.1. ***Leptosphaeria avicenniae*** A-D; A- v.s. of ascocarp, B- ascus, C- ascospores, D- ascospore 1000x).

62.2. *Leptosphaeria australiensis* (Cribb & J.W. Cribb) G.C. Hughes, 1969. *Syesis,* 2: 132. [Fig. 62.2]

Basionym: *Metasphaeria australiensis* Cribb & J.W. Cribb, *Uni. Queens. Pap. Dept. Bot.*, **3**: 79 (1955).

Ascomata: 110-200 µm high, 120-195 µm in diam, obpyriform, immersed, ostiolate, papillate, coriaceous, fuscous or light brown, lighter coloured at the base, gregarious or solitary. *Peridium*: 10-25 µm thick, 2- layered, forming a *textura angularis*; outer layer composed of irregularly rounded, thick-walled, pigmented cells; inner layer composed of trapezoid thin-walled, light coloured cells with large lumina. *Necks*: 70-225 µm long, 30-50 µm in diam, 70-75 µm at the base, conical or sub-cylindrical, fuscous or almost hyaline; ostioles lined with periphyses, 1.5-2.2 µm in diam. *Pseudoparaphyses*: 1–2.2 µm in diam, septate, somewhat branched with gelatinous walls, persistent. *Asci*: 70-105 x 10-15 µm, 8-spored, clavate-fusiform, pedunculate, indistinctly bitunicate, thick-walled when young, later relatively thin-walled with apical apparatuses (pulvillus and annulus); developing at the base of the ascomata venter. *Ascospores*: 19-26 x 5.5-9 µm, biseriate, ellipsoidal, fusiform or clavate-fusiform, 3-septate, constricted at the septa, hyaline.

Habitat: Saprobic on intertidal and mangrove wood.

Description: Based on Kohlmeyer and Kohlmeyer (1979).

Disribution: West Coast:-Gujarat: On intertidal wood of *Avicennia marina* and *Rhizophora mucronat* (Borse et al., 2000a), on intertidal wood (Patil and Borse, 2001); **Maharashtra:** On intertidal wood and proproots of *Rhizophora mucronata* (Patil and Borse, 1986; Borse, 1987d, 1988; Borse *et al.,* 2005a), on intertidal mangrove wood (Maria and Sridhar, 2002a); **Goa:** On intertidal wood (Borse et al., 1999a), on intertidal mangrove wood (Maria and Sridhar, 2002a);; **Karnataka:** On intertidal wood (Prasannarai and Sridhar, 2001, 2000-2001), on intertidal mangrove wood of *Acanthus ilicifolius, Avicennia officinalis* and *Rhizophora mucronata* (Maria and Sridhar, 2002a, 2003a, 2004), on intertidal leaf and woody litter of mangroves (Ananda and Sridhar, 2003, 2004); **Pondicherry (Mahe):** On intertidal mangrove wood (Nambiar and Raveendran, 2008a); **Kerala:** On intertidal mangrove wood (Maria and Sridhar, 2002a), on intertidal wood (Raveendran and Manimohan, 2007), on intertidal mangrove wood (Nambiar and Raveendran, 2006, 2008a, b, d, 2009a, f, g, i, 2010c; Nambiar et al., 2008a); **Lakshadweep Islands:** On intertidal wood of *Scaveola taccada* (Chinnaraj, 1992), on intertidal mangrove wood (Ananda and Sridhar, 2003).

East Coast:-Tamil Nadu: On submerged parts of *Rhizophora* spp. (Ravikumar and Vittal, 1996); **Andhara Pradesh:** On intertidal wood of *Avicennia marina, A. officinalis* and *Rhizophora apiculata* (Sarma and Vittal, 1998-99; Sarma and Vittal, 2000; Sarma and Vittal, 2001; Sarma et al., 2001; Sarma and Vittal, 2002; Sarma and Vittal, 2004; Vittal and Sarma, 2005); **Andaman-Nicobar Islands:** On intertidal wood of *Avicennia marina, A. officinalis, Rhizophora apiculata, R. mucronata* and *Sonneratia alba* (Chinnaraj, 1993a); **West Bengal:** On intertidal wood of *Avicennia marina* (Pawar et al., 2009); on mangrove woody litter (Shini et al., 2009-10); **Orissa:** On intertidal wood of *Avicennia marina, A. officinalis* and *Rhiziphora mucronata* (Borse et al., 2009).

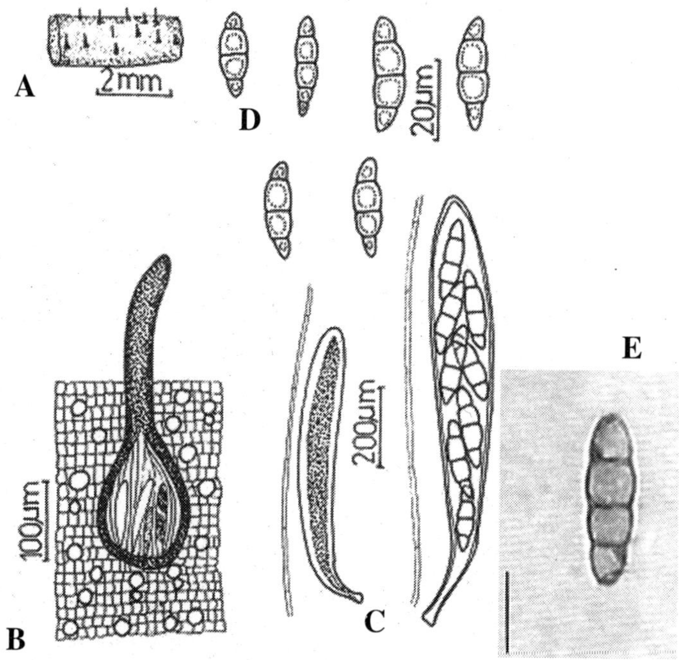

Fig. 62.2. ***Leptosphaeria australiensis,*** A-E; A- Habit, B- v.s. of ascocarp, C- asci and paraphyses, D- ascospores, E- ascospore (bar- µm).

62.3. *Leptosphaeria pelagica* E.B.G. Jones, 1962. *Trans. Br. Mycol. Soc.*, **45**: 105. [For Fig. refer original paper]

Ascomata: 115-201 µm high, 152-280 µm in diam, globose, subglobose or ellipsoidal, superficial or immersed, ostiolate, papillate, subcabonaceous, dark brown to black, some times lighter coloured bases, solitary. *Necks*: 62-103 µm long, 49-74 µm in diam, cylindrical, base dark, apex light-coloured. *Pseudoparaphyses*: 1.6–2.4 µm in diam, filiform, simple or sparingly ramose. *Asci*: 100-149 x 14.5-24 µm, 8-spored, subclavate to long cylindrical, bitunicate, thick-walled. *Ascospores*: 28-36-(40) x 8-12 µm, fusiform to clavate, straight or curved, 3-septate, constricted at the septa, apically rounded, hyaline.

Habitat: Saprobic on salt marsh plants and intertidal wood, drift wood, and test panels in temperate marine waters. It also occurs in tropical marine waters (Jones, E. B. G., personel communication).

Description: Based on Jones (1962).

Distribution: West Coast:-Goa: On intertidal wood (Prasannarai and Sridhar, 2001); **Karanataka:** On intertidal mangrove wood (Maria and Sridhar, 2002a).

62.4. *Leptosphaeria peruviana* Speg., 1881. *An. Soc. Cient. Argent.*, **12**: 179. [For Fig. refer original paper]

Habitat: Saprobic on decying stems of *Salicornia ambigua* (= *S. peruviana*), Argentina (Buenos Aires), in temperate marine waters.

Distribution: West Coast:-Karanataka: On intertidal wood of *Brugueiria gymnorrhiza* (Maria and Sridhar, 2002, 2003).

East Coast:-Andhara Pradesh: On intertidal wood of *Avicennia officinalis* and *Rhizophora apiculata* (Sarma and Vittal, 1998-99; Sarma and Vittal, 2000; Sarma and Vittal, 2001; Sarma et al., 2001; Sarma and Vittal, 2002; Sarma and Vittal, 2004; Vittal and Sarma, 2005).

Note: The present fungus is saprobic on decaying stems of *Salicornia ambigua* (= *S. peruviana*) reported from Argentina (Buenos Aires), in temperate marine waters (Kohlmeyers and Kohlmeyer, 1979; Kohlmeyer and Kohlmeyer, 1991; Jones et al., 2009a).

63. *Lignincola* Hohnk, 1955. *Veroeff. Inst. Meeresfor.Bremerhaven*, **3:** 216.

Ascomata: solitary or gregarious, subglobose or ellipsoidal, immersed or superficial, ostiolate, papillate, coriaceous, light-coloured to blackish. *Pseudoparenchyma*: of thin-walled cells with large lumina filling venter of young ascomata; eventually breaking up into catenophyses. *Asci*: 8-spored, clavate or subfusiform, pedunculate, unitunicate, thin-walled, aphysoclastic, persistent; developing at the base of the ascomata venter; breaking off at maturity, and released through the ostiole. *Ascospores*: ellipsoidal, 1-septate, hyaline, without appendages. The genus is represented by four marine species (Jones et al., 2009a; Liu et al., 2011).The genus is represented by four marine species (Jones *et al.,* 2009a Liu *et al.,* 2011).

Type species: *Lignincola laevis* Hohnk
Habitat: Saprobic on intertidal and mangrove wood.
Description: Based on Kohlmeyer and Kohlmeyer (1979).

Key to *Lignincola* species from India:

1. Ascospores thick-walled, ascus with an apical pore, ascospores 22-36 x 12-16 µm .... *Lignincola tropica*

1'. Ascospores thin-walled, ascus without an apical pore and released through the neck, ascospores 13-24 x 5-8 µm, ... *Lignincola laevis*

63.1. *Lignincola laevis* Hohnk, 1955. *Veroeff. Inst. Meeresforsch. Bremerhaven,* **3:** 216. [Fig. 63.1]

Ascomata: 125-250 µm in diam, subglobose or ellipsoidal, immersed or superficial, ostiolate, papillate, coriaceous, hyaline, light brown, fuscous or blackish, solitary or gregarious. *Peridium*: 13-16 µm thick, composed of 2 or 5 layers of elongate thick–walled cells with large lumina, forming a *textura angularis*; merging into the pseudoparenchyma of the venter. *Necks*: up to 4 µm long and 25-40 µm in diam, cylindrical, centric or eccentric; ostiolar canal at first filled with small-celled pseudoparenchyma. *Pseudoparenchyma*: of thin- walled cells with large lumina filling the venter of young ascomata; eventually breaking up into catenophyses. *Asci*: 49-69 x 15-20 µm, 8-spored, clavate or subfusiform, short pedunculate, unitunicate, thin–walled, aphysoclastic, without apical apparatuses, persistent; developing at the base of the ascomata venter on a tissue of small ascogenous cells; stipe becoming detached at ascospore maturity, and asci released through

the ostiole into the water while enclosing the ascospores; central part of ascus swelling in water. *Ascospores*: 16-24 x 6-8 mm, irregularly biseriate, ellipsoidal, 1-septate, slightly constricted at the septum, hyaline, without appendages.

Habitat: Saprobic on intertidal and mangrove wood.

Description: Based on Kohlmeyer and Kohlmeyer (1979).

Distribution: West Coast:-Daman: On intertidal wood and *Avicennia marina* (Borse et al., 2000a); **Gujarat:** On intertidal wood (Patil and Borse, 2001); **Maharashtra:** On dead pneumatophores of *Sonneratia acida*, intertidal wood and roots of *Avicennia marina* (Patil and Borse, 1983a, 1986; Borse, 1988), on intertidal mangrove wood (Maria and Sridhar, 2002a); on intertidal wood of *Avicennia alba* (Shrivastava, 1995); **Goa:** On intertidal wood (Borse et al., 1999a), on intertidal mangrove wood (Maria and Sridhar, 2002a); **Karnataka:** On intertidal wood of *Acanthus ilicifolius, Avicennia officinalis, Brugueiria gymnorrhiza, Rhizophora mucronata* and *Sonneratia caseolaris* (Maria and Sridhar, 2002a, 2003a, 2004), on intertidal leaf and woody litter of mangroves (Ananda and Sridhar, 2004); **Pondicherry (Mahe):** On intertidal wood (Nambiar and Raveendran, 2007, 2008a); **Kerala:** On intertidal mangrove wood (Maria and Sridhar, 2002a), on intertidal wood (Raveendran and Manimohan, 2007), on intertidal wood and mangrove wood (Nambiar and Raveendran, 2006, 2007, 2008a, b, d, 2009a, e, f, g, 2010c; Nambiar et al., 2006), on mangrove woody litter (Shini et al., 2009-10); **Lakshadweep Islands:** On intertidal wood of *Pemphis acidula* (Chinnaraj, 1992).

East Coast:-Tamil Nadu: On drift wood (Raghukumar, 1973), On intertidal wood (Nambiar et al., 2008b), on mangrove wood (Nambiar and Raveendran, 2008e, 2009k, l, 2010c); **Andhara Pradesh:** On intertidal wood of *Avicennia officinalis* (Sarma and Vittal, 1998-99; Sarma and Vittal, 2000; Sarma and Vittal, 2001; Sarma et al., 2001; Sarma and Vittal, 2004; Vittal and Sarma, 2005), on intertidal wood (Nambiar and Raveendran, 2009j); **West Bengal:** On intertidal wood of *Sonneratia apetala* (Pawar and Borse, 2004), on mangrove woody litter (Shini et al., 2009-10); **Andaman & Nicobar Islands:** On intertidal wood of *Avicennia marina, Rhizophora apiculata, R. mucronata* and *Sonneratia alba* (Chinnaraj, 1993aa).

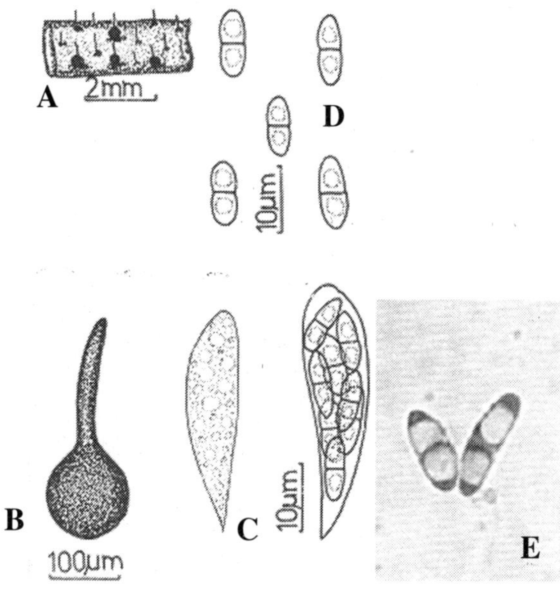

Fig. 63.1. ***Lignincola laevis,*** A-E; A- Habit, B- ascocarp, C- asci, D- ascospores, E- ascospores (1000x).

63.2. ***Lignincola tropica*** Kohlm., 1984. *Mar. Ecol. (P. S. Z. N. I.)*, **5**: 345. [For Fig. refer original paper]

Ascomata: 325-357 µm high, 250-350 µm in diam, subglobose to ellipsoidal, immersed or almost superficial, ostiolate, papillate, coriaceous, light to dark brown, solitary or gregarious. *Peridium*: 30-40 µm thick, 3- layers; outer layer 2-4 µm thick, dark brown with hyphae attached to the substrate; middle layer 15-20 µm thick, light brown, composed of irregularly interwoven cells with small lumina, forming *textura epidermoidea*; inner layer 15-20 µm thick, hyaline composed of cells with large lumina, forming a *textura angularis*, merging into the pseudoparenchyma of the centrum. *Necks*: 175-400 x 80-150 µm, cylindrical or often irregularly shaped, curved or inflated; ostiolar canal periphysate. *Pseudoparenchyma*: of thin- walled cells filling centrum of young ascomata; eventually breaking up into catenophyses. *Asci*: 130-200 x 24-34 µm, 8-spored, clavate, pedunculate, truncate above, unitunicate, thin–walled, aphysoclastic, refractive at the apexand a pore, semi-persistent, developing at the base of the

ascomatal centrum. *Ascospores*: 22-36 x 12-16 µm, ellipsoidal, 1-septate, slightly or not constricted at the septum, hyaline, without appendages.

Habitat: Saprobic on intertidal and mangrove wood.

Description: Based on Kohlmeyer (1984).

Distribution: West Coast:-Karanataka: On intertidal wood of *Avicennia officinalis* and *Brugueiria gymnorrhiza* (Maria and Sridhar, 2002a, 2003a, 2004); **Kerala:** On intertidal mangrove wood (Maria and Sridhar, 2002a), on intertidal wood (Raveendran and Manimohan, 2007; Nambiar et al., 2008a), on *Avicennia* sp. (Nambiar and Raveendran, 2009d, f, h, i), on mangrove woody litter (Shini et al., 2009-10).

East Coast:- East Coast:-Tamil Nadu: On mangrove wood (Ravikumar, 1991); **Andhara Pradesh:** On intertidal wood of *Avicennia marina, A. officinalis* and *Rhizophora apiculata* (Sarma and Vittal, 1998-99; Sarma and Vittal, 2000; Sarma and Vittal, 2001; Sarma et al., 2001; Sarma and Vittal, 2004; Vittal and Sarma, 2005).

64. *Limacospora* Jorgen Koch & E.B.G. Jones, *Can. J. bot.*, **73**: 1011 (1995).

Ascomata: deeply embedded in the wood, long neck with no periphyses. *Peridium*: 2-layered, an outer layer of prismatic, nearly isodimetric cells and an inner layer of long, thin-walled often pointed cells. *Asci*: 8-spored, clavate, tapering, pedunculate, unitunicate, early deliquescing, in a hymenial layer at the base of the ascomata. Interthecial filaments present. *Ascospores*: 2-celled, hyaline, surrounded by a mucilaginous sheath, bounded by the delimiting membrane, which is constricted at the septum. *Appendages*: variable in morphology, initially in water elliptical and "slug-like", but becoming ovoid-oblong, then spore surrounded by fibrillar mucilage drawn into long polar threads. The genus is monotypic Jones *et al*. (2009a).

Type species: *Limacospora sundica* Jorgrn Koch & E.B.C. Jones

Habitat: Saprophytic on intertidal wood.

Description: Based on Jones et al. (1995).

64.1. ***Limacospora sundica*** (Jorgen Koch & E.B.G. Jones) Jorgen Koch & E.B.G. Jones, 1995. *Can. J. bot.*, **73**: 1010. [For Fig. refer original paper]

Basionym: *Ceriosporopsis sundica* Jorgen Koch & E.B.G. Jones, *Nord. J. Bot.*, **6**: 339 (1985).

Ascomata: 270-675 µm high, 570-1080 µm broad, gregarious, immersed, deep seated, subglobose-ellipsoidal, ostiole with long neck extended to the surface of the wood, neck aperiphysate. Surrounding and bordering the ascomata there is a brown transitional zone 20-30-60 µm thick in which the substrate cells are not easily discernible. *Peridium*: 2-layered, an outer layer 16-20 µm at the base, thinning towards the apical region, consisting of hyaline, prismatic, nearly isodimetric cells (4-8 µm in diam.); an inner layer 60 µm thick at the base of the ascomata and again thinning out towards the neck region, composed of long, thin-walled often pointed cells, 20-40 x 8-10 µm, which are longer and broader towards the base of the ostiole. Melanisation of the outer peridial cells of old ascomata commences in the neck region and extend around the ascomata. *Necks*: up to 1500 µm long, 46-76 µm in diam. cylindraical, brown-black; outer layer of the neck 8-16 µm thick, consisting of slender prismatic cells (2-3 x 4-8 µm) merging with the hyaline small-celled tissue which forms the ostiolar channel. *Asci*: 90-140 x 24-28 µm, 8-spored, clavate, tapering, pedunculate, unitunicate, thin-wlled, early deliquescing, in different stages of maturity in a hymenium on a sub-hymenial layer formed as a broad, flat cushion or a flat cup in the lower part of the ascomata cavity, interthecial filaments present. Paraphyses : absent. *Ascospores*: 19-24 x 8-10 µm, 2-celled, hyaline, ellipsoidal or broadly fusiform, slightly constricted around septum, hyaline, released in masses into ascomata cavity before emerging through neck, surrounded by a mucilaginous sheath, bounded by the delimiting membrane, which is constricted at the septum. *Appendages*: spores are enveloped in a granular sheath with polar appendages at either end which appear as outgrowths of the spore wall; polar appendages initially short (ca 28 µm) then becoming long and drawn out sometimes forming fine threads; the sheath and polar appendages swell in water to form a diffuse mass, are extremely sticky and adhere to substrata; the sheath is slightly constricted at the septum with radiating striations extending from the spore wall.

Habitat: Saprophytic on intertidal wood.

Description: Based on Koch and Jones (1985).

Distribution: West Coast:-Kerala: On intertidal wood (Nambiar and Raveendran, 2007);

Karnataka: On intertidal mangrove wood (Maria and Sridhar, 2002a), on intertidal wood of *Brugueiria gymnorrhiza* and *Rhizophora mucronata* (Maria and Sridhar, 2003).

Note: The occurrence of *L. sundica* on intertidal wood and mangrove wood recorded in tropical marine waters of India is unlikely, because *L. sundica* was described on basis of samples collected on partially immersed driftwood in temperate marine waters of Denmark coast (Koch and Jones, 1985; Jones et al., 1995).

65. *Lindra* I.M. Wilson, 1956. *Trans. Br. Mycol. Soc.,* **39:** 411.

Ascomata: semiglobose or ellipsoidal, immersed or becoming exoposed, carbonaceous or subcarbonaceous, black, ostiolate, papillate or epapillate, solitary or gregarious, dark coloured. *Pseudoparenchyma*: of large, thin-walled, deliquescing cells filling venter of the young ascomata. *Asci*: cylindrical to clavate, unitunicate, thin-walled, deliquescing early, lacking apical apparatus. *Ascospores*: filiform, multiseptate septate (up to 50), not constricted at the septa, hyaline, with a golbose to semiglbose gelatinous appendage in species occurring on wood, species occurring on sea grasses are with coriaceous ascomata, immersed in the host, and lacking an apical swelling or appendages. The genus is represented by 5 obligate marine species (Jones *et al.,* 2009a).

Type species: *Lindra inflata* I.M. Wilson

Habitat: On intertidal wood, marine phanerogams, algae and hard surface (sand).

Description: Based on Kohlmeyer and Kohlmeyer (1979) and Jones et al. (2009a).

Key to *Lindra* species from India:

1. Ascomata on sea grasses, ascospores 230-390 x 3-6 mm, ... *Lindra thalassiae*

1'. Ascomata on wood, ascospores 101-188 x 3.5-5.5 mm .... *Lindra hawaiiensis*

65.1. ***Lindra hawaiiensis*** Kohlm. & Volkm.-Kohlm., 1987a, *Can. J. bot.*, **65**: 574. [For Fig. refer original paper]

Ascomata: 120-150 x 150-260 μm, perithecioid, globose, cylindrical or subctlindrical, dark brown, immersed, ostiolate, papillate. *Necks*: 140-470 x 20-40 μm, tapering towards apex. *Peridium*: 25-37.5 μm thick. *Paraphyses*: absent. Asci: 112.5-165 x 7.5-10 μm, 8-spored, elongate, cylindrical unitunicate, thin-walled, deliquescing early. *Ascosores*: 101-188 x 3.5-5.5 μm, 8-18-septate, filiform, curved, hyaline, uniform in thickness, without apical chambers, apices rounded.

Habitat: Saprobic on intertidal wood.

Description: Based on Kohlmeyer and Volkmann-Kohlmeyer (1987a) and Raveendran & Manimohan (2007).

Distribution: West Coast:-Kerala: On intertidal wood (Raveendran and Manimohan, 2007), on intertidal mangrove wood (Nambiar and Raveendran, 2006).

65.2. ***Lindra thalassiae*** Orpurt, Meyers, Boral & Simms, 1964. *Bull. Mar. Sci. Gulf Caribb.*, **14**: 406. [For Fig. refer original paper]

= *Lindra marinera* Meyers, *Mycologia*, **61**: 488 (1969).

Ascomata: 115-200 μm high (excluding necks), 127-200 μm, perithecioid, globose, subglobose or depressed, immersed, ostiolate, papillate, coriaceous, dark brown to black, gregarious. Peridium: 7.5-10 μm thick, composed of 4 or 5 layers of irregular, mostly elongate, thick-walled cells, merging into the pseudoparenchyma of the venter; outer layer of cells forming a textura epidermioidea. *Necks*: 22.5-84 μm high, 27.5- 37.5 μm in diam., conical or cylindrical; osiolar canal at first filled with a pseudoparenchyma of small, hyaline, thin-walled cells. Pseudoparenchyma: of large, thin-walled, polygonal or roundish, deliquescing cells filling venter of the young ascomata. *Asci*: 8-spored, cylindrical to subcylindrical or subclavate, curved, unitunicate, thin-walled, aphysoclastic, without apical apparatuses, deliquescing early. *Ascospores*: 230-390 x 3-6 μm, filiform, tapering toward both apices; 1-1.5 mm in diam. at the apex, curved, (s-, u-, or alpha-shaped), 14-26-septate, not or slightly constricted at the septa, hyaline; tips slightly inflated.

Habitat: Perthophytic or saprobic on *Thalassia testudinum*, vesicles of *Sargassum* sp.

Description: Based on Kohlmeyer and Kohlmeyer (1979).

Distribution: West Coast:-Goa: On *Sargassum cinereum* (Raghukumar et al., 1992a); on *Sargassum cinereum* (Sathe-Pathak et al., 1993; Sharma et al., 1994). **Karanataka:** On intertidal leaf and woody litter of mangroves (Ananda and Sridhar, 2003, 2004); **Lakshadweep Islands:** On sea grass *Thalassia testiudinum* (Chinnaraj, 1992); on *Sargassum cinereum* (Sathe-Pathak et al., 1993; Sharma et al., 1994).

66. *Lineolata* Kohlm. & Volkm.-Kohlm., 1990b. *Mycol. Res.*, **94**: 687.

Ascomata: obpyriform, almost superficial to immersed, ostiolate, periphysate, papillate, subcabonaceous to subcoraceous, black, gregarious. *Peridium*: thick, 2-lyered; outer stratum pseudostromatic, composed of irregular elongate or roundish, dark brown cells, mostly with small lumina, forming a *textura angularis*. *Hamathecium*: pseudoparaphysoid, trabeculate, with gelatinous matrix. *Asci*: 8-spored, cylindrical, short pedunculate, non-amyloid, with a multi-layered refractive ring in the endotunica around the oculus, thick-walled, fissitunicate, maturing successively at the bottom of the locule. *Ascospores*: monostichous, ellipsoidal, 1-septate, brown, with delicate costae.

Type species: *Lineolata rhizophorae* (Kohlm. & E. Kohlm.) Kohlm. & Volkm.-Kohlm.

Habitat: Saprobic on intertidal and mangrove wood.

Description: Based on Kohlmeyer and Volkmann-Kohlmeyer (1990b).

66.1. *Lineolata rhizophorae* (Kohlm. & E. Kohlm.) Kohlm. & Volkm.-Kohlm., 1990. *Mycol. Res.*, **94**: 688. [Fig. 66.1]

Basionym: *Didymosphaeria rhizophorea* Kolhm. & E. Kolhm., *Icones Fungorum Maris*, **4/5**: 62a (1967).

≡*Lojkania rhizophorea* (Kohlm. & E. Kohlm.) Barr, *N. Amer. Flora, Ser.*, 2, **13**: 58 (1990).

Ascomata: 300-490 µm high, 200-360 µm in diam., obpyriform, partly or completely embedded, ostiolate, papillate, periphysate, subcabonaceous to subcoraceous, blackish brown, gregarious, sometimes two grow together.

Peridium: 37-45 μm thick, 2-lyered; the outer, thick stratum pseudostromatic, composed of irregular elongate or roundish, dark brown cells, on the outside with a more or less recognizable hyphal structure, enclosing some decaying cells of the host; inner stratum thin, composed of 4 or 5 layers of hyaline, polygonal, elongate, thin-walled cells with large lumina, merging into the pseudoparaphyses. *Necks*: 65-95 μm in diam. at the apex, conical, emerging in immersed ascomata; ostiolar canal periphysate; periphyses 1.5-2.5 μm in diam., septate, simple. *Pseudoparaphyses*: 1-1.5 μm in diam., septate, simple or rarely branched, connecting to each other by a gelatinous outer layer, persistent. *Asci*: 150-175 x 14-17.5 μm, 8-spored, cylindrical, short pedunculate, bitunicate, thick-walled, physoclastic, with refractive apical apparatus, J-ve reaction to iodine; developing at the base of the ascomata venter. *Ascospores*: 23-32 x 9-12 μm, straight or obliquely uniseriate, ellipsoidal, 1-septate in the center, not or slightly constricted at the septum, dark brown, striate by delicate costae that run parallel or in a slight angle to the longitudinal axis of the spore; wall slightly thickened at the apex, without germ pore, germinating apically.

Habitat: Saprobic on intertidal and mangrove wood.

Description: Based on Kohlmeyer and Kohlmeyer (1979).

Distribution: West Coast:-Gujarat: On intertidal wood of *Ceriops tagal* and *Rhizophora mucronat* (Borse et al., 2000a), on intertidal mangrove wood (Patil and Borse, 2001); **Maharashtra:** On intertidal mangrove (*Ceriops tagal*) wood (Borse, 2000b; Borse et al., 2005a); **Karnataka:** On intertidal wood (Prasannarai and Sridhar, 2001. 2000-2001), on intertidal mangrove wood (Ananda and Sridhar, 2003); **Kerala:** On intertidal wood (Prasannarai and Sridhar, 2001), on mangrove woody litter (Shini et al., 2009-10).

East Coast:-Tamil Nadu: On submerged parts of *Rhizophora* spp. (Ravikumar and Vittal, 1996); **Andhara Pradesh:** On intertidal wood of *Rhizophora apiculata* (Sarma and Vittal, 2000; Sarma and Vittal, 2001; Sarma et al., 2001; Sarma and Vittal, 2002; Sarma and Vittal, 2004; Vittal and Sarma, 2005), on intertidal wood (Nambiar and Raveendran, 2009j); **Andaman & Nicobar Islands:** On intertidal wood of *Rhizophora mucronata* (Chinnaraj and Untawale, 1992), on intertidal wood of *Bruguiera gymnorhiza* and *Rhizophora mucronata* (Chinnaraj, 1993a); **West Bengal:** On mangrove woody litter (Shini et al., 2009-10).

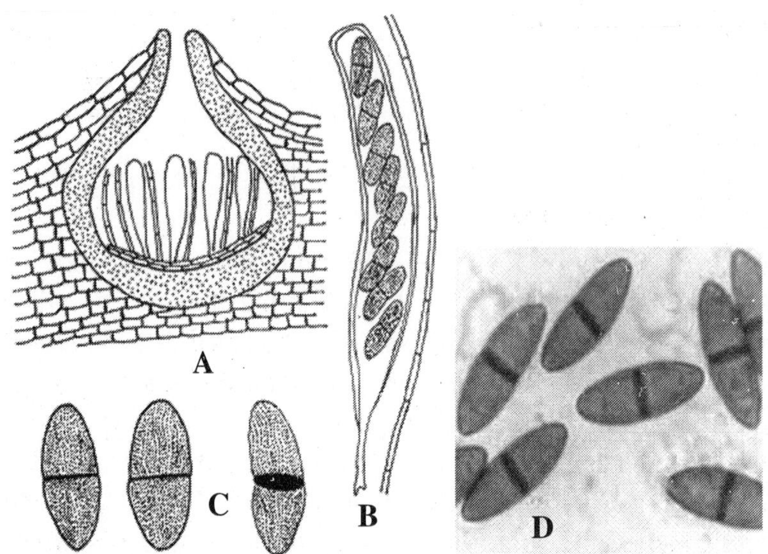

Fig. 66.1. ***Lineolata rhizophorae,*** A-D; A- v.s. of ascocarp (100x), B- asci and pseudoparaphysis (450x) , C- ascospores (1000x), D- ascospores (1000x).

67. ***Lophiostoma*** Ces. & De Not., 1863. *Comm. Sco. Crittog. Ital.,* 1: 219.

Ascomata: perithecioid, immersed or erumpent, becoming superficial by weathering of host epidermis, conical to hemisphaerical, carbonaceous, brown to black, generally with a distinct, flattened neck, opening rounded or slot-like, solitary or clustered. *Peridium*: thin to medium, comprising compressed brown-walled cells or *textura igantean* inwardly and blackened fungal cells with lumens occluded at the ostiole, wider laterally at the base. *Pseudoparaphyses*: hyphae-like, cellular, numerous, longer than asci, embedded in gelatinous matrix. *Asci*: 8-spored, clavate to cylindrical-clavate, short pedunculate, bitunicate, fissitunicate, apically rounded with a small ocular chamber and faint ring. *Ascospores*: biseriate, fusiform or narrowly fusiform, 2 to multi-celled or muriform, euseptate, hyaline or brown, smooth, verrucose or striate, surrounded by a mucilaginous sheath or with appendages.

Type species: *Lophiostoma caulium* Ces. & De Not.

Habitat: Saprobic wood.

Description: Based on Cai et al. (2006) and Jones et al. (2009a).

67.1. ***Lophiostoma armatispora*** (K.D. Hyde, Vrijmode, Chinnaraj & E.B.G. Jones) Liew, Aptroot & K.D. Hyde, 2002. *Fungi in Marine Environment* (eds.), Fungi Diversity Press, Hong Kong, pp. 106. [For Fig. refer original paper]

Basionym: *Massarina armatispora* K.D. Hyde, Vrijmode, Chinnaraj & E.B.G. Jones, *Bot. Mar.*, **35**: 325 (1992).

Ascomata: up to 260 µm high, 351-450 µm diam., perithecioid, immersed, depressed globose or almost conical, mostly with flattened base, carbonaceous, ostolae, with a neck and aring mostly singly, Necks: black, short, raised, seen as a black column on the host surface with a cylindrical almost slit-like opening, 78-150 µm high, 130-176 µm in diam. (above the host surfsce), periphysate, comprised of small dark compact cells. *Peridium*: up to 15 µm thick, brown, at the inside comprised of thicl-walled hyaline compressed cells, brown and angular toward the middle and a confused highly melanised *textura angularis* towards the outside. *Hamathecium*: hyaline, hyphae-like, unbranched, numerous, septate, weakly tapering, up to 3 µm wide, fused at the base and sides of the ascomata and joined at the top with the periphyses and upper wall. *Asci*: 112-148 x 19.6-25.5 µm, forming from the sides and base of ascomata, 8-spored, clavate, bitunicate, fissitunicate, pedunculata and with an ocular chamber and non amyloid ring. *Ascospores*: 28-39.2 x 7-9.8 µm, 2-3 seriate, 2-celled, fusiform constricted at the septum, hyaline, surrounded by a narrow mucilaginous sheath which is drawn out at the poles to form tapering appendages, 6-8 µm long.

Habitat: On intertidal mangrove wood.
Description: Based on Hyde et al. (1992).
Distribution: West Coast:-Goa: On intertidal wood of *Avicennia marina* and *A. officinalis* (Hyde et al., 1992); **Karnataka:** On intertidal wood of *Rhizophora mucronata* (Hyde et al., 1992).
East Coast:-Andaman & Nicobar Islands: On intertidal mangrove wood (Hyde et al., 1992), on intertidal wood of *Avicennia marina* and *Rhizophora mucronata* (Chinnaraj, 1993a).

68. ***Lulworthia*** G.K. Sutherl., 1916. *Trans. Br. Mycol. Soc.*, **5:** 259.

Ascomata: solitary or gregarious, globose to cylindrical depending on the host substratum, immersed or superficial, ostiolate, papillate or with a long neck, coriaceous to subcarbonaceous, pale brown red to purple to dark brown depending on species. *Paraphyses*: absent. *Asci*: 8-spored, cylindrical, clavate or fusiform,

unitunicate, thin-walled, early deliquescing. *Ascospores*: filiform, curved, hyaline, with conical end chambers formed from the outgrowth of the episporium and mesosporium, and filled with mucilage which is released through a pore at the tip of the appendage. The genus is represented by 11 obligate marine species (Jones et al., 2009a).

Type species: *Lulworthia fucicola* Sutherland

Habitat: On intertidal wood, coral or coralline algae, brown algae, molluskan shells, sea grasses.

Description: Based on Kohlmeyer and Kohlmeyer (1979) and Jones et al. (2009a).

Key to *Lulwothia* species from India:

1 Ascospores 9-12-septate, 170-240 x 4 µm .... *Lulworthia lindroidea*

1' Ascospores with no septa 2

2 (1') Ascospores over 500 µm long, 500-756 x 3-5 µm .... *Lulworthia grandispora*

2'(1') Ascospores shorter than 500 µm 3

3 (2') Ascomata dark brown, ascospores 230-320 x 3.6-5.7 µm .... *Lulworthia floridana*

3'(2') Ascomata purplish brown turning green in lactophenol, elongate, ascospores 185-265 x 3-4 µm .... *Lulworthia purpurea*

3"(2') Ascospores variable in dimensions .... *Lulworthia sp.*

68.1. ***Lulworthia floridana*** Meyers, 1957. *Mycologia*, **49:** 515.

[Fig. 68.1]

Ascomata: 187-216 x 173-216 µm, immersed or superficial, golbose to pyriform. *Necks*: 280-720 x 28-30 µm, papillate to elongate, straight, irregular or curved, hyaline at apex, concolorous with venter below, two or three, single or branched ones occasionally occurring on a single ascocarp. *Asci*: long cylindraical, deliquescent. *Ascospores*: 230-266-320 x 3.6-4.2-5.7 µm, filiform, one-celled, hyaline, each end provided with a hyaline conical appendage/chamber. *Appendages*: 7-35 –m long, predominantly 8-11 mm long.

Habitat: On intertidal wood and root stocks of *Spartina alterniflora*.

Description: Based on Meyers (1957) and Johnson and Sparrow (1961).

Distribution: West Coast:-Maharashtra: On submerged wood panels (Jones, 1968), on intertidal mangrove wood (Ramesh and Borse, 1989); **Kerala:** On submerged wood panels (Jones, 1968).

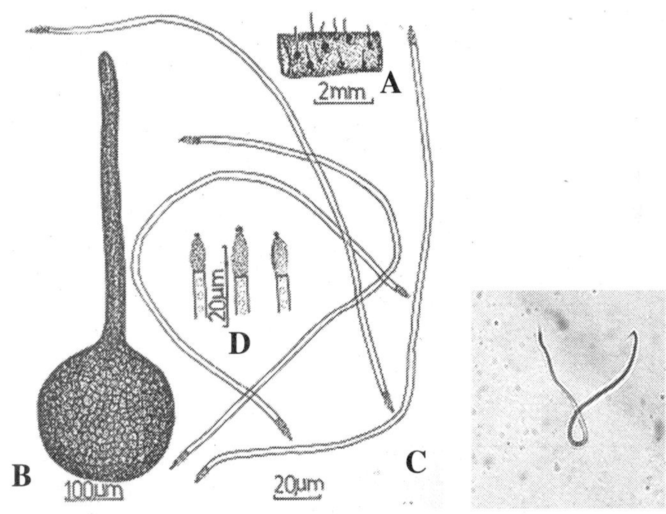

Fig. 68.1. *Lulworthia floridana,* A-D; A- Habit, B- ascocarp, C- ascospores, D- ascospore tips.

68.2. *Lulworthia grandispora* Meyers, 1957. *Mycologia,* **49:** 513. [Fig. 68.2]

Basionym: *Lulworthia grandispora* var. *apiculata* Johnson, *Mycologia*, **50:** 159 (1958).

Ascomata: 180-306 µm in diam., globose or subglobose to pyriform, immersed or superficial, ostiolate, with long neck, brown to black, solitary or gregarious. *Necks*: 75-1400 x 15-33 µm, cylindrical, straight or curved, sometimes two on one ascomata. *Paraphyses*: absent. *Asci*: 8-spored, elongate-fusiform or cylindrical, unitunicate, thin-walled, early deliquescing. *Ascospores*: 500-756 x 3-5 µm (including apical chambers), filamentous, curved, hyaline, tapering at each end into an elongate, conical process or apical chamber; processes 3.6-7 µm long, acute or rounded, filled with mucus that is released through an apical pore.

Habitat: Saprobic on intertidal and mangrove wood.

Description: Based on Kohlmeyer and Kohlmeyer (1979).

Distribution:West Coast:-Diu Island: On intertidal wood (Borse et al., 1999b); **Daman:** On intertidal wood and *Avicennia marina* (Borse et al., 2000b); **Gujarat:** On intertidal wood of *Avicennia marina* (Borse et al., 2000a), on intertidal wood and mangrove wood (Patil and Borse, 2001); **Maharashtra:** On intertidal wood of *Avicennia marina* and *Rhizophora mucronata, Sonneratia acida* (Patil and Borse, 1986, Borse, 1987a, 1988; Ramesh and Borse, 1989; Borse et al., 2005a), on intertidal wood (Shrivastava, 1995), on intertidal mangrove wood (Maria and Sridhar, 2002a); **Goa:** On intertidal wood (Nandan et al., 1993), on intertidal mangrove wood (Maria and Sridhar, 2002a); **Karnataka:** On intertidal wood and teak panels (Prasannarai et al., 1999), on intertidal wood (Prasannarai and Sridhar, 2001, 2000-2001; Ananda and Sridhar, 2003, 2004), on intertidal mangrove wood, roots of *Rhizophora mucronata* and *Sonneratia caseolaris* (Maria and Sridhar, 2002a, b, 2004), on intertidal wood of *Acanthus ilicifolius, Avicennia officinalis, Brugueiria gymnorrhiza, Rhizophora mucronata* and *Sonneratia caseolaris* (Maria and Sridhar, 2003a, 2004); **Pondecherry (Mahe):** On intertidal wood (Nambiar and Raveendran, 2007); **Kerala:** On intertidal mangrove wood (Maria and Sridhar, 2002a), on intertidal wood (Raveendran and Manimohan, 2007), on intertidal wood, wood of *Bruguiera* sp. and wood of *Rhizophora* sp. (Nambiar and Raveendran, 2006; 2007; 2008b, d; 2009a, b, d, e, f, g, h; 2010b, c; Nambiar et al., 2006, 2008a); **Lakshadweep Islands:** On intertidal wood of *Avicennia marina* (Chinnaraj, 1992); on intertidal mangrove wood (Ananda and Sridhar, 2003).

East Coast:-Tamil Nadu: On submerged parts of *Rhizophora* spp. (Ravikumar and Vittal, 1996), on intertidal wood (Nambiar et al., 2008b), on mangrove wood (Nambiar and Raveendran, 2008e, 2009l, 2010c); on intertidal woody debris (Ravikumar et al., 2009); **Pondicherry (Karaikkal):** on intertidal woody debris (Ravikumar et al., 2009); **Andhara Pradesh:** On intertidal wood of *Avicennia marina, A. officinalis, Lumnitzera racemosa, Rhizophora apiculata* and *Sonneratia apetala* (Sarma and Vittal, 1998-99; Sarma and Vittal, 2000; Sarma and Vittal, 2001; Sarma et al., 2001; Sarma and Vittal, 2002; Sarma and Vittal, 2004; Vittal and Sarma, 2005); **West Bengal:** On intertidal wood of *Aegialitis rotundifolia* and *Rhizophota mucronata* (Pawar and Borse, 2004); **Andaman & Nicobar Islands:** On intertidal wood of *Avicennia marina, A. officinalis, Bruguiera gymnorhiza, Rhizophora apiculata, R. mucronata* and *Sonneratia alba* (Chinnaraj, 1993a).

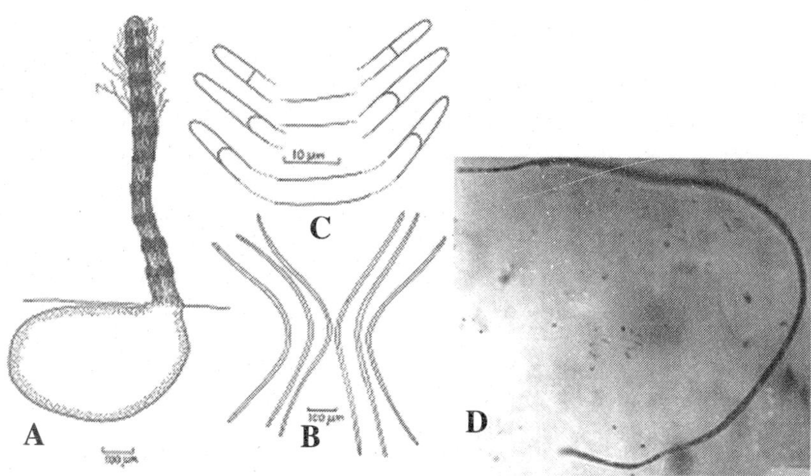

Fig. 68.2. ***Lulworthia grandispora,*** A-D; A- ascocarp, B- ascospores,C- ascospore tips, D- ascospore (1000x).

68.3. ***Lulworthia kniepii*** Kohlm., 1963. *Nova Hedwigia*, **6**: 140.

[For Fig. refer original paper]

Habitat: Parasitc on calcified members of red algae.

Distribution: West Coast:- Karanatka: On intertidal wood ((Prasannarai and Sridhar, 2001, 2000-2001), on intertidal mangrove wood (Maria and Sridhar, 2002a); **Kerala:** On intertidal mangrove wood (Maria and Sridhar, 2002a).

Note: The occurrence of *L. kniepii* Kohlmeyer on intertidal wood and mangrove wood is unlikely, because the fungus is parasitc on calcified members of red algae (coralline algae) (Kohlmeyer and Kohlmeyer, 1979; Kohlmeyer and Volkmann-Kohlmeyer, 1991; Jones et al., 2009a).

68.4. ***Lulworthia lindroidea*** Kohlm., 1980. *Bot. Mar.*, **23**: 537.

[For Fig. refer original paper]

Ascomata: 150-220 μm high, 250-450 μm in diam., ellipsoidal, half immersed in the wood, ostiolate, papillate, coriaceous, thin-walled, black, solitary. *Peridium*: 20-40 μm thick, composed of 5-8 layers of dark cells, forming a textura angularis, intimately connected with surrounding wood fibers and enclosing wood cells, merging into the hyaline pseudo-parenchyma of the venter. *Necks*: 40-90 μm long, 40-60 μm in diam., lateral; ostiolar canal 12 μm in diam., without periphyses.

Paraphyses: absent; deliquescing pseudoparenchymatous cells filling venter of immature ascocarps. *Asci*: 185-230 x 14-16 µm, 8-spored, clavate, long pedicellate, tapering toward the apex, unitunicate, thin-walled, early deliquescing, without apical apparatuses, originating opposite the ostiole and oriented horizontally in the ascocarp. *Ascospores*: 170-240 x 4 µm, filamentous, curved, 9-12 septate, not constricted at the septa, with a cylindrical chamber at each end, hyaline; end chambers 6-10 µm long, 2.5-3.0 µm in diam., slightly tapering and truncate at the tip, containing mucus that is released through an apical pore.

Habitat: Saprobic on intertidal and mangrove wood.

Description: Based on Kohlmeyer (1980).

Distribution: West Coast:-Karanataka: On intertidal mangrove wood (Maria and Sridhar, 2002a); **Lakshadweep Islands:** On intertidal mangrove wood (Ananda and Sridhar, 2003).

68.5. *Lulworthia medusa* (Ellis & Everh.) Cribb & J.W. Cribb, 1955. Univ. Queensl. Pap., Dept. Bot., **3:** 80. [For Fig. refer original paper]

Habitat: Saprobic on intertidal dead leaves of *Spartina alterniflora* Lois. Mangrove wood.

Description: Based on Kohlmeyer and Kohlmeyer (1979).

Distribution: West Coast:-Maharashtra: On intertidal wood of *Avicennia alba*, *Rhizophora mucronata, Sonneratia acida* (Borse, 1987e; Ramesh and Borse, 1989; Shrivastava, 1995)

East Coast:-Tamil Nadu: On drift and intertidal wood (Raghukumar, 1973).

Note: According to Kohlmeyer et al. (2000) the epithet *medusa* should be applied only to the *Lulworthia* occurring on *Spartina* and not be used as a catch-all for taxa with similar ascospore size from other hosts or substrates, as has been done by many marine mycologists in the past.

68.6. *Lulworthia purpurea* (I.M. Wilson) T.W. Johnson, 1958. Mycologia, 50: 154. [For Fig. refer original paper]

Ascomata:126-339 x 98-224 µm, immersed becoming superficial through erosion of the substratum, golbose, flask shaped, or flattened, black when containing spores, purple-brown when empty, wall is thin, and two-layered, outer wall layer cells are irregularly polygonal and flattened, contain a homogenous purplish-brown pigments,

inner wall layer cells are shaped like those of the outer, but are hyaline . *Necks*: up to 628 μm long, 15-27 μm wide, long, cylindrical, irregular or curved, centric, eccentric or alteral, concolourous with venter, but pale or hyaline at ostiole. *Asci*: elongate, clavate, deliquescent. *Ascospores*: 185-265 (220-350) x 3-4 μm, filiform, one-celled, hyaline, each end provided with a hyaline conical appendage/chamber. *Appendages*: 10-(18)-26 μm long, very rarely reaching only 6 mm in length.

Habitat: Saprobic on intertidal wood.

Description: Based on Johnson and Sparrow (1961).

Distribution: West Coast:-Maharashtra: On submerged wood panels (Jones, 1968), on intertidal mangrove wood (Ramesh and Borse, 1989); **Kerala:** On submerged wood panels (Jones, 1968).

68.7. ***Lulwothia*** sp.

Habitat: Saprobic on intertidal and mangrove wood.

Distributin: West Coast:-Gujarat: On intertidal wood (Borse et al., 2000b); **Maharashtra:** On intertidal wood of *Avicennia alba*, *Rhizophora mucronata, Sonneratia acida* (Patil and Borse, 1986; 1988; 340-380 x 4-5 μm, Borse et al., 2005a); on intertidal mangrove wood (350-450 μm long, Maria and Sridhar, 2002a); **Goa:** On intertidal wood (*L.* sp. 2: 245-372 x 2.5 μm, *L.* sp. 3: 255-295 x 2-2.5 μm, *L.* sp. 4: 274-333 x 2.5 μm, Prasannarai and Sridhar, 2001), on intertidal mangrove wood (350-450 μm long, Maria and Sridhar, 2002a); **Karnataka:** On intertidal wood (Prasannarai and Sridhar, 1997, *L.* sp. 1: 147-235 x 2.5 μm, *L.* sp. 3: 255-295 x 2-2.5 μm, *L.*sp. 4: 274-333 x 2.5 μm, 2001; 255-295 x 2-2.5 μm, Ananda et at., 1999), on intertidal mangrove wood (350-450 μm long, Maria and Sridhar, 2002a); **Kerala:** On intertidal wood (*L.* sp. 1: 147-235 x 2.5 μm, Prasannarai and Sridhar, 2001), on intertidal mangrove wood (350-450 m long, Maria and Sridhar, 2002a), on intertidal wood (Raveendran and Manimohan, 2007 (*L.* sp. I: 310-370 x 3-4 μm, *L.* sp. II: 250-300-500 x 3-5 μm, *L.* sp. III: 200-275 x 3-4 mm); Nambiar and Raveendran, 2008b, d, 2009d, 2010c); **Lakshadweep Islands:** On intertidal wood of *Avicennia marina* and *Pemphis acidula* (300 x 400μm long) (Chinnaraj, 1992).

East Coast:-Tamil Nadu: On intertidal wood (*L.* sp. 1: 147-235 x 2.5 μm, *L.* sp. 3: 255-295 x 2-2.5 μm, Prasannarai and Sridhar, 2001), on submerged parts of *Rhizophora* spp. (Ravikumar and Vittal, 1996), on intertidal wood (Nambiar et al., 2008b); on intertidal woody debris (Ravikumar et al., 2009); **Pondecherry**

(Karaikkal): on intertidal woody debris (Ravikumar et al., 2009); **Andhara Pradesh:** On intertidal wood of *Aegiceras corniculatum, Avicennia marina, A. officinalis, Excoecaria agallocha, Rhizophora apiculata* and *Sonneratia apetala* (Sarma and Vittal, 1998-99; Sarma and Vittal, 2000; Sarma and Vittal, 2001; Sarma et al., 2001; Sarma and Vittal, 2002; Sarma and Vittal, 2004; Vittal and Sarma, 2005); **West Bengal:** On intertidal wood of *Aegialitis rotundifolia* and *Avicennia marina* (234-275 x 3-4 µm, Pawar and Borse, 2004); **Andaman & Nicobar Islands:** On intertidal wood of *Avicennia marina, A. officinalis, Bruguiera gymnorhiza, Rhizophora apiculata, R. mucronata* and *Sonneratia alba* (Chinnaraj, 1993).

69. *Manglicola* Kohlm. & E. Kohlm., 1971. *Mycologia*, **63**: 840.

Ascomata: solitartary, obtuse clavate to fusiform, stipitate, superficial, seated in the substrate with a hypostroma, ostiolate, epapillate, coriaceous, olive-brown, periphysate. *Pseudoparaphyses*: numerous, septate, simple or reticulate. *Asci*: 8-spored, cylindrical, bitunicat, thick-walled, developing at the base of the ascomata venter between the pseudoparaphyses. *Ascospores*: fusiform, apiculate, 1-septate; apical cell large, dark brown; basal cell small, light brown; deliquescing appendages cover both apices. The genus is monotypic (Jones *et al.*, 2009a).

Type species: *Manglicola guatemalensis* Kolhm. & E. Kolhm.
Habitat: On intertidal mangrove wood.
Description: Based on Kohlmeyer and Kohlmeyer (1971).

69.1. *Manglicola* sp. 1957. *Mycologia*, **63**: 841.

Ascomata: 1000-1100 x 500-700 µm, perithecial, fusiform, superficial, papillate. *Necks*: 200-230 x 300-345 µm, cylindrical. *Paraphyses*: 300-350 x 5-7,5 µm, aseptate with forked tips. *Asci*: 630-695 x 27.5-32.5 µm, cylindrical, bitunicate, with a prominent umbrella-like apical apparatus, 8-spored with uniseriatly arranged ascospores. *Ascospores*: 90-100 x 15-18 µm, hyaline, without sheath or appendages, fusiform, immature spores aseptate.

Habitat: On intertidal wood.
Description: Based on Raveendran and Manimohan (2007).
Distribution: West Coast:-Kerala: On intertidal wood (Raveendran and Manimohan, 2007).

70. *Marinosphaera* K.D. Hyde, 1989. *Can. J. Bot.*, 67: 3080.

Ascomata: ellipsoidal to pyriform, ostiolate, papillate, periphysate, membranacious, and light to dark brown. *Necks*: long or short. *Peridium*: thin. *Paraphyses*: wide and septate. *Asci*: 8-spored, cylindrical, pedunculate, persistent, unitunicate, with subapically thickened structure. *Ascospores*: broad ellipsoidal to fusiform, 3-septate, hyaline. The genus is monotypic (Jones et al., 2009a).

Type species: *Marinosphaera mangrovei* K.D. Hyde
Habitat: On intertidal mangrove wood.
Description: Based on Hyde (1989).

70.1. *Marinosphaera mangrovei* K.D. Hyde, 1989. *Can. J. Bot.*, 67: 3080.

[For Fig. refer original paper]

Ascomata: 107-228 µm high, 98-199 µm in diam., ellipsoidal, globose, subglobose or pyriform, immersed, ostiolate, papillate, periphysate, membranacious, and light to dark brown, periphysate, solitary. *Necks*: long or short, hyaline, but dark at the base, up to 398 um long. *Peridium*: 8.5-17.4 µm thick, hyaline to brown, composed of elongate angular cells. *Paraphyses*: 1.6-6 µm wide, simple, septate. *Asci*: 88.4-112 x 18.5-22.7 µm, 8-spored, clavate, pedunculate, persistent, unitunicate, not bluing in Melzer's reagent, treatment with lactophenol reveals a subapical plate through which a faint pore may be observed. Asci develop from ascogenous tissue at the base of the ascomata. *Ascospores*: 25.5-31 x 6-8.5 µm, 2-3 seriate, broad ellipsoidal to fusiform, 3-septate, hyaline, smooth walled, hyaline in mass.

Habitat: On intertidal mangrove wood.
Description: Based on Hyde (1989).
Distribution: West Coast:-Pondecherry (Mahe): On intertidal wood (Nambiar and Raveendran, 2007); **Kerala:** On intertidal wood (Raveendran and Manimohan, 2007), on intertidal wood and mangrove wood of *Avicennia* sp. (Nambiar and Raveendran, 2006, 2008b, d, 2009a, b, d, e, f, g, h, I, 2010b, c; Nambiar et al., 2006, 2008a); **Lakshadweep Islands:** On intertidal wood of *Scaveola taccada* and pneumatophores of *Ceriops tagal* (Chinnaraj, 1992).
East Coast:-Tamil Nadu: On mangrove wood (Nambiar and Raveendran, 2008e); **Andhara Pradesh:** On intertidal wood of *Avicennia marina,* and *A. officinalis* (Sarma and Vittal, 1998-99; Sarma and Vittal, 2000; Sarma and Vittal, 2001; Sarma et al., 2001; Sarma and Vittal, 2004; Vittal and Sarma, 2005); **Andaman & Nicobar Islands:** On intertidal wood of *Avicennia marina, A.*

officinalis and *Rhizophora apiculata* (Chinnaraj, 1993a); **West Bengal:** on mangrove woody litter (Shini et al., 2009-10).

71. ***Morosphaeria*** Suetrong, E.B.G. Jones & Schoch, 2009. *Studies in Mycology*, **64:** 161.

Ascomata: solitary or gregarious, subglobose to lenticular, immersed becoming superficial, ostiolate, papillate, coriaceous, brown to black, pseudoparaphyses filamentous, anastomosing, branching and numerous. *Asci*: 8-spored, clavate to cylindrical, short pedunculate, thick-walled, bitunicate, fissitunicate, with an ocular chamber and apical apparatus, persistent. *Ascospores*: hyaline, 1-3 septate, constricted at the septa, fusiform to ellipsoidal, surrounded by a mucilaginous sheath. The genus is represented by two species (Suetrong et al., 2009).

Type species: *Morosphaeria velataspora* (K.D. Hyde & Borse) Suetron, Sakayaroj, Jones & Schoch,
Habitat: Saprobic on intertidal parts of mamgroves in marine habitats.
Discription: Based on Hyde and Borse (1986b)

71.1. ***Morosphaeria velataspora*** (K. D. Hyde & Borse) Suetron, Sakayaroj, E.B.G. Jones & Schoch, 2009. *Studies Mycology*, **64:** [Fig. 71.1]

Basionym: *Massarina velataspora* K.D. Hyde & Borse, 1986b. **27:** 163.

Ascomata: 700-1200 µm in diam., immersed, becoming erumpent, subglobose or depressed, ostolate, rarely papillate, coriaceous, brown to black, solitary or gregarious, exposed tissue sometimes covered in a fine hyphal mat. *Peridium*: upper walls composed of brown thick-walled polyhedral cells (*textura angularis*) and brown hyphae interdispersed with sparse epidermal and subepidermal tissue of the substratum. The cells are dense and darkened around the ostiole. At the rim is an area of vertical parallel brown elongate cells, wedge-shaped in section. The base is flattened and poorly developed, consisting of a thin layer of light brown pseudoparenchymatous tissue with small irregular cells. This tissue is also found in the lower region of the side walls. *Pseudoparaphyses*: 1.1-1.7 µm wide, branching, filamentous, hyaline, and numerous. *Asci*: 220-320 x 23-34 µm, 8-spored, clavate to cylindrical, short pedunculate, fissitunicate, thick-walled, with an apical apparatus which is ring-like, developing from the base of the ascomata. *Ascospores*: 45-56 x 14-19 µm, obliquely uni-seriate, fusiform to ellipsoidal, 1-3-septate, hyaline,

constricted at the septa, central cells larger, apical cells smaller and elongate, ascospores surrounded by a mucilaginous sheath, 5-22 µm wide. Spores purple in mass, after being extruded from the ascomata neck.

Habitat: On intertidal mangrove wood.

Description: Based on Hyde and Borse (1986b) and Hyde (1991b).

Distribution: West Coast:- Daman: On intertidal wood and *Avicennia marina* (Borse et al., 2000b); **Gujarat:** On intertidal wood of *Ceriops tagal* and *Rhizophora mucronat* (Borse et al., 2000a); **Maharashtra:** On intertidal prop roots of *Rhizophora mucronata* and wood of *Ceriops tagal* (Hyde and Borse, 1986b; Borse, 1988), on intertidal wood (Shrivastava, 1995); **Goa:** On intertidal prop roots of *Rhizophora mucronata* (Borse et al., 1999a), on intertidal mangrove wood (Maria and Sridhar, 2002a); **Karnataka:** On intertidal wood (Prasannarai and Sridhar, 2001), on intertidal wood of *Brugueiria gymnorrhiza* (Maria and Sridhar, 2002a, 2003a); **Lakshadweep Islands:** On intertidal wood of *Ceriops tagal* (Chinnaraj, 1992), on intertidal wood of *Avicennia marinas, Bruguiera gymnorhiza, Rhizophora mucronata* and *Sonneratia alba* (Chinnaraj, 1993).

East Coast:-Tamil Nadu: On submerged parts of *Rhizophora* spp. (Ravikumar and Vittal, 1996); **Pondecherry (Karaikkal):** on intertidal woody debris (Ravikumar et al., 2009); **Andhara Pradesh:** On intertidal wood of *Avicennia officinalis, Ezcoecaria agallocha, Rhizophora apiculata* and *Sonneraria apetala* (Sarma and Vittal, 1998-99; Sarma and Vittal, 2000; Sarma and Vittal, 2001; Sarma et al., 2001; Sarma and Vittal, 2002; Sarma and Vittal, 2004; Vittal and Sarma, 2005); **West Bengal:** On intertidal wood of *Avicennia officinalis* and *Ceriops tagal* (Pawar et al., 2005); **Andaman & Nicobar Islands:** On intertidal mangrove wood of *Avicennia marina, A. officinalis, Bruguiera gymnorhiza* and *Rhizophora mucronata* (Chinnaraj, 1993a).

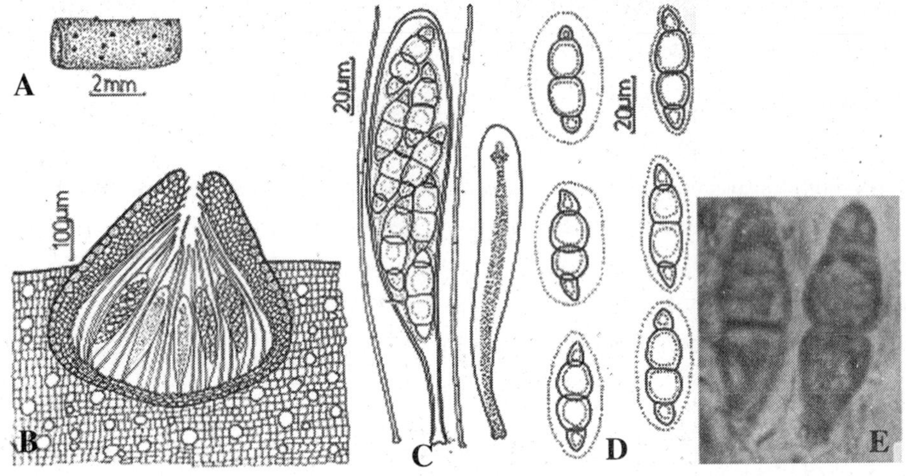

Fig. 71.1. ***Morosphaeria velataspora,*** A-E; A- Habit, B- v.s. of ascocarp, C- ascus and paraphyses, D- ascospores, E- ascospores 1000x).

72. *Mycosphaerella* Johanson, 1884. *Oefvers. Foerh. K. Sven. Vetensk.-Akad.,* **41:** 163.

Ascomata: solitary or gregarious, globose, subglobose, conical or depressed, in some species surrounded by a hyphoid stroma or even enclosed in a compact stroma, immersed or erumpent, ostiolate, epapillate or with a short papilla, thin-walled to medium thick-walled. *Peridium*: composed of 1 to 4 layers of dark, polygonal cells. *Pseudoparaphyses*: usually absent; young ascomata filled with a pseudoparenchyma of thin-walled, polygonal or rounded, deliquescing cells. *Asci*: 8-spored, elongate-cylindrical, ventricose, ovoid or rarely short clavate, bitunicate, arising ia a fascicle or parallel to each other from a small-celled ascogenous tissue at the base of the ascomata venter. *Ascospores*: bi-seriate or irregularly arranged, ellipsoidal to elongate, usually 3 times as long as wide, 1-septate near the middle, hyaline, in some species becoming brownish in age or surrounded by a gelatinous sheath. The genus is represented by species in marine environment (Jones *et al.*, 2009a).

Type species: *Mycosphaerella punctiformis* (Persoon & Fries) Starback

Habitat: Saprobic on dead substrates.

Description: Based on Kohlmeyer and Kohlmeyer (1979).

72.1. *Mycosphaerella apophlaeae* Kohlm., 1981. In: Kohlmeyer & Demoulin, *Bot. Mar.*, **24:** 13. [For Fig. refer original paper]

Distribution: East Coast:-Karnataka: On leaf and woody litter of mangroves (Ananda and Sridhar, 2004).

Note: The occurrence of *M. apophlaeae* on leaf and woody litter of mangroves is unlikely, because according to Kohlmeyer and Demoulin (1981) and Kohlmeyer and Hawkes (1983) the present fungus can be considered as mycobiont in a mycophycobiosis with marine red algae *Apophlaea lyallii* Hooker f. & Harvey and *A. sinclairii* Harvey in Hooker & Harvey.

72.2. *Mycosphaerella pneumatophorae* Kohlm., 1966. *Ber. Dtsch. Bot. Ges.*, **79**: 32. [Fig. 72.2]

Ascomata: 90-170 μm high, 110-175 μm wide, sunglobose or depressed ellipsoidal, more or less cupuliform, immersed in the bark tissue, finally apically exoposed, ostiolate, epapillate, coriaceous, brownish-black above, pallid below, solitary. *Peridium*: 15-20 μm thick, composed of irregular, polygonal, thick-walled, dark brown cells with large lumina in the upper part of the ascomat, with similar, but light-coloured cells with small lumina in the base forming a *textura angularis*, enclosing fragments of host cell walls. Ostiole: about 35 μm in diam., finally enlarged by a crumbling-away of cells of the peridium, developing into an irregular opening of about 75 μm in diam., exposing the asci. *Pseudoparaphyses*: forming an indistinct network in immature ascomata, absent in mature fruiting bodies. *Asci*: 37-60 x 15-21 μm, 8-spored, ellipsoidal or clavate, short pedunculate, bitunicate, thick-walled, physoclastic, without apical apparatuses; endoascus remaining protruding from endoascus after spores discharge and filling the ostiolar opening of the ascomata; developing at the base of the ascomat venter. *Ascospores*: 14-18 x 7-8.5 μm, bi- or tri-seriate, obovoid, 1-septate, somewhat above the middle, slightly above the middle, slightly constricted at the septum; upper cell wider than lower one; lower cell taperinh at the base; hyaline.

Habitat: Parasitic or saprobic on pneumatophores of *Avicennia* sp.

Description: Based on Kohlmeyer and Kohlmeyer (1979).

Distribution: East Coast:-Tamil Nadu: On pneumatophores of *Avicennia marina* (Ravikumar and Vittal, 1987); **Andhara Pradesh:** On pneumatophores of *Avicennia officinalis* (Sarma amd Vittal, 1998-99; Sarma and Vittal, 2000; Sarma and Vittal, 2001; Sarma et al., 2001; Sarma and Vittal, 2004; Vittal and

Sarma, 2005). **West Bengal :** On pneumatophores of *Avicennia marina* (Pawar, N.B., unpublished data). **Orissa:** On pneumatophores of *Avicennia marina* (Borse, K.N., unpublished data)

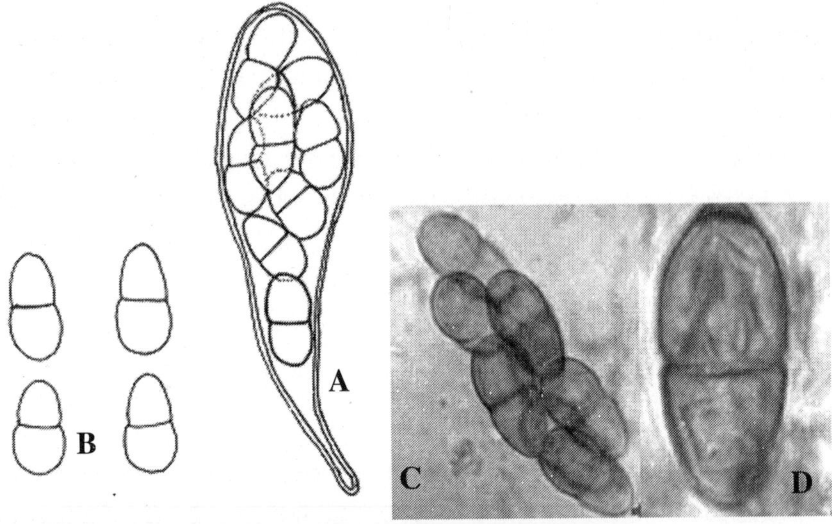

Fig. 72.2. *Mycosphaerella pneumatophorae*, A-D; A- - ascus, B- ascospores, C- Ascus, D- Ascospores (1000x).

72.3. *Mycosphaerella salicorniae* (Auersw.) Lindau, 1934. *Nova Hedwigia*, **74**: 35. [For Fig. refer original paper]

Habitat: On intertidal drying stalks and inflorescence of Chenopodiaceae (*Salicornia* and *Suaeda* sp.)

Distribution: West Coast:-Karnataka: On intertidal wood (Prasannarai and Sridhar, 2001, 2000-2001), on intertidal wood of *Brugueiria gymnorrhiza* (Maria and Sridhar, 2003a) on intertidal leaf and woody litter of mangroves (Ananda and Sridhar, 2003, 2004).

Note: The occurrence of *M. salicorniae* on intertidal wood, mangrove wood, and leaf and woody litter of mangroves is unlikely, because the fungus is a saprobic on intertidal drying stalks and inflorescence of Chenopodiaceae (*Salicornia* and *Suaeda* sp.) (Kohlmeyer and Kohlmeyer, 1979; Jones et al., 2009a).

72.4. *Mycosphaerella staticicola* (Pat.) Dias, 1970-71. *Mem. Soc. Broteriana*, **21**: 72. [For Fig. refer original paper]

Habitat: On intertidal drying inflorescence of Plumbaginaceae (*Armeria pungens* and *Limonium* sp.)

Distribution: West Coast:-Karnataka: On intertidal wood (Prasannarai and Sridhar, 2001, 2000-2001), on intertidal leaf and woody litter of mangroves (Ananda and Sridhar, 2003, 2004).

Note: The occurrence of *M. staticicola* on intertidal wood and leaf and woody litter of mangroves is unlikely, because the fungus is a saprobic on intertidal drying inflorescence of Plumbaginaceae (*Armeria pungens* and *Limonium* sp.) (Kohlmeyer and Kohlmeyer, 1979; Jones et al., 2009a).

73. *Natantispora* J. Campb., J.L. Anderson & Shearer, 2003. *Mycologia*, **95:** 543.

Ascomata: immersed to superficial, ostiolate, membranous, black. Peridium: composed of pseudoparenchyma in longitudinal section, of *textura angularis* in surface view. *Necks*: long. Cylindrical, periphysate, dark-pigmented, gradually lightening toward the apex. *Hamathecium*: contains catenophyses. *Asci*: clavate, thin-walled, lacking any thickening of the apical wall, an apical pore or an apical apparatus, deliquescing before or at spore maturity. *Ascospores*: hyaline, fusifrom to ellipsoidal, 1-septate, with a single hamate appendage at each apex. *Appendages*: thread-like, coiled into a hamate structure equal to or longer than a single spore cell, unferling in water to form a long, fine sticky, threadlike structure. The genus is represented by two species (Jones *et al.*, 2009a).

Type species: *Natantispora retorquens* (Shearer & J.L. Crane) J. Campb., J.L. Anderson & Shearer

Habitat: On intertidal and mangrove wood.

Description: Based on Campbell et al. (2003) and Jones et al. (2009a).

73.1. *Natantispora retorquens* (Shearer & J.L. Crane) J. Campb., J.L. Anderson & Shearer, 2003. *Mycologia*, **95:** 543. [Fig. 73.1]

Basionym: *Halosarpheia retorquens* Shearer and J.L. Crane.

Bot. Mar., **23:** 608 (1980).

Ascomata: 140-326 x 150-366 μm., solitary to gregarious, superficial to immersed, globose to subglobose, hyaline at first becoming black, ostiolate, Neck: 108- 564 x 14-50 μm, cylindrical, periphysate, dark at base, hyaline at apex. *Peridium*: ,embranous, pseudoparenchymatous, multilayered, the outer layer brown to black. *Centrum*: consisting of asci and catenophyses. *Asci*: 53-144 x 14-24 μm, unitunicate, clavate, thin-walled, deliquescing before or at spore maturity. *Ascospores*: 20-34 x 7-11 μm, ellipsoidal, hyaline, 1-septate, appendaged. Appendages: bipolar, composed of single, coiled or folded filament, at first hamate, finally unwinding in water to produce a long fine filament.

Habitat: On intertidal wood.

Description: Based on Shearer and Crane (1980).

Distribution: West Coast:-Maharashtra: On intertidal wood of *Rhizophora mucronata* (Patil and Borse, 1986), prop roots and seedlings of *R. mucronata* (Borse,1987d, 1988; Borse et al., 2005a); **Karnataka:** On intertidal wood (Prasannarai et al., 1999), on intertidal wood (Prasannarai and Sridhar, 2001), on intertidal mangrove wood (Maria and Sridhar, 2002a, 2004), on intertidal leaf and woody litter of mangroves (Ananda and Sridhar, 2004); **Kerala:** On intertidal mangrove wood (Maria and Sridhar, 2002a), on intertidal wood (Raveendran and Manimohan, 2007), on intertidal wood, wood of *Avicennia* sp. and *Bruguiera* sp. (Nambiar and Raveendran, 2007, 2008a, b, 2009d, h, 2010c).

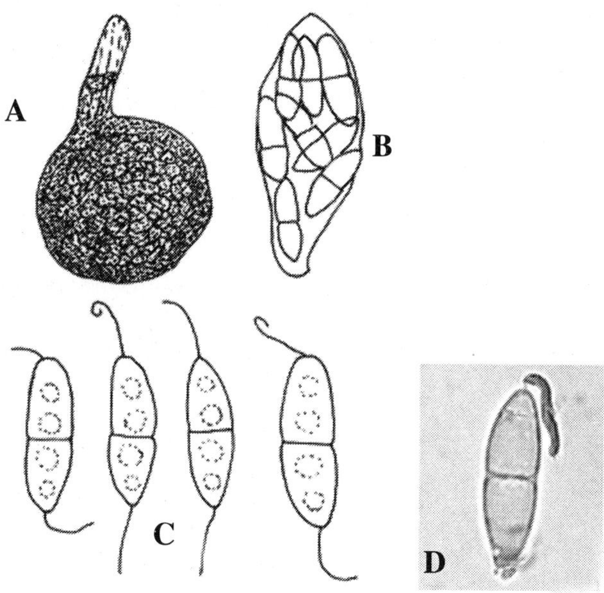

Fig. 73.1. ***Natantispora retorquens,*** A-D; A- ascocarp (100x), B- ascus (450x), C- ascospores (1000x), D- ascospore (1000x).

74. *Neomassariosphaeria* Yin, Zhang, F. Fourn. & K.D. Hyde, 2009b. *Studies in Mycology*, 64:

Aquatic. Saprobic. *Ascomata*: scattered or in small groups, immersed to erumpent, subglobose to lenticular; wall black, apex elongated. *Peridium*: thin. *Pseudoparaphyses*: trabeculate, embedded in mucilage. *Asci*: 8-spored, bitunicate, fissitunicate, clavate to broadly clavate, short pedicellate. *Ascospores*: narrowly fusiform, hyaline to reddish brown, multi-septate, constricted at the median septum, usually surrounded by an irregular, hyaline, gelatinous sheath. Ascomat or hyphae usually stain the woody substrate or cultural medium purple. The genus is represented by two species (Zangh *et al.* 2009b).

Type species: *Neomassariosphaeria typhicola* (P. Karst.) Yin, Zhang, F. Fourn. & K.D. Hyde

Habitat: On *Typha* sp. in freshwater habitats.

Description: Based on Zhang et al. (2009b)

74.1. ***Neomassariosphaeria typhicola*** (P. Karst.) Yin, Zhang, F. Fourn. & K.D. Hyde, 2009. *Studies Mycology*, **64:** [For Fig. refer original paper]

Basionym: *Leptosphaeria typhicola* P. Karst., 1873. *Bidrag Kannedom Finlands nature Folk.*, **23:** 100.

≡*Leptosphaeria baldingerae* Fautrey & F. Lamb., *Revue Mycol., Toulouse*, **19:** 53 (1897).

≡ *Leptosphaeria cladii* Cruchet, *Bull. Soc., Vaud. Sci. Nat.*, **55:** 161 (1923).

≡ *Phaeosphaeria baldingerae* (Fautrey & F. Lamb.) Hedjar., *Sydowia*, **22:** 87 (1969).

≡ *Phaeosphaeria typhicola* (P. Karst.) Hedjar., 1969. *Sydowia*, **22:** 86.

≡ *Massariosphaeria typhicola* (P. Karst.) Leuchtm., 1984. *Sydowia*, **37:** 168.

≡ *Chaetomastia typhicola* (P. Karst.) M.E. Barr, 1989. *Mycotaxon*, **34:** 514.

Ascomata: 180-220 µm high, 200-430 µm diam., subglobose to depressed-ellipsoidal, immersed or erumpent, ostiolate, epapillate or with short papillae perforating the epidermis, subcoriaceous, dark brown, solitary or gregarious. *Peridium*: 10-20 µm thick, composed of 3-5 layers of large, thin-walled cells with large lumina, forming a textura angularis. *Pseudoparaphyses*: 2-3 µm wide, septate, simple, merging apically with pseudoparenchyma, which fills the ostiolar canal before maturation of the ascomata. *Asci*: 90-140 x 18-26 um, 8-spored, cylindrical to clavate, short pedunculate, bitunicate, thick-walled; developing at the base of the ascomata venter. *Ascospores*: 34-52 x 7-10 µm, bi- or tri-seriate in the upper part of the ascus, uni-seriate below, fusiform, 7-11-septate, slightly constricted at the septa, particularly around the thickest cell (4th or 5th from the top), straight or curved, at first hyaline, becoming light brown (or golden brown) and verrucose in age, surrounded by a gelatinous, 2 to 4 µm thick sheath. *Pycnidia*: 320-400 um wide, globose, black. *Peridium*: 32-48 µm thick, covered by brown hyphae. *Conidia*: 3-4 x 2.5-3 µm, globose, ovoid or ellipsoidal, hyaline.

Habitat: Sabrobic on plant material.

Description: Based on Kohlmeyer and Kolhmeyer (1979).

Distribution: East Coast:-Andhara Pradesh: On intertidal wood of *Rhizophora apiculata* (Sarma and Vittal, 2004).

75. ***Neptunella*** K.L. Pang & E.B.G. Jones, 2003. *Mycolog. Progress,* **2:** 35.

Ascomata: subglobose or bottle-shaped, immersed or semi-immersed, ostiolate, membranous, brownish or hyaline, solitary or gregarious. *Neck*: periphysate. *Catenophyses*: present. *Asci*: unitunicate, 8-spored, ellipsoid-clavate or subfusiform, thin-walled, with short stalk, apically truncate, cytoplasm retracted below the apex, with an apical pore, semi-persistent. *Ascospores*: ellipsoidal, hyaline, thin-walled, lacking appendages but with a thin ascospore sheath or exosporium. The genus is monotypic (Jones *et al.*, 2009a).

Type species: *Neptunella longirostris* (Cribb & J.W. Cribb) K.L. Pang & E.B.G. Jones

Habitat: On intertidal wood.

Description: Based on Pang et al. (2003).

75.1. ***Neptunella longirostris*** (Cribb & J.W. Cribb) K.L. Pang & E.B.G. Jones, 2003. *Mycological progress,* **2:** 35. [Fig. 75.1]

Basionym: *Gnomonia longirostris* Cribb & J.W. Cribb, *Univ. Queen. Pap. Dept. Bot.,* **3:** 101 (1956).

≡ *Lignincola longirostris (*Cribb & J.W. Cribb) Kohlm., P.S.Z.N.I: *Marine Ecology,* **5:** 353 (1984).

Ascomata: 100-220 µm in diam., subglobose or bottle-shaped, immersed or partly immersed, ostiolate, with long necks, membranaceous, brown or almost hyaline, solitary or gregarious. *Necks*: 80-1125 x 13-40 µm, cylindrical or slightly pointed at the apex, often curved and with a variable diam, rarely branching, projecting over the surface of the substrate. *Asci*: 50-80 x 17-21 µm, 8-spored, cylindrical-clavate, ellipsoidal-clavate or subfusiform, short pedunculate, unitunicate, at first thick-walled apically, beconing thin-walled, with an apical pore; wall irregularly deliquescing. *Ascospores*: 13-19.5 x 4-7.5 mm, irregularly bi-seriate, elongate ellipsoidal or irregularly ellipsoidal, 1-septate at the middle, slightly constricted at or near the septum, hyaline.

Habitat: On intertidal wood.

Description: Based on Kohlmeyer and Kohlmeyer (1979).

Distribution: West Coast:-Gujarat: On intertidal wood (Patil and Borse, 2001); **Maharashtra:** On intertidal wood and wood of *Sonneratia acida* (Patil and Borse, 1986; Borse,1987e,1988; Shrivastava, 1995; Borse et al., 2005a); **Goa:**

on intertidal mangrove wood (Maria and Sridhar, 2002a); **Karnataka:** On intertidal mangrove wood (Maria and Sridhar, 2002a); **Pondecherry (Mahe):** On intertidal mangrove wood (Nambiar and Raveendran, 2007, 2008a, c); **Kerala:** On intertidal wood (Nair, 1970), on intertidal mangrove wood (Maria and Sridhar, 2002a), on intertidal wood (Raveendran and Manimohan, 2007), on intertidal mangrove wood, wood of *Avicennia* sp., *Bruguiera* sp. and *Rhizophora* sp. (Nambiar and Raveendran, 2008a, b, 2009b, d, e, g, h, 2010c; Nambiar et al., 2008a).

East Coast:-Tamil Nadu: On submerged parts of *Rhizophora* spp. (Ravikumar and Vittal, 1996), on mangrove wood (Nambiar and Raveendran, 2008e, 2009l); **Andhara Pradesh:** On intertidal wood of *Avicennia marina, A. officinalis* and *Rhizophora apiculata* (Sarma and Vittal, 1998-99; Sarma and Vittal, 2000; Sarma and Vittal, 2001; Sarma et al., 2001; Sarma and Vittal, 2004; Vittal and Sarma, 2005); **West Bengal:** On intertidal wood of *Sonneratia apetala* and *Phoenix poludosa* (Pawar and Borse, 2004).

76. *Nimbospora* Jorgen Koch, 1982. Nord. J. Bot., **2:** 166.

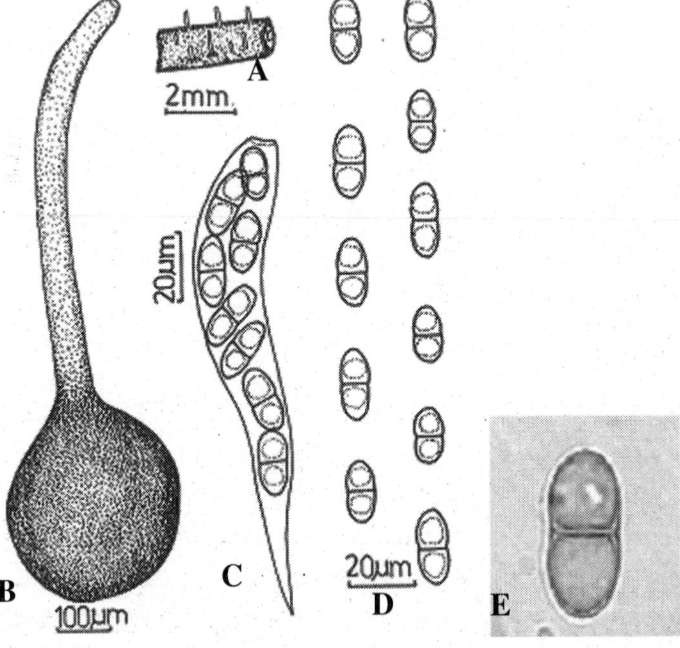

Fig. 75.1. *Neptunella longirostris,* A-E; A- Habit, B- ascocarp, C- ascus, D- ascospores, E- ascospore 1000x).

Ascomata: solitary or gregarious, partly immersed or superficial, globose to subglobose, ostiolate, hyaline. *Peridium*: membranous, pseudoparenchymatous, surface cells angular, thin-walled. *Asci*: thin-walled, clavate, deliquescing at maturity, 8-spored. *Ascospores*: ellipsoidal, 1-septate, surrounded by an excentrically developed sheet and laterally wearing a mucilaginous appendage. The genus is represented by 3 species in marine environment (Jones *et al.*, 2009a).

Type species: *Nimbospora effusa* Jorgen Koch
Habitat: On intertidal wood.
Description: Based on Koch (1982).

Key to *Nimbospora* species from India:

1 Ascospores lacking hair-like appendages, sheath with many (8) folds, 22-29 (-31) x 12-18 µm .... *Nimbospora octonae*

1' Ascospores with lateral tuft of hair-like appendages 2

2 (1') Ascospores with one lateral groop of appendages, 17-24 x 8-10 µm .... *Nimbospora effusa*

2'(1') Ascospores with 2 or more lateral groups of appendages, 18-27 x 8-12 µm .... *Nimbospora bipolaris*

76.1. ***Nimbospora bipolaris*** K.D. Hyde & E.B.G. Jones, 1985. *Can. J. Bot.*, **63**: 611. [For Fig. refer original paper]

Ascomata: 180-320 µm high, 204-380 µm diam., solitary or gregarious, immersed, globose to subglobose, ostiolate, hyaline, papillate and membranous. *Peridium*: pseudoparenchymatous, composed of two layers, an outer layer of angular cells, inner cells large and thin-walled. *Necks*: 50-180 µm long, 40-64 µm wide, cylindrical, periphysate. *Paraphyses*: absent. *Asci*: 50-80 x 16-30 µm, 8-spored, thin-walled, calvate, unitunicate, deliquescing at early, developing at different times on a dome of tissue at the base of the centrum. *Ascospores*: 18-27 x 8-12 µm, ellipsoidal, 1-septate, slightly covered, at first 1-septate, but becoming 3-septate at maturity, not constricted at the septum, hyaline and appendaged. *Appendages*: of two types (i) an elongate sheath drawn out apically and surrounding the spore, in young stages smooth, later slightly folded, especially at the centrum; (ii) two groups of equatorial, lateral appendages, initially fingerlike, 3 or 4 at opposite

sides of the septum becoming fibrillar and mucilaginous in appearance. Polar sheaths 5-12 x 10-12 µm, equatorial appendages 8-30 x 2-3 µm.

Habitat: On intertidal driftwood.
Description: Based on Hyde and Jones (1984).
Distribution: West Coast:-Karnataka: On intertidal wood (Prasannarai and Sridhar, 2001); **Kerala:** On intertidal wood (Nambiar and Raveendran, 2009f).

76.2. ***Nimbospora effusa*** Jorgen Koch, 1982. *Nordic J. Bot.*, **2**: 166.

[For Fig. refer original paper]

Ascomata: 160-236 x 72-135 µm including neck, solitary or gregarious, immersed or superficial, pyriform-subglobose, ostiolate, hyaline. *Peridium*: membranous, pseudoparenchymatous, surface cells angular, with slightly thickened walls, inner cells elongated, thin-walled, becoming large and inflated towards center of ascomata. *Necks*: up to 56 µm long, with papilliform cells. *Asci*: 56-64 x 22-28 µm, 8-spored, in a hymenium in the venter of ascomata, widely calvate with a short stalk, unitunicate, deliquescing at maturity. *Ascospores*: 28-36 x 6-8 µm, slender fusiform, 1-septate, slightly covered, at first 1-septate, but becoming 3-septate at maturity, constricted around the middle septum, hyaline to faintly grey when mature, with four conspicuous oil drops and appendaged. *Appendages*: bipolar, at first covering nearly the whole length of the spore, later folding out and swelling in a plumose appendage which ultimately can be drawn out in a long thin thread (up to 400 µm).

Habitat: On intertidal wood.
Description: Based on Koch (1982).
Distribution: West Coast:-Karnataka: On intertidal wood and teak panels (Prasannarai et al., 1999), on intertidal wood (Prasannarai and Sridhar, 2001).
East Coast:-Tamil Nadu: on intertidal woody debris (Ravikumar et al., 2009); **Pondecherry (Karaikkal):** on intertidal woody debris (Ravikumar, et al., 2009).

76.3. *Nimbospora octonae* Kohlm., 1985b. *Can. J. Bot.*, **63**: 1122.

[Fig. 76.3]

Ascomata: 850-1000 µm high, 600-1250 µm in diam., solitary, immersed, subglobose to ampulliform, ostiolate, with long necks, coriaceous, dark brown to black above, hyaline to brown below. *Peridium*: 70-100 µm thick, tow-layered; outer layer 40-60 µm thick, dark brown, irregular, growing into tracheids and fibers of the substrate; inner layer 30-40 µm thick, hyaline, composed of flat cells, forming a *textura angularis* or *prismatica*, merging with the pseudoparenchyma of the centrum. *Necks*: 200-500 µm long, 80-150 µm in diam., dark brown to black, with irregular outline, wall growing into cavities of the substrate, emerging with black tips over the wood surface, ostiolar canal periphysate; periphyses surrounded by a gelatinous mass. *Pseudoparenchyma*: of thin-walled, polygonal cells filling the centrum of young ascomata, dissolving before spore maturity. *Asci*: 110-135 x 35 µm, 8-spored, calvate, pedunculate, unitunicate, thin-walled, aphysoclastic, upper part dissolving before spores mature, developing at the base of the ascomata venter on a convex cushion of ascogenous cells, maturing successively; the ascogenous tissue is separated from the peridium by a pseudoparenchyma of thin-walled, flat cells. *Ascospores*: 22-29 x 12-16 µm, ellipsoidal, 1-septate, not or slightly constricted at the septum, hyaline, with a gelatinous sheath enclosing 6 (rarely 7) subulate equatorial, 12-20 µm long appendages and one similar appendage at each end; the equatorial appendages are evenly distributed around the septum; the sheath swells in water from 3-4 to 22 µm.

Habitat: On drift and intertidal wood.

Description: Based on Kohlmeyer (1985b).

Distribution: West Coast:-Maharashtra: On intertidal wood of *Rhizophora mucronata* (Borse, 1987a, 1988; Borse et al., 2005a).

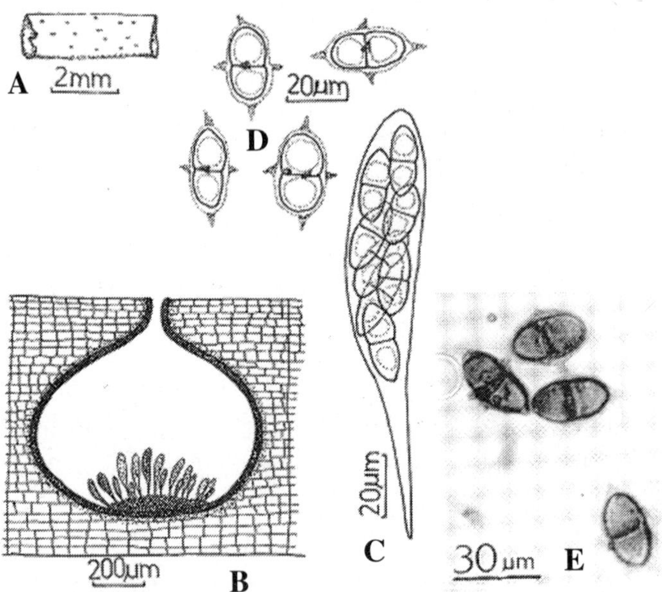

Fig. 76.3. *Nimbospora octonae,* A-E; A- Habit, B- v.s. of ascocarp, C- ascus, D- ascospores, E- ascospores.

77. *Oceanitis* Kohlm., 1977. *Rev. Mycol.*, **41**: 193.

Ascomata: gregarious, subglobose to ellipsoidal, seated on the surface of a thin hypostroma, ostiolate, periphysate, epapillate, fleshy, brownish to dull orange. *Peridium*: thick, forming a textura angularis. *Paraphyses*: absent; center of immature ascomata filled with thin-walled pseudoparenchymatous cells. *Asci*: 8-spored, clavate, unitunicate, thin-walled, ripening successively on an ascogenous tissue at the bottom of the ascomata venter, becoming detached at maturity at the base and pushed upward by young asci, eventually dissolving in the upper part of the ascomata venter. *Ascospores*: filamentous, 1-septate, hyaline, with a filiform apical appendage. The genus is represented by 4 species in the marine environment (Jones *et al.*, 2009a).

Type species: *Oceanitis scuticella* Kohlm.

Habitat: Saprobic on submerged wood in the deep sae.

Description: Based on Kohlmeyer and Kolhmeyer (1979).

Key to *Oceanitis* species from India:

1. A intertidal fungus, Ascospores 5-11-septste, cylindrical to fusiform, 34-60 x 4-5 μm .... *Oceanities cincinnatula*

1'. A deep sea fungus, Ascospores 1-septate, fusiform, 32-50 x 2(-3) μm .... *Oceanitis scuticella*

77.1. *Oceanitis scuticella* Kohlm., 1977. *Rev. Mycol.*, **41**: 194.

[For Fig. refer original paper]

Ascomata: 1.4-2.03 mm high, 1.4-1.78 mm wide, subglobose to ellipsoidal, round at the top, flat at the base, seated on the surface of a thin hypostroma, ostiolate, epapillate, fleshy, brown to dull orange, gragarios. *Hypostroma*: 30-120 μm thick, light-coloured, composed of small, polygonal to rounded cells with large lumina; masses of lilac-coloured hyphae can be found in the wood vessels under the stromata. *Peridium*: 450-470 μm thick at the apex, 240-260 μm at the sides, composed of 17-35 layers of hyaline, thick-walled, polygonal, isodimetric cells with large lumina, forming a *textura angularis*; merging toward the center into flattened cells; outer 3-7 cells filled with small yellowish globules. *Paraphyses*: absent; center of immature ascomata filled with thin-walled, polygonal, isodimetric, pseudoparenchymatous cells, which are eventually compressed by the asci. *Asci*: 70-90 x 12-17 μm, 8-spored, clavate, unitunicate, thin-walled, without apical apparatuses, ripening successively on an ascogenous tissue at the bottom of the ascomata venter, becoming detached at maturity at the base and pushed upward by young asci, eventually dissolving in the upper part of the ascomata venter. *Ascospores*: 60-80 x 4-6 μm, filiform to elongate fusiform, 1-septate, hyaline, with one apical appendage. *Appendages*: 32-50 μm long, 2(-3) μm wide at the base, filamentous, tapering toward the apex, wavy, whiplike, at first attached to the side of the ascospore, becoming detached from the wall, but adhering with its base to the apex of the ascospore.

Habitat: Saprobic on submerged wood in the deep sea.

Description: Based on Kohlmeyer and Kohlmeyer (1979).

Distribution: West Coast:-Goa: On deep sea sediments, Arabian sea (Raghukumar, 1998).

77.2. *Oceanitis cincinnatula* (Shearer & J.L.Crane) J. Dupont & E.B.G. Jones, 2009. *Mycol. Res.*, **113**: 1351. [Fig. 77.2]

Basionym: *Halosarpheia cincinnatula* Shearer & J.L. Crane, *Can, J. Bot.*, **23**: 613 (1980).

≡*Falcatispora cicinnatula* (Shearer & J.L. Crane) K.L. Pang & E.B.G. Jones, *Nova Hedwigia*, **77**: 14 (2003).

≡*Ascosalsum cincinnatulum* (Shearer & J.L. Crane) J. Campb., J.L. Anderson, & Shearer, *Mycologia*, **95**: 546 (2003).

Ascomata: 68-188-284 x 66-158-243 μm, solitary, superficial to slightly immersed, globose to subglobose, hyaline, ostiolate. *Necks*: 113-223-465 x 22-35-46 μm, cylindrical, periphysate, hyaline. *Peridium*: membranous, pseudoparenchymatous, multilayered. Centrum: lacking catenophyses. *Asci*: 41-52-62 x 11-13-15.6 μm, 8-spored, ellipsoidal to clavate, unitunicate, thin-walled, deliquescent. *Ascospores*: 33-50-60 x 3.6-4.3-4-8 μm, cylindrical to fusiform and rounded at each end, 5-11-septate, hyaline, with one apical appendage at one end. *Appendages*: apical, composed of a single coiled or folded filament at first irregular in shape, finally unwinding in water to produce a long, fine, filament.

Habitat: Saprobic on intertidal wood.

Description: Based on Shearer and Crane (1980).

Distribution: West Coast:-Maharashtra: On intertidal mangrove wood (Maria and Sridhar, 2002a); **Goa:** on intertidal mangrove wood (Maria and Sridhar, 2002a); **Karnataka:** On intertidal mangrove wood on intertidal wood of *Avicennia officinalis, Brugueiria gymnorrhiza* and *Rhizophora mucronata* (Maria and Sridhar, 2002a, 2003a, 2004), on intertidal leaf and woody litter of mangroves (Ananda and Sridhar, 2004); **Kerala:** On intertidal mangrove wood (Maria and Sridhar, 2002a).

East Coast:-Orissa: On intertidal mangrove wood (Borse et al., 2005b).

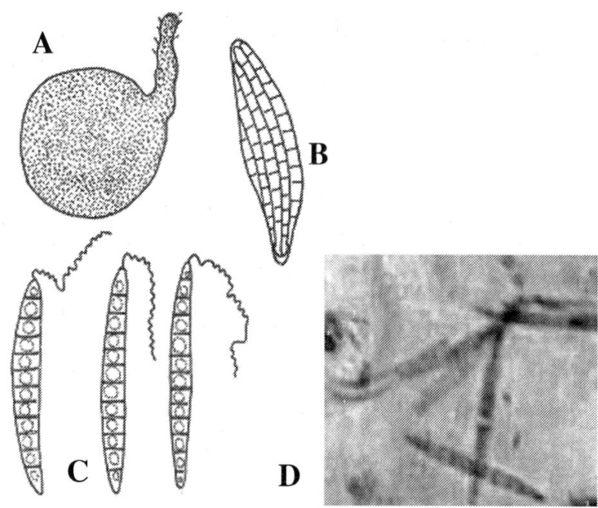

Fig. 77.2. ***Oceanitis cincinnatula,*** A-D; A- ascocarp (100x),B- ascus (450x), C-ascospores (1000x), D- ascospores (1000x).

78. ***Ocostaspora*** E.B.G. Jones, R.G. Johnson & S.T. Moss, 1983b. *Bot. Mar.*, **26**: 354.

Ascomata: globose to subglobose, hyaline, exposed part brown to black, solitary sometimes gregarious, immersed to partly immersed in the substrate, membranous, ostiolate, papillate, and lacking periphyses. *Peridium*: 4-5 layers of elongate, thin-walled cells. *Asci*: 8-spored, clavate to subcylindrical, apiculate, pedunculate, unitunicate, thin-walled, deliquescing early, asci formed on a cushion of pseudoparenchymatous cells at the base of ascomata. *Ascospores*: ellipsoidal, 1-septate, hyaline and appendaged. *Appendages*: polar and equatorial, and awl-like. The genus is monotyoic (Jones *et al.*, 2009a).

Type species: *Ocostaspora apilongissima* E.B.G. Jones, R.G. Johnson & S.T. Moss

Habitat: Saprobic on intertidal wood.

Description: Based on Jones et al. (1983b).

78.1. *Ocostaspora apilongissima* E.B.G. Jones, R.G. Johnson & S.T. Moss, 1983b. *Bot. Mar.*, **26**: 354. [Fig. 78.1]

Ascomata: 87-285 µm high, 87-270 µm wide, globose to subglobose, hyaline with short neck and upper part of centrum dark brown to black, solitary or gregarious, immersed to partly immersed in the substrate, membranous, ostiolate, papillate, and lacking periphyses. *Peridium*: 4-5 layers, cells thin-walled, elongate, cells 25-98 µm long and 25-86 µm wide. *Asci*: 46.5-54 x 10-12.3 µm, 8-spored, unitunicate, thin-walled, clavate to subcylindrical, short pedunculate, lacking apical pore and deliquescing early. Asci formed on a cushion of pseudoparenchymatous cells at the base of ascomata. Paraphyses: absent. *Ascospores*: 15-22 x 5-7.4 µm, ellipsoidal, 1-septate, not or slightly constricted at the septum, hyaline and appendaged. *Appendages*: polar appendages 10-49 µm long, awl-like up to 7.4 µm wide and tapering, 6-8 equatorial appendages awl-like 2.4-8 µm long.

Habitat: Saprobic on intertidal wood.

Description: Based on Jones et al. (1983b).

Distribution: West Coast:-Goa: On intretidal wood (Nandan et al., 1993); **Kerala:** On intertidal wood (Raveendran and Manimohan, 2007).

East Coast:- Andaman & Nicobar Islands: On intertidal wood of *Rhizophora mucronata* (Chinnaraj, 1993).

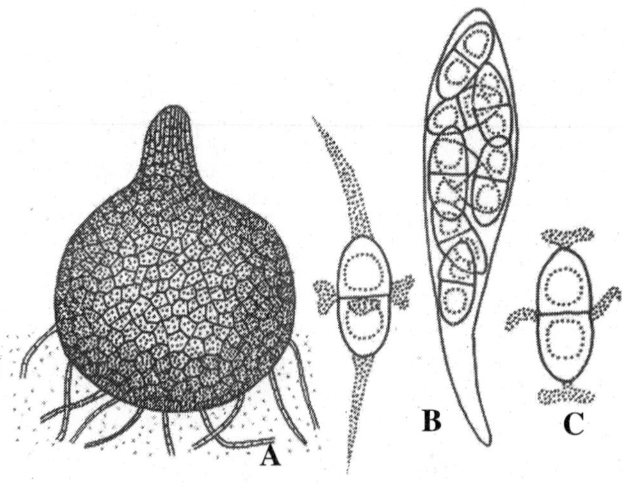

Fig. 78.1. *Ocostaspora apilongissima,* A-C; A- ascocarp (100x), B- ascus (450x), C- ascospores (1000x).

79. *Okeanomyces* K.L. Pang & E.B.G. Jones, 2004. *Bot. J. Linn. Soc.*, **146**: 228.

Ascomata: subglobose or ellipsoidal, immersed or semi-immersed, brown or black, coriaceous or carbonaceous. *Periphyses*: absent. *Asci*: clavate, with short pedunculate, thin-walled, early deliquescing. *Catenophyses*: present. *Ascospores*: cylindrical, thin-walled, with or without a single cap-like appendage. The genus is monotyoic (Jones *et al.*, 2009a).

Type species: *Okeanomyces cucullatus* (Kohlmeyer) Pang & Jones
Habitat: Saprobic on intertidal wood.
Description: Based on Pang et al. (2004).

79.1. *Okeanomyces cucullatus* (Kohlm.) K.L. Pang & E.B.G. Jones, 2004. *Bot. J. Linn. Soc.*, **146**: 228. [Fig. 79.1]

Basionym: *Remispora cucullata* Kohlmeyer, *Mycologia*, 56: 770 (1964).

≡ *Halosphaeria cucullata* (Kolhm.) Kohlm., *Can. J. Bot.*, 50: 1956 (1972).

Ascomata: 150-245 µm high, 150-250 µm wide, subglobose or ampulliform, immersed, ostiolate, papillate, coriaceous, browinish-black or brownish red, solitary or gregarious. *Peridium*: 7.5-11.5 µm thick, composed of 4 to 6 layers of thick-walled, mostly ellipsoidal cells, forming a *textura angularis*, merging toward the center into the pseudoparenchyma. *Necks*: 55-100 µm long, 28-75 µm wide, cylindrical or conical; ostiolar canal at first filled with a small-celled parenchyma. *Pseudoparenchyma*: of thin-walled, polygonal or rounded cells filling venter of young ascomata; eventually breaking up into catenophyses; cells 9-23 µm long, 6-15 µm wide. *Asci*: 45-70 x 15-30 µm, 8-spored, clavate, short pedunculate, unitunicate, thin-walled, aphysoclastic, without apical apparatuses, deliquescing very early in development, developing in the base of the ascomata venter on a small-celled ascogenous tissue. *Ascospores*: 20-68.5 x 6-15.5 µm, ctlindrical or rarely ellipsoidal, 1-septate, slightly or not constricted at the septa, hyaline, with or without cap-like, subglobose, terminal, deciduous appendage at one end, 5-8.5 µm wide.

Anamorph: *Periconia prolifica* Anastasiou
Habitat: Saprobic on intertidal wood.
Description: Based on Kohlmeyer and Kohlmeyer (1979).

Distribution: West Coast:-Maharashtra: On intertidal wood and *Avicennia marina* (Borse, 1987a, 1988; Borse et al., 2005a); **Kerala:** On intertidal wood (Raveendran and Manimohan, 2007).
East Coast:-Tamil Nadu: On mangrove wood (Ravikumar, 1991).

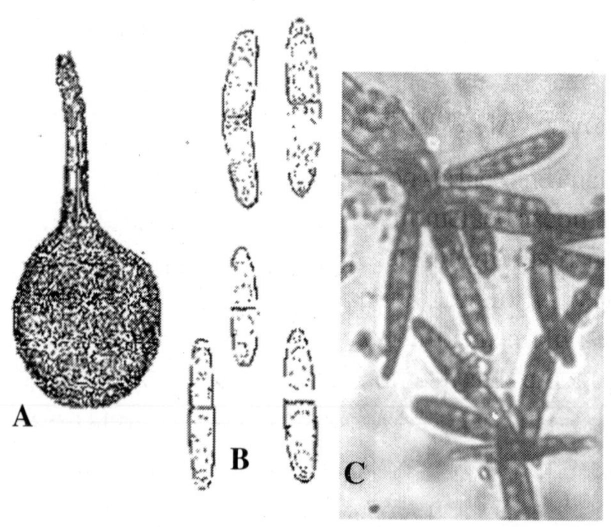

Fig. 79.1. ***Okeanomyces cucullatus,*** A-D; A- ascocarp (100x), B- ascospores (1000x), C- ascospores (1000x).

80. ***Ophiodeira*** Kohlm. & Volkm.-Kohlm., 1988b. *Can. J. Bot.*, **66**: 2062.

Ascomata: ellipsoidal, immersed under a stroma, ostolate, with long necks, light brown, single or gregarious. *Necks*: more or less laterally inserted, curved, dark brown, periphysate, piercing through the stroma. *Peridium*: thin, forming a textura angularis. *Pseudoparenchyma*: composed of thin-walled cells, filling the centrum of young ascomata, finally collapsing, catenophyses absent. *Asci*: 8-spored, clavate, pedunculate, thin-walled, unitunicate, dissolving, developing from an ascogenous tissue at the base of ascomata venter. *Ascospores*: ellipsoidal, 1-septate, hyaline, with a single appendage. *Appendages*: apical-subapical, cap-like, finally gelatinizing and transformed into long filament. The genus is monotyoic (Jones *et al.*, 2009a).

Type species: *Ophiodeira monosemeia* Kohlmeyer & Volkmann-Kolhmeyer
Habitat: Saprobic on subtidal wood of mangroves.
Description: Based on Kohlmeyer and Volkmann-Kohlmeyer (1988b).

80.1. ***Ophiodeira monosemeia*** Kohlm. & Volkm.-Kohlm., 1988b. *Can. J. Bot.*, **66**: 2062. [For Fig. refer original paper]

Ascomata: 125-270 µm high, 210-320 µm long, ellipsoidal, horizontal, immersed under a thin stroma, ostolate, with long necks, light brown, single or gregarious. *Necks*: 60-300 µm long, 25-35 µm wide, more or less laterally inserted, curved, dark brown, ostiolar canal filled with short periphyses, piercing through the stroma. *Peridium*: 15-20 µm thick, composed of 5 to 7 layers of polygonal flat cells with thick walls, forming a textura angularis, merging with a thin-walled pseudoparenchymatuos cells. *Pseudoparenchyma*: of thin-walled, large polygonal cells filling the centrum of young ascomata, finally collapsing without forming catenophyses. *Asci*: 45-65 x 15-18 µm, 8-spored, clavate, pedunculate, thin-walled, unitunicate, dissolving before spore maturity, developing successively on an ascogenous tissue at the bottom of the locule, mostly arranged horizontally, parallel to the wood surface. *Ascospores*: 16-21 x 6-8 µm, ellipsoidal, 1-septate, not constricted at the septum, hyaline, with a single appendage. *Appendages*: 1 µm thick, 7-11 µm long, probably originating from an apical pore, attached to the apex and side of the spore, at first stiff and homogenous, in water becoming soft and banner-like, eventually transforming into a coil of delicate fibers that uncoil and form long, sticky filament, remaining attached to the spore with their bases.

Habitat: Saprobic on subtidal wood of mangroves.
Description: Based on Kohlmeyer and Volkmann-Kohlmeyer (1988b).
Distribution: East Coast:-Andhara Pradesh: On intertidal wood of *Avicennia marina,* and *A. officinalis* (Sarma and Vittal, 1998-99; Sarma and Vittal, 2000; Sarma and Vittal, 2001; Sarma et al., 2001; Sarma and Vittal, 2004; Vittal and Sarma, 2005).

81. ***Panorbis*** J. Campb., J.L. Anderson & Shearer, 2003. *Mycologia*, **95**: 544.

Ascomata: globose to subglobose, ostiolate, at first hyaline becoming black with age. *Peridium*: 10-13 cells wide, membranous, cells thin-walled forming a textura angularis. *Necks*: long, cylindrical, periphysate, hyaline at the apex and pigmented at the base. *Hamathecium*: absent or present as catenophyses. *Asci*: 8-spored, ellipsoidal to clavate, thin-walled, persistent or deliquescent, separating from the ascogenous hyphae, lacking an apical pore and apical apparatus. *Ascospores*: hyaline, 1-septate, ellipsoidal, tapered or rounded at the apices, often flattened on one side, with an apical appendage at each end. *Appendages*: small, hamate at

first, less than or equal to ½ the spore length, unfurling in water to form long, sticky, threadlike structures.

Type species: *Panorbis viscosus* (Schmidt) Campbell, Anderson & Shearer
Habitat: Saprobic on intertidal wood.
Description: Based on Kohlmeyer and Kohlmeyer (1979).

81.1. ***Panorbis viscosus*** (I. Schmidt) J. Campb., J.L. Anderson & Shearer, 2003. *Mycologia*, **95**: 544. [Fig. 81.1]

Basionym: *Halosphaeria viscosa* I. Schmidt, *Natur und Naturschutz in Mecklenburg*, **12**: 70 (1974) 1979 and *Mycotaxon*, **24**: 420 (1985).

≡ *Halosarpheia viscosa* (I. Schmidt) Shearer & J.L. Crane, *Bot. Mar.*, **23**: 608 (1980).

Ascomata: 68-248-455 x 68-250-385 µm, globose to subglobose, at first hyaline becoming black with age, osiolate, immersed to superficial. *Necks*: 73-262-648 x 18-36-54 µm, long cylindrical, periphysate, dark at base becoming hyaline at apex. *Peridium*: membranous, pseudoparenchymatic. *Catenophyses*: apparently absent. *Asci*: 36-65-114 x 9-13-23 µm, 8-spored, ellipsoidal to clavate, thin-walled, unitunicate, deliquescent. *Ascospores*: 13-20-26 x 4.4-6.6-8.4 µm (16.5-23 x 6.5-12.5 µm, Schmidt, 1974; 21-28-31 x 8-11 µm, Kohlm. & Volkm.-Kohlm., 1991), hyaline, 1-septate, ellipsoidal, appendaged. *Appendages*: bipolar, composed of a single, coiled or folded filament, at first hamate to irregular, finally unwinding in water to produce a long, fine filament.

Habitat: Saprobic on intertidal wood
Description: Based on Shearer and Crane (1980).
Distribution: West Coast:-Maharashtra: On intertidal wood of mangroves (*Rhizophora mucronata*) (Borse, 2000b; Borse et al., 2005a); **Karnataka:** On intertidal wood (Sridhar and Prasannarai, 1993; Prasannarai et al., 1999), on intertidal wood (Prasannarai and Sridhar, 2001), on intertidal leaf and woody litter of mangroves (Ananda and Sridhar, 2004); **Kerala:** On intertidal wood (Raveendran and Manimohan, 2007; Nambiar et al., 2008a), on intertidal wood, wood of *Avicennia* sp. and *Bruguiera* sp. (Nambiar and Raveendran, 2009e, f, h).

East Coast:-Tamil Nadu: On intertidal wood (Prasannarai and Sridhar, 2001), on intertidal wood (Nambiar and Raveendran, 2010c) ; **Andhara Pradesh:** On

intertidal wood of *Avicennia marina* (Sarma and Vittal, 2000; Sarma and Vittal, 2001; Sarma et al., 2001; Sarma and Vittal, 2004; Vittal and Sarma, 2005).

Fig. 81.1. *Panorbis viscosus,* A-E; A- Habit, B- v.s. of ascocarp, C- ascus, D- ascospores, E- ascospore 1000x).

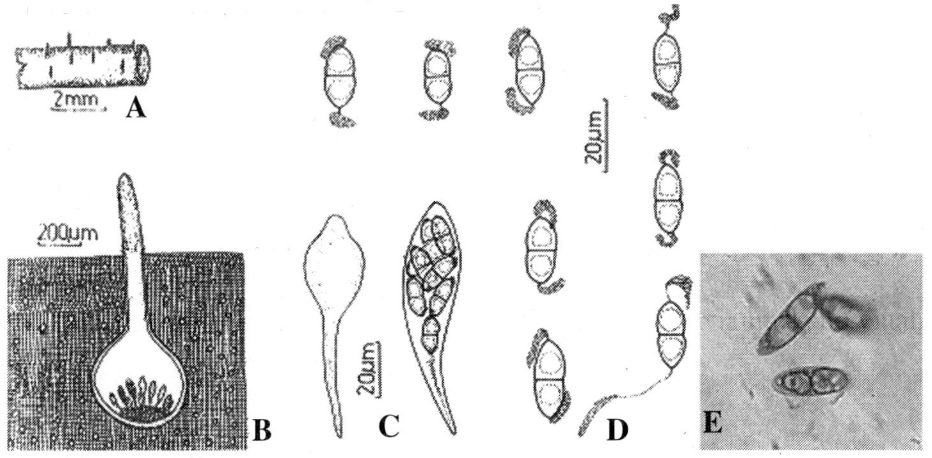

82. *Patellaria* Fr., 1822. *Syst. Mycol.* (Lundae), **2**: 158.

Ascomata: apthecium, sessile, superficial, carbonaceous, black, discoid, opening widely, sub-gelatinous, hard. *Excipulum*: formed by paraphyses, which are hyaline becoming brown towards tips, branched at tips. *Asci*: cylindrical, bitunicate, pedunculate, 4-8-spored, dehiscing by shedding the upper portion of the outer wall. *Ascospores*: hyaline, ellipsoidal to fusoid, transversely multi-septate. Only one facultative marine species is known among several terrestrial species (Jones *et al*., 2009a).

Type species: *Patellaria atrata* (Hedw.) Fries
Habitat: Saprobic on wood.
Description: Based on Pande (2008).

82.1. *Patellaria atrata* (Hedw.) Fr., 1822. *Syst. Mycol.*, **2**: 158.

[For Fig. refer original paper]

= *Lecanidion atratum* (P. Karst.) Sacc., *Syll. Fung.*, **8**: 796 (1889).

Ascomata: 225-690 µm in diam., appothecioid, sessile, typically patelliform, smooth, carbonaceous when dry and subgelatinous under moist conditions. Saucer shaped, black, disc flat or slightly convex with a low margin. *Pseudoparaphyses*: slender, branched, tips olive brown, swollen to 3-5 um. *Asci*: 74-145 x 16-24 µm, 8-spored, irregularly bi-seriate, narrowly clavate, thick-walled especially in upper part, J-. *Ascospores*: 33-45 x 8-10 µm, hyaline, 5-11-septate (mostly 7-septate), occasionally horizontally septate, fusiform-clavate, hyaline, slightly curved,

Habitat: Saprobic on intertidal mangrove wood

Description: Based on Sarma and Vittal (2004) and Pande (2008).

Distribution: West Coast:-Andhra Pradesh: On intertidal wood of *Avicennia marina, A. officinalis* and *Rhizophora apiculata* (Sarma and Vittal, 1998-99; Sarma and Vittal, 2000; Sarma and Vittal, 2001; Sarma et al., 2001; Sarma and Vittal, 2002; Sarma and Vittal, 2004; Vittal and Sarma, 2005).

83. *Paraliomyces* Kohlm., 1959. *Nova Hedwigia*, **1:** 81.

Stromata: black, immersed. *Ascomata*: solitary, subglobose to pyriform, subiculate or nonsubiculate, immersed or erumpent, ostiolate, papillate or epapillate, carbonaceous, black. *Pseudoparaphyses*: filamentous, numerous, persistent. *Asci*: 8-spored, cylindrical, pedunculate, bi-tunicate, thick-walled. *Ascospores*: ellipsoidal to subfusiform, 1-septate, hyaline, surrounded by a gelatinous sheath that contracts to form a lateral flat appendage. The genus is monotypic (Jones et al., 2009a).

Type species: *Paraliomyces lentiferus* Kohlmeyer

Habitat: Saprobic on intertidal and mangrove wood.

Description: Based on Kohlmeyer and Kohlmeyer (1979).

83.1. *Paraliomyces lentiferus* Kohlm., 1959. *Nova Hedwigia*, **1:** 81.

[For Fig. refer original paper]

Stromata: black, immersed, penetrating into the substrate with dark brown hyphae. *Ascomata*: 353-707 µm high, 306-840 µm wide, solitary, subglobose to pyriform, subiculate or nonsubiculate, immersed or erumpent, ostiolate, papillate or epapillate, carbonaceous, thick-walled, black. *Necks*: 63-159 µm high, 56-153 µm wide; ostiolar canal ca. 70 µm in diam., periphysate. *Pseudoparaphyses*: 140-160 x o.5-1.5 µm, filamentous, numerous, persistent, septate, slightly swollen at the tips.

Asci: 85-150 x 13-17.7 µm, 8-spored, cylindrical, short pedunculate, bi-tunicate, thick-walled, persistent; developing in a hymenium from the base and sides in the lower half of the ascomata, with an ocular chamber at the apex; at maturity the ectoascus ruptures and the endoascus elongates to 3 to 4 times its original length. *Ascospores*: 17.7-28.6 x 8-13.6 µm, ellipsoidal to subfusiform, 1-septate below the middle, constricted at the septum, hyaline; at first surrounded by a gelatinous sheath, at maturity sheath becomes less distinct and the lateral appendages apparent. *Appendages*: single, lateral, spherical to lentiform at the septum, viscous, 6.5-13.6 µm in long, 1-4.3 µm wide.

Habitat: Saprobic on wood (fishing boats) and mangrove wood.
Description: Based on Kohlmeyer and Kohlmeyer (1979) and Read et al. (1992).
Distribution: East Coast:-Tamil Nadu: On intertidal wood of fishing boats (Kohlmeyer, 1959).

84. ***Passeriniella*** Berl., 1891. Icon. Fung. (Abellini), **1:** 51.

Ascomata: globose to subglobse, immersed, ostiolate, papillate, dark brown to black, coriaceous, solitary or gregarious, periphysat. *Pseudoparaphyses*: hyaline, branched, septate. *Asci*: cylindrical, pedunculate, bi-tunicate, with an ocular chamber. *Ascospores*: ellipsoidal, versicolour, smooth-walled, 3-septate, constricted at the septa, central cells larger and brown, end cells small and hyaline, lacking a sheath or appendages. The genus is represented by 2 species in marine environment among several terrestrial species (Jones *et al.*, 2009a).

Type species: *Passeriniella incarcerata* (Berl. & Cooke) Berl.
Habitat: Saprobic on wood.
Description: Based on Jones et al. (2009a).

Key to *Passeriniella* species from India:

1. Asci with 4 ascospores, A'spores, 64-88 x 24-28 µm .... *P. savoryellopsis*

1'. Asci with 8 ascospores, A'spores, 44-62 x17-24 µm .... *P. mangrovei*

84.1. ***Passeriniella mangrovei*** Maria & K.R. Sridhar, 2002b. *Ind. J. Forestry*, **25**: 319. [For Fig. refer original paper]

Ascomata: 450-650 µm high, 550-800 µm wide, globose to subglobose, immersed, solitary or gregarious, black, coriaceous, ostiolate, papillate. *Necks*: 200-300 µm long, 200-220 µm wide, periphysate. *Pseudoparaphyses*: 2.4-3.6 µm wide, hyaline, numerous, branched, septate. *Peridium*: 41-72 µm thick, composed of 2 layers, outer layer consists of irregular cells amongst host cells and inner layer is thick-walled irregular cells forming *textura angularis*. *Asci*: 360-450 x 24-36 µm, 8-spored, cylindrical, pedunculate, bi-tunicate, with ocular chamber, at maturity ascus pushes out into the ostiolar canal. *Ascospores*: 44-62 x 17-24 µm, 1-seriate, smooth-walled, ellipsoidal, 3-septate, constricted at the middle septum, central cells large, brown, end cells small and hyaline.

Habitat: Saprobic on intertidal wood

Description: Based on Maria and Sridhar (2002b).

Distribution: West Coast:-Kerala: On intertidal mangrove wood (Maria and Sridhar, 2002a); **Karanataka:** On intertidal wood of *Avicennia officinalis, Brugueiria gymnorrhiza* and *Rhizophora mucronata* (Maria and Sridhar, 2002b, 2003a); *Avicennia officinalis, Bruguiera gymnorrhiza* and *Rhizophora mucronata* (Maria and Sridhar, 2002a, 2004).

84.2. *Passeriniella savoryellopsis* K.D. Hyde & Mouzouras, 1988. *Trans. Br. Mycol. Soc.*, **91**: 179. [For Fig. refer original paper]

Ascomata: 500-700 µm high, 800-1300 µm wide, globose to subglobose, immersed in the substratum beneath a thin stroma, solitary or gregarious, black, coriaceous, ostiolate, papillate. *Necks*: 300-600 µm long, 200-350 µm wide, periphysate. *Pseudoparaphyses*: 3.9-5.9 µm wide, hyaline, numerous, filamentous, branched, faintly septate. *Peridium*: 80-100 µm thick, composed of 2 layers, an outer layer consists of irregular malanized hyphae amongst host cells and inner layer of thick-walled irregular cells forming textura angularis. *Asci*: 280-440 x 24-32 µm, 4-spored, cylindrical, pedunculate, bi-tunicate, thick-walled, with an ocular chamber, at maturity endoascus swelling in water, stretching and pushing the ascus into the ostiolar canal, arising from a basal ascogenous tissue. *Ascospores*: 64-88 x 24-28 µm, 1-seriate, smooth-walled, ellipsoidal, 3-septate, constricted at the central septum, central cells brown, large, end cells small, conical and hyaline.

Habitat: Saprobic on intertidal mangrove wood.

Description: Based on Hyde and Mouzouras (1988).

Distribution: East Coast:-Andaman-Nicobar Islands: On intertidal wood of *Rhizophora mucronata* (Chinnaraj and Untawale, 1992), on intertidal wood of *Avicennia marina, Bruguiera gymnorhiza, Rhizophora mucronata* and *Sonneratia alba* (Chinnaraj, 1993).

85. ***Payosphaeria*** W.F. Leong, 1990. *Bot. Mar.*, **33**: 511.

Ascomata: globose to pyriform, superficial, papillate, ostiolate, lacking periphyses, membranous, hyaline, solitary to gregarious. *Necks*: cylindrical, hyaline and stout. *Peridium*: composed of a single layer of thin-walled elongate cells. *Paraphyses*: few, branched, septate and hyaline. *Asci*: 8-spored, long-cylindrical, short pedunculate, persistent, uni-tunicate, thin-walled and developing at the base of the ascomata. *Ascospores*: 1-seriate, round to oval, 1-celled, hyaline, thin-walled, smooth and lacking appendages. The genus is monotypic (Jones et al., 2009a).

Type species: *P. minuta* W.F. Leong
Habitat: Saprobic on intertidal mangrove wood.
Description: Based on Leong et al. (1990).

85.1. ***Payosphaeria minuta*** W.F. Leong, 1990. *Bot. Mar.*, **33**: 511.

[For Fig. refer original paper]

Ascomata: 119-206 µm high, 115-182 µm wide, globose to flask-shaped, superficial on natural substrate, papillate, ostiolate, membranous, hyaline, subhyaline to yellowish brown and solitary to gregarious. *Necks*: 43.4-146.5 µm long, 24-55 µm wide, conical to elongate, cylindrical, and subhyaline to yellowish brown. *Peridium*: 8 um thick, 1-3 layered and composed of thin-walled elongate cells. *Paraphyses*: 1-1.5 µm thick, few, branched, septate and hyaline. *Asci*: 46-72 x 5-8 µm, 8-spored, cylindrical, pedunculate, uni-tunicate, thin-walled and developing at the base of the ascomata. *Ascospores*: 6-10 x 4-6 µm, 1-seriate, globose to pyriform, 1-celled, hyaline, thin-walled, smooth without appendages or gelatinous sheath, with 1 to 4 conspicuous refractive globules in the center or at the apices of each spore.

Habitat: Saprobic on intertidal mangrove wood, blocks and drift wood.
Description: Based on Leong et al. (1990).

Distribution:West Coast:-Kerala: On intertidal wood (Raveendran and Manimohan, 2007, 2008b), on intertidal wood (Nambiar and Raveendran, 2010c); **Lakshadweep Islands:** On intertidal wood of *Avicennia marina* and *Ceriops tagal* (Chinnaraj, 1992).

East Coast:-Andaman & Nicobar Islands: On intertidal wood of *Avicennia marina, A. officinalis, Rhizophora mucronata* and *Sonneratia alba* (Chinnaraj, 1993).

86. *Pedumispora* K.D. Hyde & E.B.G. Jones, 1992. *Mycol. Res.*, **96**: 78.

Pseudostroma: brown to black, covering the wood surface and enclosing elements of wood, with erumpent pustules containg 1-4 immersed ascomata. *Ascomata*: brown to black, subglobose, coriaceous, aggregated in groups, ostiolate, papillate, periphysate and with paraphyses. *Asci*: 8-spored, fusiform, pedunculate, unitunicate and apically truncate. *Ascospores*: filiform, multi-septate, curved, longitudinally striate, with tapering poles, one or both ends crook-like. The genus is monotypic (Jones *et al.*, 2009a).

Type species: *P. rhizophorae* Hyde & Jones
Habitat: Saprobic on submerged mangrove wood.
Description: Based on Hyde and Jones (1992).

86.1. *Pedumispora rhizophorae* K.D. Hyde & E.B.G. Jones, 1992. *Mycol. Res.*, **96**: 78. [For Fig. refer original paper]

Pseudostroma: up to 130 µm thick, brown to black, covering the wood surface and enclosing elements of wood, with erumpent pustules containg 1-4 immersed ascomata, central region containg 1-4 short necks. *Ascomata*: 560-720 µm high, 660-1040 µm wide, brown to black, subglobose, coriaceous, ostiolate, papillate, sparse pseudostromatic tissues surrounding each ascoma. *Necks*: collectively erumpent and united, up to 320 µm wide, with periphyses. *Peridium*: 100-165 µm thick, an inner hyaline layer of elongate cells and an outer layer composed of malanized thick-walled hyphae forming a *textura epidermioidea*, incorporating host cells, the outside fusing with the host tissue. *Paraphyses*: 3-8 um thick, filiform and simple. *Asci*: 200-265 x 17-26 µm, 8-spored, irregularly fusiform, pedunculate, uni-tunicate, thin-walled, apically truncate without an apical apparuses. *Ascospores*: 160-240 x 5-7 µm, filiform, tapering towards both ends, curved with a characteristic hook-shape at one end, 13-17-septate, not constricted at the septa, with 7-9 longitudinal striations running the length of the spore, tips 15-30 x 1.5-2 µm, lacking cytoplasm, slightly inflated and yellow-yellowish brown.

Habitat: Saprobic on submerged mangrove wood.

Description: Based on Hyde and Jones (1992).

Distribution: East Coast:-Andhara Pradesh: On intertidal wood of *Rhizophora apiculata* (Sarma and Vittal, 1998-99; Sarma and Vittal, 2000; Sarma and Vittal, 2001; Sarma et al., 2001; Sarma and Vittal, 2002; Sarma and Vittal, 2004; Vittal and Sarma, 2005).

87. *Phaeosphaeria* I. Miyake, 1909. *Bot. Mag.*, **23**: 93, in Japanese.

J. Coll. Agric., Yokyo Imp. Univ., **2**: 246 (1910), in German.

Ascomata: solitary gregarious. Subglobose to pyriform, in some species with a tendency to form dothideaceous stromata, immersed or subepidermal, ostolate, papillate or with a long neck, generally thin-walled, smooth or covered by hyphae. *Peridium*: pseudoparenchymatous, composed of small rounded or flattened cells with rather thin walls. *Pseudoparaphyses*: numerous, filamentous, hyaline and distinctly septate. *Asci*: 8-spored, clavate to cylindrical, short pedunculate, bi-tunicate, in some species with apical appatuses. *Ascospores*: 2 to 3 seriate, fusoid to ellipsoidal, 3 to multi-septate, often with one distinctly inflated cell, generally yellowish brown, rarely hyaline. Only 11 of several terrestrial species are known from estuarine and marine habitats (Jones et al., 2009a).

Type species: *Phaeosphaeria oryzae* I. Miyake

Habitat: Saprobic rarely parasitic on Monocotyledons rarely on Dicotyledons or Cryptogams.

Description: Based on Kohlmeyer and Kohlmeyer (1979) and Jones et al. (2009a).

87.1. *Phaeosphaeria halima* (T.W. Johnson) Shoemaker & C.E. Babc., 1989. *Can. J. Bot.*, **67**: 1514. [For Fig. refer original paper]

Basionym: *Leptosphaeria halima* T.W. Johnson, Mycologia, 48: 502 (1956).

Ascomata: 90-252 µm high, 100-216) µm wide, subglobose to pyriform, immersed or superficial, ostiolate, papillate, subcarbonaceous or subcoriaceous, black, solitary or gregarious. *Peridium*: 10-12.5 µm thick, composed of 4 or 5 layers of ellipsoidal or irregularly rounded, thick-walled cells with large lumina, forming a *textura angularis*. *Ostiolar canal*: periphysate. *Pseudoparaphyses*: 2-2.5 µm wide, septate, ramose. *Asci*: 64-108 x 8-14 µm, 8-spored, subclavate to subcylindrical, short pedunculate, bi-tunicate, thick-walled, physoclastic, without apical apparatuses; developing at the base of the ascomata venter. *Ascospores*: 12-18 x 5-7(-8) µm, biseriate, oblong-ellipsoidal, subcylindrical or subfusiform, 3-septate, constricted at the septa, yellow-brown, surrounded by a gelatinous sheath, the second cell from the top is largest.

Habitat: Saprobic dead culms of salt marsh grass *Spartina alterniflora*, test panels and driftwood.

Description: Based on Kohlmeyer and Kohlmeyer (1979).

Distribution: West Coast:-Kerala: On intertidal mangrove wood (Shini et al., 2009-10).

87.2. ***Phaeosphaeria neomaritima*** (R.V. Gesnner & Kohlm.) Shoemaker & C.E. Babc., 1989. *Can. J. Bot.*, **67**: 1572. [For Fig. refer original paper]

Habitat: Saprobic on salt marsh grass *Juncus* sp.

Distribution: West Coast:-Maharashtra: On intertidal wood of *Rhizophora mucronata* and *Sonneratia apetala* (Ramesh and Borse, 1989; Borse et al., 2005a); **Karnataka:** On intertidal wood (Sridhar and Prasannarai, 1993).

Note: The occurrence of *P. neomaritima* on intertidal wood and mangrove wood from India is unlikely, because the fungus occurs on salt marsh grass *Juncus* sp. (Kohlmeyer and Kohlmeyer, 1979; Kohlmeyer and Volkmann-Kohlmeyer, 1991; Jones et al., 2009a).

87.3. ***Phaeosphaeria orae-maris*** (Linder) Khashn. & Shearer, 1996. *Mycol. Res.*, **100**: 1351. [For Fig. refer original paper]

Basionym: *Leptosphaeria orae-maris* Linder, *Farlowia*, **1**: 413 (1944).

Ascomata: 125-240 µm high, 135-319(-372) µm wide, subglobose, ellipsoidal or broadly ovoid, partly or totally immersed, ostiolate, short papillate or epapillate, carbonaceous or subcabonaceous, black, base often lighter coloured, solitary or gregarious. *Peridium*: 20-25 µm thick at the top, 15-20 µm at the sides and base, composed of 6 to 8 layers of irregular or elongated polygonal thick-walled cells with large lumina, forming a *textura angularis*. Ostiolar canal: periphysate.

Pseudoparaphyses: 1-1.5 µm wide, septate, ramose, anastomosing. *Asci*: 96-141 x 8-12 µm, 8-spored, cylindrical or subclavate, short pedunculate, bi-tunicate, thick-walled, physoclastic, without apical apparatuses; developing at the base of the ascomata venter. *Ascospores*: 17-29 (-32) x (4-)5-8 µm, biseriate, fusiform or rarely oblong-ellipsoidal or subclavate, straight or slightly curved, 1 or 3-septate, strongly constricted at the central septum, less so at the others, pale brown, becoming darker in age.

Habitat: Saprobic on intertidal wood.

Description: Based on Kolhmeyer and Kolhmeyer (1979).

Distribution: West Coast:-Maharashtra: On intertidal wood (Borse, 1987a); **Goa:** On intertidal wood (Raghukumar, 1988); **Karnataka:** On intertidal leaf and woody litter of mangroves (Ananda and Sridhar, 2004).

East Coast:-Tamil Nadu: On wood of *Avicennia marina* (Ravikumar and Vittal, 1987).

87.4. ***Phaeosphaeria spartinae*** (Ellis & Everh.) Shoemaker & C.E. Babc., 1989. *Can. J. Bot.*, **67**: 1573. [For Fig. refer original paper]

Basionym: *Leptosphaeria spartinae* Ellis & Everh., J. Mycol., 1: 43 (1885).

= *Leptosphaeria sticta* Ellis & Everh., J. Mycol., 1: 43 (1885).

Habitat: Saprobic on salt marsh grass *Spartina* sp.

Distribution: East Coast:-Kerala: On intertidal mangrove wood (Shini et al., 2009-10).

Note: The occurrence of *P. spartinae* on mangrove wood is unlikely, because the fungus occurs on salt marsh grass *Spartina* sp. (Kohlmeyer and Kohlmeyer, 1979; Kohlmeyer and Volkmann-Kohlmeyer, 1991; Jones et al., 2009a).

87.5. ***Phaeosphaeria spartinaecola*** Leuchtm, 1991. *Mycotaxon*, **41**: 1. [For Fig. refer original paper]

Habitat: Saprobic on salt marsh grass *Juncus* sp.

Distribution: East Coast:-Tamil Nadu: On intertidal wood (Prasannarai and Sridhar, 2001).

Note: The occurrence of *P. spartinacola* on intertidal wood is unlikely, because the fungus occurs on salt marsh grass *Juncus* sp. (Kohlmeyer and Kohlmeyer, 1979; Kohlmeyer and Volkmann-Kohlmeyer, 1991; Jones et al., 2009a).

87.6. *Phaeosphaeria typharum* (Desm.) L. Holm., 1957. *Symb. Bot. Ups.*, **14**: 126. [For Fig. refer original paper]

Basionym: *Sphaeria scripicola* var. *typharum* Desm., Platentes Crypt. France ed. **2**: 1778 (1849).

≡ *Sphaeria typharum* (Desm.) Raben., Herb. Myc. Ed. **2**: 731 (1858).

≡ *Pleospora typharum* (Desm.) Fuckel, Symb. Mycol., 137 (1870).

≡ *Leptosphaeria typharum* (Desm.) Karst., Bidr. Kann. Finl. Nat. Folk, **23**: 100 (1873).

= *Sphaeria perpusilla* var. *typharum* Auers., Rabenhorst Fungi Europaeae: 831 (1865).

Habitat: Saprobic on *Typha* sp. in freshwater habitats.

Distribution: East Coast:-Kerala: On intertidal mangrove wood (Shini et al., 2009-10).

Note: The occurrence of *P. typharum* on intertidal mangrove wood is unlikely, because the fungus occurs on *Typha* sp. in freshwater habitats (Kohlmeyer and Kohlmeyer, 1979; Kohlmeyer and Volkmeyer-Kohlmeyer, 1991; Jones et al., 2009a).

88. *Pharcidia* Koerber, 1865. *Parerga Lichenologiea*, Breslau, pp. 469-470.

= *Epicymatia* Fuckel, *Symb. Mycol.*, pp. 118 (1869).

Ascomata: solitary, small, globose to ellipsoidal, often apically flattened, ostolate, periphysate, epapillate, dark. *Peridium*: partially thick around the ostiole, composed of thick-walled, subglobose, or near the center, flattened cells. *Paraphyses*: numerous, often hanging downward into ascomata cavity. *Asci*: 4-8-spored, elongate-clavate, weakly pedunculate, thick-walled, uni-tunicate. *Ascospores*: ellipsoidal or elongate, one-septate in the center, hyaline, with or without appendages (gelatinous cap-like and at both ends). The genus contains some 90 names, but many are referred to *Stigmidium,* or lichen genera such as *Arthopyrenia* and *Lichenodiplis.* The genus is represented by three marine species (Jones *et al.*, 2009a).

Type species: *Pharecidia congesta* Koerber

Habitat: Marine species occur on brown alga, *Laminaria digitata* or shells of marine animals.

Description: Based on Kohlmeyer and Kohlmeyer (1979) and Jones et al. (2009a).

88.1. *Pharecidia rhachiana* Kohlm., 1973. *Bot. mar.*, **16**: 210.

[For Fig. refer original paper]

Habitat: Symboitic in lichenoid association with epiphytic blue-green and brown alga *Laminaria digitata.*

Distribution: East Coast:-West Bengal: On intertidal mangrove wood (Shini et al., 2009-10).

Note: The occurrence of *P. rhachiana* on intertidal mangrove wood is unlikely, because the fungus occurs as in lichenoid association with epiphytic blue-green and brown alga *Laminaria digitata* (Kohlmeyer and Kohlmeyer, 1979; Kohlmeyer and Volkmeyer-Kohlmeyer, 1991; Jones et al., 2009a).

89. *Pleospora* Raben. ex Ces. & De Not., 1863. Com. Soc. Crittogam. Ital., **1:** 217.

Type species: *Pleospora harbarum* (Pers. & Fr.) Raben.

Habitat: Saprobic rarely parasitic on Monocotyledons rarely on Dicotyledons or Cryptogams.

89.1. *Pleospora pelagica* T.W. Johnson, 1956. *Mycologia*, **48**: 504.

[For Fig. refer original paper]

Habitat: Saprobic on dead culms of salt marsh grass *Spartina alterniflora*

Distribution: West Coast:-Maharashtra: On intertidal mangrove wood (Ramesh and Borse, 1979); **Kerala:** On intertidal wood (Raveendran ang Manimohan, 2007, Nambiar and Raveendran, 2008b, 2009e, g, I, 2010c; Nambiar et al., 2008a).

Note: The occurrence of *P. pelagica* on intertidal wood and mangrove wood is unlikely, because the fungus occurs on dead culms of salt marsh grass *Spartina alterniflora* (Kohlmeyer and Kohlmeyer, 1979; Kohlmeyer. and Volkmann-Kohlmeyer, 1991; Jones et al., 2009a).

89.2. *Pleospora triglochinicola* J. Webster, 1969. *Trans. Br. Mycol. Soc.*, **53**: 481. [For Fig. refer original paper]

Habitat: Saprobic on old leaves and inflorescences of salt marsh plant *Triglochin igantea.*

Distribution: West Coast:-Kerala: On intertidal mangrove wood (Shini et al., 2009-10).

Note: The occurrence of *P. triglochinicola* on intertidal mangrove wood is unlikely, because the fungus occurs as saprobic on old leaves and inflorescences of salt marsh plant *Triglochin maritima* (Kohlmeyr and Kohlmeyer, 1979; Kohlmeyer and Volkmeyer-Kohlmeyer, 1991; Jones et al., 2009a).

90. *Quintaria* Kohlm. & Volkm.-Kohlm., 1991. *Bot. Mar.*, **34**: 35.

Ascomata: obpyriform, somewhat compressed laterally, completely immersed, ostolate, papillate, carbonaceous to subcoriaceous, black. *Peridium*: 2-layered, outer stratum of hyphoid cells, inner stratum of flattened cells, forming a *textura angularis*. *Ostiolar canal*: filled with a network of thin hyphae, embedded in a gelatinous matrix, black incrustations lining the sides. *Pseudoparaphyses*: trabeculate, above the asci, embedded in gelatinous matrix. *Asci*: 8-spored, cylindrical, pedunculate, thick-walled, fissitunicate, with refractive apical plate. IKI negative. *Ascospores*: bi-seriate, fusiform, 5-septate, hyaline. The genus is monotypic (Jones *et al.*, 2009a).

Type species: *Quintaria lignatilis* (Kohlm.) Kohlm. & Volkm.-Kohlm.
Habitat: Saprobic on intertidal wood and bark of mangroves.
Description: Based on Kohlmeyer and Volkmann-Kohlmeyer (1991).

90.1. *Quintaria lignatilis* (Kohlm.) Kohlm. & Volkm.-Kohlm., 1991. *Bot. Mar.*, **34**: 35. [Fig. 90.1]

Basionym: *Trematosphaeria lignatilis* Kohlm., *P.S.Z.N.I: Mar. Ecol.*, **5**: 365 (1984).

Ascomata: 340-670 μm high, 350-600 μm in diam., obpyriform, somewhat compressed laterally, completely immersed, ostolate, papillate, carbonaceous to coriaceous, black, gregarious. *Peridium*: 40-50 μm thick, 2-layered; outer stratum 20-30 μm thick, composed of irregular hyphoid cells mixed with decomposing wood or bark paricles; inner stratum 16-20 μm thick, composed of flattened cells, forming a *textura angularis,* with oil droplets in the lumina and black pigment deposits in the cell walls. *Necks*: 250-460 µm long, 160-280 µm in diam., irregularly cylindrical; ostiolar canal 90 um in diam., filled with a network of thin, hyaline hyphae, embedded in a gelatinous matrix, black incrustations lining the sides of the canal. *Pseudoparaphyses*: 1.5-3.5 µm in diam., more or less parallel between the asci and unbranched, becoming trabeculate (branched and anastomosing), above the asci and merging with the periphysoid filaments in the ostiolar canal, embedded in gelatinous matrix which often adheres to asci in squash mounts. *Asci*: 230-290 x 28-37 µm, 8-spored, cylindrical, pedunculate, thick-walled, fissitunicate, with an inconspicuous, often igantean refractive apical plate, not bluing in IKI (IKI-negative); arising from a basal ascogenous tissue. *Ascospores*: 51-80 x 14-20 µm, biseriate, fusiform, 5-septate, slightly constricted at the centra septum, less so at the others, hyaline, rarely pale yellowish, smooth-walled, filled with many oil droplets.

Habitat: Saprobic on immersed wood and bark of mangroves.

Description: Based on Kohlmeyer (1984).

Distribution: West Coast:-Gujarat: On intertidal wood (Patil and Borse, 2001); **Maharashtra:** On prop roots of *Rhizophora mucronata,* wood of *Avicennia alba* (Borse and Shrivastava,1988; Borse et al., 2005a); **Lakshadweep Islands:** On intertidal wood of *Ceriops tagal* and *Pemphis acidula* (Chinnaraj, 1992).

East Coast:-Tamil Nadu: On wood of *Rhizophora mucronata* and *R. apiculata* (Ravikumar and Vittal, 1987, 1996); **Andhara Pradesh:** On intertidal wood of *Aegiceras corniculatum, Ceriops decandra* and *Rhizophora apiculata* (Sarma and Vittal, 1998-99; Sarma and Vittal, 2000; Sarma and Vittal, 2001; Sarma et al., 2001; Sarma and Vittal, 2004; Vittal and Sarma, 2005); **West Bengal:** On intertidal wood of *Avicennia marina, A. officinalis* and *Ceriops tagal* (Pawar et al., 2005); **Andaman & Nicobar Islands:** On intertidal wood of *Avicennia marina, A. officinalis, Bruguiera gymnorhiza, Rhizophora apiculata, R. mucronata* and *Sonneratia alba* (Chinnaraj, 1993a).

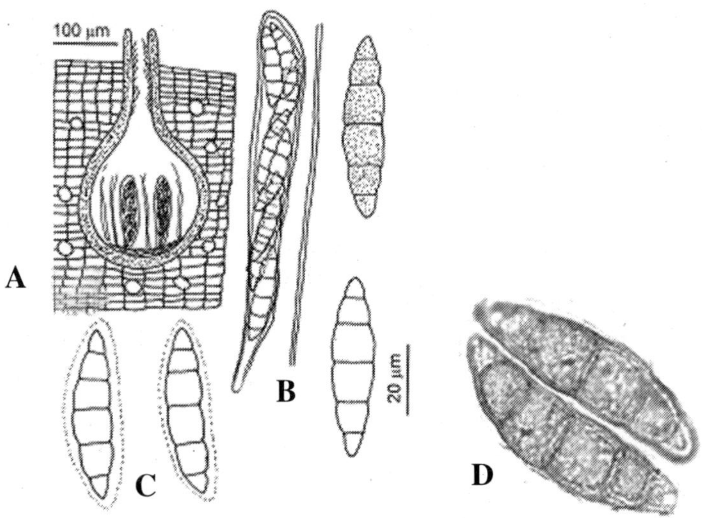

Fig. 90.1. ***Quintaria lignatilis,*** A-D; A- v.s. of ascocarp, B- ascus and pseudoparaphyses (450x) , C- ascospores, D- ascospores (1000x).

91. ***Remispora*** Linder, 1944. *Farlowia*, **1**: 409.

Ascomata: globose, subglobose, ovoid, immersed or superficial, osiolate, papillate, coriaceous, hyaline to dark brown, solitary or gregarious, catenophysate, necks periphysate, peridium composed of one type of cells. *Asci*: clavate, broadly fusoid, pedunculate, apiculate, unitunicate, thin-walled, without an apical apparatuses, deliquescing. *Ascospores*: 1-septate, ellipsoidal to rhomboidal, hyaline, with polar appendages. *Appendages*: morphologically variable depending on the species. The genus is represented by 7 obligate marine species (Jones et al., 2009a; Suetrong et al. (2009).

Type species: *Remispora maritima* Linder
Habitat: On intertidal wood.
Description: Based on Jones et al. (2009a).

91.1. ***Remispora quadri-remis*** (Hohnk) Kohlm., 1960. *Nova Hedwigia*, **2:** 332. [For Fig. refer original paper]

Basionym: *Palomyces quadri-remis* Hohnk, *Veroff. Inst. meeresforsch. Bremerhaven,* **3:** 213 (1955).

≡ *Arenariomyces quadri-remis* (Hohnk) Meyers, *Mycologia*, **49:** 505 (1957).

≡ *Halosphaeria quadri-remis* (Hohnk) Kohlm., *Can. J. Bot*., **50:** 1957 (1972).

Ascomata: 135-400 μm in diam., globose or subglobose, immersed or superficial, osiolate, papillate, membranaceous or coriaceous, cream-coloured, grayish or brown, sometimes the top darker than the base, solitary or gregarious. *Peridium*: 12-22 μm thick, composed of cells with large lumina, merging into the pseudoparenchyma of the venter. *Necks*: up to 600 μm long, 24-58 μm in diameter, cylindrical or eccentric, one or several occurring on one ascocarp. *Catenophyses*: of thin walled polygonal to ellipsoidal cells filling venter of young ascocarp; eventually breaking up into catenophyses. *Asci*: 70-100 x 20-40 μm, 8-spored, clavate or subfusiform, short pedunculate, unitunicate, thin-walled, aphysoclastic, without an apical apparatuses, early deliquescing; developing at the base of the ascocarp venter. *Ascospores*: 18-30(-34) x 8-12(-15.5) μm (excluding appendages), one-septate, ellipsoidal, not or slightly constricted at the septum, hyaline, appendaged; *Appendages*: at each end four radiating appendages, developing by fragmentation of the exospore; appendages 12-21.5 μm long, 2.5-4 μm in diameter at the base, terminal, obclavate, curved, attenuate, semirigid, inconspicuously striate by fiberlike elements embedded in the subgelatinous matrix.

Habitat: Saprobic on intertidal wood, drift wood and, test panels.
Description: Based on Kohlmeyer and Kohlmeyer (1979).
Distribution: West Coast:-Kerala: On driftwood (Khan and Manimohan, 2011).

92. *Rhizophila* K.D. Hyde & E.B.G. Jones, 1989b. *Mycotaxon*, **34**: 527.

Ascomata: large, globose to subglobose, immersed in the substratum, ostiolate, papillate, periphysate, coriaceous, dark brown to black, solitary or gregarious. *Paraphyses*: like structures irregular in morphology, septate. *Asci*: 8-spored, clavate, pedunculate, unitunicate, thin-walled, with an indistinct apical thickening, without an apical apparatus, developing from ascogenous tissue lining the base and sides of the ascomata wall. *Ascospores*: 2-3 seriate, ellipsoidal or fusiform, one-celled, yellowish to yellowish brown. The genus is monotypic (Jones et al., 2009a).

Type species: *Rhizophila marina* Hyde & Jones
Habitat: On intertidal mangrove roots.
Description: Based on Hyde and Jones (1989b).

92.1. *Rhizophila marina* K.D. Hyde & E.B.G. Jones, 1989b. *Mycotaxon*, **34**: 527. Fig. 91.1]

Ascomata: 500-960 µm in diam., 425-730 µm high, globose to subglobose, immersed in the substratum, ostiolate, papillate, periphysate, coriaceous, dark brown to black, solitary or gregarious. *Peridium*: 45-112 µm thick, 2-layered, the outer layer is thick, composed of a highly melanized layer of irregular hyphae interdispersed at the outside with the host cells, the inner layer, thin, hyaline, composed of irregular thin-walled elongate cells which fuse with the paraphyses-like structures. *Necks*: up to 515 µm long, 113-198 µm in diam., black, of similar structure to the ascomata outer wall layer. *Paraphyses*: like structures 3-11 µm wide, consisting of irregular shaped cells, septate. *Asci*: 74-101 µm long, 13-24 µm in diam., 8-spored, 2-3 seriate, clavate, pedunculate, unitunicate, thin-walled, with an indistinct apical thickening, slightly truncate, without an apical apparatus, developing from ascogenous tissue lining the base and sides of the ascomata wall. *Ascospores*: 20-28 x 7-10 µm, ellipsoidal or fusiform, one-celled, hyaline when immature, yellowish to yellowish brown at maturity.

Habitat: On prop root and branches of *Rhizophora mucronata*.
Description: Based on Hyde and Jones (1989b).
Distribution: West Coast:-Maharashtra: On intertidal mangrove (*Rhizophora mucronata*) wood (Borse, 2000b; Borse et al., 2005a); **Goa:** On intertidal mangrove wood (Maria and Sridhar, 2002a); **Karnataka:** On intertidal wood of *Rhizophora mucronata* (Chinnaraj and Untawale, 1992), on intertidal mangrove wood (Maria and Sridhar, 2002a); **Kerala:** On intertidal mangrove wood (Maria and Sridhar, 2002a).
East Coast:- Tamil Nadu: On mangrove wood (Ravikumar, 1991);
Andhara Pradesh: On intertidal wood of *Rhizophora apiculata* (Sarma and Vittal, 1998-99; Sarma and Vittal, 2000; Sarma and Vittal, 2001; Sarma et al., 2001; Sarma and Vittal, 2002; Sarma and Vittal, 2004; Vittal and Sarma, 2005);
Andaman & Nicobar Islands: On intertidal wood of *Rhizophora apiculata* and *R. mucronata* (Chinnaraj, 1993a).

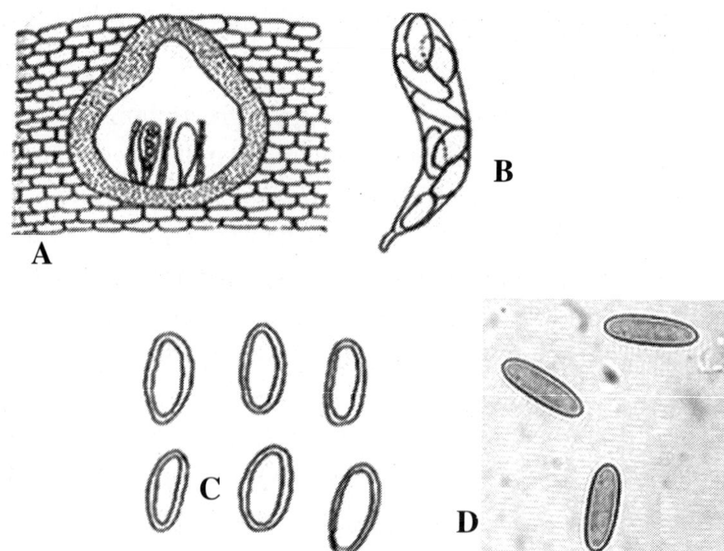

Fig. 92.1. ***Rhizophila marina,*** A-D; A- v.s. of ascocarp (100x), B- ascus (450x), C- ascospores (1000x), D- ascospores (1000x).

93. ***Rimora*** Kohlm., Volkm.-Kohlm., Suetrong, Sakayaroj & E.B.G. Jones, 2009. *Studies in Mycology*, **64**: 166.

Ascomata: erumpent, with flat tops, elongated, apapillate, opening with a periphysate cleft-like ostiole, black, gregarious. *Peridium*: of thick-walled cells, forming a *textura angularis. Hamathecium*: of branched pseudoparaphyses. *Asci*: 8-spored, cylindrical, pedunculate, thick-walled, fissitunicate, without apical apparati. *Ascospores*: biseriate, fusiform, 3-septate, hyaline, surrounded by an evanescent sheath.

Type species: *Rimora mangrovei* (Kohlm. & Vittal) Kohlm.,Volkm- Kohlm., Suetrong, Sakayaroj & Jones

Habitat: On intertidal bark and wood of mangrove trees.

Description: Based on Suetrong et al. (2009).

93.1. ***Rimora mangrovei*** (Kohlm. & Vittal) Kohlm., Volkm.-Kohlm., Suetrong, Sakayaroj & E.B.G. Jones, 2009. *Studies in Mycology*, **64**: 167.

[Fig. 93.1]

Basionym: *Lophiostroma mangrovei* Kohlm. & Vittal, 1986, *Mycologia*, 78: 487.

≡ *Astrosphaeriella mangrovei* (Kohlm. & Vittal) Aptroot & Hyde, 2002, *Fungi in Marine Environments*, Fungal Diversity Press, 7: 106.

Ascomata: 260-400 µm high, 775-950 µm wide, 450-535 µm thick, broadly oblong with flat top in frontal view, conoidal to truncate in sagittal section, elongated, laterally compressed, developing on a thin black stromatic crust, at first immersed, later erumpent by removal of the covering host tissue, opening with an elongated cleft-like ostiolum at the flattened top, epapillate, carbonaceous, black, gregarious. *Peridium*: 60-150 µm thick at the sides, 5-18 µm at the base, encrusted with melanin particles, thick-walled cella forming a *textura angularis*, cells of the host are enclosed in the lower half of the peridium. *Ostioles*: 10-60 mm wide, ostiolar canal filled with hyaline periphyses, 1.8-2.2 mm thick. *Pseudoparaphyses*: 2 mm in diameter, branched, merging with the periphyses. *Asci*: 155-218 x 13-19 mm, 8-spored, cylindrical, short pedunculate, thick-walled, fissitunicate, without apical apparatus, arising in parallel from a basal ascogenous tissue. *Ascospores*: 37-55 x 7-11 mm, biseriate, fusiform, 1-septate, upon germination 3-septate, constricted at the central primary septum, hyaline, surrounded by an evanescent gelatinous sheath.

Habitat: On intertidal mangrove wood.

Description: Based on Kohlmeyer and Vittal (1986).

Distribution: West Coast:-Gujarat: On intertidal mangrove wood (Patil and Borse, 2001); **Maharashtra:** On bark of *Avicennia officinalis* (Kohkmeyer and Vittal, 1986), on mangrove wood (Borse, 1988), on intertidal wood of *Avicennia alba* (Shrivastava, 1995; Borse et al., 2005a), on intertidal mangrove wood (Maria and Sridhar, 2002a); **Goa:** On intertidal mangrove wood (Maria and Sridhar, 2002a); **Karnataka:** On intertidal wood (Prasannarai and Sridhar, 1997, 2001, 2000-2001), on intertidal wood of *Acanthus ilicifolius, Rhizophora mucronata* and *Sonneratia caseolaris* (Maria and Sridhar, 2002a, 2003a, 2004); **Kerala:** On intertidal mangrove wood (Maria and Sridhar, 2002a); **Lakshadweep Islands:** On intertidal bark of *Avicennia marina* (Chinnaraj, 1992).

East Coast:-Tamil Nadu: On dead seedling of *Rhizophora mucronata* (Kohlmeyer and Vittal, 1986), on submerged parts of *Rhizophora* spp. (Ravikumar and Vittal, 1996); on intertidal woody debris (Ravikumar et al., 2009); **Pondicherry (Karaikkal):** on intertidal woody debris (Ravikumar et al., 2009); **Andhara Pradesh:** On intertidal wood of *Avicennia marina, A. officinalis, Ceriops decandra, Lumnitzera recemosa* and *Rhizophora apiculata* (Sarma and Vittal, 1998-99; Sarma and Vittal, 2000; Sarma and Vittal, 2001; Sarma et al., 2001; Sarma and Vittal, 2002; Sarma and Vittal, 2004; Vittal and Sarma, 2005); **West Bengal:** On intertidal wood of *Avicennia marina* (Pawar et al., 2009); **Orissa:** On intertidal wood of *Avicennia marina, Phoinix poludosa, Rhizophora apiculat* and *Tamarix troupii* (Borse et al., 2009); **Andaman & Nicobar Islands:** On intertidal wood of *Avicennia marina, A. officinalis, Bruguiera gymnorhiza, Rhizophora apiculata, R. mucronata* and *Sonneratia alba* (Chinnaraj, 1993a).

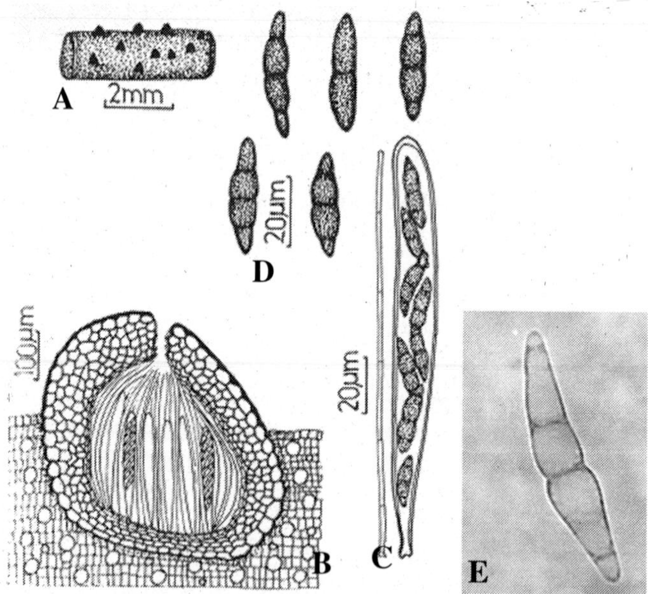

Fig. 93.1. *Rimora mangrovei,* A-E; A- Habit, B- v.s. of ascocarp, C- ascus and paraphysises, D- ascospores, E- ascospore 1000x).

94. *Saagaromyces* K.L. Pang & E.B.G. Jones, 2003. *Mycolo. Progress*, **2**: 35.

=*Littispora* J. Campb., J.L. Anderson & Shearer, *Mycologia,* **95:** 547, 2003.

Ascomata: globose, subglobose to ellipsoidal, immersed, ostiolate, papillate, membranaceous to coriaceous, hyaline, cream-coloured, brown to black. *Necks*: periphydate. *Catenophyses*: present. *Asci*: 8-spored, cylindrical-clavate, pedunculate, unitunicate, thick–walled, persistent. *Ascospores*: oval, hyaline, heavily guttulated, with or without unfurling appendages. Three species have been assigned to the genus from marine environment (Jones *et al.*, 2009a).

Type species: *Saagaromyces ratnagiriensis* (S.D. Patil & Borse) K.L. Pang & E.B.G. Jones
Habitat: Saprobic on intertidal mangrove wood.
Description: Based on Pang et al. (2003).

Key to *Saagaromyces* species from India:

1 Ascopspores less than 20 um in diam., Ascospores 33-47 x 14-22 µm,.......... .... *S. abonnis*

1' Ascopspores more than 20 um in diam., ...2

2 Ascospores 42-59 x 21- 31 µm, ... *S. glitra*

2' Ascospores 48-64(-72) x 22-28 µm, ... *S. ratnagiriensis*

2" Ascospores 33-47 x 14-22 µm, ... *S. abonnis*

94.1. *Saagaromyces abonnis* (Kohlm.) K.L. Pang & E.B.G. Jones, 2003. *Mycolo. Progress*, **2**: 35. [Fig. 94.1]

Basionym: *Halosarpheia abonnis* Kohlm., *Mar. Ecol. (P.S.Z.N.I.),* **5**: 339 (1980).

≡ *Littispora abonnis* (Kohlm.) J. Campb. *et al., Mycologia*, **95**: 549 (2003).

Ascomata: 450-600 µm high, 300-450 µm in diam., ellipsoidal, immersed, ostiolate, papillate, coriaceous, light brown, main axis sometimes horizontal with a lateral, recurved neck. *Peridium*: 20-45 µm thick, 2-layered, forming a *textura angularis*; outer stratum composed of small, subglobose cells that are intimately bound to the softened, decomposing wood; inner stratum composed of larger, elongate and flattened cells that merge with the pseudoparenchyma of the centrum.

Necks: 350-450 µm long, 80-180 µm in diam., cylindrical, centrally or laterally inserted, sometimes curved and irregular; ostiolar canal periphysate. *Pseudoparenchyma*: of thin-walled, large polygonal cells filling the centrum of young ascomata before the development of ascogenous cells, breaking up to form catenophyses; cells of pseudoparenchyma and peridium are filled with lipid globules. *Asci*: 210-190 x 30-54 µm, 8-spored, clavate, pedunculate, unitunicate, thin-walled, thicker at the apex, persistent, apical apparatuses not observed, developing at the base of the ascomata venter on a small-celled ascogenous tissue that is separated from the peridium by thin-walled, polygonal cells of the sterile pseudoparenchyma; asci mature successively; in squash mounts breaking off basally from the ascogenous tissue without releasing the ascospores. *Ascospores*: 33-47 x 16-22 µm (excluding igantean), ellipsoidal, 1-septate, slightly constricted at the septum, hyaline, with apical-subapical appendages. *Appendages*: one cap-like, stiff and homogenous is attached to each apex and run partially along the side of the spore, 5-8 µm thick, 16-20 µm long, 10 µm wide, at maturity becoming soft and scooplike, eventually transforming into a coil of delicate fibers that uncoil and form long, sticky filaments, but remain attached to the ascospore apices with their bases.

Habitat: Saprobic on immersed mangrove wood.

Description: Based on Kohlmeyer (1984).

Distribution: West Coast:-Gujarat: On intertidal wood (Patil and Borse, 2001); **Maharashtra:** On intertidal wood and mangrove wood of *Avicennia marina* (Borse, 1988; Borse and Shrivastava, 1988; Borse et al., 2005a); **Kerala:** On intertidal mangrove wood (Maria and Sridhar, 2002a), on intertidal mangrove wood and wood of *Avicennia* sp. (Nambiar and Raveendran, 2008a, b, 2009e, h, 2010c; Nambiar et al., 2008a); **Lakshadweep Islands:** On intertidal pneumatophores of *Avicennia marina* (Chinnaraj, 1992).

East Coast:-Andhara Pradesh: On intertidal wood of *Avicennia marina* and *Rhizophora apiculata* (Sarma and Vittal, 1998-99; Sarma and Vittal, 2000; Sarma and Vittal, 2001; Sarma et al., 2001; Sarma and Vittal, 2004; Vittal and Sarma, 2005); **West Bengal:** On intertidal wood of *Avicennia marina* (Pawar and Borse, 2004); **Andaman & Nicobar Islands:** On intertidal wood of *Rhizophora apiculata* (Chinnaraj, 1993a).

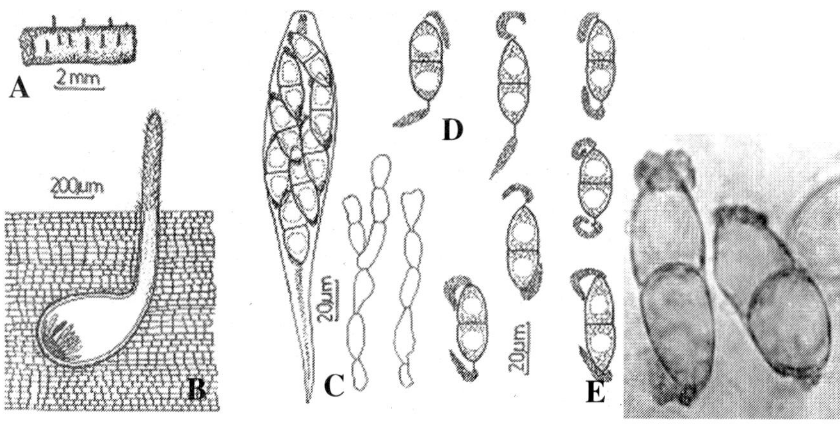

Fig. 94.1. ***Saagaromyces abonnis,*** A-E; A- Habit, B- v.s. of ascocarp, C- ascus and catenophyses, D- ascospores, E- ascospores (1000x).

94.2. ***Saagaromyces glitra*** (J.L. Crane & Shearer) K.L. Pang & E.B.G. Jones, 2003. *Mycological Progress*, **2**: 35. [Fig. 94.2]

Basionym: *Nias glitra* J.L. Crane & Shearer, *T. B. M. S.*, **86**: 509 (1986).

Ascomata: 400-544 µm in diam., globose to subglobose, membranous, cream-coloured becoming black with age. *Peridium*: composed of textura epidermiodia with thick-walled dark brown cells. *Necks*: 544-1780 x 148-197 µm, cylindrical, ostolate, dark brown at base becoming light brown to subhyaline at apex, periphysate. *Periphyses*: numerous, filamentous and branched, loosely interwoven, becoming compact, at times protruding from the neck in the form of a penicillus, centrum aparaphysate. Developing asci surrounded by pseudoparrenchyma which breaks up into rudimentary catenophyses. *Asci*: 240-312 x 43-62 µm, 8-spored, biseriate, unitunicate, clavate to ellipsoid, long-stalked, thin-walled, with an apical pore, deliquescing at maturity. *Ascospores*: 42-59 x 21-31 µm, oblong to obovate, hyaline, 1-septate, slightly constricted and with granular deposits at the septum which stain with cotton blue, walls up to 1.0 µm thick, apical cell may be either equal in length and width or longer (1.o-15.8 µm) and broader (1.0-6.9 µm) than the basal cell.

Habitat: Saprobic on submerged mangrove wood.

Description: Based on Crane and Shearer (1986).

Distribution: West Coast:-Goa: On intertidal wood (Borse et al., 1999a); **Kerala:** On mangrove woody litter (Manimohan et al. 2011).

East Coast:-Tamil Nadu: On prop roots of *Rhizophora mucronata* (Ravikumar and Vittal, 1987), on submerged parts of *Rhizophora* spp. (Ravikumar and Vittal, 1996); **Andhara Pradesh:** On intertidal wood of *Avicennia officinalis* (Sarma and Vittal, 1998-99; Sarma and Vittal, 2000; Sarma and Vittal, 2001; Sarma et al., 2001; Sarma and Vittal, 2004; Vittal and Sarma, 2005); **Andaman & Nicobar Islands:** On intertidal wood of *Rhizophora mucronata* and *Sonneratia alba* (Chinnaraj, 1993a).

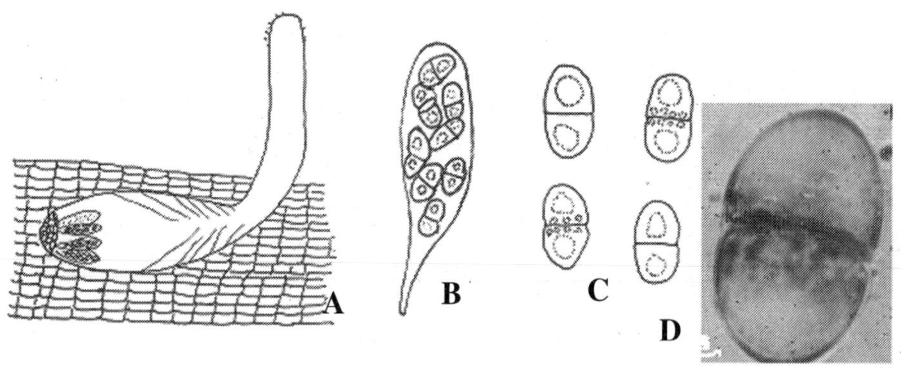

Fig. 94.2. *Saagaromyces glitra,* A-D; A- v.s. of ascocarp (100x), B- ascus (450x), C- ascospores (1000x), D- ascospore (1000x).

94.3. *Saagaromyces ratnagiriensis* (S.D. Patil & Borse) K.L. Pang & E.B.G. Jones, 2003. *Mycological Progress*, **2**: 35. [Fig. 94.3]

Basionym:Halosarpheia ratnagiriensis (S.D. Patil & Borse), *Ind. Bot. Reptr.* **1:** 102 (1982).

≡*Littispora ratnagiriensis* (S.D. Patil & Borse) J. Campb. *et al., Mycologia,* **95:** 549 (2003).

Ascomata: 300-650 µm high, 400-480 µm in diam., ellipsoidal, immersed, ostiolate, papillate, coriaceous, hyaline, main axis sometimes horizontal with a lateral, recurved neck. *Peridium*: 25-40 µm thick, cells forming a *textura angularis,* mergeing with the pseudoparenchyma of the centrum. *Necks*: about 1mm long, 100-160 µm in diam., cylindrical, emerging with whitish, fuzzy tips over the wood surface, centrally or laterally inserted; ostiolar canal periphysate.

Pseudoparenchyma: of thin-walled, large polygonal cells filling the centrum of young ascomata, breaking up to form catenophyses. *Asci*: 250-340 x 50-68 μm, 8-spored, clavate, pedunculate, unitunicate, thin-walled, thicker below the apex, persistent; with a refractive lens-shaped, apical plate, 3-6 μm in diam., about 0.8 μm thick; developing at the base of the ascomata venter, mature successively; in squash mounts breaking off basally from the ascogenous tissue without releasing the ascospores. *Ascospores*: 48-64(-71.5) x 22-28 μm (excluding igantean), ellipsoidal, 1-septate, slightly constricted at the septum, hyaline, with apical-subapical appendages. *Appendages*: one cap-like, stiff and homogenous is attached to each apex and run partially along the side of the spore, 4-7 μm thick, 25-34 μm long, 16-18 μm wide, at maturity becoming soft and scooplike, eventually transforming into a coil of delicate fibers that uncoil and form long, sticky filaments of about 0.7 μm in diam., but remains attached to the ascospore apex with its 1.0 μm thick base.

Habitat: Saprobic on immersed wood and bark of mangroves.

Description: Based on Kohlmeyer (1984) and Patil and Borse (1982).

Disribution: West Coast:-Gujarat: On intertidal wood (Patil and Borse, 2001); **Maharashtra:** On intertidal wood (Patil and Borse, 1982; Shrivastava, 1995), intertidal wood, proproots, seedlings of *Rhizophora mucronata*, pneumatophores of *Sonneratia acida* (Patil and Borse, 1986; Borse, 1987d, 1988; Borse et al., 2005a), on intertidal mangrove wood (Maria and Sridhar, 2002a); **Goa:** On intertidal prop roots of *Rhizophora mucronata* (Borse et al., 1999b), on intertidal wood (Prasannarai and Sridhar, 2001), on intertidal mangrove wood (Maria and Sridhar, 2002a); **Karnataka:** On intertidal wood (Prasannarai and Sridhar, 2001, 2000-2001), on intertidal mangrove wood on intertidal wood of *Acanthus ilicifolius, Avicennia officinalis, Rhizophora mucronata* and *Sonneratia caseolaris* (Maria and Sridhar, 2002a, 2003a), on intertidal leaf and woody litter of mangroves (Ananda and Sridhar, 2003, 2004); **Kerala:** On intertidal mangrove wood (Maria and Sridhar, 2002a), on intertidal wood (Raveendran and Manimohan, 2007), on intertidal mangrove wood of *Rhizophora* sp. (Nambiar and Raveendran, 2006, 2008a, 2009d, i; Nambiar et al., 2008a).

East Coast:-Tamil Nadu: On intertidal wood (Nambiar and Raveendran, 2010c); on intertidal woody debris (Ravikumar et al., 2009); **Pondicherry (Karaikkal):** on intertidal woody debris (Ravikumar et al., 2009); **Andhara Pradesh:** On intertidal wood of *Avicennia marina, A. officinalis* and *Rhizophora apiculata* (Sarma and Vittal, 1998-99; Sarma and Vittal, 2000; Sarma and Vittal, 2001;

Sarma et al., 2001; Sarma and Vittal, 2002; Sarma and Vittal, 2004; Vittal and Sarma, 2005); **West Bengal:** On intertidal wood of *Avicennia marina* and *Ceriops tagel* (Pawar and Borse, 2004); **Andaman & Nicobar Islands:** On intertidal wood of *Avicennia marina, Rhizophora apiculata* and *R. mucronata* (Chinnaraj, 1993a).

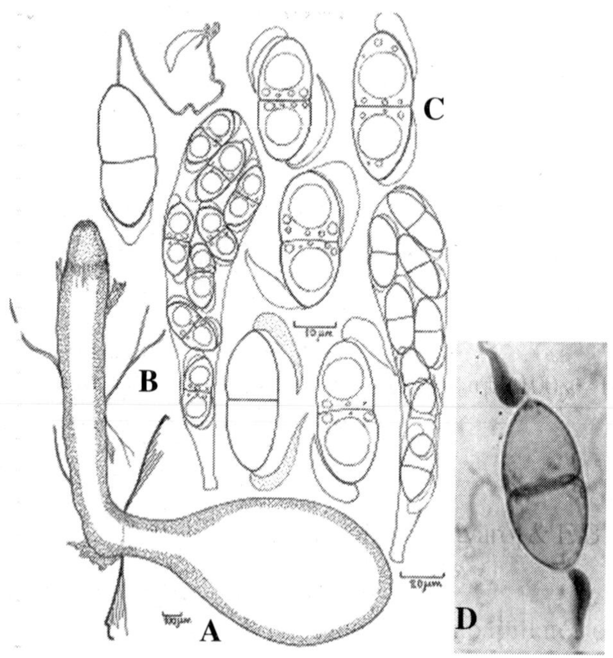

Fig. 94.3. ***Saagaromyces ratnagiriensis,*** A-D; A- v.s. of ascocarp, B- ascus, C- ascospores, D- ascospore (1000x).

95. *Saccardoella* Speg., 1879. Michelia, **1:** 461.

Ascomata: large or small, globose to subglobose, immersed or erumpent, ostiolate, erumpent papillate, coriaceous, carbonaceous, solitary or gregarious, dark brown to black, clypeate, no periphyses, Paraphyses: numerous, filamentous, hyphae-like, branching and anastomosing in gelatinous matrix. *Asci*: long-cylindrical, short pedunculate, thin-walled, unitunicate, apically flattened with a subapical ring. *Ascospores*: uniseriate, fusiform, 3-9-septate, not constricted at the septa, with acute end, variable septation, hyaline, with a thin to well developed gelatinous sheath or polar appendages. Three species have been assigned to the genus from marine environment (Jones *et al.*, 2009a).

Type species: *Saccardoella montellica* Speg.
Habitat: Saprobic on wood in aquatic and terrestrial environments.
Description: Based on Hyde (1992b) and Jones et al. (2009a).

Key to *Saccardoella* species from India:

1.Ascospores with (4-)5(-6)-septate, 19-26 x 6-8 µm .... *S. rhizophorae*

1' Ascospores with 3-septate, 25-31 x 7.5-10 µm .... *S. marinosphora*

95.1. *Saccardoella marinospora* K.D. Hyde, 1992b. *Mycolo.*, **84**: 803.

[For Fig. refer original paper]

Ascomata: 780-1040 µm high, 650-960 µm in diam., golbose to subglobose, immersed under a clypeus, ostiolate, short papillate, coriaceous, brown-black, solitary or in groups of tow to three. *Clypeus*: extending outwards around the papilla, composed of host cells filled with highly melanised fungal hyphae, becoming very dense and blackened neares the ostiole. *Necks*: black, blunt conical, up to 240 µm high, ostiole rounded, periphyse not observed. *Peridium*: of equal thickness throughout, except at the top where it becomes indistinguishable from the clypeus, up to 90 µm thick, comprised of two strata, an inner layer of brown relatively thick-walled cells. These cells are small and rounded to angular at the base and irregularly elongate or rounded at the sides. The outer layer is composed of host cells and fungal hyphae in the form of *textura igantean*. *Hamathecium*: hyphae-like, filamentous, up to 2 µm thick, septate, rarely branched, numerous, gelatinous and attached to the base and top of the ascomata. *Asci*: 190-240 x 10-12 µm, 8-spored, cylindrical, thin-walled, unitunicate, short pedunculate, apically rounded or flattened, with a small ring-like subapical apparatus. *Ascospores*: 25-31 x 7.5-10 µm, overlapping uniseriate, hyaline at maturity, cylindrical with acute poles, 3-septate, not constricted at the septa, surrounded by a thin sheath which swells in water to become large and indistinct.

Habitat: Saprobic on intertidal mangrove wood.
Description: Based on Hyde (1992b).
Distribution: West Coast:-Maharashtra: On intertidal wood of *Rhizophora mucronata* (Borse *et al.*, 2005a).
East Coast:-Tamil Nadu: On mangrove wood (Ravikumar, 1991);

Andhara Pradesh: On intertidal wood of *Avicennia officinalis* (Sarma and Vittal, 1998-99; Sarma and Vittal, 2000; Sarma and Vittal, 2001; Sarma et al., 2001; Sarma and Vittal, 2004; Vittal and Sarma, 2005); **West Coast:** On intertidal wood of *Avicennia marina* (Pawar et al., 2009); **Andaman & Nicobar Islands:** On intertidal mangrove wood (Chinnaraj, 1993b).

95.2. ***Saccardoella rhizophorae*** K.D. Hyde, 1992b. *Mycologia*, **84**: 806.

[For Fig. refer original paper]

Ascomata: 325-455 µm high, up to 364 µm in diam., golbose to subglobose, immersed, ostiolate, ostiole round in section, short papillate, coriaceous, brown-black, neck surrounded by a small clypeus, solitary or . *Clypeus*: extending outwards around the papilla, composed of host small dense pigmented cells filled with highly melanised fungal hyphae, becoming very dense and blackened neares the ostiole. *Peridium*: equal throughout, up to 58 um thick, comprised of two strata, an inner stratum of thin-walled brown angular cells forming a textura angularis, individual cells up to 12 x 3 um, becoming rounded towards the outside and eventually composed of hyphae interdispersed with the host cells forming a *textura igantean*. *Hamathecium*: hyphae-like, filamentous, up to 2 µm thick, septate, rarely branched, numerous ina gel. *Asci*: 135-160 x 8-10 µm, 8-spored, cylindrical, thin-walled, unitunicate, short pedunculate, apically rounded or flattened, with a small ring-like subapical apparatus. *Ascospores*: 19-26 x 6-8 µm, overlapping uniseriate, young spores 2-celled, becoming (4-) 5 (-6)-septate, when mature, hyaline when mature, ellipsoidal, surrounded by a thin sheath which swells in water.

Habitat: Saprobic on intertidal mangrove wood.
Description: Based on Hyde (1992b).
Distribution: East Coast:-Tamil Nadu: On mangrove wood (Ravikumar, 1991); **Andhara Pradesh:** On intertidal wood of *Rhizophora apiculata* (Sarma and Vittal, 1998-99; Sarma and Vittal, 2000; Sarma and Vittal, 2001; Sarma et al., 2001; Sarma and Vittal, 2002; Sarma and Vittal, 2004; Vittal and Sarma, 2005); **Andaman & Nicobar Islands:** On intertidal mangrove wood (Chinnaraj, 1993b).

96. *Salsuginea* K.D. Hyde, 1991a. *Bot. Mar.*, **34**: 316.

Ascomata: immersed, fusiform, conical or subglobose, clypeate, appearing as a raised darkened area with a central ostiole and short neck, solitary. *Peridium*: thin and hyaline at the base, of rounded cells at the sides, fusing with the clypeus at the top, clypeus of dark fungal tissue. At the region between the side and base is a area of elongate parallel cells in the form of a wedge. *Hamathecium*: thin, anastomosing in a gel. Asci: 8-spored, fissitunicate, pedunculate, with an ocular chamber and ring. *Ascospores*: irregularly 2-celled, brown to black, obovoid with hyaline apical germ pores, highly constricted at the septa. The genus is monotypic (Jones *et al.*, 2009a).

Type species: *Salsuginea ramicola* K.D. Hyde
Habitat: Saprobic on intertidal mangrove wood.
Description: Based on Hyde (1991a).

96.1. *Salsuginea ramicola* K.D. Hyde, 1991a. *Bot. Mar.*, **34**: 316.

[Fig. 96.1]

Ascomata: 1040-2600 µm high, 455–1430 µm in diam., fusiform, conical or subglobose, with or without a flattened base, immersed under a darkened clypeus, appearing as a raised darkened area on the wood surface with a central ostiole and short neck, solitary. *Peridium*: up to 39 µm thick of round cells (in cross section) at the sides, fusing at the top with the clypeus, thin at the base, clypeus composed of host cells interdispersed with darkened fungal tissue and rounded melanin particles, at the margin between the side and base is wedge-shaped region of elongate parallel cells. *Hamathecium*: 0.9-1.8 µm in diam., thin, numerous, anastomosing in a gel. *Asci*: 440–512 x 29-34 µm, 8-spored, fissitunicate, cylindrical, pedunculate, with an apical apparatus consisting of a large disitinctive ocular chamber and prominent ring. *Ascospores*: 59-72 x 24-30 µm, uniseriate, unequally 2-celled, obovoid, highly constricted at the septum, brown to black, smooth-walled with hyaline apical germ pores, lacking a mucilaginous sheath.

Habitat: Saprobic on intertidal mangrove wood.
Description: Based on Hyde (1991a).
Distribution: West Coast:-Daman: On intertidal wood and *Avicennia marina* (Borse et al., 2000b); **Gujarat:** On intertidal mangrove wood (Patil and Borse, 2001); **Maharashtra:** On intertidal dead and decaying wood of *Avicennia alba, A. marina, Sonneratia acida,* and *S, apetala* (Patil and Borse, 1985, 1986;

Borse, 1988; Borse et al., 1988; Ramesh and Borse, 1989; Srivastava, 1989); **Goa:** On intertidal wood of *Sonneratia alba* (Borse et al., 1999a); **Pondecherry (Mahe):** On intertidal wood (Nambiar and Raveendran, 2007); **Kerala:** On intertidal wood (Raveendran and Manimohan, 2007), on intertidal wood and *Avicennia* sp. (Nambiar and Raveendran, 2006, 2007, 2009a, b, d, e, h, i; Nambiar et al., 2006, 2008a).

East Coast:-Tamilnadu: On mangrove wood (Nambiar and Raveendran, 2008e, 2010c); **Orissa:** On intertidal wood of *Avicennia marina, A. officinalis, Ceriops tagal* and *Heritiera foams* (Borse et al., 2005b); **West Coast:** On intertidal wood of *Phoenix poludosa* (Pawar et al., 2009); **Andaman & Nicobar Islands:** On intertidal wood of *Avicennia marina* and *Sonneratia alba* (Chinnaraj, 1993a).

Note: The fungus was earlier reported by many researchers from India as *Helicascus kanaloanus.* All the records included as *H. kanaloanus* may be considered as *S. ramicola* in the previous papers (Patil and Borse, 1985, 1986; Borse, 1988; Borse et al., 1988; Ramesh and Borse, 1989; Srivastava, 1989).

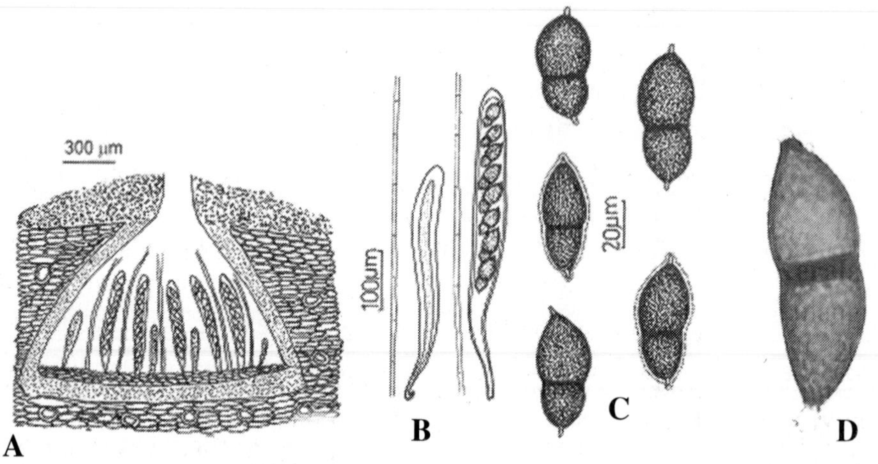

Fig. 96.1. ***Salsuginea ramicola,*** A-D; A- v.s. of ascocarp, B- asci and pseudoparaphyses, C- ascospores, D- ascospore (1000x).

97. ***Savoryella*** E.B.G. Jones & R.A. Eaton, 1969. *Trans. Br. Mycol. Soc.*, **52**: 162.

Ascomata: solitary to gregarious, immersed, partly immersed to superficial, ostiolate, periphysate, papillate, membranous and brown, Peridium of *textura angularis* when viewed from the surface and in section composed of several layers of angular cells. Paraphyses presents in young ascomata, wide and septate. Asci: 2 to 8-spored, cylindrical to clavate, short pendunculate, unitunicate, persistent, with a non-amyloid apical thickening containing a pore. Ascospores: ellipsoidal, 3-septate, central cells brown, end cells hyaline, with or without polar appendages.

Type species: *Savoryella lignicola* E.B.G. Jones & R.A. Eaton
Habitat: On submerged wood.
Description: Based on Jones and Eaton (1969) and Jones and Hyde (1992).

Key to *Savoryella* species from India:

1. Asci 2-spored, 36-60 x 12-16.5 mm .... *S. paucispora*

1' Asci 8-spored 2

2. (1') Ascospores appendaged, 26-35 x 11-16 mm............*S. appendiculata*

2'(1') Ascospores not appendaged 3

3. (2') Ascospores 24-36 x 8-12 mm .... *S.lignicola*

3'(2') Ascospores 33.5-46.5 x 7.5-12 m .... *S.longispora*

97.1. ***Savoryella appendiculata*** K.D. Hyde & E.B.G. Jones, 1992. *Bot. Mar.*, **35**: 84. [Fig. 97.1]

Ascomata: 280-340 µm high, 140-250 µm in diam., subglobose to ellipsoidal, immersed, partly immersed to superficial, often horizontal to the substrate surface, ostiolate, papillate, membranous, brown-black, solitary or gregarious. *Necks*: 120-300 µm long, 50-70 µm in diam., brown to black withlighter tips, periphysate. *Peridium*: in surface view, a *textura epidermioidea*, in section composed of several layers of thick-walled elongate cells. *Paraphyses*: up to 8 um wide. *Asci*: 150-225 x 14-24 µm, 8-spored, cylindrical, stalked, unitunicate, persistent, apices truncate, with a non-amyloid apical thickening containing a pore. *Ascospores*: 26-35 x 11-16 µm, uniseriate, ellipsoidal, 3-septate, constricted at the septa;

central cells brown, apical cells smaller and hyaline, wall minutely verrucose, appendaged (appendages 30 µm long).

Habitat: Saprobic on drift wood, ascospores accumulate in foam along shore.

Description: Based on Jones and Hyde (1992).

Distribution: West Coast:-Kerala: on intertidal wood (Nambiar and Raveendran, 2010a).

East Coast:-Tamil Nadu: On intertidal wood (Nambiar et al., 2008b), on intertidal wood (Nambiar and Raveendran, 2009l, 2010c); **Andhra Pradesh:** On intertidal wood (Nambiar and Raveendran, 2009j); **West Bengal:** On intertidal wood (Borse. 2000a); **Orissa:** Ascospores in foam samples ((Borse et al., 2001b).

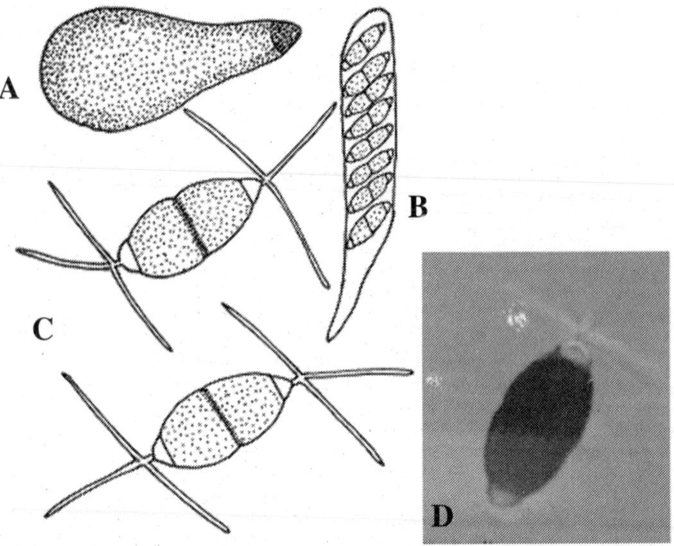

Fig. 97.1. ***Savoryella appendiculata,*** A-D; A- ascocarp (100x), B- ascus (450x), C- ascospores (1000x), D- ascospore (1000x).

97.2. ***Savoryella lignicola*** E.B.G. Jones & R.A. Eaton, 1969. *Trans. Br. Mycol. Soc.*, **52:** 161. [Fig. 97.2]

Ascomata: 170-350 µm high, 120-250 µm in diam., globose, subglobose or ellipsoidal, immersed, partly immersed or superficial, ostiolate, papillate, membranous and pale to dark brown. *Necks*: (48)-80-165 µm long and up to 72 µm in diam., brown with periphyses. *Peridium*: brown, a *textura angularis* when viewed from the surface, while in section composed of several layers of thick-walled angular cells. *Paraphyses*: present but sparse. *Asci*: 100 (128)-180 x 16-24 µm, 8-spored, cylindrical or clavate, short- stalked, unitunicate, persistent, with an apical truncate non-amyloid apical thickening containing a pore. *Ascospores*: 24-36 x 8-12 µm (The length breadth ratio 2.9 : 1.0), uni or biseriate, ellipsoidal, 3-septate, not markedly constricted at the septa; central cells brown (10.6-16 µm), apical cells smaller and hyaline (2.6-6 µm).

Habitat: On intertidal wood.

Description: Based on Jones and Eaton (1969) and Jones and Hyde (1992).

Distribution: West Coast:-Daman: On intertidal wood and *Avicennia marina* (Borse et al., 2000b); **Gujarat:** On intertidal wood (Patil and Borse, 2001); **Maharashtra:** On intertidal wood and *Rhizophora mucronat* (Borse, 1987e, 1988; Shrivastava, 1995; Borse et al., 2005a); on intertidal mangrove wood (Maria and Sridhar, 2002a); **Goa:** On intertidal wood (Borse et al., 1999a), on intertidal wood (Prasannarai and Sridhar, 2001), on intertidal mangrove wood (Maria and Sridhar, 2002a); **Karnataka:** On intertidal wood (Prasannarai et al., 1999), on intertidal wood (Prasannarai and Sridhar, 2001, 2000-2001; Ananda and Sridhar, 2004), on intertidal wood of *Acanthus ilicifolius, Avicennia officinalis, Brugueiria gymnorrhiza, Rhizophora mucronata* and *Sonneratia caseolaris* (Maria and Sridhar, 2002a, 2003a. 2004), on intertidal mangrove wood (Ananda and Sridhar, 2003); **Pondecherry (Mahe):** On intertidal wood (Nambiar and Raveendran, 2007, 2008a, c); **Kerala:** On intertidal mangrove wood (Maria and Sridhar, 2002a), on intertidal wood (Raveendran and Manimohan, 2007), on intertidal wood, mangrove wood of *Avicennia* sp., *Bruguiera* sp. and *Rhizophora* sp. (Nambiar and Raveendran, 2006, 2007, 2008a, b, d, 2009a, b, d, e, f, h, I, 2010c; Nambiar et al., 2006, 2008a), on mangrove woody litter (Shini et al., 2009-10); **Lakshadweep Islands:** On intertidal wood of *Ceriops tagal* and *Pemphis acidula* (Chinnaraj, 1992), on intertidal mangrove wood (Ananda and Sridhar, 2003).

East Coast:-Tamil Nadu: On drift wood (Raghukumar, 1973), On intertidal wood (Nambiar et al., 2008b), on mangrove wood (Nambiar and Raveendran, 2008e, 2009k, l, 2010c); on intertidal woody debris (Ravikumar et al., 2009); **Pondicherry:** On intertidal wood (Nambiar and Raveendran, 2008c); **Pondicherry (Karaikkal):** on intertidal woody debris (Ravikumar et al., 2009); **Andhara Pradesh:** On intertidal wood of *Avicennia officinalis, Lumnitzera recemosa* and *Rhizophora apiculata* (Sarma and Vittal, 1998-99; Sarma and Vittal, 2000; Sarma and Vittal, 2001; Sarma et al., 2001; Sarma and Vittal, 2002;

Sarma and Vittal, 2004; Vittal and Sarma, 2005); **West Bengal:** On intertidal wood (Borse. 2000a), on mangrove woody litter (Shini et al., 2009-10); **Andaman &Nicobar Islands:** On intertidal wood of *Avicennia marina, Rhizophora apiculata, R. mucronata* and *Sonneratia alba* (Chinnaraj, 1993a).

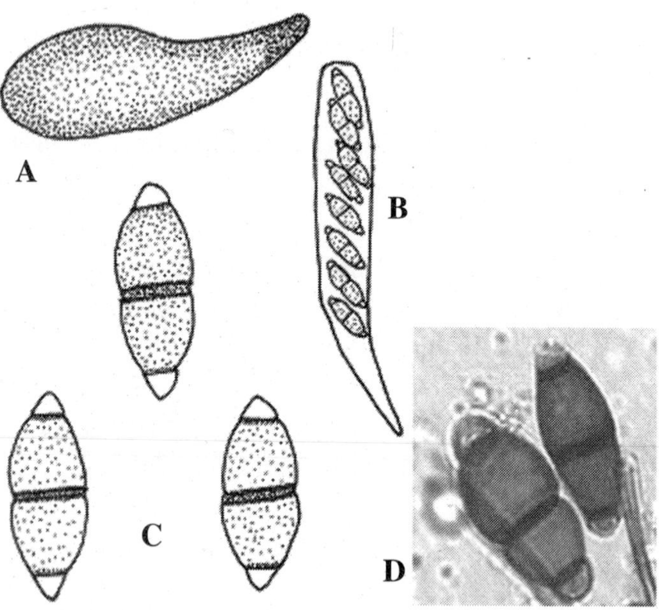

Fig. 97.2. ***Savoryella lignicola*** **,** A-D; A- ascocarp (100x), B- ascus (450x) , C- ascospores (1000x), D- ascospores (1000x).

97.3. ***Savoryella longispora*** E.B.G. Jones & K.D. Hyde, 1992. *Bot. Mar.*, **35**: 83. [Fig.97.3]

Ascomata: 200-300 µm high, 120-235 µm in diam., globose, subglobose or ellipsoidal, immersed, partly immersed or superficial, ostiolate, papillate, membranous and pale to dark brown. *Necks*: 80-160 µm long and up to 70 µm in diam., brown with periphyses. *Peridium*: brown, a *textura angularis* when viewed from the surface, while in section composed of several layers of thick-walled angular cells. *Paraphyses*: present but sparse. *Asci*: 100-160 x 16-21 µm, 8-spored, cylindrical or clavate, short- stalked, unitunicate, persistent, with an apical truncate non-amyloid apical thickening containing a pore. *Ascospores*: 33.5-46.5 x 7.5-12 m (The length breadth ratio 4.1: 1.0), uni or biseriate, ellipsoidal,

3-septate, not markedly constricted at the septa; central cells brown, apical cells smaller and hyaline.

Habitat: On intertidal wood.

Description: Based on Jones and Hyde (1992).

Distribution: West Coast:-Karanataka: On intertidal wood of *Acanthus ilicifolius, Avicennia officinalis, Brugueiria gymnorrhiza* and *Rhizophora mucronata* (Maria and Sridhar, 2002a, 2003a); **Kerala:** On intertidal mangrove wood (Maria and Sridhar, 2002a).

East Coast:-Orissa: On intertidal wood of *Ceriops tagal* (Borse et al., 2005b).

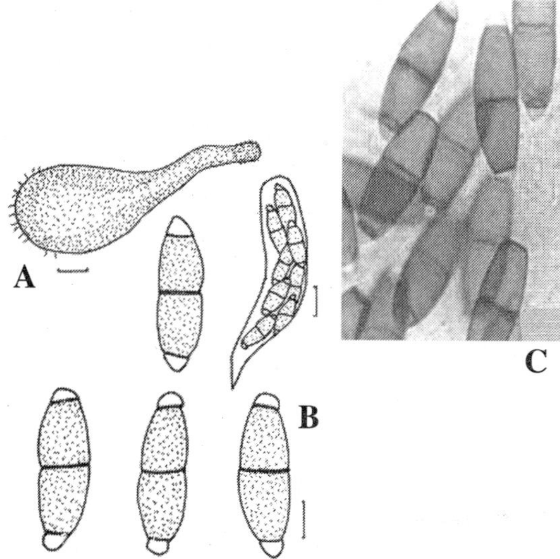

Fig. 97.3. ***Savoryella longispora,*** A-D; A- ascocarp, B,C- ascospores.

97.4. ***Savoryella paucispora*** (Cribb & J.W. Cribb) Jorgen Koch, 1982. *Nordic. J. Bot.*, **2:** 169. [Fig. 97.4]

Basionym: *Leptosphaeria paucispora* Cribb & J.W. Cribb, Pap. Dept. Bot., Uni. Qd., **4:** 41, (1960)

Ascomata: 84-140 µm diam., flask-shaped to ellipsoidal, immersed or superfial, ostiolate, papillate, membranous, cream-coloured to broon, solitary or in groups. *Peridium*: in surface view textura epidermioidea, in section composed of 3-6 layers of thick-walled angular to elongate cells. *Necks*: 70-100 µm long, 40-

55μm in diam., stout, pallid, lighter in colour at the tip. *Pseudoparaphyses*: up to 1.0 μm in diam., scarce, simple or racemose. *Asci*: 80-106 x 16-23 μm, 2-spored, cylindrical or clavate, short stalked, unitunicate, persistent, with non-amyloid apical thickening containing a pore. *Ascospores*: (36)-44-50(60) x 12-16.5 μm, fusoid-ellipsoidal, 3-septate, slightly constricted at the septa, central cells brown, apical cells hyaline, without appendages.

Habitat: Saprobic on drift wood and mangrove wood.

Description: Based on Kohlmeyer and Kohlmeyer (1979) and Jones and Hyde (1992).

Distribution: West Coast:-Maharashtra: On intertidal wood of *Avicennia marina* and prop roots of *Rhizophora mucronata*. (Borse, 2000a; Borse et al., 2005a), on intertidal mangrove wood (Maria and Sridhar, 2002a); **Goa:** On intertidal wood (Prasannarai and Sridhar, 2001), on intertidal mangrove wood (Maria and Sridhar, 2002a); **Karnataka:** On intertidal wood (Prasannarai and Sridhar, 2001), on intertidal wood of *Acanthus ilicifolius, Avicennia officinalis, Brugueiria gymnorrhiza, Rhizophora mucronata* and *Sonneratia caseolaris* (Maria and Sridhar, 2002a, 2003a, 2004), on intertidal leaf and woody litter of mangroves (Ananda and Sridhar, 2004); **Pondicherry (Mahe):** On intertidal wood (Nambiar and Raveendran, 2007); **Kerala:** On intertidal mangrove wood (Maria and Sridhar, 2002a), on intertidal wood (Raveendran and Manimohan, 2007), on intertidal mangrove wood, wood of *Avicennia* sp., *Bruguiera* sp. and *Rhizophora* sp. (Nambiar and Raveendran, 2006, 2007, 2008b, d, 2009a, b, d, e, f, g, h, I, 2010c; Nambiar et al., 2008a), on mangrove woody litter (Shini et al., 2009-10); **Lakshadweep Islands:** On intertidal wood (Chinnaraj, 1992).

East Coast:-Tamil Nadu: On intertidal wood (Nambiar et al., 2008b), on mangrove wood (Nambiar and Raveendran, 2008e, 2009k, l, 2010c); on intertidal woody debris (Ravikumar et al., 2009); **Pondicherry (Karaikkal):** on intertidal woody debris (Ravikumar et al., 2009); **West Bengal:** On intertidal mangrove wood (Pawar and Borse, 2005b); **Andaman & Nicobar Islands:** On intertidal wood of *Avicennia marina, A. officinalis, Rhizophora mucronata* and *Sonneratia alba* (Chinnaraj, 1993a).

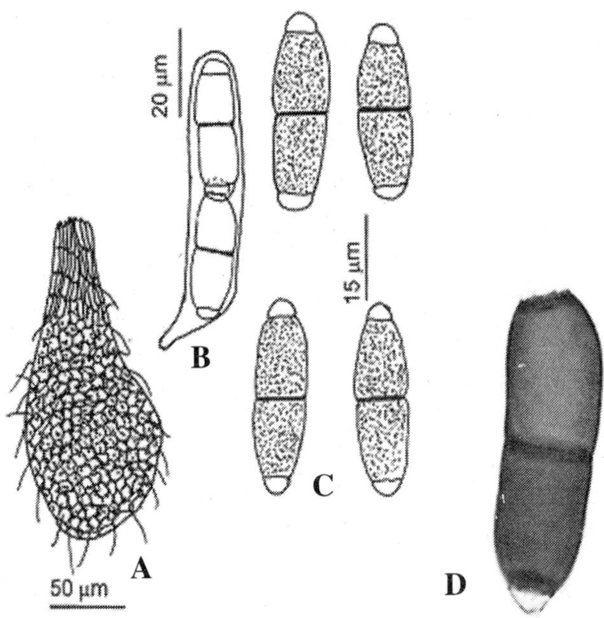

Fig. 97.4. ***Savoryella paucispora,*** A-D; A- ascocarp, B- ascus,C- ascospores, D- ascospore (1000x).

98. ***Swampomyces*** Kohlm & Volkm.- Kohlm., 1987a. *Bot. Mar.*, **30**: 200.

Ascomata: pyriform, subglobose, globose, centrun apricot coloured, solitary, immersed, ostiolate, necks with periphysate, papillate or epapillate, clypeate, coriaceous, black in the clupeus, dark brown to black, single. *Peridium*; thin. *Paraphyses*: simple, numerous, rarely branched, embedded in a gelatinous matrix, between and above the asci, mearging with periphyses. *Asci*: 8-spored, cylindrical, oblong, pedunculate, persistent, unitunicate, apically thickened, J-, arising from a thin ascogenous tissue at the base and along the sides of the locule. *Ascospores*: clavate, ellipsoidal, 3-septate, hyaline, slightly to constricted at the septa, no appendage or sheath. The genus is represented by 4 species in marine habitat (Jones *et al.*, 2009a).

Type species: *Swampomyces armeniacus* Kohlm. & Volkm.- Kohlm.

Habitat: Saprobic on mangrove wood.

Description: Based on Kohlmyer and Volkmann- Kohlmeyer (1987a).

Key to *Swampomyces* species from India:

1 Ascospores 1-septate, 13-20 x 6-9 µm, ... *Swampomyces armeniacus*

1' Ascospores 3-setate, 18-25 x 8-11 µm, ... *Swampomyces triseptatus*

98.1. ***Swampomyces armeniacus*** Kohlm. & Volkm.- Kohlm., 1987a. *Bot. Mar.*, **30**: 200. [For Fig. refer original paper]

Ascomata: 370-450 µm high, 290-370 µm in diam., subglobose, totally immersed in the substrate, ostiolate, periphysate, epapillate, clypeate, coriaceous, black in the clupeus, light brown, intimately connected with heavily deco,posed wood particles, single. *Ostiolar canal*: 40-75 µm in diam., conical, filled with 2-7 µm thick periphyses. *Clypeus*: 45-100 µm thick, black, surrounding the ostiole, enclosing deteriorating host cells. *Peridium*; 14-15 µm thick, almost hyaline, composed of 5-8 layers of flat cells with narrow lumina. *Paraphyses*: 1.3-1.8 µm thick, simple, rarely branched, embedded in a gelatinous matrix, between and above the asci, mearging with periphysesin the ostiolar canal. *Asci*: 99-132 x 13-17 µm, 8-spored, cylindrical, short pedunculate, persistent, unitunicate, J-, arising from a thin ascogenous tissue at the base and along the sides of the locule. *Ascospores*: 13.3-19.5 x 6-8.9 µm, uni- to biseriate, broad ellipsoidal, 1-septate, hyaline to yellowish, not constricted at the septum, appearing light apricot-coloured in masses, smooth, no appendage or sheath.

Habitat: Saprobic on intertidal mangrove wood.

Description: Based on Kohlmeyer and Volkmann-Kohlmeyer (1987a).

Distribution: West Coast:-Karnataka: On intertidal wood (Prasannarai and Sridhar, 2001, 2000-2001).

East Coast:-Tamil Nadu: On submerged parts of *Rhizophora* spp. (Ravikumar and Vittal, 1996); **Andaman & Nicobar Islands:** On intertidal wood of *Avicennia marina* and *Rhizophora mucronata* (Chinnaraj and Untawale, 1992), on intertidal wood of *Avicennia marina* and *Rhizophora mucronata* (Chinnaraj, 1993a).

98.2. ***Swampomyces triseptatus*** K.D. Hyde & Nakagiri, 1992. *Sydowia*, **44**: 122. [Fig. 98.2]

Ascomata: 240-320 µm high, 175-200 µm in diam., subglobose or pyriform, immersed, with long axis horizontal, oblique or vertical to the host surface, brown-black, ostiolate, periphysate, coriaceous, contents apricot coloured in mass, occurring mostly singly, developing under a thin darkened superficial pseudostroma, up to 20 mm thick, covering area where ascomata develop and composed of host cells with darkened fungal hyphae. *Necks*: up to 70 µm in diam., dark, with periphyses. *Peridium*; up to 25 µm thick, composed of 6-8 layers of brown flattened angular cells, these cells with narrow lumina. *Paraphyses*: 1.3-1.8 µm wide, branching, appearing to anastomosing, numerous, filamentous, hyaline, embedded in a gelatinous matrix and fusing with periphyses in the neck. *Asci*: 120-150 x 10-12 µm, 8-spored, cylindrical, thin-walled, short pedunculate, apically thickened with indication po pore and non-amyloid. *Ascospores*: 18-25 x 8-11 µm, overlapping uniseriate, hyaline to yellowish, ellipsoidal, 3-septate, weakly constricted at the septum, granular ornamentations under SEM,

Habitat: Saprobic on intertidal mangrove wood.

Description: Based on Hyde and Nakagiri (1992).

Distribution: West Coast:-Gujarat: On intertidal wood (Patil and Borse, 2001); **East Coast:-West Bengal:** On intertidal mangrove wood (Pawar and Borse, 2005b).

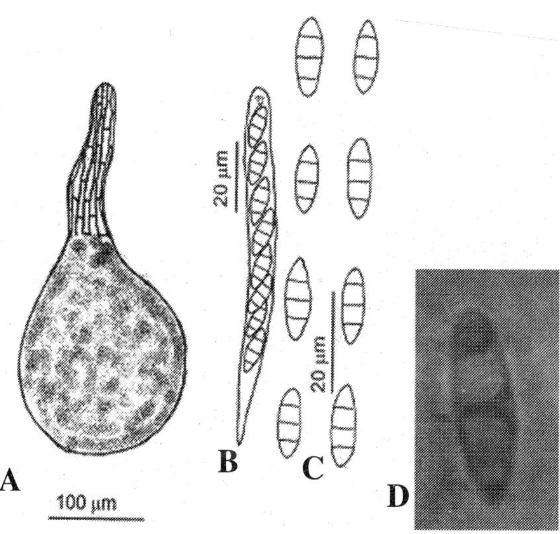

Fig. 98.2. ***Swampomyces triseptatus,*** A-D; A- ascocarp, B- ascus, C- ascospores, D- ascospore (1000x).

99. *Thalossogena* Kohlm. & Volkm.- Kohlm., 1987c. *Syst. Ascom.*, **6**: 225.

Ascomata: subglobose, immersed to superficial, ostiolate, periphysate, papillate, cream-coloured, single. *Peridium*: thin, forming a *textura angularis*. *Catenophyses*: developing from, the thin-walled pseudoparenchyma. *Asci*: 8-spored, clavate, pedunculate, with an ascul pore, non-amyloid, thin-walled, unitunicate, persitent or deliquescent, maturing successively on an ascogenous tissue at the bottom of the locule. *Ascospores*: subglobose, one-celled, hyaline, without sheath or appendages. The genus is monotypic (Jones *et al.*, 2009a).

Type species: *Thalossogena sphaerica* Kohlm. & Volkm.- Kohlm.
Habitat: Saprobic on subtidal or intertidal wood.
Description: Based on Kohlmeyer and Volkmann-Kohlmeyer (1987c).

99.1. *Thalossogena sphaerica* Kohlm. & Volkm.-Kohlm., 1987c. *Syst. Ascom.*, **6**: 225. [Fig. 99.1]

Ascomata: 150-260 µm in diam., subglobose, immersed to superficial, ostiolate, periphysate, papillate, cream-coloured, single, surrounded by brown hyphae. *Necks*: 230-570 µm long, 50-85 µm wide, ostolar canal filled with periphyses. *Peridium*: 10-16 µm thick, composed of 4-5 layers of polygonal cells, forming a *textura angularis*, mearing with the thin-walled pseudoparenchymatous cells. *Catenophyses*: 3-11 µm diam., developing from, the thin-walled pseudoparenchyma that originally fills the central cavity. *Asci*: 72-89 x 22-27 µm, 8-spored, clavate, pedunculate, flattened at the apex and with a pore, thin-walled, unitunicate, persitent or deliquescent, maturing successively on an ascogenous tissue at the bottom of the locule. *Ascospores*: 12.5-16.5 µm, subglobose, rarely ellipsoidal, one-celled, hyaline, without sheath or appendages, with one large lipid globule surrounded by numerous small non-lipoid igantea. The genus is monotypic (Jones *et al.*, 2009).

Habitat: Saprobic on subtidal or intertidal wood.
Description: Based on Kohlmeyer and Volkmann-Kohlmeyer (1987c).
Distribution: East Coast:-West Bengal: On intertidal wood of *Excoecaria agallocha* (Pawar and Borse, 2005b).

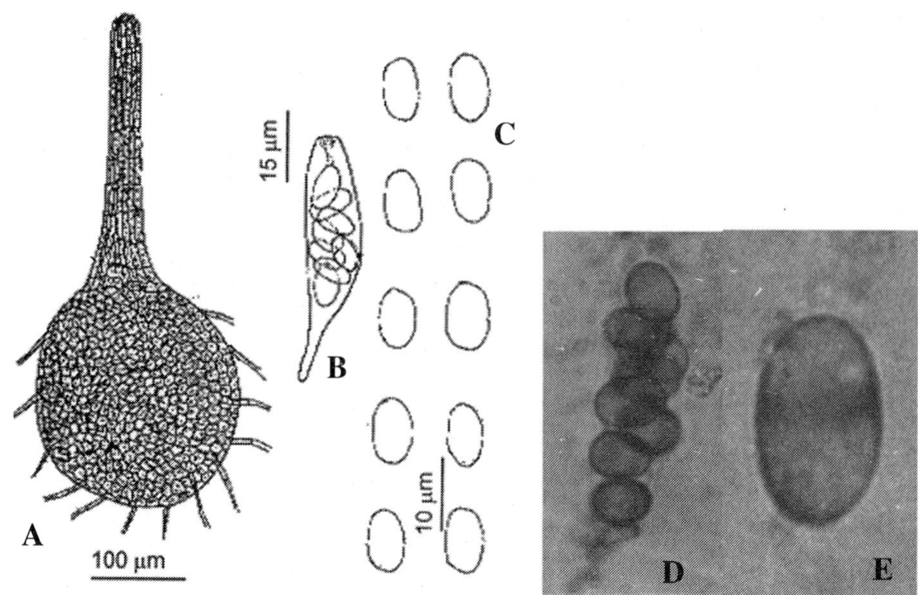

Fig. 99.1. ***Thalossogena sphaerica,*** A-D; A- ascocarp, B- ascus, C- ascospores, D- ascus (450x), E-ascospore (1000x).

100. *Tirispora* E.B.G. Jones & Vrijmoed, 1994. *Can. J. Bot.*, **72**: 1373.

Ascomata: globose to subglobose, solitary, superficial, ostiolate, periphysate, papillate, pale to dark brown. *Peridium*: thin. *Catenophyses*: present. *Asci*: 8-spored, clavate, pedunculate, with a ring and apical plate, thin-walled, unitunicate, persitent, maturing successively on an ascogenous cushion at the base of ascoma. *Ascospores*: ellipsoidal, bicelled, thick-walled, hyaline, with a single appendage at one pole that is initially adpressed to spore wall but unfurls to form a long filamentous thread. The genus is represented by 2 obligate marine species (Jones *et al.*, 2009a).

Type species: *Tirispora unicaudata* E.B.G. Jones & Vrijmoed
Habitat: Saprobic on intertidal mangrove wood.
Description: Based on Jones et al. (1994).

Key to *Tirispora* species from India:

1.Ascospores 24-32 x 8-12 µm .... *Tirispora unicaudata*

1' Ascospores 15-22 x 8-12 µm .... *Tirispora mandoviana*

100.1. ***Tirispora unicaudata*** E.B.G. Jones & Vrijmoed, 1994. *Can. J. Bot.*, **72**: 1373. [For Fig. refer original paper]

= *Aniptodera indica* Ananda & K.R. Sridhar, *J. Environ. Biol.*, **22**: 283 (2001).

Ascomata: 90-300 µm long, 66-216 µm in diam., globose to subglobose, solitary, superficial, ostiolate, periphysate, papillate, pale to dark brown. *Necks*: 22-48 µm long, 12-46 µm wide. *Peridium*: thin, 4-6 layeres of cells. *Catenophyses*: present. *Asci*: 40-80 x 14-28 µm, 8-spored, clavate, pedunculate, with a ring and apical plate, thick-walled, unitunicate, persitent, maturing successively on an ascogenous cushion at the base of ascoma. *Ascospores*: 24-32 x 8-12 µm, thick-walled, ellipsoidal, bicelled, hyaline, with a single appendage at one pole, which is initially adpressed to spore wall but unfurls to form a long filamentous thread.

Habitat: Saprobic on intertidal mangrove wood.

Description: Based on Jones et al. (1994) and Ananda and Sridhar (2001).

Distribution: West Coast:-Maharashtra: On intertidal mangrove wood (Maria and Sridhar, 2002a); **Goa:** On mangrove wood (Maria and Sridhar, 2002); **Karnataka:** On intertidal wood and leaves of mangroves (Ananda and Sridhar, 2001, 2003), on intertidal mangrove wood of *Acanthus ilicifolius, Avicennia officinalis, Brugueiria gymnorrhiza, Rhizophora mucronata* and *Sonneratia caseolaris* (Maria and Sridhar, 2002a, 2003a); **Kerala:** On intertidal mangrove wood (Maria and Sridhar, 2002a).

East Coast:-Tamil Nadu: On intertidal wood (Nambiar et al., 2008b), on intertidal wood (Nambiar and Raveendran, 2009k, l, 2010c).

100.2. ***Tirispora mandoviana*** V.V. Sarma & K.D. Hyde, 2000. *Aust. Mycologist*, **19**: 52. [Fig. 100.2]

Ascomata: 264-420 µm long, 250-430 µm in diam., globose to subglobose, solitary, superficial or semi-immersed, hyaline to pale brown, ostiolate, papillate, pale to dark brown. *Necks*: 220-466µm long, 60-90 µm wide, periphysate. *Peridium*: 15-22 µm thick, hyaline to light brown, of 5-8 layeres of elongated, compressed cellsPseudoparenchyma: compressed in mature ascomata. *Catenophyses*: absent. *Asci*: 80-115 x 19-31 µm, eight-spored, clavate, pedunculate, with an indistinct apical pore, thin-walled, unitunicate, persitent, maturing successively on an ascogenous cushion at the base of ascoma. *Ascospores*: 15-22 x 8-12 µm, thick-walled, ellipsoidal, bicelled with larege oil globules, hyaline, with a single appendage at one pole, 2-4.5 μm wide, 1.5-3.5 μm high, apical, cap-like, appressed, arranged randomly in elation to the apex of the ascus.

Habitat: Saprobic on intertidal mangrove wood.

Description: Based on Sarma and Hyde (2000b).

Distribution: West Coast:-Goa: On intertidal wood of *Rhizophora mucronata* (Sarma and Hyde, 2000); **Karanataka:** On intertidal wood of *Avicennia officinalis* and *Rhizophora mucronata* (Maria and Sridhar, 2004).

East Coast:-West Bengal: On intertidal wood of *Avicennia marina* (Pawar and Borse, 2005b).

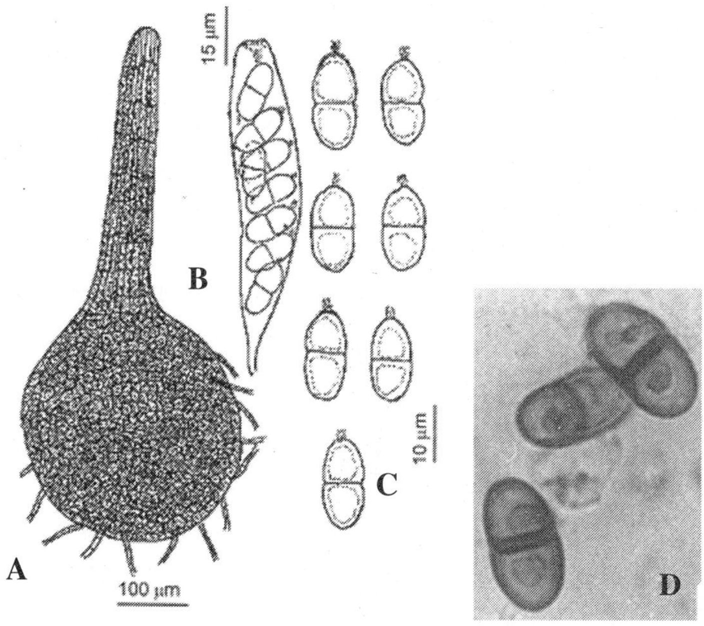

Fig. 100.2. ***Tirispora mandoviana,*** A-D; A- ascocarp, B- ascus,C- ascospores, D- ascospore (1000x).

101. ***Torpedospora*** Meyers, 1957. *Mycologia*, **49**: 496.

Ascomata: solitary, subglobose to ellipsoidal, immersed or superficial, ostiolate, papillate or epapillate, subcarbonaceous to subcoriaceous, dark- coloured. *Paraphyses*: ramose, growing irregularly through the ascomata venter; persistent or deliquescing. *Asci*: 8-spored, clavate or ellipsoidal, unituncate, thin-walled, aphysoclastic, early deliquescing. *Ascospores*: cylindrical to elongate-ellipsoidal,

3-septate, hyaline but pale orange in mass, with several radiating appendages at one or both ends. The genus is represented by two obligate marine species (Jones et al., 2009a).

Type species: *Torpedospora radiata* Meyers
Habitat: Saprobic on intertidal wood.
Description: Based on Meyers (1957).

101.1. *Torpedospora radiata* Meyers, 1957. *Mycologia*, **49**: 496.

[Fig. 101.1]

Ascomata: (75-)100-340 µm high, 100-361 µm in diam., subglobose to pyriform, immersed or superficial, ostiolate, papillate, subcoriaceous, fuscous or dark brown above, subhyaline grey or brownish below, gregarious. *Peridium*: 15-25 µm thick, 2-layered; outer layer composed of subglobose, thick-walled cells with small lumina; inner layer composed of elongate thick-walled cells with larger lumina. *Necks*: up to 225 µm long, 18–32 µm in diam., cylindrical, fuscous; ostiolar canal at first filled with small thin-walled pseudoparenchymatous cells. *Paraphyses*: 0.7-1.3 µm in diam., septate, ramose, growing irregularly through the venter of the ascomata. *Asci*: 103-181 x 12-16 µm, 8-spored, clavate or oblonge-ellipsoidal, sessile or short pedunculate, unitunicate, thin-walled, aphysoclastic, without apical apparatuses, early deliquescing, developing along the inner wall of the lower half of the ascomata. *Ascospores*: 30.5-52 x 4-9 µm, cylindrical or clavate, broader at the apex, 3-septate, not or slightly constricted at the speta, hyaline, appendaged. Appendages: 3 to 4 radiating on the lower end, 19-39.5 x 1.5-4.5 µm, semirigid, straight or slightly curved, with a thick base, tapering towards the apex.

Habitat: Saprobic on intertidal wood and mangrove wood, ascospores accumulate in foam along shore.
Description: Based on Meyers (1957).
Distribution: West Coast:-Gujarat: On intertidal wood of *Avicennia marina* (Borse et al., 2000a), on intertidal wood (Patil and Borse, 2001); **Maharashtra:** On intertidal wood and wood of *Avicennia marina* (Borse, 1987e, 1988; Shrivastava, 1995; Borse et al., 2005a); **Goa:** On submerged wood panels of *Mangifera indica* (Raghukumar et al., 1988a), on intertidal wood (Borse et al., 1999b), on intertidal wood (Prasannarai and Sridhar, 2001), on intertidal mangrove wood (Maria and Sridhar, 2002a), Ascospores in foam samples (Borse and Tuwar, 2006); **Karnataka:** On intertidal wood of *Bambusa* sp. (Prasannarai and Sridhar,

1997, 2001, 2000-2001), on intertidal wood and teak panels (Prasannarai et al., 1999), on intertidal leaf and woody litter of mangroves (Ananda and Sridhar, 2003, 2004); **Pondicherry (Mahe):** Ascospores in foam samples (Borse and Pawar, 2005); **Kerala:** On intertidal wood (Prasannarai and Sridhar, 2001), on wood of *Avicennia* sp. and *Bruguiera* sp. (Nambiar and Raveendran, 2009h, 2010c); **Lakshadweep Islands:** On intertidal of *Pemphis acidula* (Chinnaraj, 1992), on intertidal mangrove wood (Ananda and Sridhar, 2003).

East Coast:-Tamil Nadu: On intertidal wood (Raghukumar, 1973), on intertidal wood (Prasannarai and Sridhar, 2001), on intertidal wood (Nambiar and Raveendran, 2009l); **West Coast:** On intertidal wood of mangroves (Pawar et al., 2009); **Orissa:** On intertidal wood of *Avicennia marina* (Borse et al., 2009).

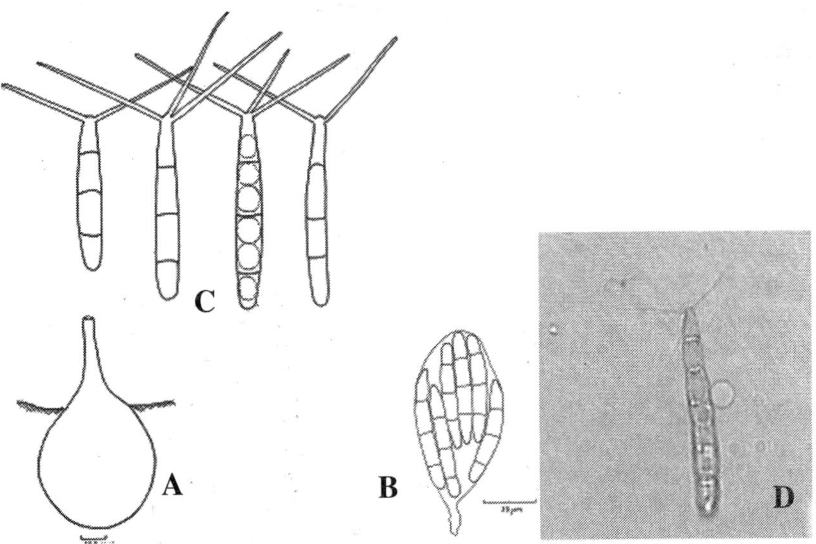

Fig. 101.1. ***Torpedospora radiata,*** A-D; A- ascocarp, B- ascus,C- ascospores, D- ascospore (1000x).

102. ***Trematosphaeria*** Fuckel, 1869-70. *Jahrb. Nassau. Ver. Naturkd.*, **23-24**: 161.

Ascomat: subglobose, obpyriform, immersed, ostiolate, papillate, coriaceous to carbonaceous, black, solitary or gregarious. *Necks*: with hyaline thin hyphae. *Pseudoparenchyma*: unbranched, trabeculate, in gelatinous matrix. *Asci*: cylindrical, pedunculate, thick-walled, fissitunicate, with an eccentric apical plate, J-. *Ascospores*: fusiform, multi-septate, slightly constricted to constricted at the septa, generally more or less cinnamon-coloured, frequently lighter at apices, smooth-walled, with or without sheath or appengages. The genus is represented by 3 obligate marine species (Jones *et al.,* 2009a).

Type species: *Trematosphaeria pertusa* Fuckel
Habitat: Saprobic on intertidal and mangrove wood.
Description: Based on Kohlmeyer and Kohlmeyer. (1979) and Jones et al. (2009a).

Key to *Trematosphaeria* species from India:

1 Ascospores versicolour with hyaline end-cells, 34-48 x 7-10 μm,

....*Trematosphaeria lineolatispora*

1' Ascospores uniformly coloured, 30-41 x 10-16 μm,

.... *Trematosphaeria mangrovei*

102.1. ***Trematosphaeria lineolatispora*** K.D. Hyde, 1992c. *Mycol. Res.*, **96**: 28. [Fig. 102.1]

Ascomata: 90-180 μm high, 210-360 μm diam., conoid to subglobose, immersed with a flattened base, ostiolate, papillate, as darkened spots on wood surface, clypeate, solitary or gregarious. *Necks*: up to 150 μm long, 75-100 μm diam., periphysate, brown. *Peridium*: up to 25 μm thick, at the sides composed of one layer of pseudoparenchymatous cells of *textura angularis*, becoming elongate towards the ostiole and fusing with the clypeus. The peridium is thin at the base. *Pseudoparaphyses*: 2-4 μm wide, straight, cellular, septate, appearing trabeculate, in a gel, forming a thick packing tissue between the asci and extending into the neck. *Asci*: 120-204 x 14-18 μm, 8-spored, cylindrical-subclavate, bitunicate, thick-walled, pedunculate, with an ocular chamber and faint ring structure, arising from the base and lower sides of the ascoma. *Ascospores*: 34-48 x 7-10 μm, 1-2-seriate, fusiform, mostly 5-septate, third cell from the top the largest, cinnamon-brown in center with lighter end cells, hyaline when young, surface covered in striations and surrounded by a mucilaginous sheath.

Habitat: Saprobic on intertidal mangrove wood.
Description: Based on Hyde (1992c).
Distribution: West Coast:-Karnataka: On intertidal wood (Prasannarai and Sridhar, 1997); **Kerala:** On mangrove woody litter (Shini et al., 2009-10).

East Coast:-Orissa: On intertidal wood (Borse and Borse, 2001); **West Bengal:** On intertidal mangrove wood (Pawar et al., 2005).

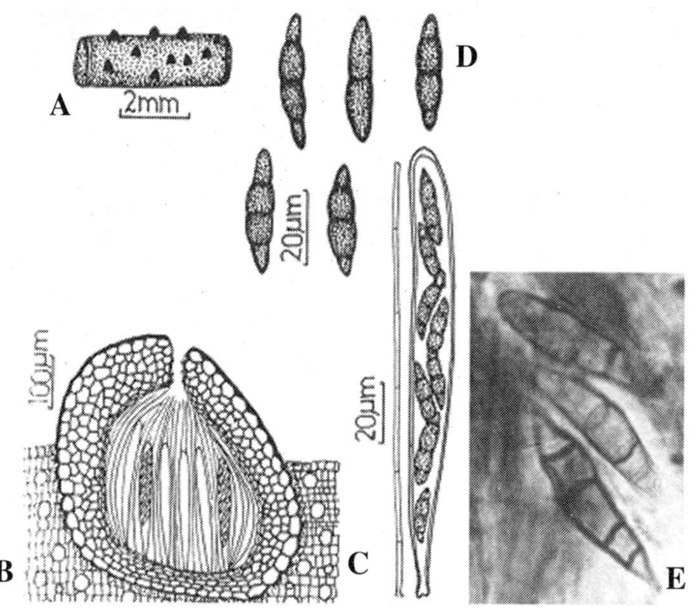

Fig. 102.1. ***Trematosphaeria lineolatispora,*** A-E; A- Habit, B- v.s. of ascocarp, C- ascus and paraphysis, D- ascospores, E- ascospores 1000x).

102.2. ***Trematosphaeria mangrovei*** Kohlm., 1968. *Mycopath. Mycol. Appl.*, **34**: 1. [For Fig. refer original paper]

Ascomata: 380-750 µm high (including necks), 450-800 µm diam., conical or semiglobose, flattened at the base base, seated partly in the substrate with subicula that consists of dark, plectenchymatous layers and downward of brown hyphae; ostiolate, short papillate, carbonaceous, black, gregarious, sometimes confluent. *Peridium*: 64-88 µm thick, 2-layered, enclosing cells of host; outer black layer 56-84 µm thick, scleroplectenchymatous, composed of thick-walled, brown cells with irregular, small lumina; inner hyaline layer 8-12 µm thick, composed of elongate, thin-walled cells with large lumina at the sides and in the angle between cover and basis of the ascomata, mearging into periphyses and pseudoparaphyses; cells of inner layer isodimetric under the asci. *Necks*: 160-220 µm long, 160-180 µm diam., conical, ostiolar canal at first closed at the apex by polygonal, light brown cells with small lumina, 3-4.5 µm in diam.; filled with delicate, periphysoid hyphae.

Pseudoparaphyses: 1.6- 2.2 µm wide, septate, simple ot rarely reticulate. *Asci*: 190-220 x 20-22 µm, 8-spored, cylindrical, bitunicate, thick-walled, pedunculate, without apical apparatuses; arising from the base ascomata venter. *Ascospores*: 30-41 x 10-13(-16.5) µm, obliquely 1-2-seriate, broad fusiform or ellipsoidal, sometimes slightly apiculate, somewhat curved, 3-septate, slightely constricted at the septa, fuscous, sometimes lighter coloured at the extreme apices.

Habitat: Saprobic on intertidal mangrove wood.

Description: Based on Kohlmyer and Kohlmeyer (1979).

Distribution:-West Coast:-Karnataka: On intertidal wood (Sridhar and Prasannarai, 1993).

103. *Verruculinia* Kohlm. & Volkm.-Kohlm., 1990b. *Mycol. Res.,* **94:** 689.

Ascomata: subglobose, ampulliform or ellipsoidal, partly or incompletely immersed, ostiolate, periphyaste, papillate, clypeate, carbonaceous, black, single. *Necks*: short, conical, surrounded by the clypeus, mostly immersed; ostiolar canal obturbinate. *Peridium*: composed of six or more layers of irregular rounded or flattended dark brown cells, forming a *textura angularis*, with hyphae or stromatic structures embedded in the substrate. *Hamathecium*: pseudoparaphysoid, trabeculate, with gelatinous matrix. *Asci*: eight–spored, cylindrical, short pedunculate, non-amyloid, without apical apparatuses, with ocular chamber, thick-walled, fissitunicate, developing at the bottom of the locule. *Ascospores*: monostichous, ellipsoidal, one-septate, dark brown, verrucose to verruculose. The genus is monotypic (Jones *et al.,* 2009a).

Type species: *Verruculinia enalia* (Kohlm.) Kohlm. & Volkm.-Kohlm.

Habitat: Saprobic on intertidal and mangrove wood.

Description: Based on Kohlmeyer and Volkmann-Kohlmeyer (1990b).

103.1. *Verruculinia enalia* (Kohlm.) Kohlm. & Volkm.-Kohlm., 1990b. *Mycol. Res.*, **94**: 689. [Fig. 103.1]

Basionym: *Didymosphaeria enalia* Kohlm, *Ber. Destch. Bot. Ges.* **79:** 28, 1966.

Ascomata: 295-480 µm high (including papilla), 140-520 µm in diam., subglobose, ampulliform or depressed ellipsoidal, partly or completely immersed, ostiolate, papillate, clypeate, carbonaceous, black, solitary. *Peridium*: 12.5-17.5 µm thick, composed of about 6 or more layers of irregular roundish or elongate, thick-walled cells with usually small lumina, forming a *textura angularis*; merging on the outside into thick, brown hyphae or stromatic structures. *Necks*: 80-145 µm long, 140-300 mm in diam. (including clypeus), conical, surrounded by blackish-brown clypeus; ostiolar canal obturbinate, filled with long delicate, hyaline periphyses; the pore is closed by somewhat thicker, shorter, hyaline hyphae. *Pseudoparaphyses*: 1.5-2 µm in diam., septate, rarely branched, attached at both ends and reaching into the ostiolar canal; connected to each other by a gelatinous outer layerand thereby forming a compact ball that firmly encloses the asci. *Asci*: 117-135 x 12.5-15.5 µm, 8-spored, cylindrical, pedunculate, bitunicate, thick-walled, physoclastic, without apical apparatuses; developing at the base of the ascomata venter. *Ascospores*: (15.5-) 16.5-23 x (6.5-) 7.5-10(-11) µm, obliquely uniseriate, ellipsoidal, 1-septate, constricted at the septum, dark brown, verrucose to verruculose.

Habitat: Saprobic on intertidal and mangrove wood.

Description: Based on Kohlmeyer and Kohlmeyer (1979).

Distribution: West Coast:-Diu Island: On intertidal wood (Borse et al., 1999b); **Daman:** On intertidal wood and *Avicennia marina* (Borse et al., 2000b); **Gujarat:** On intertidal wood of *Avicennia marina* and *Rhizophora mucronat* (Borse et al., 2000a), on intertidal wood and mangrove wood (Patil and Borse, 2001); **Maharashtra:** On intertidal wood and dead proproots of *Rhizophora mucronata,* intertidal wood, intertidal dead roots of *Avicennia alba* (Borse,1985, 1986, 1988; Srivastava, 1989), intertidal wood of Ceriops *tagal* and *Rhizophora mucronata* (Patil and Borse, 1986; Borse et al., 2005a), on intertidal mangrove wood (Maria and Sridhar, 2002a); **Goa:** On intertidal wood (Borse et al., 1999b), on intertidal wood (Prasannarai and Sridhar, 2001, 2000-2001), on intertidal mangrove wood (Maria and Sridhar, 2002a); **Karnataka:** On intertidal wood (Prasannarai and Sridhar, 1997, 2001), on intertidal wood of *Acanthus ilicifolius, Avicennia officinalis, Brugueiria gymnorrhiza, Rhizophora mucronata* and *Sonneratia caseolaris* (Maria and Sridhar, 2002a, 2003a, 2004), on intertidal leaf and woody litter of mangroves (Ananda and Sridhar, 2003, 2004); **Pondicherry (Mahe):** On intertidal wood (Nambiar and Raveendran, 2007, 2008c); **Kerala:** On intertidal mangrove wood (Maria and Sridhar, 2002a), on intertidal wood (Raveendran and Manimohan, 2007), on intertidal wood, mangrove wood of *Avicennia* sp., *Bruguiera* sp. and *Rhizophora* sp. (Nambiar and Raveendran, 2006, 2008b, d, 2009a, b, d, e, f, g, h, I, 2010c; Nambiar et al., 2006, 2008a), on mangrove woody litter (Shini et al., 2009-10); **Lakshadweep Islands:** On intertidal wood of *Pemphis acidula* and *Scaveola taccada* (Chinnaraj, 1992) on intertidal mangrove wood (Ananda and Sridhar, 2003). **East Coast:-Tamil Nadu:** On submerged parts of *Rhizophora* spp. (Ravikumar and Vittal, 1996), on intertidal wood (Nambiar et al., 2008b), on mangrove wood (Nambiar and Raveendran, 2008e, 2009k, 2010c); on intertidal woody debris (Ravikumar et al., 2009); **Pondicherry (Karaikkal):** on intertidal woody debris (Ravikumar et al., 2009); **Andhara Pradesh:** On intertidal wood of *Acanthua ilicifolius, Aegiceras corniculatum, Avicennia marina, A. officinalis, Ceriops decandra, Excoecaria agallocha, Lumnitzrea recemosa, Rhizophora apiculata* and *Sonneratia apetala* (Sarma and Vittal, 1998-99; Sarma and Vittal, 2000; Sarma and Vittal, 2001; Sarma et al., 2001; Sarma and Vittal, 2002; Sarma and Vittal, 2004; Vittal and Sarma, 2005), on intertidal wood (Nambiar and Raveendran, 2009j); **West Coast:** On intertidal wood of *Avicennia marina* (Pawar et al., 2009), on mangrove woody litter (Shini *et al.,* 2009-10); **Orissa:** On intertidal wood of *Acanthus ilicifolius, Acrostichum aureum, Aegiceras corniculatum, Avicennia marina, A. officinalis, Ceriops decandra, Excoecaria agallocha, Heritiera fomes, Hibiscus teliaceous, Kandelia candel, Nypa frutians, Sonneratia griffithii* and *Tamarix troupii* (Borse et al., 2009); **Andaman & Nicobar Islands:** On intertidal wood of *Avicennia marina, A. officinalis, Bruguiera gymnorhiza, Rhizophora apiculata, R. mucronata* and *Sonneratia alba* (Chinnaraj, 1993a).

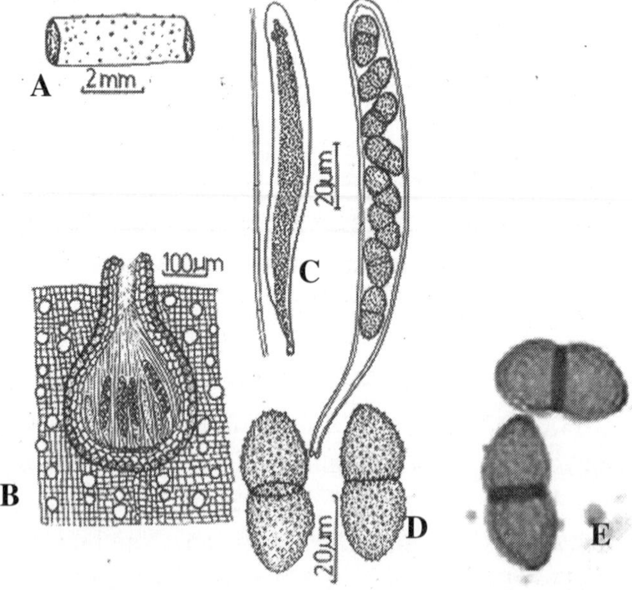

Fig. 103.1. ***Verruculinia enalia,*** A-E; A- Habit, B- v.s. of ascocarp, C- asci and paraphysis, D- ascospores, E- ascospores (1000x).

104. ***Zopfiella*** G. Winter, 1884. Kryptogamenflora, **1:** 56.

Ascomata: solitary or gregarious, globose to subglobose, superficial or rarely immersed, nonostiolate, thin-walled, irregularly dehiscing, covered with hairs. *Peridium*: pseudoparenchymatous, membranaceous, cephalothecoid in some species, cells forming a *textura angularis*. *Paraphyses*: early deliquescing, indistinct or absent. *Asci*: 4 to 8-spored, clavate to cylindrical or rarely subglobose, pedunculate, unitunicate, deliquescing, in some species with an apical ring, fasciculate or irregularly arranged. *Ascospores*: uni-, bi-, or triseriate, at first 1-celled, hyaline, becoming 1-septate, hyaline in the lower third, forming a large ellipsoidal, dark upper cell and a small, mostly cylindrical, hyaline, often collapsing basal cell; the upper cell may become divided by a horizontal septum in some species; with an apical or subapical germ pore. The genus is represented by 2 species in marine habitats (Jones et al., 2009a).

Type species: *Zopfiella tabulata* Winter
Habitat: Saprobic on intertidal and mangrove wood.
Description: Based on Kohlmeyer and Kohlmeyer (1979).

Key to *Zopfiella* species from India:

1.Ascospores with subapical germ pore, upper cell 16-22(-25) x 10-13(-15) mm, olivaceous to brown; basal cell 4-8(-9) x 3.5-7 mm, hyaline .... *Z. latipes*

1' Ascospores with apical germ pore, upper cell 14-20 x 10-14 mm, olivaceous to dark broen, basal cell up to 4 mm diam., hyaline.......................................
.... *Z. marina*

104.1. ***Zopfiella latipes*** (N. Lundq.) Maloch & Cain, 1971. *Can. J. Bot.*, **49**: 876. [Fig. 104.1]

Ascomata: 120-700 µm in diam., globose to subglobose, superficial or immersed, nonostiolate, coriaceous, irregularly dehiscing, dark brown, covered with hyaline to grayish- or yellowish-brown, septate, branched hairs, 1.5-4 µm in diam.; solitary. *Peridium*: 40-50 µm thick, semitransparent, composed of 3 or 4 layers of irregular or angular, thin-walled cells of 5-12 µm in diam., forming a *textura angularis*. *Paraphyses*: up to 12 µm in diam., composed of vesicular cells, early deliquescing. *Asci*: 80-120 x 12-18 µm, 8-spored, clavate, broadest in the middle, short pedunculate, apicallty truncate, unitunicate, deliquescing, with a simple apical ring, 2.1 µm wide; fasciculate. *Ascospores*: biseriate, ellipsoidal, becoming 1-septate in the lower third; slightly constricted at the septum; larger upper cell 16-22(-25) x 10-13(-15) µm, ellipsoidal, apex conical or abonate, base truncate, olivaceous to brown, thin-walled, smooth, with a apical germ pore, 1 um in diam.; smaller lower cell 4-8(-9) µm long, 3.5-7 µm in diam., broadly cylindrical, apex truncate, base broadly rounded, hyaline, at maturity without cytoplasm; the base and one side of the lower cell thin-walled, collapsing, and giving it a cuplike shape; collapsed lower cell appearing triangular in lateral view.

Habitat: Saprobic on intertidal and mangrove wood.

Description: Based on Kohlmeyer and Kohlmeyer (1979).

Distribution: West Coast:-Gujarat: On intertidal wood of *Avicennia marina* (Borse et al., 2000b); **Maharashtra:** On intertidal wood and submerged wood of *Avicennia alba* and *A. marina* (Borse and Shrivastava, 1988; Borse et al., 2005a); **Karnataka:** On intertidal wood of *Avicennia officinalis, Brugueiria gymnorrhiza,* and *Sonneratia caseolaris* (Maria and Sridhar, 2003a, 2004); **Pondecherry (Mahe):** On intertidal wood (Nambiar and Raveendran, 2007), **Kerala:** On intertidal wood (Kolhmeyer and Kolhmeyer, 1979), on intertidal mangrove wood (Nambiar and Raveendran, 2008b, 2009g, 2010c), on mangrove woody litter (Shini et al., 2009-10).

East Coast:-Tamil Nadu: On intertidal wood (Nambiar et al., 2008b), On mangrove wood (Nambiar and Raveendran, 2008e, 2010c); **Andhara Pradesh:** On intertidal wood (Kolhmeyer and Kolhmeyer, 1979), on intertidal wood of *Avicennia officinalis* (Sarma and Vittal, 1998-99; Sarma and Vittal, 2000; Sarma and Vittal, 2001; Sarma et al., 2001; Sarma and Vittal, 2004; Vittal and Sarma, 2005); **West Bengal:** On intertidal wood of mangroves (Pawar et al., 2009), on mangrove woody litter (Shini et al., 2009-10); **Orissa:** On intertidal wood of *Avicennia marina* (Borse et al., 2009).

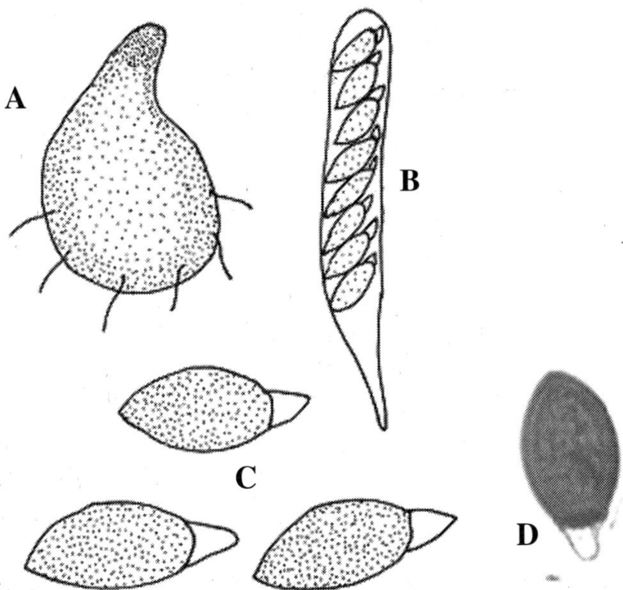

Fig. 104.1. *Zopfiella latipes,* A-D; A- ascocarp (100x), B- ascus (450x) , C- ascospores (1000x), D- ascospore (1000x).

104.2. *Zopfiella marina* Furuya & Udagawa, 1975. *J. Jpn. Bot.*, **50**: 149.

[For Fig. refer original paper]

Ascomata: 180-450 µm in diam., globose to subglobose, superficial or immersed, nonostiolate, coriaceous, irregularly dehiscing, black, almost glabrous or loosely covered by some curved, hyaline, septate, simple hyphae, 2-2.5 µm in diam. at the base; solitary. *Peridium*: 15-28 µm thick, semitransparent, composed of about 4 layers of angular, thin-walled cells of 5-14 µm in diam., forming a *textura angularis*. *Paraphyses*: composed of vesicular cells, early deliquescing. *Asci*: 75-90 x 14-20 µm, 8-spored, clavate, broadest in the middle, short pedunculate, unitunicate, deliquescing, with an indistinct apical ring. *Ascospores*: biseriate, clavate, becoming 1-septate in the lower third; slightly constricted at the septum; larger upper cell (14-)15-20 x 10-13(-14) µm, ellipsoidal, apex slightly abonate, base truncate, olivaceous to dark brown, thin-walled, smooth, with a apical germ pore, about 1 µm in diam.; smaller lower cell 8-10 µm long, 3-4 µm in diam., elongate cylindrical, apex truncate, base rounded, straight or slightly curved, hyaline, collapsed lower cell appearing more or less truncate in lateral view.

Habitat: Isolated on marine sediments at depth of 120 m. in East China Sea (Pacific Ocean).

Description: Based on Kohlmeyer and Kohlmeyer (1979) and Furuya and Udagawa (1975)

Distribution: East Coast:- Andhara Pradesh: On intertidal wood of *Avicennia officinalis* (Sarma and Vittal, 1998-99; Sarma and Vittal, 2000; Sarma and Vittal, 2001; Sarma et al., 2001; Sarma and Vittal, 2004; Vittal and Sarma, 2005).

Phylum: Basidiomycota

Subphylum: **Agaricomycotina**

Class: **Agaricomycetes**

Subclass: **Agaricomycetidae**

Order: **Agaricales**

Family: **Lachnellaceae**

Key to marine Basidiomycetes from India:

1. Basidioma sub-globose, puff ball-like, no stalk, basidiospores with appendages, 11-12 x 7-8 µm .... *Nia vibrissa* Moore & Meyers

2. Basidioma cyphalloid, white, surface hairy smooth, basidiospres 8-10 x 8-9.5 µm .... *Halocyphina villosa* Kohlm. & Kohlm.

3. Basidioma yellowish to ochre yellow, cup-like, wall with surface hairs, brownish, dextrinoid, minutely encrusted, basidiospores 8- 12 x 6-10 µm .... *Calathella mangrovei* Jones & Agerer

105. ***Calathella*** D.A. Reid, 1964. *Persoonia,* **3**: 122.

Basidiomes: cyphalloid, pedunculate, superficial on wood, yellowish with a reddish tinge, older specimens ocher-yellow, leathery, tomentose, solitary or gregarious, mycelium with clamp connections, sterile hairs from a ring around the mouth of the basidiocarp that are simple, brownish, non-septate, dextrinoid, round and minutely encrusted. *Basidia*: suburniform to cylindrical, hyaline, with 4 sterigmata.

Basidiospores: hyaline, unicellular, smooth-walled, ellipsoidal and accumulate at the tip of the basidiocarp.

Type species: *Calathella mangrovei* E.B.G. Jones & Agerer
Habitat: A wide spread tropical monotypic genus, often occurring on freshly cut and exposed branches of mangrove tree *Bruguiera.*
Description: Based on Jones et al. (2009a).

105.1. ***Calathella mangrovei*** E.B.G. Jones & Agerer, 1992. *Bot. Mar.*, **35**: 259. [For Fig. refer original paper]

Basidiomes: cyphelloid, 500-2520 µm high, 300-1080 µm in diameter at the apex, 243-450 µm in diameter in the stalk region, 720-810 µm in the mid region, turbinate to club-shaped, tube-shaped, becoming horn-like in shape, pedunculate, superficial on the wood, yellowish with a reddish colouring, old basidiomes ochre-yellow, leathery, tomentose, solitary or gregarious. *Mycelium*: with clamp connections. Cup wall surface hairs 60-520 X 203 µm forming a ring at the top around the opening of the basidiome, simple, brownish, non-septate, dextrinoid, round and minutely encrusted. *Hymenium*: lining the inner walls of the basidiome. *Basidia*: 60-80 x 5-8.5 µm, suburniform, sometimes cylindrical, tapering with a long tail, non-septate, hyaline, with 4 sterigmata, 4 µm long. *Basidiospores*: 8-12 x 6-10 µm, one-celled, smooth, hyaline, elliptical and accumulating at maturity at the tip of the basidiomes.

Habitat: Saprophytic on mangrove litter.
Description: Based on Jones and Agerer (1992).
Distribution: West Coast:-Maharashtra: On intertidal prop roots and seedlings of *Rhizophora mucronata* (Borse, 2000b; Borse et al., 2005a); **Goa:** On intertidal wood (Prasannarai and Sridhar, 2001); **Karnataka:** On intertidal wood (Prasannarai and Sridhar, 1997, 2001, 2000-2001).

106. ***Halocyphina*** Kohlm. & E. Kohlm., 1965. *Nova Hedwigia*, **9**: 100.

Basidiomes: cyphelloid, solitary or gregarious, initially turbinate or clavate becoming funnel-shaped or cupulate, pedunculate, superficial, white or yellowish, soft, thin-walled, tomentous, basidiome apex covered by sterile hairs. *Mycelium*: with clamp connections. *Hymenium*: lining the inner wall of the funnel. *Basidia*: 4-spored, clavate to cylindrical with a narrow base, non-septate, hyaline, with two to four evanescent sterigmata. *Basidiospores*: subglobose, one-celled, smooth, hyaline, non-amyloid, accumulating at the tip of the basidiome and washed away by the incoming tide. The genus is monotypic.

Type Species: *Halocyphina villosa* Kohlm. & E. Kohlm.
Habitat: Saprophytic on intertidal wood and mangrove substrates.
Description: Based on Kohlmeyer and Kohlmeyer (1965, 1979) and Jones et al. (2009a).

106.1. *Halocyphina villosa* Kohlm. & E. Kohlm, 1965. *Nova Hedwigia*, **9**: 100. [Fig. 106.1]

Basidiomes: cyphelloid, 350-500 µm high, 310-440 µm in diam in the apex, 100-150 µm in diam in the stalk, solitary or gregarious, turbinate or clavate, eventually funnel-shaped or cupulate, pedunculate, superficial in protected part of the wood, whitish or light-coloured, soft, thin-walled, tomentous. Peridium: 25-40 µm thock, composed of closely packed hyaline, rarely branched hyphae with thin walls and rare clamp connections, 3-6 µm in diam; external hairs up to 100 µm long, 4-5 µm in diam at the base, 1.5 µm at the apex, simple or dichotomously branched near the tip, hyaline. *Hymenium*: lining the inner wall of the funnel-shaped basidiome. *Basidia*: 13-22 x 5-10 µm, 4-spored, clavate or cylindrical with a narrow base, nonseptate, hyaline, with two to four evanescent sterigmata, about 3.5 µm long. *Basidiospores*: 8-10 x 8-9.5 µm, subglobose, one-celled, smooth, hyaline, nonamyloid, accumulating at maturity in the opening of the basidiome.

Habitat: Saprophytic on intertidal wood and mangrove substrates.
Description: Based on Kohlmeyer and Kohlmeyer (1965, 1979) and Ginns and Malloch (1977).
Distribution: West Coast:-Gujarat: On intertidal wood of *Avicennia marina* (Borse et al., 2000a), on intertidal wood (Patil and Borse, 2001); **Maharashtra:** On intertidal dead roots of *Avicennia alba* (Patil and Borse, 1983b, 1986; Shrivastava, 1995), intertidal wood of *Rhizophora mucronata* and *Sonneratia acida*, (Patil and Borse, 1986, Borse, 1988; Borse et al., 2005a); **Goa:** On intertidal wood of *Rhizophora mucronata* (Nandan et al., 1993), on intertidal animal substrates (Ananda et al., 1998), On intertidal wood (Prasannarai and Sridhar, 2001), on intertidal mangrove wood (Maria and Sridhar, 2002a); **Karnataka:** On intertidal wood (Prasannarai and Sridhar, 1997, 2001, 2000-2001), on intertidal animal substrates (Ananda et al., 1998), on intertidal wood of *Acanthus ilicifolius, Avicennia officinalis* and *Rhizophora mucronata* (Maria and Sridhar, 2002a, 2003a), on intertidal mangrove wood (Ananda and Sridhar,

2003); **Pondecherry (Mahe):** On intertidal mangrove wood (Nambiar and Raveendran, 2008a); **Kerala:** On intertidal mangrove wood (Maria and Sridhar, 2002a), on intertidal wood (Raveendran and Manimohan, 2007), on intertidal wood and mangrove wood of *Avicennia* sp. and *Bruguiera* sp. (Nambiar and Raveendran, 2006, 2007, 2008a, b, d, 2009a, b, d, e, f, g, h, i, j, 2010b, c; Nambiar et al., 2006, 2008a); **Lakshadweep Islands:** On intretidal wood of *Ceriops tagal* (Chinnaraj, 1992).

East Coast:-Tamil Nadu: On submerged parts of *Rhizophora* spp. (Ravikumar and Vittal, 1996), on intertidal wood (Nambiar et al., 2008b), on intertidal wood (Nambiar and Raveendran, 2009l, 2010c); on intertidal woody debris (Ravikumar et al., 2009); **Pondecherry (Karaikkal):** on intertidal woody debris (Ravikumar et al., 2009); **Andhara Pradesh:** On intertidal wood of *Avicennia marina, A. officinalis, Excoecaria agallocha,* and *Rhizophora apiculata* (Sarma and Vittal, 1998-99; Sarma and Vittal, 2000; Sarma and Vittal, 2001; Sarma et al., 2001; Sarma and Vittal, 2002; Sarma and Vittal, 2004; Vittal and Sarma, 2005), on mangrove wood (Nambiar and Raveendran, 2008e); **West Coast:** On intertidal wood of mangroves (Pawar et al., 2009); **Orissa:** On intertidal wood of *Avicennia marina, A. officinalis, Ceriops tagal, Hibiscus teliaceous* and *Sonneratia griffithii* (Borse et al., 2009); **Andaman & Nicobar Islands:** On intertidal wood of *Avicennia marina, A. officinalis, Bruguiera gymnorhiza, Rhizophora apiculata, R. mucronata* and *Sonneratia alba* (Chinnaraj, 1993a).

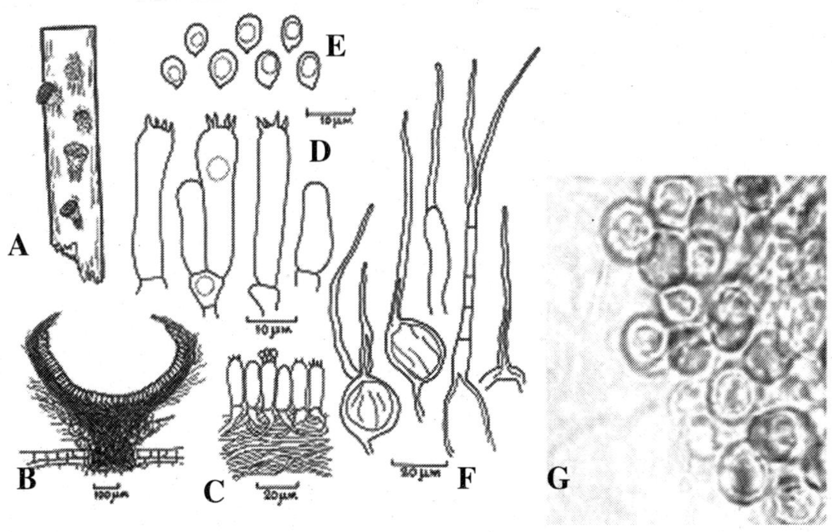

Fig. 106.1.: *Halocyphina villosa,* A-G: A- habit of Basidiocarp, B- v.s. of basidiocarp, C- v. s. of hymenium, D- basidia, E- basidiospores, F- external hyphae on peridium of basidiocarp, G- basidiospores (1000x).

107. ***Nia*** R.T. Moore & Meyers, 1959. *Mycologia*, **51**: 874.

Basidioms: solitary or gregarious, subglobose, superficial, pedicellate, light coloured becoming yellow, pink or orange, soft, thin-walled, villous or smooth, containing homogenous gleba, opening by irregular rupture of the peridium. *Hyphae*: septate, ramose, with clamp connections, hyaline. *Basidia*: 4- to 8-spored, hyaline without sterigmata. *Basidiospores*: ovoid or ellipsoidal, one-celled, hyaline, with several slender appendages, one terminal the others lateral. The genus is represented by three marine species (Rossello *et al.*, 1993).

Type Species: *Nia vibrissa* R.T. Moore & Meyers
Habitat: Intertidal dead wood.
Description: Based on Moore and Meyers (1959), Kohlmeyer and Kohlmeyer (1965, 1979), Ginns and Malloch, (1977) and Jones et al. (2009a).

107.1. ***Nia vibrissa*** R.T. Moore & Meyers, 1959. *Mycol.*, **51**: 874. [Fig. 107.1]

Basidiomes: 1-3 mm in diam, solitary or gregarious, subglobose, superficial, anchored in the substrate with an inconspicuous, cylindrical pedicel up to 80 µm in diam, whitish, yellowish, pinkish, and finally orange-coloured, soft, thin-walled, villose or smooth, containing homogenous gleba, opening by irregular rupture of the peridium. *Hyphae*: 2-3 µm in diam, septate, ramose, with clamp connections, hyaline. *Peridium*: 10-15 µm, bearing on the outside long hairs up to 275 µm long, 4-7 µm in diam., thick-walled, straight or curved, sometimes slightly curled apically and incinate. *Gleba*: at first composed of hyphae with inflated tips of 15-25 µm in diam (protocyst), later replaced by basidia developing at random throughout the venter. *Basidia*: 35-50 x 2 µm, with an apical subglobose inflation, 10-12 µm in diam, bearing at the round apex four- to eight basidiospores; hyaline without sterigmata. *Basidiospores*: 9-15 x 6-11 µm, ovoid or ellipsoidal, one-celled, hyaline, with several slender appendages. *Appendages*: at the apex provided with a single, slender, flexible, attenuate, hyaline appendage, 20-47 x 1 µm, terminally slightly inflated; four (rarely 3 or 5) similar, sub-terminal, radiating appendages around the base, 20-32 µm long; at the point of attachment to the basidium with a short cylindrical projection.

Type Species: *Nia vibrissa* R.T. Moore & Meyers

Habitat: Intertidal dead wood and mangrove substrates.

Description: based on Moore and Meyers (1959), Ginns and Malloch, (1977).

Distribution: West Coast:-Gujarat: On intertidal wood and mangrove wood (Patil and Borse, 2001)**; Maharashtra:** On intertidal wood of *Avicennia marina* and *Rhizophora mucronata* (Borse, 1985, 1988; Borse et al., 2005a); **Goa:** On intertidal wood (Nandan et al., 1993), on intertidal wood (Prasannarai and Sridhar, 2001); **Karnataka:** On intertidal wood (Prasannarai and Sridhar, 2001); **Kerala:** On intertidal wood (Prasannarai and Sridhar, 2001), on intertidal wood (Nambiar and Raveendran, 2007, 2008b, 2009b, 2010a, c).

East Coast:-Tamil Nadu: On intertidal mangrove wood (Ravikumar, 1991), on intertidal wood (Nambiar and Raveendran, 2009k, l); **Andhra Pradesh:** On intertidal wood (Nambiar and Raveendran, 2009j); **West Bengal:** On intertidal wood of *Excoecaria agallocha* (Pawar and Borse, 2005a, Pawar et al., 2009); **Orissa:** On intertidal wood of *Avicennia marina* and *Ceriops tagal* (Borse et al., 2009).

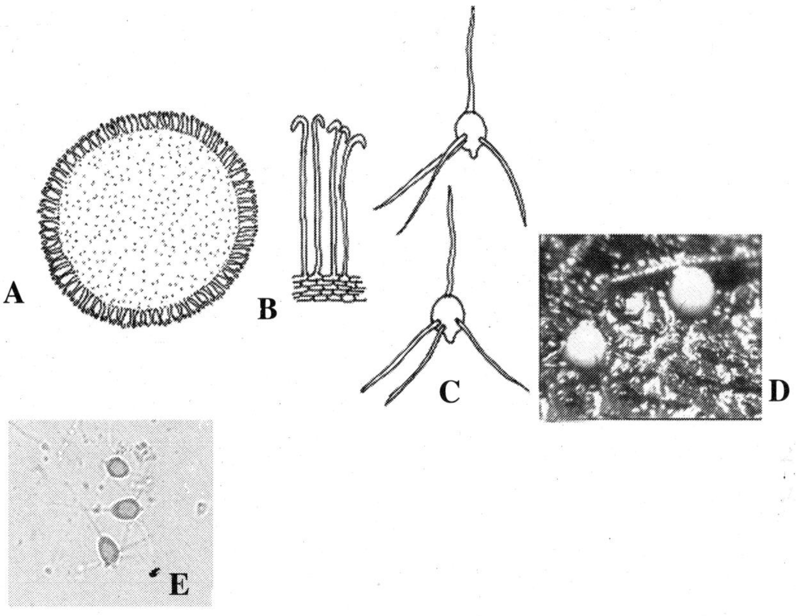

Fig. 107.1. ***Nia vibrissa,*** A-B: A- Basidiocarp (100x), B- appendages of basidiocarp (450x), C-basidiospores (1000x), D-basidiocarp, E-basidiospores (1000x).

Anamorphic (Mitosporic) Fungi

Hyphomycetes:

108. ***Alleschieriella*** Henn.

Hedwigia, **36**: 244 (1897).

Hyphae: interwoven, anastomosing, superficial or immersed, forming sporodochia. *Conidiophores*: semimacronematous, mononematous, branched, straight or curved, hyaline to light reddish-brown, smooth. *Conidiogenous cells*: monoblastic, integrated, terminal on stipe and branches or discrete, determinate, cylindrical. *Conidia*: acrogenous, solitary, ellipsoidal to fusiform, one-celled, thick-walled, smooth, pale-brown. The genus is represented by one species in marine environment (Jones *et al*., 2009a).

Type species: *Alleschieriella crocea* (Mont.) Hughes
Habitat: Saprobic on wood.
Discription: Based on Ellis (1971) and Kohlmeyer and Kohlmeyer (1979).

108.1. ***Alleschieriella bathygena*** Kohlm., 1977. *Rev. Mycol.*, **41**: 199.

[For Fig. refer original paper]

Hyphae: 2-5 µm in diam, septate, brown, immersed, forming sporodochia. *Stroma*: 40-50 µm high, 80-130 µm in diam, pulvinate, partly immersed, pseudoparenchymatous, composed of polygonal, thin-walled cells with large lumina, forming a *textura angularis*, producing conidia on the surface. *Conidiogenous cells*: 20-4 µm, monoblastic, terminal, determinate, cylindrical, straight, hyaline to light brown, arising from the upper cells of the stroma. *Conidia*: 17-24 x 6.5-9 µm, acrogenous, solitary, ellipsoidal, to fusiform with rounded ends, one-celled, thick-walled (especially at the apices), smooth, light-brown.

Habitat: Saprobic on submerged wood in deep sea.
Discription: Based on Kohlmeyer (1977).
Distribution: West Coast:-Goa: On deep sea sediments, Arabian Sea (Raghukumar, C. & Raghukumar, S. 1998).

109. ***Alternaria*** Nees, 1816. *Das System der Pilze und Schwamme, Wurzburg*, p. 72.

Conidiophores: cylindrical, septate, simple or irregularly branched, straight or curved, basal cell occasionally swollen, smooth, bearing conidia at the uniperforate apex, yellowish to brown, singly or fascicled. *Conidia*: enteroblastic-tretic, ovoid, obclavate, obpyriform or ellipsoidal, with a basal pore, taperinh or not into an apical beak, muriform, constricted at the septa, smooth or rough, olivaceous to brown, produced singly or in acropetal succession, forming simple or branched chains.

Type species: *Alternaria alternata* (Fries) Keissler

Habitat: Saprophytic or parasitic on plant material.

Discription: Based on Ellis (1971) and Kohlmeyer and Kohlmyer (1979).

109.1. ***Alternaria*** sp.

No obligate marine species is known, but species of this genus are occasionally found on detritus along the shores and conidia in foam samples. (Kohlmeyer and Volkmann-Kohlmeyer, 1991; Jones et al., 2009a).

Habitat: On intertidal wood and mangrove wood.
Distribution:-
West Coast:-Karanataka: On intertidal wood (Prasannarai and Sridhar, 2001, 2000-2001), on intertidal wood of *Acanthus ilicifolius, Avicennia officinalis, Brugueiria gymnorrhiza, Rhizophora mucronata* and *Sonneratia caseolaris* (Maria and Sridhar, 2002a, 2003a, 2004), on intertidal leaf and woody litter of mangroves (Ananda and Sridhar, 2003, 2004); **Kerala:** On intertidal mangrove wood (Maria and Sridhar, 2002a).

East Coast:-Orissa: On intertidal fronds of *Acrostichum aureum* (Borse and Borse, 2005); **West Bengal:** On intertidal wood of Mangroves (Pawar and Borse, 2005b).

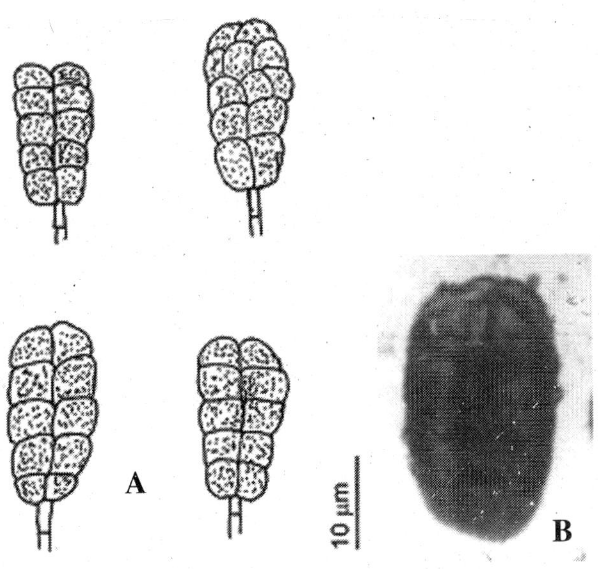

Fig. 109.1. ***Alternaria*** sp.A-B: A-conidia, B-conidium (1000x)

110. *Anguillospora* Ingold, 1942. *Trans. Br. Mycol. Soc.*, **25**: 401.

The genus characterized by having *Mycelium*: hyaline, branched and septate. *Conidiophores*: hyaline, septate and simple or branched. *Conidiogenous Cells*: hyaline, monoblastic, terminal, precurrent, without a separating cell. *Conidia*: hyaline, solitary, filiform, unicellular, curved or sigmoid, usually with right and sharp angles and borne terminally. The genus is represented by several freshwater and one marine species (Jones *et al.*, 2009a).

Type species: *Anguillospora longissima* Ingold
Habitat: Saprophytic on submerged plant material.
Discription: Based on Ingold (1942).

110. 1. ***Anguillospora marina*** Nakagiri & Tubaki, 1983. *Mycologia*, **75**: 488. [For Fig. refer original paper]

Hyphae: 2-4.8 µm in diam, septate, barely constricted at the septa, hyaline. *Conidiphores*: 35-187.5 x 1.8- 3.5 µm, simple or branched, hyaline. *Conidiogenous cells*: hyaline, monoblastic, terminal, precurrent, without a separating cell. *Conidia*: 150-255 x 2.5-4 µm, solitary, filiform, straighr or curved, 9-13(-19)-septate, swollen at both ends, (especially at the basal end), hyaline. Phialoconidia absent.

Habitat: Saprophytic on submerged plant material.
Discription: Based on Nakagiri and Tubaki (1983).
Distribution: West Coast:-Karanataka: on intertidal wood of *Rhizophora mucronata* (Maria and Sridhar, 2003a).

111. ***Arthrobotrys*** Corda, 1839. *Pracht-Fl. Eur. Schimmebild.*, **43**:

Colonies white to cream to yellow, fast growing. Mycelium is hyaline, septate and branched. Conidiphores erect, septate, rarely producing side branches, proliferating and bearing numerous conidia. Conidia are hyaline, septate, variously shaped, elongate, ellipsoidal, pyriform, rarely constricted at the septa. The genus is represented by 12 species in marine environment (Jones *et al.*, 2009a).

Type species: *Arthrobotrys superba* Corda
Habitat: Predacious on nematodes.
Discription: Based on Jones et al. (2009a)

111.1. ***Arthrobotrys oligospora*** Fresen., 1850. *Beitr. Mykol.*, **1**: 18.

[For Fig. refer original paper]

≡ *Arthrobotrys superba* var. *oligospora* (Fresenius) Coem.

≡ *Didimozoophaga oligospora* (Fresen.) Soprunov & Galiulina, Microbiol., Reading, **20**: 493 (1951).

Conidiophores: up to 440 µm long. *Conidia*: predominantly 1-septate, septum below the middle, obovoid to pyriform, 17-29 x 10-15 µm.

Habitat: Saprophytic on submerged plant material.
Discription: Based on Jones et al. (2009a).

Distribution:- West Coast:-Karanataka: on intertidal mangrove wood (Maria and Sridhar, 2004).

112. ***Aspergillus*** Link, 1809. *Observ. Mycol. (Lipstae)*, **1**: 16.

A number of *Aspergillus* species have been isolated from intertidal plant materials, but only one of these has been considered to be marine (*Jones* et al., 2009).

112.1. ***Aspergillus sydowii*** (Bainier & Sartory) Thom & Church, 1926. The Aspergilli, pp. 147. [For Fig. refer original paper]

Basionym: *Sterigmatocystis sydowii* Bainier & Sartory, *Annls. Mycol.*, **11**: 25 (1913).

Colonies: on Czapek at 25 C attaining a diam. of 3-4 cm in 2 weeks, velvety sometimes floccose, usually consisting of a dence felt of intensely blue green (Delft blue) conidiophores. *Conidial heads*: blue green, radiate; exudates reddish brown often abundant. Reverse mostly in shades of reddish brown to dark red (marron). *Conidiophore stipe*: up to 500 μm in length, hyaline sometimes slightly coloured, smooth-walled. Heads biseriate, vesicles globose, spathulate to subclavate, hyaline up to 20 μm in diam. Phialides borne on metulae, 5-7(-10) x 2-3.5 μm. Metulae 4-6(-7) x 2-3.5 μm. *Conidia*: globose, (2.5)3-4 um in diam., hyaline, conspicuously rough-walled.

Distribution: West Coast:-Isolated from deep sea sediments core from the Chagos, Indian Ocean (Raghukumar et al., 2004).
East Coast:-Tamil Nadu: on intertidal woody debris (Ravikumar et al., 2009); **Pondicherry (Karaikkal):** on intertidal woody debris (Ravikumar et al., 2009); **West Bengal:** Isolated from mangrove mud (Rai et al., 1979, Chowdhery and Rai, 1980).

112.2. ***Aspergillus*** sp.

Distribution: West Coast:-Deep sea: *A. ustus* isolated from deep sea sediments of the Indian Ocean (Raghukumar & Raghukumar, 1998). *A. terres* isolated from deep sea sediments of the central Indian Ocean basin (Damare et al., 2006); **Maharashtra:** On intertidal mangrove wood (Maria and Sridhar, 2002a); **Goa:** On intertidal mangrove wood (Maria and Sridhar, 2002a); **Karanataka:** On intertidal mangrove wood (Ananda and Sridhar, 2002, 2004; Maria and Sridharar

2002a, 2003, 2004), on animal substrate (Ananda and Sridhar, 2001); **Kerala:** On intertidal mangrove wood (Maria and Sridhar, 2002a).

East Coast:-Tamilnadu: On seagrass Halophila ovalis (Devarajan et al., 2002), on macroalgae (Surayanarayanan et al., 2010); **West Bengal:** Isolated from mangrove mud (Rai et al., 1979, Chowdhery and Rai, 1980).

113. *Bactrodesmium* Cooke, 1883. *Grevillea,* **12:** 35.

The genus is characterized by having *Sporodochia*: scattered brown or black. Mycelium: mostly immersed, composed of branched and septate. *Conidiophores*: micronematous, thin-walled hyaline or thick-wallwd brown, fasciculate, broad slightly near the apex, simple or branched, smooth, septate. *Conidiogenous cells*: holoblastic, integrated, terminal or intercalary, smooth, cylindrical, determinate. *Conidia*: solitary, subglobose to obpyriform, 1-3 septate, without constriction, apical cell larger, dark brown to black, basal and sub-basal cells smaller, light brown, wall unequal in height, hence the base of the conidia become curved. The genus is represented by one species in marine environment (Jones et al., 2009a).

Type species: *Bactrodesmium abruptum* Cooke
Habitat: Saprophytic on plant material.
Discription: Based on Ellis (1971, 1976) and Jones et al. (2009a).

113.1. *Bactrodesmium linderii* (J.L. Crane & Shearer) Palm & Stewart, 1982. *Mycotaxon,* **15:** 319. [Fig. 113.1]

Basionym: *Trichocladium linderi* J.L. Crane & Shearer, *Mycologia,* **70**: 866 (1978).

Mycelium: composed of branched, septate, at first hyaline, latter sub hyaline to light brown hyphae, *Conidiophores*: macronematous, mononematous, smooth, thin-walled and hyaline or thick-walled and brown. *Conidiogenous cells*: holoblastic, integrated, terminal or intercalary, smooth, cylindrical, determinate. *Conidia*: solitary, subglobose to obpyriform, 1-2 septate, without constriction, 20-33.6 x 14.5-20.5 μm, becoming 3-6 μm wide at base, apical cell larger, dark brown to black, 11-16 μm high, basal and sub-basal cells smaller, light brown, wall unequal in height, hence the base of the conidia become curved.

Habitat: Saprophytic on interdidal and mangrove wood.

Discription: Based on Crane and Shearer (1978).

Distribution: West Coast:-Maharashtra: On intertidal wood (Borse, 1984, 1988), intertidal wood of *Rhizophora mucronata* (Patil and Borse, 1986; Borse et al., 2005a); **Goa:** On intertidal wood (Bhat, 2010); **Karnataka:** On intertidal wood of *Acanthus ilicifolius, Avicennia officinalis, Brugueiria gymnorrhiza, Rhizophora mucronata* and *Sonneratia caseolaris* (Maria and Sridhar, 2002a, 2003a, 2004), on intertidal wood (Bhat, 2010); **Kerala:** On intertidal mangrove wood (Maria and Sridhar, 2002a), on mangrove woody litter (Shini et al., 2009-10).

East Coast:-Andhara Pradesh: On intertidal wood of *Rhizophora apiculata* (Sarma and Vittal, 1998-99; Sarma and Vittal, 2000; Sarma and Vittal, 2001; Sarma et al., 2001; Sarma and Vittal, 2002; Sarma and Vittal, 2004; Vittal and Sarma, 2005); **West Bengal:** On intertidal wood of *Phoenix poludosa* (Pawar and Borse, 2005b), on mangrove woody litter (Shini et al., 2009-10).

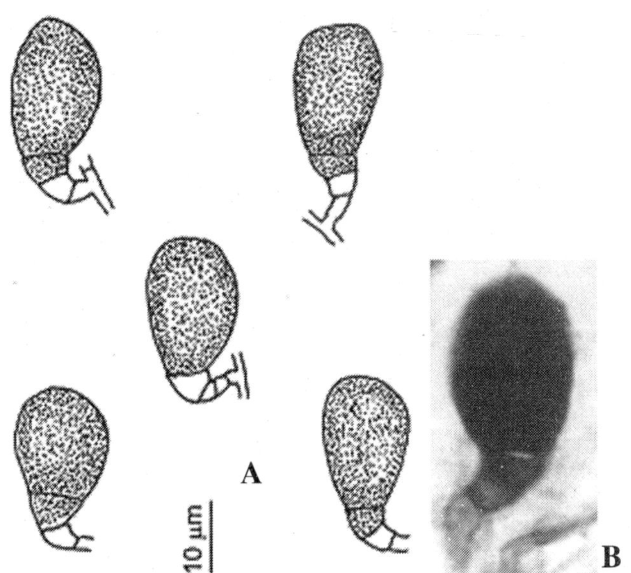

Fig. 113.1. ***Bactrodesmium linderii***. A-B: A-conidia, B-conidium (1000x)

114. ***Blastobotrys*** von Klopetek emend. C.P. Kurtzman & C.J. Robnett 1967, *Archiv. Microbiol.*, **58:** 92-96; *FEMS Yeast Res.*, **7:** 141-151 (2007).

Colonies: are white to cream-coloured, sometimes slightly glistering and butyrous, but often dull, powdery, and mycelial. Reproduction is usually by multilateral budding and less commonly by arthroconidia formed by disarticulation of hyphae. *Blastoconidia*: are commonly formed on denticles that may occur singly on hyphae or in sympodially arranged clusters at the ends of conidiophores. Primary conidia may produce secondary conidia. Conidia may develop long setae. Hyphae and pseudohyphae are usually abundantly present. Sugars may be fermented and nitrate may be assimilated.

Type species: *Blastobotrys nivea* von Klopetek
Discription: Based on Kurtzman and Robnett (2007).

114.1 *Blastobotrys* sp. [Fig. 114.1]

Colonies on seawater embedded MEA slow growing, attaining a diam of 2-2.5 cm in 10 days, flat, dull-white, slimy on the surface. Hyphae with pseudomycelium, cells exhibiting multilateral budding. Blastoconidial cells spherical, thick-walled, hyaline, 2-4 μm diam.; in mass, dull coloured.

Habitat: On submerged wood.
Distribution: West Coast:-Karnatak: off Honnavar coast (Prof Bhat, D. J., Unpublished).

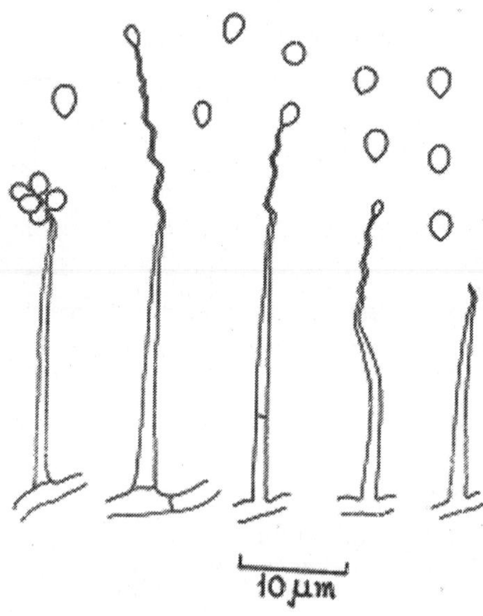

Fig. 114.1 *Blastobotrys* sp., Conidiophores with conidia.

115. *Botryophialophora* Linder, 1944. *Farlowia*, **1**: 403.

Hyphae: septate, branched, hyaline to pale brown, *Conidiophores*: subgloboser, lateral or apical on the main hyphae, bears clusters of conidiogenous cells. *Conidiogenous cells*: phialidic, flask-shaped, apex narrow, trumet-shaped, simple, hyaline, on conidiophores or rarely produced singly on hypha. *Conidia*: enteroblastic-phialidic, globose or subglobose, one-celled, hyaline, no sheath or appendage. The genus is monotypic.

Type species: *Botryophialophora marina* Linder
Habitat: Saprobic on wood, test panels, sand and sediments.
Discription: Based on Ellis (1971) and Jones et al. (2009a).

115.1. *Botryophialophora marina* Linder , 1944. *Farlowia*, **1**: 404.

[Fig. 115.1]

Hyphae: 4-10 µm in diam, septate, simple or branched, hyaline or light coloured, *Conidiophores*: subglobose or irregularly polygonal, lateral or apical on the main hyphae, bearing clusters of uo to 20 or more conidiogenous cells. *Conidiogenous cells*: 3.5-8.5 µm long, 1.5-3.3 µm in diam, phialidic, flask-shaped, apex narrow, trumet-shaped, simple, hyaline, on conidiophores or rarely produced singly on hyphae. *Conidia*: enteroblastic-phialidic, 2-3.3 µm in diam, globose or subglobose, one-celled, hyaline, is containing a large vacuole.

Habitat: Saprophytic on submerged plant material.
Discription: Based on Kohlmeyer and Kohlmeyer (1979).
Distribution: East Coast:-West Bengal: On intertidal wood of *Sonneratia apetala* (Pawar and Borse, 2005b).

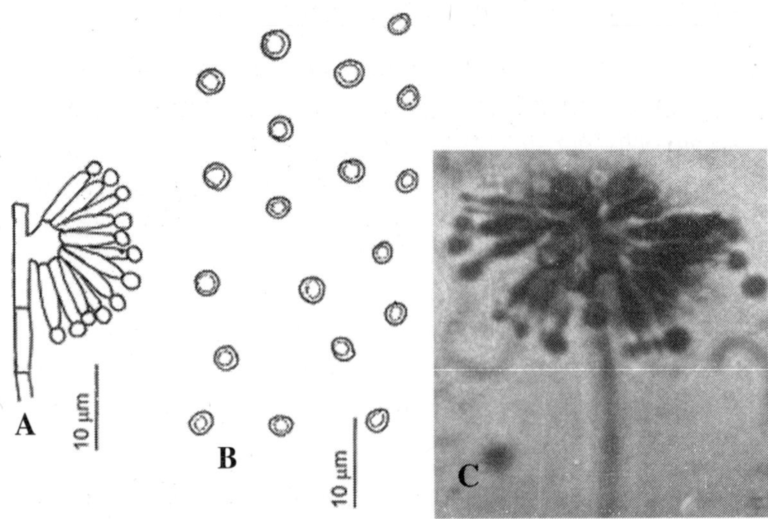

Fig. 115.1. ***Botryophialophora marina***.A-C: A-conidiophore with conidia, B-conidia, C- conidiophore with conidia (1000x).

116. *Cirrenalia* Meyers & R.T. Moore, 1960. *Am. J. Bot.,* **47:** 346.

The genus is characterized by having, *Conidiophores*: Present or lacking, cylindrical, septate or non-septate; acrogenous or laterally on the hyphae, hyaline or light brown. *Conidiogenous cells*: monoblastic, integrated, terminal, determinant. *Conidia*: acrogenous, solitary, igantean, septate, constricted or not constricted at the septa, brown; diam and pigmentation increasing from base to apex; cells dissimilar. The morphology and dimensions of the apical cell determine the species. The genus is represented by four marine among several terrestrial species (Jones *et al.,* 2009a; Somrithipal *et al*., 2002; Zhao and Liu, 2005; Abdel-Wahab *et al*., 2010).

Type species: *Cirrenalia macrocephala* (Kohlm.) Meyers & R.T. Moore
Habitat: On intertidal and mangrove wood.
Discription: Based on Kohlmeyer and Kohlmeyer (1979).

Key to *Cirrenalia* species from India:

1.Apical cell 16-20 μm, 3-6-septate, fuscous to grey brown.. *Cirrenalia pseudomacrocephala*

1'. Apical cell 9-16 μm, 3-5-septate, pale brown .... *Cirrenalia basiminuta*

1". Apical cell 6-14(-17) μm, 2-6-septate, reddish-brown .... *Cirrenalia macrocephala*

116.1. ***Cirrenalia basiminuta*** Raghukumar & Zainal, 1988. *Mycotaxon*, **31**: 163. [Fig. 116.1]

Hyphae: 2.4-5.5 μm in diam, septate, hyaline to pale brown. *Conidiophores*: terminal, integrated, monoblastic and determinate, 8.2-28.1 x 1.4-2.5 μm. *Conidia*: borne acrogenously, rarely borne laterally, and directly on conidiophore, solitary, helicoids, semi-contorted, 29-41 μm high and 18-34 μm in diam. Conidia: 3-5-septate, constricted at the septa. Conidial cells increasing in size from base to apex, pale brown. Apical cell 9-16 x 9-13.7 μm, subglobose, basal cell cylindrical and tapering, 4.9-17.1 x 2.5-7 μm. Pigmentation of the cells increasing from base to apex, the apical cell light brown with a reddish tinge.

Habitat: On intertidal wood, mangroves and shoreline plants.

Discription: Based on Raghukumar et al. (1988b).

Distribution: West Coast:-Gujarat: On intertidal mangrove wood (Patil and Borse, 2001); **Maharashtra:** On intertidal dead seedilings of *Rhizophora mucronata* (Borse, 1984; Borse et al., 2005a); **Goa:** Isolated from *Rhizophora mucronata* (Raghukumar et al., 1988b); **Karnataka:** on intertidal leaf and woody litter of mangroves (Ananda and Sridhar, 2004); **Kerala:** On intertidal wood (Raveendran and Manimohan, 2007), on intertidal mangrove wood of *Avicennia* sp. (Nambiar and Raveendran, 2006, 2009d, g), on mangrove woody litter (Shini et al., 2009-10); **Lakshadweep Islands:** On intertidal frond of *Cocos nucifera* (Chinnaraj, 1992).

East Coast:-Tamil Nadu: On submerged parts of *Rhizophora* spp. (Ravikumar and Vittal, 1996); **Andhara Pradesh:** On intertidal wood of *Rhizophora apiculata* (Sarma and Vittal, 1998-99; Sarma and Vittal, 2000; Sarma and Vittal, 2001; Sarma et al., 2001; Sarma and Vittal, 2004; Vittal and Sarma, 2005); **Orissa:** On intertidal fronds of *Acrostichum aureum* (Borse and Borse, 2005).

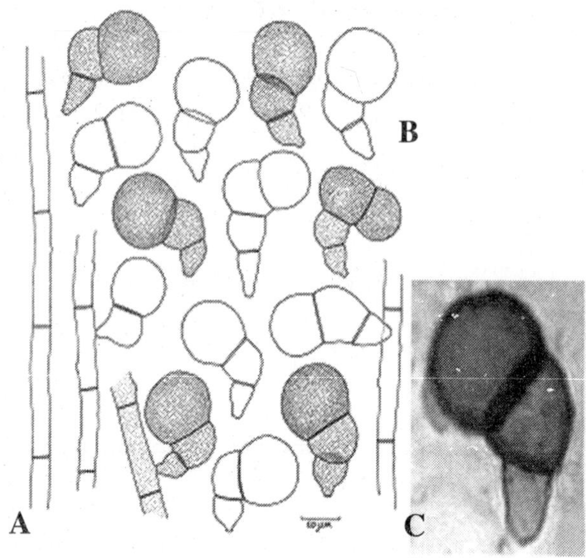

Fig. 116.1. ***Cirrenalia basiminuta***. A-C: A-conidiophore with conidia, B-conidia, C- conidium (1000x).

116.2. ***Cirrenalia macrocephala*** (Kohlm.) Meyers & R.T. Moore, 1960. *Am. J. Bot.*, **47**: 347. [For Fig. refer original paper]

Basionym: *Helicoma macrocephala* Kohlm., Ber. Dtsch. Bot. Ges., 71:99 (1958).

Hyphae: 1.5-2 µm in diam, septate, ramose, hyaline to dark brown. *Conidiophores*: 3.5-25 µm long, 2-5 µm in diam, cylindrical, occasionally apically somewhat inflated, zero- to three-septate, simple, straight or curved, hyaline to yellowish. *Conidiogenous cells*: monobalstic, integrated, terminal, determinate. *Conidia*: acrogenous, solitary, igantean, rarely straight, ¼ to 1 time contorted, 2-6-septate, strongly constricted at the septa, multiguttulate, pale brown; cells increasing in diameter and pigmentation from base to apex, distinctly dissimilar; spirals 12-31 µm high, 12-23 µm in diam; terminal cell 6-14 (17) µm high, 6.5-14 µm in diam, subglobose, basely flattened, reddish-brown, largest and darkest of all cells; basal cells 1.5-6 µm high, 2.5-7 µm in diam, semiglobose or obtusely conical, hyaline; central cells subglobose, cylindrical or doliiform, brownish.

Habitat: On intertidal and mangrove wood.

Discription: Based on Meyers and Moore (1960) and Kohlmeyer and Kohlmeyer. (1979).

Distribution: West Coast:-Gujarat: On intertidal wood (Patil and Borse, 2001); **Maharashtra:** On submerged wood panels (Jones, 1968), on intertidal dead seedilings of *Rhizophora mucronata* (Borse, 1984, 1988; Patil and Borse, 1986; Shrivastava, 1994; Borse et al., 2005a), on lignite beds (Kumaran et al., 2004a, b); **Karnataka:** On intertidal wood of *Acanthus ilicifolius, Avicennia officinalis, Rhizophora mucronata* and *Sonneratia caseolaris* (Maria and Sridhar, 2002a, 2003a); **Kerala:** On submerged wood panels (Jones, 1968), on intertidal wood and wood of *Avicennia* sp. (Raveendran and Manimohan, 2007, Nambiar and Raveendran, 2008b, 2009d, e, h, 2010c; Nambiar et al., 2008a).

East Coast:-Tamil Nadu: On submerged parts of *Rhizophora* spp. (Ravikumar and Vittal, 1996); on intertidal woody debris (Ravikumar et al., 2009); **Pondicharry (Karaikkal):** on intertidal woody debris (Ravikumar et al., 2009); **Andhara Pradesh:** On intertidal wood of *Rhizophora apiculata* (Sarma and Vittal, 1998-99; Sarma and Vittal, 2000; Vittal and Sarma, 2005).

Note: A report of *C. macrocephala* appears to be doubtful and some collections identified as *M. macrocephala* might actually belong to *C. basiminuta* (Borse, 2000, Sarma and Vittal, 2004).

116.3. *Cirrenalia pseudomacrocephala* Kohlm., 1968. *Mycologia,* **60:** 266. [Fig. 116. 3]

Hyphae: 2.5-5 µm in diameter, septate, superficial or immersed, brown. *Conidiophores*: 23-30 x 3-5 µm, cylindrical zero- to two-septate, simple, straight or curved, acrogenous or laterally on hyphae, often remaining connected with detached conidia, sometimes obsolete, light brown. *Conidiogenous cells*: monoblastic, integrated, terminal, determinate. *Conidia*: acrogenous, solitary, igantean, semicontorted to 1¼ times contorted, 3-6-septate, slightly or strongly constricted at the septa, fuscous to gray-brown (upper 3 to 5 cells dark, lower one to three cells light-colored); cells increasing in diameter and pigmentation from base to apex, distinctly dissimilar; spirals 27.5-38.5 µm in diameter; terminal cell 16-20 µm, subglobose to ellipsoidal, basally flattened, mostly largest and darkest of all cells; basal cells 5.5-10 x 4.5-5.5 µm in diam, obtusely conical or cylindrical, light brown; central cells subglobose or obtusely conical.

Habitat: On intertidal and mangrove wood.

Discription: Based on Kohlmeyer (1968).

Distribution: West Coast:-Maharashtra: On lignite beds (Kumaran et al., 2004a, b); **Goa:** On intertidal wood of *Avicennia marina* (Bhat, 2010); **Karnataka:** On intertidal wood (Prasannarai and Sridhar, 2001, 2000-2001), on intertidal mangrove wood (Ananda and Sridhar, 2003); **Lakshadweep Islands:** On intertidal mangrove wood (Ananda and Sridhar, 2003).

East Coast:-Tamil Nadu: on intertidal woody debris (Ravikumar et al., 2009); **Pondecharry (Karaikkal):** on intertidal woody debris (Ravikumar et al., 2009); **Orissa:** On intertidal wood of *Avicennia marina* (Borse and Borse, 2005); **Andaman & Nicobar Islands:** On intertidal wood of *Avicennia marina* (Chinnaraj, 1993a).

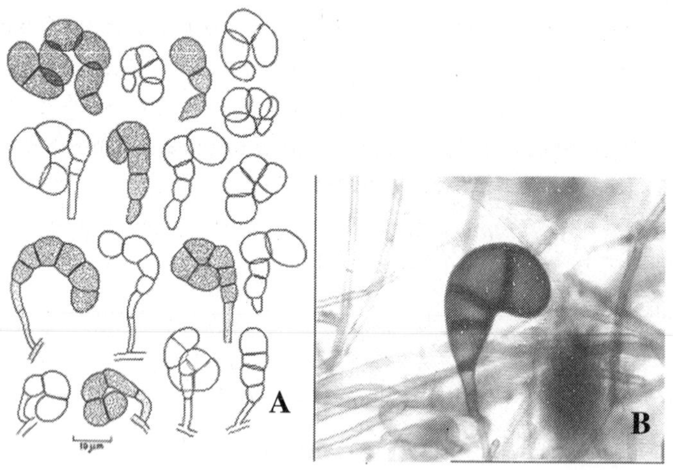

Fig. 116.3. ***Cirrenalia pseudomacrocephala***. A-C: A-conidia, B- conidia (1000x).

117. ***Cladosporium*** Link ex Fr., 1816. Link, *Mag. Ges. Naturf. Freunde*, Berlin, **7**: 37. Fries, *Systema Mycologicum, Gryphiswaldiae*, **1**: 46 (1821).

Hyphae: septate, branched, pale brown. *Conidiphores*: macronematous, cylindrical, simple or only branched apically, straight or curved, smooth or verrucose, olive-brown or brown. *Ramo-conidia*: frequently occurring. *Conidiogenous cells*: polyblastic, mostly integrated, terminal, intercalary or rarely discrete, sympodial, cylindrical, cicatrized with prominent scars. *Conidia*: simple, cylindrical, doliiform, fusiform, ellipsoidal or subglobose, with a protuberant basal scar or with scar at each end, zero to three (or more)-septate, smooth, verruculose or echinulate, olive-brown or brown, usually catenate, also solitary, chains often branched, acropleurogenous. The genus is represented by one marine among several terrestrial species (Jones *et al.,* 2009a).

Type species: *Cladosporium herbarum* (Pers.) Link ex S. F. Gray

Habitat: Saprophytic or parasitic on the substrate.

Discription: Based on Kohlmeyer and Kohlmeyer (1979).

117.1. *Cladosporium algarum* Cooke & Massee, 1888. *Grevillea*, **16**: 80. [For Fig. refer original paper]

Conidiphores: 140-175 um long, 8-13 um in diam at the base, 4 um at the tip, macronematous, cylindrical, septate, simple, straight or somewhat curved, smooth, brown. *Ramo-conidia*: up to 34 um long, 6 um in diam, three or four-septate, cylindrical or elongate-ellipsoidal, bearing up to four conidia on small protuberances, smooth, brown. *Conidiogenous cells*: polyblastic, integrated, terminal or intercalary, more or less, cylindrical, cicatrized with prominent scars. *Conidia*: 8-24 x 4.5-8 um, ellipsoidal, with a protuberant basal scar or with scars at each end, 0-2(-3)-septate, not or slightely constricted at the septa, smooth, thick-walled, olive-brown, solitary, rarely in short chains, acropleurogenous.

Habitat: On drift brown algae and sea grasses.
Discription: Based on Kohlmeyer and Kohlmeyer (1979).
Distribution: West Coast:-Karanataka: On intertidal wood (Prasannarai and Sridhar, 1997, 2000-2001); **Kerala:** On intertidal wood (Prasannarai and Sridhar, 2001, 2000-2001).
East Coast:-Tamil Nadu: On intertidal wood (Prasannarai and Sridhar, 2001); on intertidal woody debris (Ravikumar et al., 2009); **Pondicherry (Karaikkal):** on intertidal woody debris (Ravikumar et al., 2009).

118. *Clavatospora* Sv. Nilsson ex Marvanova & Sv. Nilsson, 1971. *Trans. Br. Mycol. Soc.*, **57**: 531.

The genus is characterized by *Hyphae*: septate, hyaline, without sporodochia. *Conidiophores*: simple or sparsely branched, hyaline. *Conidiogenous cells*: phialidic, hyaline. *Conidia*: enteroblastic-phialidic, one-celled, tetraradiate, consisting of a clavate, straight, basal main axis, from which three narrow cylindrical or conical, elongate branches arise at the apex; hyaline. The genus is represented by one marine and four fresh water species (Ingold, 1975; Jones *et al.,* 2009a).

Type species: *Clavatospora longibrachiata* (Ingold) Sv. Nilsson
Habitat: Saprobic on submerged plant substrates.
Discription: Based on Nilsson (1964) and Kohlmeyer and Kohlmeyer (1979).

118.1 ***Clavatospora bulbosa*** (Anastasiou) Nakagiri & Tubaki, 1985. *Bot. Mar.*, **28**: 489. [Fig. 118.1]

Basionym: *Clavariopsis bulbosa* Anastasiou, *Mycologia*, **53**: 11 (1962).

Conidiophores: 10-300 x 2-5 µm, cylindrical, septate, simple or branched, hyaline. *Conidia*: tetraradiate, septate, slightly constricted at the septa, hyaline to light brown, developing by transformation of the inflated apex of the conidiophore; basal arm one–septate; proximal cell 10-20 x 5-11.5 µm, ellipsoidal or ovoid, truncate at the base, light brown; distal cell 6.5-13 x 6.5-13 µm , cylindrical or shortly three branched, fuscous; three divergent arms arising simultaneously from the inflated distal cell of basal arm, 20-70 x 4-6 µm, cylindrical, one- to seven-septate, light brown. *One armed conidia*: Conidia consisted of only single arm also observed, these are grey brown, 5-10 celled, 54-60 x 6.5-9 µm, constricted at the septa, basal and apical cells are with lighter colour.

Teleomorph: *Corollospora pulchella* Kohlm., Schmidt & Nair
Habitat: On intertidal and mangrove wood, conidia accumulate in foam along shore.

Discription: Based on Anastasiou (1961) and Kohlmeyer and Kohlmeyer (1979).
Distribution: West Coast:-Gujarat: On intertidal wood (Patil and Borse, 2001); **Maharashtra:** On intertidal wood (Borse, 1984, 1988, Shrivastsva, 1984), on intertidal wood of *Rhizophora mucronata, Sonneratia acida* and *S. alba* (Patil and Borse, 1986; Borse et al., 2005a); **Goa:** On submerged wood panels of *Anogeissus latifolia* (Raghukumar et al., 1988a), on intertidal wood (Nandan et al., 1993), on intertidal animal substrates (Ananda et al., 1998), on intertidal mangrove wood (Maria and Sridhar, 2002a), Conidia in foam samples (Borse and Tuwar, 2006), on intertidal wood of *Avicennia marina* and *Rhizophora apiculata* (Bhat, 2010); **Karnataka:** On intertidal leaf litter (Sridhar and Kaveriappa, 1988, 1991), on intertidal wood and teak panels (Prasannarai and Sridhar, 1997, 2001, 2000-2001), On intertidal animal substrates (Ananda et al., 1998), on intertidal wood of *Avicennia officinalis* and *Rhizophora mucronata* (Maria and Sridhar, 2002a, 2003a, 2004,), on intertidal leaf and woody litter of mangroves (Ananda and Sridhar, 2003, 2004); **Pondicherry (Mahe):** Conidia in foam samples (Borse and Pawar, 2005); **Kerala:** on intertidal wood (Nambiar and Raveendran, 2007, 2008b, 2009b, e, I, 2010a, c; Nambiar et al., 2008a); **Lakshadweep Island:** On intertidal mangrove wood (Ananda and Sridhar, 2003). **East Coast:-Tamil Nadu:** On drift wood (Raghukumar, 1973), on intertidal wood (Prasannarai and Sridhar, 2001; Nambiar et al., 2008b), on mangrove

wood (Nambiar and Raveendran, 2008e, 2009k, l, 2010a, c); **Pondicherry:** On intertidal wood (Nambiar and Raveendran, 2008c); **Andhra Pradesh:** On intertidal wood (Nambiar and Raveendran, 2009j); **West Bengal:** Conidia in foam samples (Borse et al, 2001a); **Andaman & Nicobar Islands:** On intertidal wood of *Avicennia marina* (Chinnaraj, 1993a).

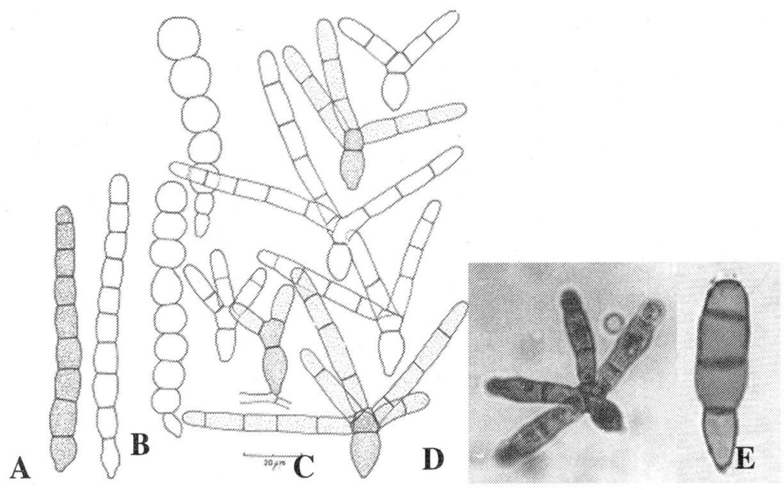

Fig. 118.1: ***Clavatospora bulbosa***, A-D: A- one armed conodia, B- chlamydospores, C- normal conidia, D- normal conidium (1000x), E- one armed conidium(1000x).

119. *Cumulospora* I. Schmidt, 1985. *Mycotaxon,* **24:** 420.

The genus is represented by two species (Chatmala *et al.*, 2004).The genus is iganteaned by having septate, branched, superficial or immerged, pale brown mycelium. *Conidiophores*: absent. *Conidiogenus cells*: holoblastic, integrated, terminal, determinate. *Conidia*: dark grey to fuscous, solitary, scattedered or gregarious, muriform a complex cluster of subglobose cells in knot like group of cells. The genus is represented by two marine species (Jones *et al.,* 2009a).
Type species: *Cumulospora marina* I. Schmidt

Habitat: Saprobic on submerged plant substrates.
Discription: Based on Chatmala et al. (2004).

119.1. *Cumulospora marina* I. Schmidt, 1985. *Mycotaxon*, **24**: 421.

[Fig. 119.1]

Basionym: *Vesicularia marina* I. Schmidt, Natur Naturschutz Mecklenberg, **12**: 117 (1974).

≡ *Basramyces marinus* (I. Schmidt) Abdullah, Abdulk & Goos, *Intern. J. Mycol. & Lichenol.*, **4**: 183 (1989).

Hyphae: septate, branched, superficial or immersed, pale brown. *Conidiophores*: absent. *Conidiogenous cells*: holoblastic, integrated, terminal, determinate. *Conidia*: 52-91 x 40-71 µm, solitary, scattered or gregarious, with up to 20 or more globose cells and up to 30 µm diam or more of each cell, and with a dark brown, sheath-like structure around the conidia.

Habitat: On intertidal wood and Mangrove substrates.

Discription: Based on Chatmala et al. (2004).

Distribution: West Coast:-Karanataka: On intertidal wood of *Avicennia officinalis* and *Rhizophora mucronata* (Maria and Sridhar, 2003b, 2004); **Kerala:** On intertidal mangrove wood (Nambiar and Raveendran, 2008b, 2009e, I, 2010c; Nambiar et al., 2008a).

East Coast:-Tamil Nadu: On intertidal wood (Nambiar et al., 2008b), on mangrove wood (Nambiar and Raveendran, 2008e, 2010c); **West Bengal:** On mangrove woody litter (Shini et al., 2009-10).

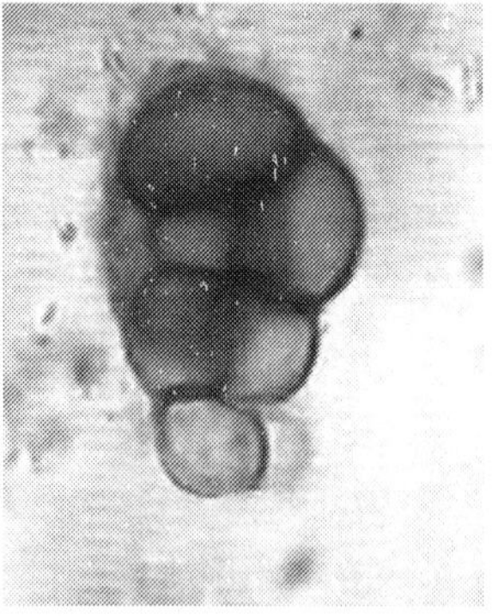

Fig. 119.1. *Cumulospora marina*m, Conidium(1000x)

120. *Dendryphiella* Bubak & Ranoj., 1914. In: Ranojevic, *Ann. Mycol.*, **12**: 417.

Hyphae: septate, branched, pale brown. *Conidiophores*: macronematous, septate, simple or irregularly branched near the apex, smooth or verruculose, bearing conidia at the swollen tip and at intercalary swellings, light to dark or reddish brown, solitary, showing indefinite growth. *Conidiogenous cells*: polytretic, integrated, terminal but becoming intercalary, sympodial, subglobose to clavate, cicatrized on the nodose swellings with scars close to each other. *Conidia*: simple, cylindrical or oblong with rounded ends, or ellipsoidal, 1-11-septate, smooth or verruculose, light brown to olive-brown, solitary, acropleurogenous, constricted at the septa, with short stalk but not denticulate. The genus is represented by two marine among several terrestrial species (Jones et al., 2009a).

Type species: *Cumulospora vinosa* (Berk. & Curt.) Reisinger
Habitat: Saprobic on plant litter.
Discription: Based on Ellis (1976) and Kohlmeyer and Kohlmeyer (1979).

120.1. *Dendryphiella salina* (G.K. Suther.) G.J.F. Pugh & Nicot., 1964. *Trans. Br. Mycol. Soc.*, **47**: 266. [For Fig. refer original paper]

Basionym: *Cercospora salina* Sutherl., New Phytol., **15**: 43 (1918).

Hyphae: 2-8 µm in diam, septate, hyaline, light brown or lightolivaceous. *Conidiophores*: 15-60 µm long, 4-4.5 µm in diam, macronematous, cylindrical, septate, simple or rarely branched, straight or curved, apically or subapically somewhat swollen, hyaline, light olivaceous or pale brown. *Conidia*: 14-75 x 6-10 µm, cylindrical or sub-ellipsoidal, with a basal scar, 2-9-septate, predominantly 3- to 5-septate, slightly or not constricted at the septa, straight or somewhat curved, smooth, subhyaline, pale brown or olivaceous, catenate or solitary, acropleurogenous.

Habitat: On intertidal wood.
Discription: Based on Ellis (1971) and Kohlmeyer and Kohlmeyer (1979).
Distribution: West Coast:-Pondecherry (Mahe): on intertidal wood (Nambiar and Raveendran, 2007, 2008c); **Kerala:** On intertidal wood (Raveendran and Manimohan, 2007), on intertidal mangrove wood of *Avicennia* sp. (Nambiar and Raveendran, 2006, 2007, 2008b, 2009h, I, k, 2010c; Nambiar et al., 2006, 2008a).

East Coast:-Tamil Nadu: On intertidal wood (Nambiar et al., 2008b; Nambiar and Raveendran, 2009k, l, 2010c).

Note: The present fungus is a temperatewater species. The one armed conidia (illustrated by Koch, 1982) of *Clavatospora bulbosa* (Anastasiou) Nakagiri and Tubaki can be misidentified as conidia of *Dendryfiella salina*

121. *Dictyosporium* Corda, 1836. *Beitr. Zur Gesammten Natur-Heilwiss.*, Prag, **1**: 87.

Conidiphores: micronematous, mononematous, septate, irregularly branched, smooth, hyaline to brown. *Conidiogenous cells*: monoblastic, integrated, terminal or rarely intercalary, determinate, cylindrical, doliform, globose or subglobose. *Conidia*: solitary, holoblastic, cheiroid, branched, branches arising from a single basal cell, mostly flattened in one plane, multiseptate, mostly constricted at the septa, smooth, olive to brown, acrogenous or rarely pleurogenous. Among several species two occur in marine waters (Jones et al., 2009a).

Type species: *Dictyosporium elegans* Corda
Habitat: Saprobic on plant litter.
Discription: Based on Ellis (1971) and Kohlmeyer and Kohlmeyer (1979).

121.1. *Dictyosporium pelagicum* (Linder) G.C. Hughes ex T.W. Johnson & F.K. Sparrow, 1961. Fungi in Oceans and Estuaries, Cramer, p. 391.

[For Fig. refer original paper]

Basionym: *Speira pelagica* Linder, *Farlowia*, **1**: 407 (1944).

= *Speira littoralis* Hohnk, *Veroff. Inst. Meeresforsch.* Bremerhaven, **3**: 221 (1955).

Hyphae: 1.3-3.1 μm in diam, septate, ramose, and hyaline to fuscous. *Conidiphores*: 10-35 μm long, 1.5-2.5 μm in diam, simply or rarely septate, sometimes swollen at the apex, hyaline to light brown. *Conidia*: 12-66 x 9-28 (-36) μm, ellipsoidal, ovoid, cylindrical, clavate or cheiroid sometimes branched, multiseptate, constricted at the septa; rows of cells aring from a single basal cell; cells initially in three, rarely two rows, then forming three to eight irregular rows; basal cells 2.5-7.5 μm high, 1.5-9.5 μm in diam; apical cells 4-10 μm high, 3.5-8.5 μm in diam; smooth, dark to almost black.

Habitat: Saprobic on intertidal and mangrove wood.

Discription: Based on Ellis (1971) and Kohlmeyer and Kohlmeyer (1979).

Distribution: West Coast:-Maharashtra: On intertidal wood and submerged wood of *Avicennia alba* and *A. marina* (Borse et al., 1988; Borse et al., 2005a), on intertidal roots of *Avicennia alba* (Shrivastava, 1994); **Goa:** On intertidal wood (Prasannarai and Sridhar, 2001); **Karnataka:** On intertidal wood (Prasannarai and Sridhar, 1997, 2001, 2000-2001); **Kerala:** On intertidal wood (Prasannarai and Sridhar, 2001).

East Coast:-Tami Nadu: on intertidal woody debris (Ravikumar et al., 2009); **Pondicharry (Karaikkal):** on intertidal woody debris (Ravikumar et al., 2009); **Orissa:** Conidia in foam samples (Borse et al., 2001b).

122. *Fusarium* sp.

A number of *Fusarium* species have been isolated from sediments, sand dunesand intertidal plant materials, but none of these have been considered to be marine (Jones et al., 2009a).

Distribution: West Coast:-Karanataka: On intertidal mangrove wood (Ananda and Sridhar, 2002, 2004; Maria and Sridhar, 2002a); **Kerala:** On intertidal mangrove wood (Ananda and Sridhar, 2002; Maria and Sridhar, 2002a).

East Coast:-Tamil Nadu: On seagrass Halophila ovalis (Devarajan et al., 2002), on macroalgae (Surayanarayanan et al., 2010).

123. *Halazoon* Abdel-Aziz, Abdel-Wahab & Nagahama, 2010. *Mycol. Progress*, **9**: 545.

Conidiphores: present or obsolete, micronematous, hyaline to yellow-brown in colour, cylindrical, septate. *Conidiogensis*: is holoblastic. *Conidia*: terminal or lateral, forming single coil or branched to form two to three coils, coiled in two or three dimensions, semi-contorted to several coils, conidial cells increase in size and pigmentation from base to apex. *Note*: The genus is represented by two species in marine environment (Abdel-Wahab et al., 2010).

Type species: *Halazoon melhae* Abdel-Aziz, Abdel-Wahab & Nagahama

Habitat: Occurring on intertidal wood.

Discription: Based on Abdel-Wahab et al. (2010).

123.1. *Halazoon fuscus* (I. Schmidt) Abdel-Wahab, K.L. Pang, Nagahama, Abdel-Aziz & E.B.G. Jones, 2010. *Mycol. Progress*, **9**: 547.

[For Fig. refer original paper]

Basionym: *Cirrenalia fusca* I. Schmidt, *Fedds Reperttorium*, **80**: 107-112 (1969).

Hyphae: 2.5-3.3 µm in diam, septate, ramose, superficial or immersed in the substrate, yellow or brown. *Conidiophores*: up to 23 µm long, 2 µm in diam, cylindrical, one- or two-septate, simple, sometimes obsolete, yellow or brown. *Conidiogenous cells*: monobalstic, integrated, terminal, determinate. *Conidia*: acrogenous, solitary, regularly or irregularly igantean, conidia semicontorted to 1½ times contorted, 2-4-septate, constricted at the septa, brown to dark brown (never reddish); cells increasing in diam and pigmentation from base to apex, distinctly dissimilar; spirals 20-36 µm high, 20-30 µm in diam; terminal cell 11-22 µm high, 11-22.5 µm in diam, reniform or ellipsoidal, fuscous; basal cells obtusely conical; central cells irregularly ellipsoidal or subglobose. *Note*: The apical cell in *H. fuscus* resembles that of *Hydea pygmea* but lacks the dark pigmentation of the latter (Abdel-Wahab *et al*., 2010).

Habitat: On intertidal mangrove wood.
Discription: Based on Kohlmeyer and Kohlmeyer (1979).
Distribution: West Coast:-Pondicherry (Mahe): on intertidal mangrove wood (Nambiar and Raveendran, 2008a); **Kerala:** on mangrove wood (Nambiar and Raveendran, 2009f).

124. *Halenospora* E.B.G. Jones, 2009. *Fungal Diversity*, **35**: 154.

Anamorphic Ascomycota (Leotiaceae). Conidiophores: simple or absent. Conidia: dark brown to black, cells formed in a chain, becoming tortous and appearing as terminal dictyospores. The genus is monotypic.

Type species: *Halenospora varia* (Anastasiou) E.B.G. Jones
Habitat: Occurring on intertidal wood, submerged leaves and seedings of mangroves.
Discription: Based on Jones et al. (2009a).

124.1. ***Halenospora varia*** (Anastasiou) E.B.G. Jones, 2009. *Fungal Diversity*, **35**: 154. [Fig. 124.1]

Basionym: *Zalerion varium* Anastasiou, *Can. J. Bot.*, **41**: 1136 (1963).

Hyphae: septate, branched, immersed, hyaline, *Conidiophores*: up to 30 µm long, 2-3.5 µm in diam, micronematous, simple, cylindrical, septate, sometimes absent, superficial, hyaline to light olive coloured. *Conidia*: 15-65 x 13.5-56 µm, solitary, irregularly igantean or coiled in three planes, forming a knot or ball of about 10 to 30 cells; Conidial filament lateral, thick–walled, smooth, brown to dark brown, appearing black in mass; cells 5-13 x 4-10.5 µm.

Habitat: On intertidal wood, animal substrates and mangrove wood.

Discription: Based on Ellis (1971) and Kohlmeyer and Kohlmeyer (1979).

Distribution: West Coast:-Daman: On intertidal wood and *Avicennia marina* (Borse et al., 2000b); **Gujarat:** On intertidal wood and mangrove wood (Patil and Borse, 2001); **Maharashtra:** On Intertidal wood of *Avicennia marina* and *A. officinalis* (Borse, 1984, 1988; Shrivastava, 1994; Borse et al., 2005a), on wood of *Sonneratia acida* (Patil and Borse, 1986) on intertidal mangrove wood (Maria and Sridhar, 2002a), on lignite beds (Kumaran et al., 2004a); **Goa:** On intertidal wood (Nandan et al., 1993), on intertidal animal substrates (Ananda et al., 1998), On intertidal wood (Prasannarai and Sridhar, 2001), on intertidal mangrove wood (Maria and Sridhar, 2002a), on lignite beds (Kumaran et al., 2004a, b); **Karnataka:** Conidia in foam and sand samples (Sridhar and Kaveriappa, 1991), on intertidal animal substrates (Ananda et al., 1998), on intertidal wood (Prasannarai et al., 1999), on intertidal wood (Prasannarai and Sridhar, 2001, 2000-2001), on intertidal wood of *Acanthus ilicifolius, Avicennia officinalis, Brugueiria gymnorrhiza, Rhizophora mucronata* and *Sonneratia caseolaris* (Maria and Sridhar, 2002a, 2003a, 2004), on intertidal leaf and woody litter of mangroves (Ananda & Sridhar, 2003, 2004); **Pondecherry (Mahe):** On intertidal mangrove wood (Nambiar and Raveendran, 2008a); **Kerala:** On intertidal wood (Prasannarai and Sridhar, 2001) on intertidal mangrove wood (Maria and Sridhar, 2002a), on intertidal wood (Raveendran and Manimohan, 2007), on intertidal wood, mangrove wood of *Rhizophora* sp. and *Bruguiera* sp. (Nambiar and Raveendran, 2006, 2007, 2008a, b, d, 2009a, b, d, e, f, h, 2010a, c; Nambiar et al., 2006); **Lakshadweep Islands:** On intertidal wood (Chinnaraj, 1992), on intertidal mangrove wood (Ananda and Sridhar, 2003).

East Coast:-Tamil Nadu: On drift wood (Raghukumar, 1973), on submerged parts of *Rhizophora* spp. (Ravikumar and Vittal, 1996), on intertidal wood (Prasannarai and Sridhar, 2001; Nambiar et al., 2008b), on intertidal wood

(Nambiar and Raveendran, 2009l, 2010a, c); on intertidal woody debris (Ravikumar et al., 2009); **Pondecherry:** On intertidal mangrove wood (Nambiar and Raveendran, 2010a); **Pondicherry (Karaikkal):** on intertidal woody debris (Ravikumar et al., 2009); **Andhara Pradesh:** On intertidal wood of *Avicennia marina, Ceriops decandra* and *Rhizophora apiculata* (Sarma and Vittal, 1998-99; Sarma and Vittal, 2000; Sarma and Vittal, 2001; Sarma et al., 2001; Sarma and Vittal, 2002; Sarma and Vittal, 2004; Vittal and Sarma, 2005), on mangrove wood (Nambiar and Raveendran, 2008e); **Orissa:** On intertidal wood of *Avicennia marina* (Borse and Borse, 2005); **West Bengal:** On intertidal wood of *Avicennia marina* (Pawar and Borse, 2005b); **Andaman & Nicobar Islands:** On intertidal wood of *Avicennia marina, A. officinalis, Bruguiera gymnorhiza, Rhizophora apiculata, R. mucronata* and *Sonneratia alba* (Chinnaraj, 1993); on intertidal wood of *Avicennia marina, A. officinalis, Bruguiera gymnorhiza, Rhizophora apiculata, R. mucronata* and *Sonneratia alba* (Chinnaraj, 1993a).

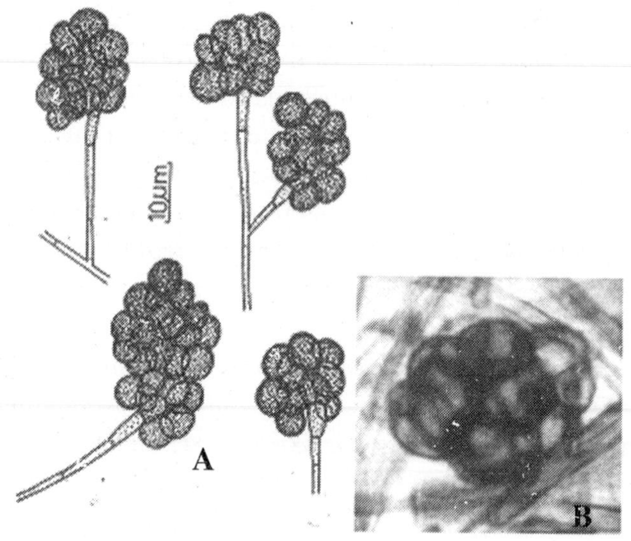

Fig. 124.1. ***Halenospora varia***, A-B: A- conidia, B- conidium (1000x).

125. *Hydea* K.L. Pang & E.B.G. Jones, 2010. In: Abdel-Wahab *et al., Mycol. Progress*, **9**: 549.

Hyphae: septate, ramose, fuscous. *Conidiphores*: obsolete. *Conidia*: acrogenous, solitary, helicoids, ½-1 times contorted, 3- to 4-septate, black or fuscous, not or slightly constricted at the septa, cells increasing in diameter and pigmentation from the base to the apex. *Note*: The genus is monotypic (Abdel-Wahab *et al*., 2010).

Type species: *Hydea pygmea* (Kohlm.) K.L. Pang & E.B.G. Jones
Habitat: Occurring on decayed driftwood in the intertidal zone, especially on mangrove wood.
Discription: Based on Abdel-Wahab et al. (2010).

125.1. *Hydea pygmea* (Kohlm.) K.L. Pang & E.B.G. Jones, 2010. In: Abdel-Wahab *et al., Mycol. Progress*, **9**: 549. [Fig. 125.1]

Basionym: *Cirrenalia pygmea* Kohlm., *Ber. Drsch. Bot. Ges.,* **79:** 35 (1966).

Hyphae: 2.2-4.5 µm in diam, septate, ramose, fuscous. *Conidiophores*: obsolete. *Conidia*: acrogenous, solitary, igantean, contorted ½ or 1 time contorted, 3-4-septate, not or slightly constricted at the septa, hooked appearance, black or fuscous, fulgent (upper three cells dark, lower two or three cells light-coloured); cells increasing in diameter from base to apex, distinctly dissimilar; spirals 25.5-31 x 28.5-34 µm; terminal cell 16-23 µm in diam, subglobose to reniform, basely flattened; basal cells 3.5-5.5µm in diam; central cells irregularly conical or almost wedge-shaped. *Note*: It is a common species on mangrove wood, especially *Rhizophora* species, growing on the bark, with slow groth in culture but sporulates readily. It differs from all othe *Cirrenalia*-like species by the dark-brown to black hooked nature of the conidia. It is also characterized by internal proliferation of the conidia (Kohlm. & Kohlm., 1979; Abdel-Wahab *et al*., 2010).

Habitat: Occurring on decayed driftwood in the intertidal zone, especially on mangrove wood.
Discription: Based on Kohlmeyer and Kohlmeyer (1979).
Distribution:West Coast:-Gujarat: On intertidal wood of *Rhizophora mucronat* (Borse et al., 2000a), on intertidal mangrove wood (Patil and Borse, 2001); **Maharashtra:** On intertidal wood (Borse, 1984, 1988), intertidal wood of *Ceriops tagal* and *Rhizophora mucronata* (Patil and Borse, 1986; Borse et al., 2005a), on intertidal mangrove wood (Maria and Sridhar, 2002a), on lignite beds (Kumaran et al., 2004a, b); **Goa:** On intertidal prop roots of *Rhizophora*

mucronata (Nandan et al., 1993); **Karnataka:** On intertidal mangrove wood and roots of *Rhizophora mucronata* (Maria and Sridhar, 2002a), on intertidal wood of *Acanthus ilicifolius, Avicennia officinalis, Brugueiria gymnorrhiza, Rhizophora mucronata* and *Sonneratia caseolaris* (Maria and Sridhar, 2003a, 2004), on intertidal leaf and woody litter of mangroves (Ananda and Sridhar, 2004); **Pondecherry (Mahe):** On intertidal wood (Nambiar and Raveendran, 2007, 2008c); **Kerala:** On intertidal wood (Raveendran and Manimohan, 2007), on intertidal wood, mangrove wood of *Avicennia* sp., *Bruguiera* sp. and *Rhizophora* sp. (Nambiar and Raveendran, 2006, 2007, 2008b, d, 2009a, d, e, f, g, h, 2010a, c; Nambiar et al., 2006, 2008a); **Lakshadweep Islands:** On pneumatophores of *Ceriops tagal* (Chinnaraj, 1992).

East Coast:-Tamil Nadu: On submerged parts of *Rhizophora* spp. (Ravikumar and Vittal, 1996), on intertidal wood (Nambiar et al., 2008b), on mangrove wood (Nambiar and Raveendran, 2008e, 2009k, l, 2010a, c); **Pondicherry:** On intertidal wood (Nambiar and Raveendran, 2008c); **Andhara Pradesh:** On intertidal wood of *Rhizophora apiculata* (Sarma and Vittal, 1998-99; Sarma and Vittal, 2000; Sarma and Vittal, 2001; Sarma et al., 2001; Sarma and Vittal, 2002; Sarma and Vittal, 2004; Vittal and Sarma, 2005), on intertidal wood (Nambiar and Raveendran, 2009j); **Orissa:** On intertidal wood of *Avicennia marina* (Borse and Borse, 2005); **West Bengal:** On intertidal wood of mangroves (Pawar and Borse, 2005b); **Andaman & Nicobar Islands:** On intertidal wood of *Avicennia marina, A. officinalis, Bruguiera gymnorhiza, Rhizophora apiculata, R. mucronata* and *Sonneratia alba* (Chinnaraj, 1993a).

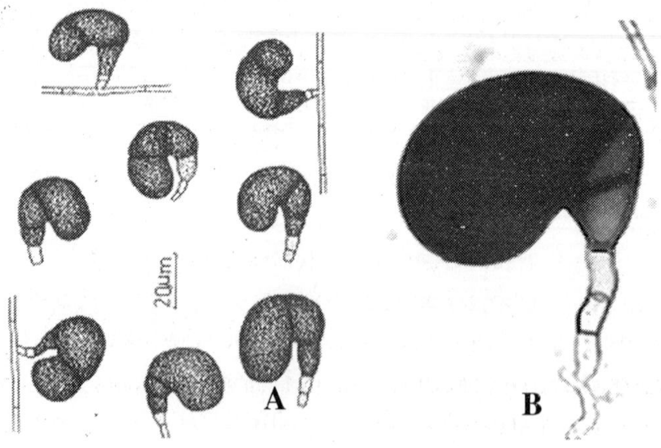

Fig. 125.1. ***Hydea pygmea***, A-B: A- conidia, B- conidium (1000x).

126. *Matsusporium* E.B.G. Jones & K.L. Pang, 2010. In: Abdel-Wahab et al., *Mycol. Progress*, **9**: 550.

Hyphae: septate, superficial or immersed, light-brown. *Conidiophores*: obsolete or present, when present micronematous, cylindrical, light-brown, cylindrical, septate, acrogenous or lateral. *Conidia*: acrogenous or lateral, solitary, helicoids, up to 12-septate, immature conidia are surrounded by sheath, dark-brown to black, not constricted at the septa, cells increasing in diameter and pigmentation from the base to the apex. *Note*: The genus is monotypic. It differs from other helicoids genera by having 12-septate conidia that usually have a long stalk (Abdel-Wahab et al., 2010).

Type species: *Matsusporium tropicale* (Kohlm.) E.B.G. Jones & K.L. Pang
Habitat: Occurring on decayed submerged and intertidal wood, especially on mangrove wood.
Discription: Based on Abdel-Wahab et al. (2010).

126.1. *Matsusporium tropicale* (Kohlm.) E.B.G. Jones & K.L. Pang, 2010. In: Abdel-Wahab et al., *Mycol. Progress*, **9**: 550. [Fig. 126.1]

Basionym: *Cirrenalia tropicalis* Kohlm., *Mycologia*, **60**: 267 (1968).

Hyphae: 2.5-5 µm in diam, septate, superficial or immersed, light brown. *Conidiophores*: 25-42 x 2.5-5 µm, cylindrical, zero- to four-septate, simple, acrongenous or laterally on hyphae, often remaining connected with detached conidia, sometimes obsolete, straight or curved, light brown. *Conidiogenous cells*: monoblastic, integrated, terminal, determinate. *Conidia*: acrogenous, solitary, regularly or irregularly igantean, mostly 1 to 1 ½ times contorted, rarely semicontorted, 5-12-septate, not or slightly constricted at the septa, reddish brown; cells increasing in diameter from base to apex, distinctly dissimilar; spirals 23-30µm in diameter; terminal cell 8-14 µm high, 10-13 um diam, subglobose to ellipsoidal, basally flattened; basal cells 8-17 um high, 3-5 µm in diam, cylindrical; central cells subglobose, obtusely conical or dolliform. *Note*: It is a widely distributed species in tropical and subtropical mangroves, growing on the bark of mangrove trees. The fungus differs from other marine helicoids species by the repeated coiled nature of the conidia that dark brown in colour and surrounded by sheath when young (Abdel-Wahab *et al.*, 2010).

Habitat: Occurring on decayed submerged and intertidal wood, especially on mangrove wood.

Discription: Based on Kohlmeyer (1968) and Somrithipal et al. (2002).

Distribution: West Coast:-Maharashtra: On intertidal mangrove, *Rhiziphora mucronata* wood (Borse, 2000b; Borse et al., 2005a), on lignite beds (Kumaran et al., 2004a, b); **Goa:** On intertidal wood of *Cocos nucifera* (Bhat, 2010); **Karnataka:** on intertidal wood of *Acanthus ilicifolius, Avicennia officinalis, Brugueiria gymnorrhiza* and *Rhizophora mucronata* (Maria and Sridhar, 2002a, 2003a, 2004), on intertidal leaf and woody litter of mangroves (Ananda and Sridhar, 2004); **Kerala:** on intertidal mangrove wood (Maria and Sridhar, 2002a), on intertidal wood (Nambiar and Raveendran, 2009b, k, 2010a, c), on mangrove woody litter (Shini et al., 2009-10); **Lakshadweep Islands:** On drift wood (Raghukumar, 1988).**East Coast:-Tamil nadu:** On intertidal wood (Nambiar and Raveendran, 2010c); on intertidal woody debris (Ravikumar et al., 2009); **Pondicherry (Karaikkal):** on intertidal woody debris (Ravikumar et al., 2009); **Andhara Pradesh:** On intertidal wood of *Avicennia officinalis* and *Rhizophora apiculata* (Sarma and Vittal, 1998-99; Sarma and Vittal, 2000; Sarma and Vittal, 2001; Sarma et al., 2001; Sarma and Vittal, 2004; Vittal and Sarma, 2005), on intertidal wood (Nambiar and Raveendran, 2009j); **Orissa:** On intertidal fronds of *Acrostichum aureum* (Borse and Borse, 2005); **West Bengal:** On intertidal wood of *Rhizophora mucronata* (Pawar and Borse, 2005b), on mangrove woody litter (Shini et al., 2009-10); **Andaman & Nicobar Islands:** On intertidal wood of *Sonneratia alba* (Chinnaraj, 1993a).

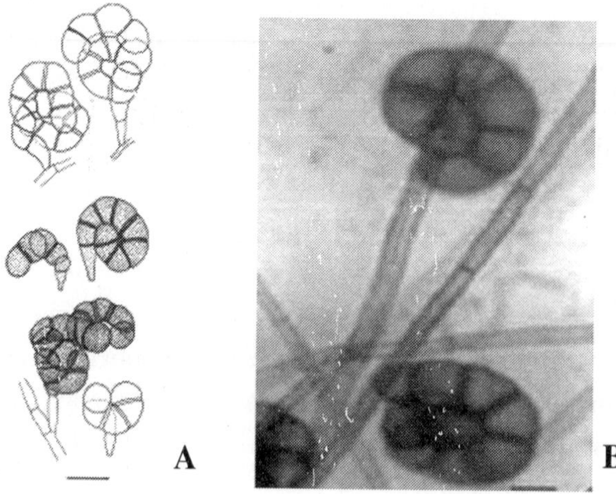

Fig. 126.1. ***Matsusporium tropicale***, A-B: A- conidia (bar= μm), B- conidia (1000x).

127. ***Monodictys*** S.J. Hughes, 1958. *Can. J. Bot.*, **36**: 785.

Hyphae: septate, branched, subhyaline or dark brown, immersed, partly immersed or superficial. *Conidiophores*: micronematous or semimicronematous, mononematous, subcylindrical, cells occasionally swollen, septate, simple or irregularly branched, straight or curved, smooth, hyaline to brown, single or sometimes forming pustules. *Conidiogenous cells*: monoblastic, integrated, terminal, determinate, cylindrical, doliform or subglobose. *Conidia*: solitary, dry, acrogenous, oblong with rounded ends, subpyriform, clavate, ellipsoidal, subglobose or irregular, occasionally spirally twisted, muriform, sometimes constricted at the septa, smooth or verrucose, brown to black, acrogenous, solitary; basal cell sometimes inflated and lighter coloured than the others. Among several species two occur in marine waters (Jones et al., 2009a).

Type species: *Monodictys putredinis* (Wallr.) S.J. Hughes
Habitat: Saprobic on plant litter.
Discription: Based on Hughes (1958) and Ellis (1971).

127.1. ***Monodictys pelagica*** (T.W. Johnson) E.B.G. Jones, 1963. *Trans. Br. Mycol. Soc.*, **46**: 138. [Fig. 127.1]

Basionym: *Piricauda pelagica* T.W. Johnson, *J. elisha Mitchell Sci. Soc.*, **74**: 42 (1958).

= *Piricauda arcticoceanorum* R.T. Moore, *Rhodora*, **61**: 95 (1959).

Hyphae: septate, rarely branched, dark brown or fuscous. *Conidiophores*: short or lacking, cylindrical. 0- to 2-septate, simple, lateral, dark brown, Conidiogenous cells: monoblastic, integrated, terminal, determinate, cylindrical, doliform or subspherical. *Conidia*: 15-41 μm long, 12.5-37 μm in diam, obpyriform, ovoid or rarely subglobose, muriform, not or slightly constricted at the septa, smooth, black, with one to three yellowish or light brown basal cells, solitary, dry, acrogenous, basal cell sometimes inflated, paler and thin-walled than the other cells; the septation is only visible in immature conidia.

Teleomorph: *Nereiospora cristata* (Kohlm.) E.B.G. Jones, R.G. Johnson & S.T. Moss

Habitat: Saprobic on driftwood.
Distributin: West Coast:-Karnatak: Off Mangalore coast (Prof. Bhat, unpublished).

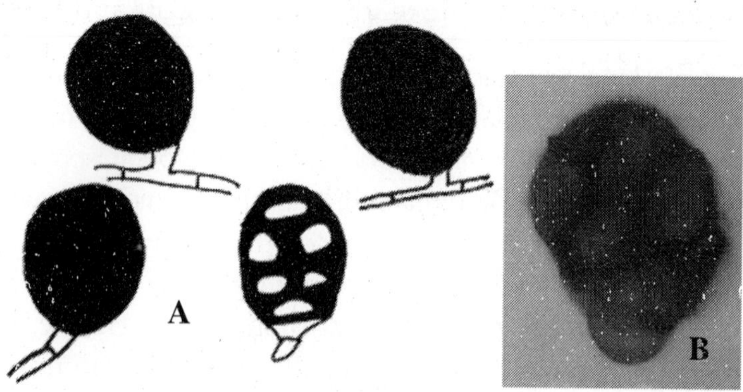

Fig. 127.1. ***Monodictys pelagica***, A-B: A- conidia (1000x), B- conidia (1000x).

128. ***Paecilomyces*** G. Bainier, 1907. *Bull. Soc. Mycol. Fr.*, **23:** 26-27.

Genus related to *Penicillium* and *Gliocladium*, distinguished by phialids which are short tubular, or more or less swollen in the basal portion and tapering into long slender conidium-bearing tubes which are mostly curved or bent slightly away from the main axis of the cell; phialids variously arranged, partly in verticils and branching systems suggesting *Penicillium*, partly irregularly arranged upon short branchlets, partly sessile on and arising singly from the fertile hyphae; conidia in chains, but, in some species, especially when grown under very moist conditions, tending to slip sideways, or even to slime down into balls as in *Gliocladium*; colonies white, or even pale coloured, or in dull brownish shades, occasionally with a greenish tone but never true green. Perfect state, where known, *Byssochlamys* Westling

Type species: *Paecilomyces varoti* G. Bainier,
Habitat: Saprobic on plant parts.
Discription: Based on Brown and Smith (1957).

128.1. ***Paecilomyces ramosus*** Samson & Evans, Stud. Mycol. 6: 44 (1974). [Fig. 128.1]

Colonies on seawater embedded MEA 4-5 cm in diam after 10 days, flat, with white mycelium. Hyphae branched, septate, hyaline, smooth-walled. Conidiophores mononematous, elongated, sparsely branched, 50-110 μm long, 2.5-4 μm wide, terminating in a cylindrical to swollen, fertile conidiogenous cell. Conidiogenous cells phialidic, 8.0-20 x 2.5-3.5 μm, tapering at the tip. Conidia pyriform, hyaline, thick-walled, smooth-walled, 3.5-5 x 1.5-3 μm, accumulating in dry masses.

Habitat: Saprobic on driftwoods,
Distributin:- off Mangalore coast
West Coast:-Goa: (Prof. Bhat, unpublished).

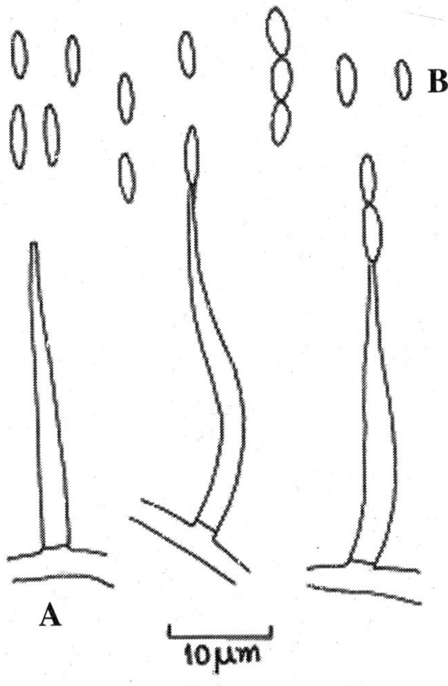

Fig. 128. ***Paecilomyces* sp.**, A-B: A- conidiophore, B- conidia

129. ***Periconia*** Tode ex Fr., 1791. Tode, *Fungi Mecklenburgenses Selecti,* Luneburg, **2**: 2-3. Fries, *Systema Mycologicum, Gryphiswaldiae,* **1**: 47 (1821).

The genus is characterized by having *Stroma*: often present, brown, pseudoparenchymatous. *Conidiophores*: macronematous, occasionally micronematous, mononematous, macronematous ones stalked with a globose head, with or without branches, stipe straight or curved, brown or appearing black, smooth, rarely verrucose; the apex may be sterile and setiform. *Conidiogenous cells*: monoblastic or polyblastic, discrete on the stipe and branches, determinate, ellipsoidal, globose or subglobose. *Conidia*: globose, subglobose or rarely ellipsoidal, oblong or broad cylindrical, one-celled, verruculose, echinulate or rarely smooth, light to dark brown, catenate, often in branched chains, developing acropetally at one or several points on the surface of the conidiogenous cell. Among several species two occur in marine waters (Jones et al., 2009a).

Type species: *Periconia lichenoides* Tode ex Merat
Habitat: Saprobic or parasitic on plant parts.
Discription: Based on Ellis (1971) and Kohlmeyer and Kohlmeyer (1979).

Key to *Periconia* species from India:

1.Conidia 6-13 µm diam., light to dark brown, littoral species .... *Periconia prolifica*

1'. Conidia 16-20 µm diam., brown, deep sea species.................................... .... *Periconia abyssa*

129.1. *Periconia abyssa* Kohlm., 1977. *Rev. Mycol.*, **41**: 202.

[For Fig. refer original paper]

Hyphae: 3-6 um in diam, septate, light brown, superficial; some unbranched hyphae perpendicular to the surface may occur. *Conidiogenous cell*: 5 um in diam, cylindrical to clavate, semimacronematous, hyaline to light brown. *Conidia*: 16-20 µm in diam, globose to subglobose, single or catenate, smooth, thin-walled, brown, with pore between adjoining cells; often forming clumps on the surface of the substrate.

Habitat: Saprobic on submerged wood in deep sea.
Discription: Based on Kohlmeyer and Kohlmeyer (1979).
Distributin: West Coast:-Goa: Deep sea sediments, Arabian sea (Raghukumar,1998).

129.2. *Periconia prolifica* Anastasiou, 1963. *Nova Hedwigia*, **6**: 260.

[Fig. 129.2]

Condidiophores: 2-200 x 2.5 µm; cylindrical, septate, simple or branched, hyaline, often forming pustules on the surface of the substrates. *Conidiogenous cell*: ellipsoidal or ovoid, hyaline, produced acrogenously. *Conidia*: 6-13 µm in diameter, one-celled, subglobose or ovoid, smooth, thick-walled, light brown with a reddish tint or dark brown, developing basipetally, catenulate, cells finally separating.

Teleomorph: *Okeanomyces cucullatus* (Kohlm.) Pang & Jones

Habitat: On intertidal and mangrove wood.

Discription: Based on Anastasiou (1963) and Kohlmeyer and Kohlmeyer (1979).

Distribution: West Coast:-Diu Island: On intertidal wood (Borse et al., 1999b); **Daman:** On intertidal wood (Borse et al., 2000b); **Gujarat:** On intertidal wood of *Avicennia marina*, *Ceriops tagal* and *Rhizophora mucronat* (Borse et al., 2000a), on intertidal wood and mangrove wood (Patil and Borse, 2001); **Maharashtra:** On intertidal wood (Borse, 1984, 1988), on intertidal wood of *Avicennia marin* , *Ceriops tagal* and *Sonneratia alba* (Patil and Borse, 1986; Borse et al., 2005a), on intertidal roots of *Avicennia alba* (Shrivastava, 1994), on intertidal mangrove wood (Maria and Sridhar, 2002), on lignite beds (Kumaran et al., 2004a, b); **Goa:** On submerged wood panels of *Anogeissus latifolia*, *Acrocarpus fraxinifolius*, *Carapa molluscensis*, *Acacia auriculiformis* (Raghukumar et al., 1988a), on intertidal endocarp of *Cocos nucifera* fruit (Nandan et al., 1993), On intertidal wood (Prasannarai and Sridhar, 2001), on intertidal mangrove wood (Maria and Sridhar, 2002a), on intertidal wood of *Avicennia marina* and *Rhizophora apiculata* (Bhat, 2010); **Karnataka:** On intertidal wood and teak panels (Prasannarai and Sridhar, 1997, 2001, 2000-2001; Prasannarai et al., 1999), on intertidal wood of *Avicennia officinalis* and *Sonneratia caseolaris* (Maria and Sridhar, 2002a, 2003a, 2004), on intertidal leaf and woody litter of mangroves (Ananda and Sridhar, 2003, 2004); **Pondecherry (Mahe):** on intertidal wood (Nambiar and Raveendran, 2007, 2008a, c); **Kerala:** On intertidal wood (Prasannarai and Sridhar, 2001), on intertidal mangrove wood (Maria and Sridhar, 2002a), on intertidal wood (Raveendran and Manimohan, 2007), on intertidal wood, mangrove wood of *Avicennia* sp., *Rhizophora* sp. and and *Bruguiera* sp. (Nambiar and Raveendran, 2006, 2007, 2008a, b, d, 2009a, d, e, f, g, h, i, k, 2010c; Nambiar et al., 2006, 2008a); **Lakshadweep Islands:** On intertidal wood of *Pemphis acidula* (Chinnaraj, 1992) on intertidal mangrove wood (Ananda and Sridhar, 2003).

East Coast:-Tamil Nadu: On drift wood (Raghukumar, 1973), on submerged parts of *Rhizophora* spp. (Ravikumar and Vittal, 1996), on intertidal wood (Prasannarai and Sridhar, 2001; Nambiar et al., 2008b),), on mangrove wood (Nambiar and Raveendran, 2008e, 2009l. 2010c); on intertidal woody debris (Ravikumar *et al.,* 2009); **Pondicherry:** On intertidal wood (Nambiar and Raveendran, 2008c); **Pondicherry (Karaikkal):** on intertidal woody debris (Ravikumar et al., 2009); **Andhara Pradesh:** On intertidal wood of *Avicennia* *marina, A. officinalis* and *Rhizophora apiculata* (Sarma and Vittal, 1998-99; Sarma and Vittal, 2000; Sarma and Vittal, 2001; Sarma et al., 2001; Sarma and Vittal, 2002; Sarma and Vittal, 2004; Vittal and Sarma, 2005), on intertidal wood (Nambiar and Raveendran, 2009j); **Orissa:** On intertidal wood of *Kandelia candel*(Borse and Borse, 2005); **West Bengal:** On intertidal wood of *Acanthus ilicifolius, Avicennia marina, A. officinalis, Rhizophora mucronata, Suaeda maritime* and *Sonneratia apetala* (Pawar and Borse, 2005b); **Andaman & Nicobar Islands:** On intertidal wood of *Avicennia marina, A. officinalis, Bruguiera gymnorhiza, Rhizophora apiculata, R. mucronata* and *Sonneratia alba* (Chinnaraj, 1993a).

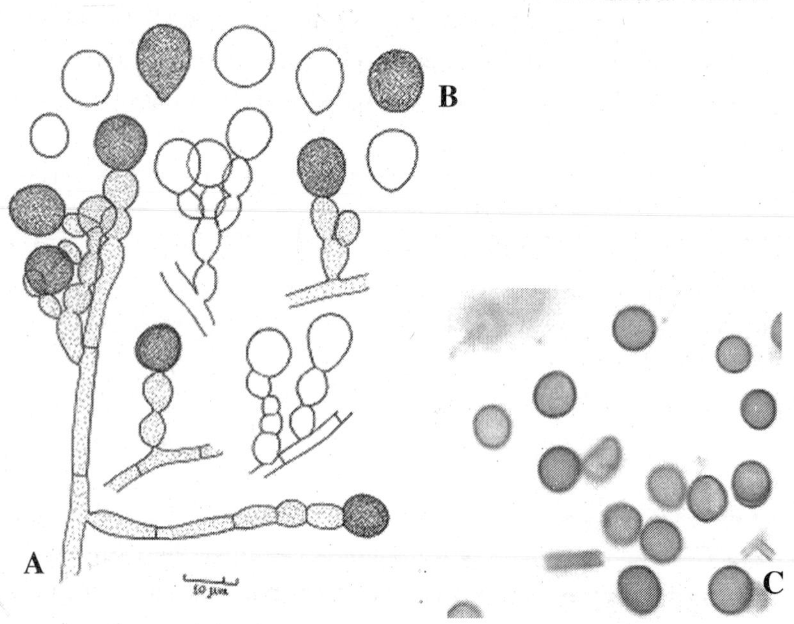

Fig. 129.2. *Periconia prolifica*, A-C: A- conidiophore with conidia, B- conidia, C- conidia (1000x).

130. *Scolecobasidium* Abbott, 1927. *Mycologia*, **19**: 29-31.

Colony: effuse grey to brown flat or funiculose. Mycelium: superficial or immersed septate, branched, sometimes producing chlamydospores, straight or flexous 2-3 celled, cylindrical. Conidiophores: macronematous. *Conidiogenous cells*: polyblastic, integrated, terminal or intercalary, sympodic, cylindrical or clavate or cuneiform, denticulate, denticles prominent. *Conidia*: holoblastic, fusiform, ellipsoidal rounded at ends, sometimes T or Y shaled, smooth or echinulate, 0-3 septate, pale or mid brown.

Type species: *Scolecobasidium terreum* Abbott
Habitat: Saprobic on organic matter.
Discription: Based on Rao et al. (2004).

130.1. *Scolecobasidium* sp.

This fungus parasite was found to cause necrotic patches on corals. Such patches renged from 9% in the coral *Montipora tuberculosa* to 54% in *Goniophora* sp. in the Andaman Islands (Raghukumar and Raghukumar, 1991). A detailed investigstion for the present fungus in healthy, partially dead, bleached and pink-line syndrome affected scleractanian corals in the Laksadweep Islands has been carried out by Ravindran et al. (2001).

Habitat: On corals.
Distributin: West Coast:-Lakshadweep Islands: On coral *Porites lutea* (Ravindran et al., 2001).
East Coast:-Andaman & Nicobar Islands: On corals *Porites lutea*, *P. lichen*, *Montipora tuberculosa*, *Goniopora* sp. and *Goniastra* sp. (Raghukumar and Raghukumar, 1991).

131. *Trichocladium* Harz., 1871. *Bull. Soc. Imp. Nat. Moscou,* **44:** 125-127.

The genus is characterized by having *Conidiophores*: mostly inconspicuous, poorely differentiated, forming short pedicels, 0-3-septate, simple or rarely brached, straight or curved, smooth, bearing conidia at the apex, hyaline or pale brown. *Conidiogenous cells*: manoblastic or polyblastic, integrated, terminal, determinate, cylindrical, doliiform. *Conidia*: acrogenous or acropleurogenous, solitary, clavate, obovoid, pyriform or cylindrical, uni- to multisepate, mostly thick-walled, smooth or verrucose, light to dark brown; cells may be unequally pigmented; germ pores occur in some or all cells. *Phialides*: may be formed in some pleomorphic species producing small, hyaline, one-celled conidia. Among several species seven occur in marine waters (Jones *et al.*, 2009a).

Type species: *Trichocladium asperum* Harz
Habitat: On intertidal and mangrove wood.

Discription: Based on Ellis (1971) and Kohlmeyer and Kohlmeyer (1979).

Key to *Trichocladium* species from India:

1.Conidia 1-2(3)-septate 2

1'.Conidia more than 3-septate 3

2. Conidia 10-22(-38) x 8-18 µm, apical cell 8.5-15.5 x 7-12 µm, fuscous .... *Trichocladium alopallonellum*

2'. Conidium 6.5-14 x 3.5-9 µm, apical cell 6.5-13 x 4-9 µm, dark brown.........

.... *Trichocladium melhae*

3. Conidia with 2-5(-6)-septate, 25-32 x 12-17 µm, slightly constricted at the septa, light brown .... *Trichocladium lignicola*

3'. Conidia 2-4-septate 4

4. Conidia 25-47 x 8-20 µm, subglobose, markedly constricted at the septa, reddish-brown .... *Trichocladium constrictum*

4'. Conidia (15-)20-34(-45) x (8-)10-24 µm, often sporodochial, less constricted at the septa, dark brown .... *Trichocladium achrasporum*

131.1. ***Trichocladium achrasporum*** (Meyers and R.T. Moore) Dixon, 1971. In: Shearer & Crane, *Mycologia,* **63:** 344. [Fig. 131.1]

Basionym: *Culcitalna achraspora* Meyers & R.T. Moore, *Am. J. Bot.*, **47**: 349 (1960).

Sporodochia: occationaly found, superficial, compact, fuscous to black. *Conidiophores*: absent or short, 2-3-septate, simple, formed laterally on hyphae, hyaline to light brown or fuscous. *Conidia***:** (blastoconidia) 20-34 x 10-24 µm, clavate, ovoid or obpyriform, 2-4-septate, less constricted at the septa, straight or slighty curved, igantean in diameter from base to apex, formed singly on the conidiophores; apical cells subglobose, dark brown; basal cells conical or subcylindrical, subhyaline to light brown or fuscous..

Habitat: On intertidal and mangrove wood.
Discription: Based on Ellis (1971) and Kohlmeyer and Kohlmeyer (1979).
Distribution: West Coast:-Gujarat: On intertidal wood of *Avicennia marina* and *Rhizophora mucronat* (Borse et al., 2000a), on intertidal wood (Patil and

Borse, 2001); **Maharashtra:** On intertidal wood (Borse, 1984, 1988, Shrivastava, 1994; Borse et al., 2005a), on intertidal wood of *Rhizophora mucronata* (Patil and Borse, 1986), on lignite beds (Kumaran et al., 2004a, b); **Goa:** On intertidal wood (Nandan et al., 1993); **Karnataka:** On intertidal wood (Prasannarai and Sridhar, 1997, 2001, 2000-2001), on intertidal wood of *Acanthus ilicifolius, Avicennia officinalis, Brugueiria gymnorrhiza, Rhizophora mucronata* and *Sonneratia caseolaris* (Maria and Sridhar, 2002a, 2003a, 2004), on intertidal leaf and woody litter of mangroves (Ananda and Sridhar, 2003, 2004); **Pondecherry (Mahe):** On intertidal wood (Nambiar and Raveendran, 2007); **Kerala:** On intertidal wood (Prasannarai and Sridhar, 2001, 2000-2001), on intertidal mangrove wood (Maria and Sridhar, 2002a), on intertidal wood (Nambiar and Raveendran, 2007, 2008b, 2009b, g, I, k, 2010a, c; Nambiar et al., 2008a); **Lakshadweep Islands:** On pneumatophores of *Ceriops tagal* (Chinnaraj, 1992). **East Coast:-Tamil Nadu:** On submerged parts of *Rhizophora* spp. (Ravikumar and Vittal, 1996), on intertidal wood (Nambiar et al., 2008b),), on mangrove wood (Nambiar and Raveendran, 2008e, 2009l, 2010c); **Andhara Pradesh:** On intertidal wood of *Aegiceras corniculatum, Avicennia marina, A. officinalis, Ceriops decandra, Lumnitzrea recemosa* and *Rhizophora apiculata* (Sarma and Vittal, 1998-99; Sarma and Vittal, 2000; Sarma and Vittal, 2001; Sarma et al., 2001; Sarma and Vittal, 2002; Sarma and Vittal, 2004; Vittal and Sarma, 2005); **Orissa:** On intertidal wood of *Avicennia marina* (Borse and Borse, 2005); **West Bengal:** On intertidal wood of *Bruguiera gymnorrhiza* and *Sonneratia apetala* (Pawar and Borse, 2005b): **Andaman & Nicobar Islands:** On intertidal wood of *Avicennia marina, A. officinalis, Rhizophora apiculata, R. mucronata* and *Sonneratia alba* (Chinnaraj, 1993a).

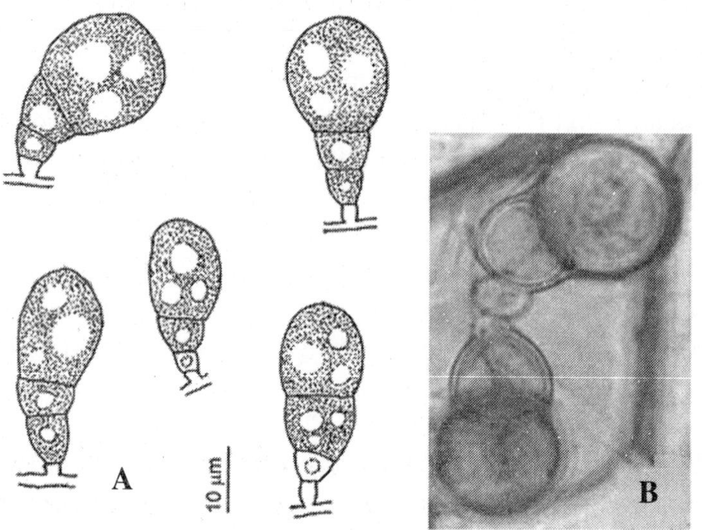

Fig. 131.1. ***Trichocladium achrasporum***, A-B: A- conidia, B- conidia (1000x).

131.2. *Trichocladium alopallonellum* (R.T. Moore & Meyers) Kohlm. & Volkm.-Kohlm., 1995. *Mycotaxon*, **53**: 352. [Fig. 131.2]

Basionym: *Humicola alopallonella* Meyers & R.T. Moore, *Am. J. Bot.*, **47**: 346 (1960).

Hyphae: septate, subhyaline to light brown. *Conidiophores*: 3.5-6.5 x 3.5-6 µm, macronematous, simple, one- to two-celled, smooth, subhyaline to light brown, lateral, short, sometimes indistinct. *Conidia*: 10-22 x 8-18 µm, obpyriform, ovoidal or subglobose, one- to two-celled, fuscous; when two-celled, apical cell larger (8.5–15.5 x 7-12 µm), ovoid, fuscous, basal cell smaller, obconical to cylindrical, light brown, conidiogenous cell usually remaining connected with the conidium.

Habitat: On intertidal and mangrove wood.

Discription: Based on Moore and Meyers (1963) and Kohlmeyer and Kohlmeyer (1979).

Distribution: West Coast:-Gujarat: On intertidal wood of *Rhizophora mucronat* (Borse et al., 2000a), on intertidal wood and mangrove wood (Patil and Borse, 2001); **Maharashtra:** On intertidal dead prop roots and seedlings of *Rhizophora mucronata* (Borse, 1984, 1988), on intertidal wood of *Rhizophora mucronata,* and *Sonneratia acida* (Patil and Borse, 1986), on intertidal wood and wood of *Avicennia alba* (Shrivastava, 1994); **Goa:** On submerged wood panels of *Acacia auriculiformis* and *Hevea braziliensis* (Raghukumar et al.,

1988a), on intertidal wood (Nandan et al., 1993), on intertidal animal substrates (Ananda et al., 1998), on intertidal wood (Prasannarai and Sridhar, 2001), on intertidal mangrove wood (Maria and Sridhar, 2002a); **Karnataka:** On intertidal wood (Prasannarai and Sridhar, 1997, 2001, 2000-2001; Prasannarai et al., 1999), on intertidal animal substrates (Ananda et al., 1998; Ananda and Sridhar, 2001), on intertidal mangrove wood and roots of *Rhizophora mucronata* (Maria and Sridhar, 2002a), on intertidal wood of *Avicennia officinalis* and *Rhizophora mucronata* (Maria and Sridhar, 2003a), on intertidal leaf and woody litter of mangroves (Ananda and Sridhar, 2003, 2004); **Pondicherry (Mahe):** On intertidal wood (Nambiar and Raveendran, 2007, 2008c); **Kerala:** On intertidal wood (Raveendran and Manimohan, 2007), on intertidal wood, mangrove wood of *Avicennia* sp. and *Bruguiera* sp. (Nambiar and Raveendran, 2006, 2007, 2008b, d, 2009b, d, e, f, g, h, k, 2010c; Nambiar et al., 2006, 2008a); **Lakshadweep Islands:** On pneumatophores of *Ceriops tagal* (Chinnaraj, 1992) on intertidal mangrove wood (Ananda and Sridhar, 2003).

East Coast:-Tamil Nadu: On submerged parts of *Rhizophora* spp. (Ravikumar and Vittal, 1996), on intertidal wood (Prasannarai and Sridhar, 2001; Nambiar et al., 2008b), on mangrove wood (Nambiar and Raveendran, 2008e, 2009l, 2010c); **Andhara Pradesh:** On intertidal wood of *Avicennia officinalis* and *Rhizophora apiculata* (Sarma and Vittal, 1998-99; Sarma and Vittal, 2000; Sarma and Vittal, 2001; Sarma et al., 2001; Sarma and Vittal, 2002; Sarma and Vittal, 2004; Vittal and Sarma, 2005), on intertidal wood (Nambiar and Raveendran, 2009j); **Orissa:** On intertidal wood of *Sonneratia griffithii* (Borse and Borse, 2005); **West Bengal:** On intertidal mangrove wood (Pawar and Borse, 2005b); **Andaman & Nicobar Islands:** On intertidal wood of *Avicennia marina, Rhizophora mucronata* and *Sonneratia alba* (Chinnaraj, 1993a).

Note: The conidia of the present fungus are very similar to *T. melhae* E.B.G. Jones, Abdel-Wahab & Vrijmode (2001) and records from India before discovery *T. melhae* are based on misidentification (Borse, 1984, 1988).

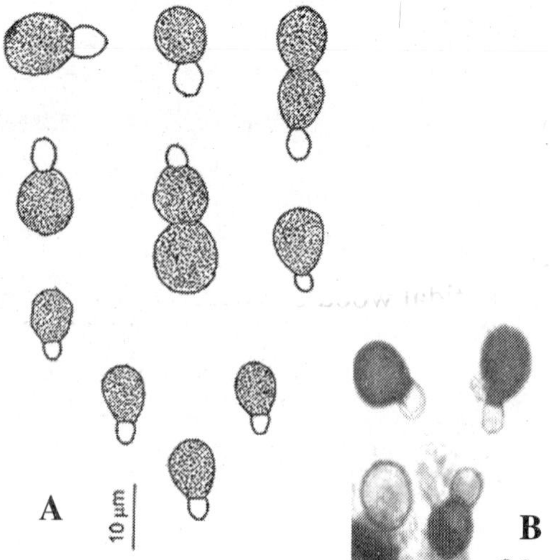

Fig. 131.2. ***Trichocladium alopallonellum***, A-B: A- conidia, B- conidia (1000x).

131.3. ***Trichocladium constrictum*** I. Schmidt, 1985. *Mycotaxon*, **24**: 419.

[Fig. 131.3]

Hyphae: sub-hyaline to light brown, septate and branched. *Conidiophores*: macronematous, simple, light brown, lateral (apical), and short. *Conidia*: 25-47 x 8-20 µm, apical cell 13-20 µm in diam, 2-4 septate, redish-brown, markedly constricted at the septa, subglobose.

Habitat: On intertidal mangrove substrates.

Discription: Based on Kohlmeyer and Kohlmeyer (1991) and Jones et al. (2009a).

Distribution: West Coast:-Kerala: On intertidal wood (Prasannarai and Sridhar, 2001), on wood of *Rhizophora* sp. (Nambiar and Raveendran, 2009d).

East Coast:-Orissa: On intertidal fronds of *Nypa fruticans* (Borse and Borse, 2005); **West Bengal:** On intertidal wood of *Aegiceras corniculatum* and *Ceriops tagal* (Pawar and Borse, 2005b).

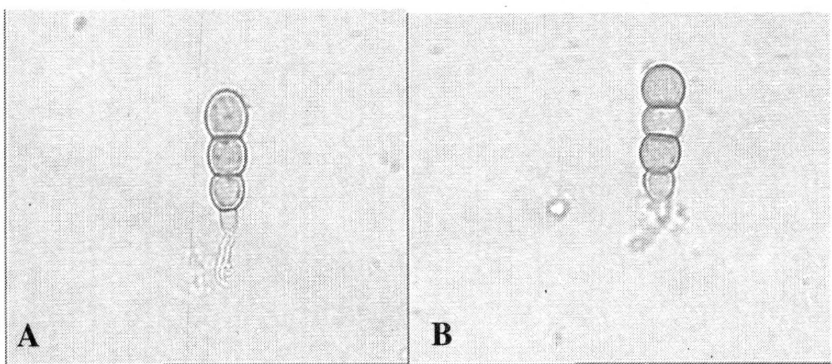

Fig. 131.3. ***Trichocladium constrictum***, A-C: Conidia (1000X)

131.4. ***Trichocladium lignicola*** I. Schmidt, 1985. *Mycotaxon*, **24**: 419.

[Fig. 131.4]

Hyphae: sub-hyaline to light brown, septate and branched. *Conidiophores*: macronematous, simple, light brown, lateral (apical), and short. *Conidia*: 25-47 x 8-20 µm, apical cell 13-20 µm in diam, 2-4 septate, redish-brown, markedly constricted at the septa, subglobose.

Habitat: On intertidal mangrove substrates.

Discription: Based on Kohlmeyer and Kohlmeyer (1991) and Jones et al. (2009a).

Distribution: West Coast:-Gujarat: On intertidal wood (B. D. Borse, unpublished).

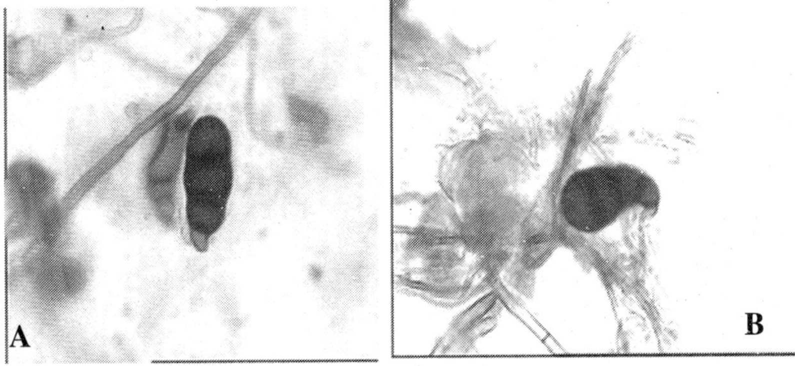

Fig. 131.4. ***Trichocladium lignicola***, A-C: Conidia (1000x)

131.5. *Trichocladium melhae* E.B.G. Jones, Abdel-Wahab & Vrijmode, 2001. *Fungal Diversity*, **7:** 50. [Fig. 131.5]

Hyphae: sub-hyaline to light brown, septate and branched. *Conidiophores*: 5-20 x 2-5 µm, macronematous, simple, one (two) celled, light brown, lateral (apical), and short. *Conidia*: 6.5-14 x 3.5-9 µm, dark brown to fuscous, and constricted at the septa,. Apical cell: 6.5- 13 x 4-9 µm, elongated to ellipsoidal in shape, Middle cell: 4-13.5 x 2.5-6 µm, while Third cell: 5-7 x 4.5- 5 µm. Solitary conidia sometimes born directly from the mycelium.

Habitat: On intertidal mangrove substrates.

Discription: Based on Jones et al. (2001).

Distribution: West Coast:-Karanataka: On intertidal wood of *Rhizophora mucronata* (Maria and Sridhar, 2004).

East Coast:-Andhra Pradesh: On intertidal wood (Nambiar and Raveendran, 2009j); **Orissa:** On intertidal wood of *Sonneratia griffithii* (Borse and Borse, 2005); **West Bengal:** On intertidal wood of *Avicennia marina* (Pawar and Borse, 2005b).

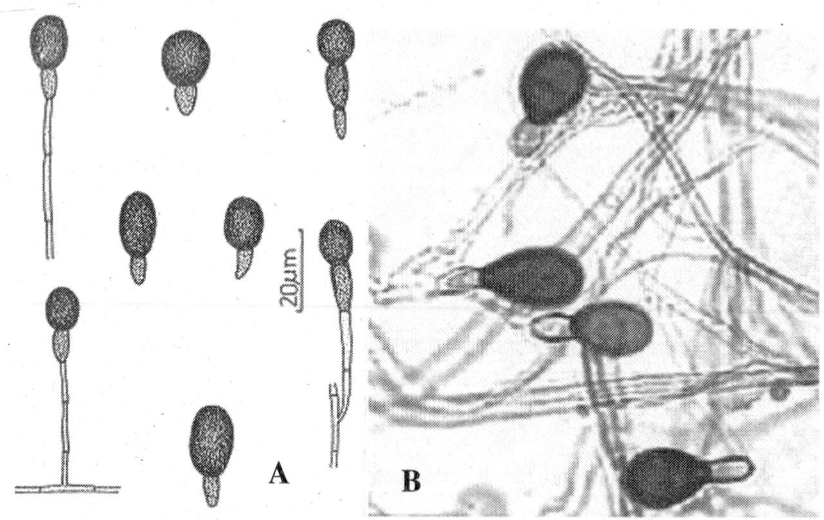

Fig. 131.5. *Trichocladium melhae*, A-B: A- conidia, B- conidia (1000x).

132. *Zalerion* R.T. Moore & Meyers, 1962. *Can. J. Microbiol.*, **8:** 408.

Conidiophores: present or obsolete, cylindrical, septate, hyaline, or grayish. *Conidiogenous cells*: monoblastic, integrated, terminal, determinate. *Conidia*: acrogenous, solitary, igantean, septate, constricted at the septa, subhyaline to fuscous; diameter and pigmentation of conidial filament more or less equal throughout; conidia coiled regularly or irregularly in three dimensions, simple or rarely branched, often producing a ball or knot of cells. Among several species one occurs in marine waters (Jones *et al.*, 2009a).

Type species: *Zalerion nepura* (Linder) Anastasiou
Habitat: On intertidal and mangrove wood.
Discription: Based on Ellis (1971) and Kohlmeyer and Kohlmeyer (1979).

132.1. *Zalerion maritium* (Linder) Anastasiou, 1963. *Can. J. Bot.*, **41**: 1136. [Fig. 132.1]

Basionym: *Helicoma maritimum* Linder, *Farlowia,* **1**: 405 (1944).

= *Helicoma salinum* Linder, *Farlowia*, **1**: 406 (1944).

= *Zalerion nepura* R.T.Moore & Meyers, *Can. J. Bot.*, **8**: 413 (1962).

= *Zalerion eistla* R.T. Moore & Meyers, *Can. J. Bot.*, **8**: 413 (1962).

= *Zalerion xylestris* R.T. Moore & Meyers, *Can. J. Bot.*, **8**: 414 (1962).

= *Zalerion raptor* R.T. Moore & Meyers, *Can. J. Bot.*, **8**: 415 (1962).

Hyphae: 2-4 µm in diam, septate, branched, hyaline to fuscous. *Conidiophores*: 10-115 x 3.5-5 µm. micronematous or semimacronematous, slightly increasing in diam from base to apex, simple or rarely ramose, straight or curved, cylindrical, septate, rarely absent, superficial, light to dark fuscous. *Conidia*: 19.5-50 µm in diam, pluricellular, monacrogenous, single or branched, solitary, blastic, coiled in tow or three dimensions, forming 1-5 more or less irregular coils; conidial filament 5.5-12 µm in diam, 4-27-septate, slightly or strongly constricted at the septa, fuscous or almost black.

Habitat: On intertidal and mangrove wood.
Teleomorph: *Lulwoana uniseptata* (Nakagiri) Kohlm., Volkm.-Kohlm, Campb., Spatafora & Graf., Mycol. Res., 109: 62 (2006).
Discription: Based on Ellis (1971) and Kohlmeyer and Kohlmeyer (1979).

Distributin: West Coast:-Diu Island: On intertidal wood (Borse et al., 1999b); **Daman:** On intertidal wood and *Avicennia marina* (Borse et al., 2000b); **Gujarat:** On intertidal wood and mangrove wood (Patil and Borse, 2001); **Maharashtra:** On intertidal wood,of *Rhizophora mucronata* (Borse, 1984, 1988, Shrivastava, 1994; Borse et al., 2005a); on intertidal mangrove wood (Maria and Sridhar, 2002); **Goa:** On intertidal wood (Nandan et al., 1993), on intertidal mangrove wood (Maria and Sridhar, 2002a); **Karnataka:** Conidia in foam and sand samples (Sridhar and Kaveriappa, 1991), on intertidal wood (Prasannarai et al., 1999), On intertidal wood (Prasannarai and Sridhar, 2001, 2000-2001), on intertidal mangrove wood and roots of *Acanthus ilicifolius* (Maria and Sridhar, 2002a), on intertidal wood of *Acanthus ilicifolius, Avicennia officinalis, Brugueiria gymnorrhiza, Rhizophora mucronata* and *Sonneratia caseolaris* (Maria and Sridhar, 2003a, 2004), on intertidal leaf and woody litter of mangroves (Ananda and Sridhar, 2003, 2004); **Kerala:** On intertidal wood (Prasannarai and Sridhar, 2001), on intertidal mangrove wood (Maria and Sridhar, 2002a), on intertidal wood (Raveendran and Manimohan, 2007), on intertidal wood, wood of *Avicennia* sp. and *Bruguiera* sp. (Nambiar and Raveendran, 2006, 2007, 2008b, 2009d, e, g, h, 2010a, c; Nambiar et al., 2006); **Lakshadweep Islands:** On frond of *Cocos nucifera* (Chinnaraj, 1992) on intertidal mangrove wood (Ananda and Sridhar, 2003).

East Coast:-Tamil Nadu: On drift wood (Raghukumar, 1973), on mangrove wood (Nambiar and Raveendran, 2008e, 2009l, 2010a, c); **Orissa:** On intertidal wood of *Heritiera foams* (Borse and Borse, 2005); **West Bengal:** On intertidal wood of *Aegiceras corniculatus* (Pawar and Borse, 2005b).

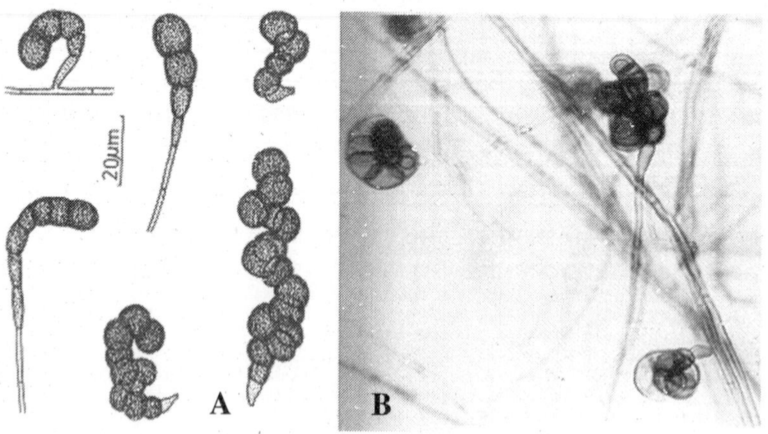

Fig. 132.1. ***Zalerion maritimum***, A-B: A- conidia (1000x), B- conidia (1000x).

Coelomycetes:

133. ***Ascochyta*** Lib. *Pl. Crypt. Arduenna, Fasc.* **1**, 8 (1830).

Pycnidia: separate, immersed, subglobose to lens-shaped, unilocular, ostiolate, brown, glabrous, sometimes hairy. *Conidiogenous cells*: phialidic, flask-shaped to pyriform, hyaline. *Conidia*: blastic, ellipsoidal, obovate or cylindrical, 1-septate, slightly constricted at the septum, smooth-walled, hyaline to to yellowish to pale brown. The genus is represented by two marine species among several terrestrial species (Jones et al., 2009a).

Type species: *Ascochyta pisi* Libert
Habitat: Parasitic or saprophytic on the salt marsh plants.
Discription: Based on Kohlmeyer and Kohlmeyer (1979) and Sutton (1980).

133.1. ***Ascochyta salicorniae*** Magnus, 1902, *Schr. Naturwiss. Ver. Schleswing-Holstein,* **12**: 345. [Fig. 133.1]

≡*Stagnosporopsis salicorniae* (Magnus) Diedicke, *Ann. Mycol.*, **10**: 141-142 (1912).

= *Ascochyta salicorniae* var. *salicorniae-patulae* Trotter, *Ann. Mycol.*, **3**: 30 (1905).

= *Ascochyta salicorniae* var. *salicorniae-patulae* (Trotter) Melnic, *Nov. Sist. Niz. Rast.*, **12**: 205 (1975).

= *Ascochyta salicorniae* Trotter, *Ann. Mycol.*, **2**: 536 (1904), nom. Illegit

= *Diplodia salicorniae* Japp, *Verh. Bot. Ver. Prov. Brandenberg*, **49**: 16 (1907).

Pycnidia: 90-180 x 90-210 µm, subglobose, ellipsoidal or pyriform, immersed, ostolate, epapillate to papillate, coriaceous, thin-walled, brown to black, solitary or gregarious. *Peridium*: 8-12 µm thick, usually composed of two layers of elongate, lenticular, thick-walled cells, dark on the outside, light-coloured near the venter, forming a *textura angularis*. *Papillae*: 25-30 um high, 45-50 µm in diam, conical or cylindrical, centric or eccentric, projecting above the surface of the host. *Ostioles*: 20-30 µm in diam, ostolar canal filled with mucilage and some small hyaline, periphysoid cells broken off the wall. *Conidiogenous cells*: phialidic, flask-shaped, hyaline, originating along the inner wall of the peridium. *Conidia*: 10-19 x 4-7 µm, blastic, cylindrical with rounded ends to ellipsoidal, one-septate in the middle, rarely with an additional septum in the lower or upper half, very seldom three-septate, not or slightly constricted at the septa, smooth, hyaline to yellowish or light brown, enclosed inan indistinct, irregular, mucilaginous sheath.

Habitat: Parasitic or saprophytic on the salt marsh plants.

Discription: Based on Kohlmeyer and Kohlmeyer (1979).

Distribution: West Coast:-Gujarat: On intertidal stem of *Salicornia brachiata* (Borse et al., 2000a; Patil and Borse, 2001); **Maharashtra:** On intertidal dead stem of *Salicornia brachiata* (Borse, 1984, 1988, Patil and Borse, 1986; Borse et al., 2005a).

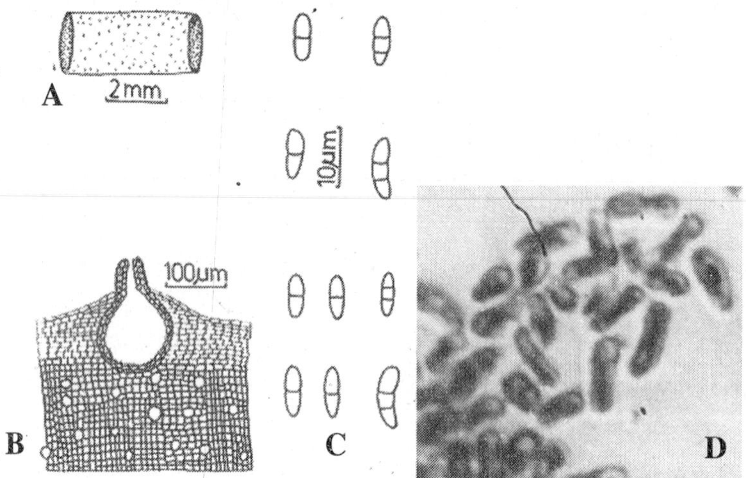

Fig. 133.1. ***Ascochyta salicorniae***, A-D: A- Habit, B- v.s.of pycnidium, C- conidia, D- conidia (1000x).

134. ***Camarosporium*** Schulzer von Muggenburg, 1870, Verh. Zool.- Bot. Ges. Wien, **20:** 641.

Pycnidia: solitary or gregarious, subglobose or ellipsoidal, unilocular, immersed or erumpent, ostiolate, papillate or epapillate, thin- or thick-walled, dark brown. *Conidiogenous cells*: holoblastic, annellated, cylindrical. *Conidia*: irregular in shape, muriform, thick-walled, smooth, dark brown. Among several species tow occur in marine waters (Jones et al., 2009a).

Type species: *Camarosporium quaternatum* (Hazsl.) Schulzer

Habitat: Parasitic or saprophytic on the salt marsh plants.

Discription: Based on Kohlmeyer and Kohlmeyer (1979) and Sutton (1980).

Key to *Camarosporium* species from India:

1. Conidia without sheath, 3-transverse-septate, 1-longi-septate, 10-20 x 7-13 µm .... *Camarosporium rumeguerii*

1'. Conidia with sheath, 5-transverse-septate, 1-longi-septate, 20-34 x 9-20 µm .... *Camarosporium palliatum*

134.1. *Camarosporium palliatum* Kohlm. & E. Kohlm., 1979. *Marine Mycology: The Higher marine fungi, Academic Press* p. 509.
[For Fig. refer original paper]

Pycnidia: 120-190 x 100-190 µm, subglobose to pyriform, immersed, ostiolate, epapillate, coriaceous, light or dark brown to black, solitary or gregarious. *Ostioles*: 10-12 µm in diam, at first closed by hyaline, ellipsoidal cells, which are exuded together with the conidia. *Peridium*: 12-14 µm thick, composed of about 3 layers of thin-walled cells with large lumina, forming a *texture angularis*. *Paraphyses*: 30-40 x 3-4 µm, simple, rarely forked, filiform, hyaline, non-septate. *Conidiogenous cells*: phialidic, flask-shaped, simple, hyaline, originating along the inner wall of the pycnidium. *Conidia*: 20-34 x 9-20 µm, enteroblastic, monophialidic, ellipsoidal or oblong with rounded ends, muriform, with (3-) 5 (-6) transverse and 3 or 4 longitudinal or oblique septa, slightly constricted at the septa, smooth, at first yellowish-brown, becoming dark brown, surrounded by gelatinous sheath, which slowly swell in water and eventually dissolves.

Habitat: Saprophytic or perthophytic on the salt marsh and mangrove plants.

Discription: Based on Kohlmeyer and Kohlmeyer (1979).

Distribution: West Coast:-Maharashtra: On intertidal wood , wood of *Avicennia alba, Salicornia brachiata* and *Sonneratia acida* (Borse et al., 1988; Borse et al., 2005a), on intertidal wood (Shrivastava, 1994).

134.2. *Camarosporium roumeguerii* Sacc., 1880. *Michelia*, **2**: 112.

[Fig. 134.2]

= *Camarosporium obiones* Jaap, *Verh. Bot. Ver. Prov. Branenburg*, **47**: 97 (1905).

Pycnidia: 90-210 x 85-260 µm, globose to ellipsoidal or lenticular, immersed, ostiolate, short papillate, coriaceous, yellow brown to olivacious, darkes at the ostiole, solitary or gregarious; venter filled with a mucilage, especially in the ostiolar canal. *Peridium*: 7-12 µm thick at the base, 12- 20 µm at the ostiole, composed of up to 4 layers of thin-walled cells with large lumina, forming a *texture angularis*. *Paraphyses*: simple, filiform, hyaline, non-septate. *Conidiogenous cells*: phialidic, flask-shaped, simple, hyaline, originating along the inner wall of the pycnidium. *Conidia*: 10-20 x 7-13 µm, enteroblastic, monophialidic, initiall one-celled, hyaline, mature conidia subglobose, ovoid, ellipsoidal or irregular, muriform, with (1-) 3-transverse and 1 or 2 longitudinal or oblique septa, slightly constricted at the septa, composed of 2 to 8 cells, smooth, gold-, ochraceous-, or olive-brown, without appendages or gelatinous sheath.

Habitat: Saprophytic or perthophytic on the salt marsh and mangrove plants.
Discription: Based on Kohlmeyer and Kohlmeyer (1979).
Distribution: West Coast:-Gujarat: On intertidal mangrove wood (Patil and Borse, 2001); **Maharashtra:** On intertidal dead stem of *Salicornia brachiata* (Patil and Borse, 1986, Borse, 1988, Borse et al., 1988; Borse et al., 2005a), on intertidal wood (Shrivastava, 1994).

East Coast:-Tamil Nadu: On intertidal mangrove wood (Ravikumar, 1991), **Andhara Pradesh:** On intertidal wood of *Avicennia marina,* and *A. officinalis* (Sarma and Vittal, 1998-99; Sarma and Vittal, 2000; Sarma and Vittal, 2001; Sarma et al., 2001; Sarma and Vittal, 2004; Vittal and Sarma, 2005); **West Bengal:** On intertidal wood of *Avicennia alba* (Pawar and Borse, 2005b), on mangrove woody litter (Shini et al., 2009-10).

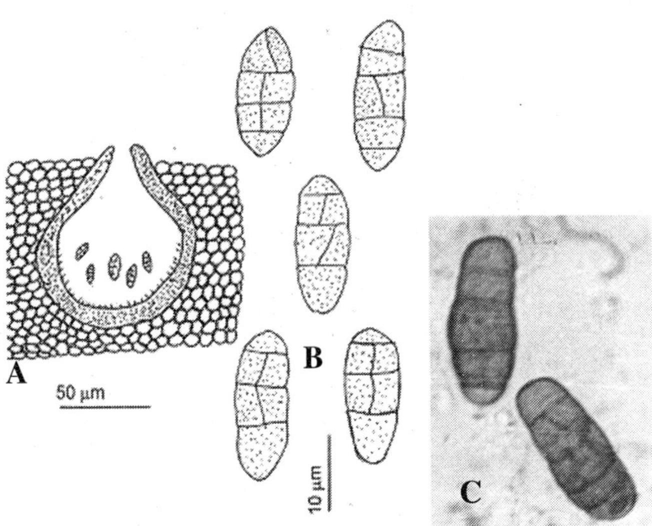

Fig. 134.2. ***Camarosporium roumeguerii***, A-C: A- v.s.of pycnidia, B- conidia, C- conidia (1000x).

135. ***Coniothyrium*** Corda, 1840. *Icon. Fung*. **4**, p. 8, 38.

Pycnidia: separate or gregarious, subglobose or depressed, unilocular, immersed, later erumpent or almost superficial, ostiolate, papillate, dark brown, membranaceous. *Conidiogenous cells*: holoblastic, phialidic, flask-shaped, annellidic, doliiform, with 1 to 3 annellations, hyaline. *Conidia*: cylindrical to oval, one-celled (rarely two-celled), thick-walled, smooth, olivaceous to pale brown. Among several species one occurs in marine waters (Jones *et al.*, 2009a).

Type species: *Coniothyrium palmarum* Corda
Habitat: Parasitic or saprophytic on the salt plants.
Discription: Based on Kohlmeyer and Kohlmeyer (1979) and Sutton (1980).

135.1. ***Coniothyrium obiones*** Japp, 1907. *Schr. Naturwiss. Ver. Schleswing-Holstein,* **14**: 29. [For Fig. refer original paper]

Pycnidia: 140-190 x 150-230 µm, gregarious, subglobose, at first immersed, finally reumpent, ostiolate, short papillate, coracious, thin-walled, brown. *Peridium*: 10-12 µm thick; on the outside composed of small, brown, thick-walled cells that merge toward the venter with hyaline, thin-walled, flattened cells; cells of inner stratum bearing conidiophores; forming a *textura angularis. Conidiogenous cells*: up to 7 μm long, holoblastic, phialidic, flask-shaped, tapeing at the apex, hyaline. *Conidia*: 4-8 μm long, 3-6 μm in diam, ellipsoidal, ovoid or subglobose, one-celled, thick-walled, smooth, olivaceous to pale brown, originating in basipetal succession on the apex of the conidiogenous cell; hyaline when released, maturing in the pycnidial venter.

Habitat: Saprophytic on the salt marsh plants.
Discription: Based on Kohlmeyer and Kohlmeyer (1979).
Distribution: East Coast:-Orissa: On intertidal wood of *Sonneratia griffithii* (Borse and Borse, 2005b).

136. *Cytospora* Ehrenb. Ex Fries, 1823. *Systema Mycologicum, Gryphiswaldiae*, **2**, p. 540.

Pycnidia: eustromatic, complex, immersed, cavity convoluted, divided, with a single ostiole. *Conidiophores*: septate, branched. *Conidiogenous cells*: enteroblastic, monophialidic. *Conidia*: allantoid, one-celled, hyaline, discharged as cirri or beads. Among several species two occur in marine waters (Jones *et al.*, 2009a).

Type species: *Cytospora chrysosperma* (Pers.) Fries
Habitat: Parasitic or saprobic on higher plants.
Discription: Based on Kohlmeyer and Kohlmeyer (1979).

136.1. *Cytospora rhozophorae* Kohlm. & E. Kohlm., 1971. *Mycologia*, **63**: 847. [For Fig. refer original paper]

Pycnidia: 0.5- 0.8 mm high, 1.5-1.7 mm in diam, of irregular shape, composed of several locules, immersedunder the host cortex, ostoilate, coriaceous, brown, solitary or gregarious, surrounded by thick, white hyphae. Locules: 370-500 μm high, 100-310 μm in diam, irregular ampulliform, opening into a common ostiole. *Peridium*: 10-25 μm thick. *Necks*: cylindrical, central, sometimes clypeate, breaking through the lenticels or the bark of the host; ostiole about 30 μm in diam; canal formed schizogenously. *Conidiophores*: 13-20 μm long, 1-2 μm in diam, cylindrical, filiform, attenuate, simple or rarely branched, septate, producing conidia at the apex, covering the inner wall of the pycnidia. *Conidia*: 3-6 x 1-1.5 μm, allantoid or ellipsoid-cylindrical, straight or slightly curved, one-celled, hyaline or light yellowish, discharged as cirri or beads.

Habitat: Parasitic on roots and seedlings of *Rhizophora* sp. and causing dieback of young plants.
Discription: Based on Kohlmeyer and Kohlmeyer (1971).
Distribution: West Coast:-Gujarat: On seedlings of *Rhizophora mucronat* (Borse et al., 2000a); **Maharashtra:** On seedlings of *Rhizophora mucronata* (Borse, 2000b; Borse et al., 2005a); **Goa:** On intertidal mangrove wood (Maria and Sridhar, 2002a); **Karnataka:** On intertidal wood of *Brugueiria gymnorrhiza* (Maria and Sridhar, 2003); **Kerala:** On intertidal mangrove wood (Maria and Sridhar, 2002a).
East Coast:-Tamil Nadu: On intertidal mangrove wood (Ravikumar, 1991); **Andhara Pradesh:** On intertidal wood of *Ceriops decandra* and *Rhizophora apiculata* (Sarma and Vittal, 1998-99; Sarma and Vittal, 2000; Sarma and Vittal, 2001; Sarma et al., 2001; Sarma and Vittal, 2002; Sarma and Vittal, 2004; Vittal and Sarma, 2005); **Andaman & Nicobar Islands:** On intertidal wood of *Avicennia marina* and *Rhizophora mucronata* (Chinnaraj, 1993a).

137. *Macrophoma* (Sacc.) Berl. & Voglino., 1886. *Atti Soc. Veneto-Trentina Sci. Nat.,* **10**: 4.

Pycnidia: separate or gregarious, globose, subglobose, ampulliform, obpyriform or irregular, unilocular, immersed or erumpent, ostiolate, papillate or epapillate, thin-walled, dark coloured, glabrous or sometimes hairy or setose; occasionally with more than one ostiole. *Peridium*: pseudo-parenchymatous, prosenchymatous, or rarely pseudosclerenchymatous; cells on the outside dark and thick-walled, toward the centrum hyaline and isodimetric. *Conidiogenous cells*: monoblastic, enteroblastic, annelidic, ampulliform, similar to the inner cells of the peridium, hyaline. *Conidia*: globose, obovoid, ellipsoidal or clavate, usually once or twice as long as widw, one-celled, smooth, and hyaline.

Type species: *Macrophoma sapinea* (Fries) Petrak
Habitat: Saprobic on higher plants.
Discription: Based on Kohlmeyer and Kohlmeyer (1979).

137.1. *Macrophoma* sp.

Coelomycetes resembiling *Macrophoma* are frequently encountered in marine and estuarine environment on a variety of substrates such as intertidal wood, mangrove roots, marsh plants, seawater, and sediments. However few have been fully identified. Some of them may be anamorph of Ascomycetes, others may be spermagonia. No effort has been made so far to separate and describe the different species (Kohlm. & Kohlm., 1991; Jones *et al.*, 2009).

Distribution: East Coast:-Orissa: On intertidal wood of *Ceriops decandra* (Borse & Borse, 2005b).

138. *Phialophorophoma* Linder, 1944. *Farlowia*, **1**: 403.

Pycnidia: septate, subglobose or ellipsoidal, unilocular, immersed, ostiolate, papillate or epapillate, subcarbonaceous, black, glabrous. Peridium: parenchymatous; cells on the outside dark, toward the centrum hyaline and bearing conidiophores. *Conidiophores*: cyalindrical, branched, septate, with terminal and lateral phialides, hyaline, produced on the inner cells of the peridium. *Conidiogenous cells*: monophialidic, enteroblastic. *Conidia*: obovoid-ellipsoidal to clavate, one-celled, smooth-walled, hyaline, no sheath or appendages. The genus is monotypic.

Type species: *Phialophorophoma litoralis* Linder

Habitat: Saprobic on drift and intertidal wood, bark and dead roots of *Avicennia marina*.

Discription: Based on Kohlmeyer and Kohlmeyer (1979).

138.1. *Phialophorophoma litoralis* Linder, 1944. *Farlowia*, **1**: 403.

[For Fig. refer original paper]

Pycnidia: up to 200 μm high, 120-200 um in diam, separate, subglobose or ellipsoidal, unilocular, immersed, ostiolate, papillate or epapillate, subcarbonaceous, brown to black, glabrous. Peridium: composed of 2 to 5 layers, pseudoparenchymatous; cells on the outside dark, toward the centrum hyaline and bearing conidiophores. *Conidiophores*: 10-25 x 2- 2.5 μm, cyalindrical, branched, septate, with terminal and lateral phialides, hyaline, produced on the inner cells of the peridium. *Conidiogenous cells*: 7-11 x 3-3.5 μm, cylindrical, branched, monophialidic, enteroblastic, tapering upward to a cylindrical nack or cylindrical throughout; lateral phialides produced immediately below transverse septa of the conidiophore. *Conidia*: 2.5-4.5 x1.5-2 μm, obovoid-ellipsoidal to clavate, one-celled, smooth-walled, hyaline, no sheath or appendages.

Habitat: Saprobic on intertidal wood.

Discription: Based on Kohlmeyer and Kohlmeyer (1979).
Distribution: West Coast:-Kerala: On intertidal wood (Raveendran and Manimohan, 2007).
East Coast:-Tamil Nadu: On intertidal mangrove wood (Ravikumar, 1991).

139. ***Phoma*** Sacc., 1880. *Michelia,* **2**: 4.

Pycnidia: separate or gregarious, globose, subglobose, ampulliform, obpyriform or irregular, unilocular, immersed or erumpent, ostiolate, papillate or epapillate, thin-walled, dark coloured, glabrous or sometimes hairy or setose; occasionally with more than one ostiole. *Peridium*: pseudo-parenchymatous, prosenchymatous, or rarely pseudosclerenchymatous; cells on the outside dark and thick-walled, toward the centrum hyaline and isodimetric. *Conidiogenous cells*: monoblastic, enteroblastic, annelidic, ampulliform, similar to the inner cells of the peridium, hyaline. *Conidia*: globose, obovoid, ellipsoidal or clavate, usually once or twice as long as widw, one-celled, smooth, and hyaline. Among several species three occur in marine waters (Jones et al., 2009a).

Type species: *Phoma herbarum* Westened
Habitat: Saprobic or parasitic on plants.
Discription: Based on Kohlmeyer and Kohlmeyer (1979).

139.1. ***Phoma suaedae*** Japp, 1907. *Schr. Naturwiss. Ver. Schleswing-Holstein,* **14**: 27. [Fig. 139.1]

Pycnidia: 150 µm high, 150-230 µm in diam, solitary or gregarious, subglobose to ellipsoidal, immersed or erumpent, ostiolate, short papillate or epapillate, coriaceous, light to dark brown, darkest around the ostiole. *Peridium*: composed of an outer layer with thick-walled, dark cells and an inner layer with hyaline cells, bearing conidiogeneous cells. Papillae: about 15 µm in diam, ostiolar canal about 10 µm in diam. *Conidiogenous cells*: 5-8 µm long, phialidic, flask-shaped, originating along the inner wall of the peridium, similar to the inner cells of the peridium, hyaline. *Conidia*: 6-8 x 3-4 µm, ellipsoidal to subglobose, one-celled, often biguttulate, hyaline, originating in basipetal succession.

Habitat: Saprobic on leaves and stems of *Suaeda* sp.
Discription: Based on Kohlmeyer and Kohlmeyer (1979).
Distribution: East Coast:-West Bengal: On intertidal wood of *Suaeda igantea* (Pawar and Borse, 2005b).

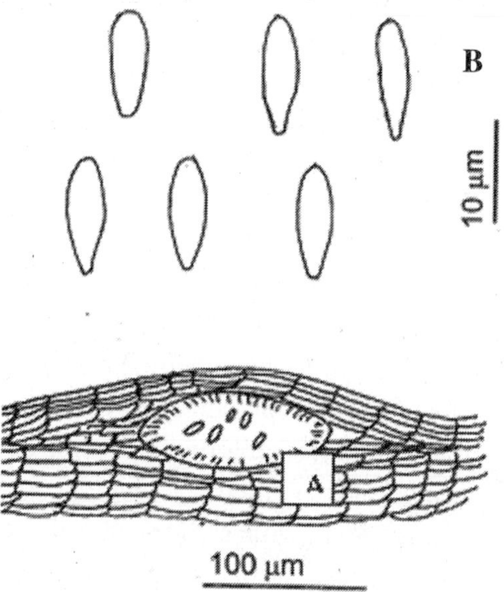

Fig. 139.1. ***Phoma suaedae***, A-B: A- v.s.of conidiomata, B- conidia.

139.2. ***Phoma*** sp. [Fig. 138.2]

Coelomycetes resembiling *Phoma* are frequently encountered in marine and estuarine environment on a variety of substrates. Some of them may be anamorph of Ascomycetes, others may be spermagonia. No effort has been made so far to separate and describe the different species (Kohlm. & Kohlm., 1991). *Phoma* species are ubiquitous, geographically widespread, occurring in variety of environments, endophytes, as biodeteriogens and parasitic on planta and animals. Because of difficulties in their identification, most are simply reffered to as *Phoma* species. Pawar *et al.* (1967) isolated and described *P. capitulam*, *P. multispora* and *P. ostiolata* from marine soils of Mumbai (Maharashtra, India) and thus not be regarded as obligately marine species (Jones *et al.*, 2009).

Habitat: Saprobic on intertidal and mangrove wood.

Distribution: West Coast:-Gujarat: On intertidal wood of *Avicennia marina* and *Rhizophora mucronat* (Borse et al., 2000a), on intertidal igantea wood (Patil and Borse, 2001); **Maharashtra:** On intertidal wood of mangroves (Borse et al., 2005a); **Karnataka:** On intertidal wood of *Acanthus ilicifolius, Avicennia officinalis, Brugueiria gymnorrhiza, Rhizophora mucronata* and *Sonneratia caseolaris* (Maria and Sridhar, 2003a), on intertidal leaf and woody litter of mangroves (Ananda and Sridhar, 2004); **Pondicherry (Mahe):** On intertidal mangrove wood (Nambiar and Raveendran, 2008a, c); **Kerala:** On intertidal wood (Raveendran and Manimohan, 2007), on intertidal mangrove wood of *Avicennia* sp. (Nambiar and Raveendran, 2008a, 2009f, g, h, i; Nambiar et al., 2008a).

East Coast:-Tamil Nadu: On intertidal mangrove wood (Ravikumar, 1991), on intertidal wood (Nambiar and Raveendran, 2009l); **Pondicherry:** On intertidal wood (Nambiar and Raveendran, 2008c); **Andhara Pradesh:** On intertidal wood of *Avicennia marina, A. officinalis, Lumnitzera recemosa* and *Rhizophora apiculata* (Sarma and Vittal, 1998-99; Sarma and Vittal, 2000; Sarma and Vittal, 2001; Sarma et al., 2001; Sarma and Vittal, 2002; Sarma and Vittal, 2004; Vittal and Sarma, 2005),), on mangrove wood (Nambiar and Raveendran, 2008e); **Orissa:** On intertidal fronds of *Acrostichum aureum* (Borse and Borse, 2005); **West Bengal:** On intertidal wood of mangroves (Pawar and Borse, 2005b); **Andaman & Nicobar Islands:** On intertidal wood of *Avicennia marina, A. officinalis, Rhizophora mucronata* and *Sonneratia alba* (Chinnaraj, 1993a).

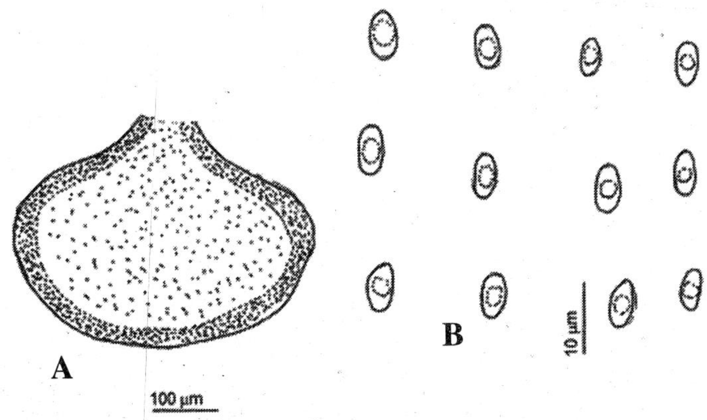

Fig. 139.2. ***Phoma*** sp. A-B: A- v.s.of Conidiomata, B- conidia.

140. ***Phomopsis*** (Sacc.) Bubak, 1905. *Ost. Bot. Z.,* **55**: 78.

Conidiomata: immersed, ellipsoidal or subglobose, ostiolate, solitary or gregarious, black. *Conidiophores*: branched, stout to filiform, septate, hyaline. *Conidiogenous cells*: phialidic, cylindrical, with a terminal collarette. *Conidia*: holoblastic, fusiform or ellipsoidal, straight, unicellular, hyaline, rounded apically, truncate at the base, no sheath or appendages. Among several species one occurs in marine waters (Jones *et al.*, 2009).

Type species: *Phomopsis lamii* Sacc. & D. Sacc. Or *P. pritchardiae* (Cke. & Harkn.) Sacc.
Habitat: Saprobic on plants.
Discription: Based on Jones et al. (2009a).

140.1. *Phomopsis mangrovei* K.D. Hyde, 1991. *Mycol. Res.*, **95**: 1149.

[For Fig. refer original paper]

Conidiomata: ellipsoid or subglobose, 250-400 μm long, 130-195 μm wide and 162-260 μm high, eustromatic, immersed, black, separate or aggregated in groups of 2-3, unilocular, thick-walled, of *textura angularis*, darker in the upper region and lined with a layer of highly melanized amaller celled tissue. *Ostiole*: single, circular, non-papillate. *Conidiophores*: up to 52 μm long, branched, stout or filiform, septate at the base with 0-3 septa above, hyaline, formed from the inner cells of the locular walls. *Conidiogenous cells*: phialidic, determinate, integrated hyaline, cylindrical, collarette, channel and periclinal thickening minute. *Conidia*: holoblastic, of one type only (alpha), 11-18 x 3-4 μm, hyaline, fusiform or ellipsoid, straight, 0-3 guttulate, aseptate, rounded at one end and with a truncate scar at the base.

Habitat: On intertidal mangrove wood and prop roots of *Rhizophora* spp.
Discription: Based on Hyde (1991).
Distribution: West Coast:-Gujarat: On intertidal mangrove wood (Patil and Borse, 2001); **Maharashtra:** On intertidal prop roots of *Rhizophora mucronata* (Borse, 2000b; Borse et al., 2005a).
East Coast:-Andhara Pradesh: On intertidal prop roots of *Rhizophora apiculata* (Sarma and Vittal, 1998-99; Sarma and Vittal, 2000; Sarma and Vittal, 2001; Sarma et al., 2001; Sarma and Vittal, 2004; Vittal and Sarma, 2005); **West Bengal:** On intertidal wood of *Avicennia marina* (Pawar and Borse, 2005b). **Andaman & Nicobar Islands:** On prop roots of *Rhizophora mucronata* (Chinnaraj, 1993a).

141. *Robillarda* Sacc., 1880, *Michelia*, **2:** 8

Pycnidia: immersed or erumpent, ellipsoidal, unilocular, epapillate, subcoriaceous, black, glabrous, solitary or gregarious. *Conidiogenous cells*: conical or cylindrical, hyaline. *Conidia*: ellipsoidal, 1-septate, slightly constricted at the septum, smooth-walled, hyaline, with 3-4 apical radiating appendages.

Type species: *Robillarda sessilis* (Sacc.) Sacc.
Habitat: Seedlings, live and dead proproots of *Rhizophora mangle*.
Discription: Based on Jones et al. (2009a).

141.1. *Robillarda rhizophorae* Kohlm., 1969. *Can. J. Bot.*, **47:** 1483.

[For Fig. refer original paper]

Pycnidia: 89-200 μm high, 248-770 μm long, ellipsoidal, unilocular, at first immersed, later erumpent, epapillate, opening apically with a longitudinal slit, subcoriaceous, black, glabrous, solitary or gregarious, rarely confluent. Peridium: 7-24 μm thick, composed of four to six layers of ellipsoidal or irregular, thick-walled cells, dark on the outside, hyaline near the venter, forming a textura angularis. *Conidiogenous cells*: 4.5-8.5 μm high, 2.5-6 μm in diam, conical or cylindrical, hyaline, originating along the inner wall of the peridium. *Conidia*: 9.5-13.5 x 3-4.5 μm, single, ellipsoidal, one-septate in the middle, slightly constricted at the septum, smooth-walled, hyaline, provided with 3-4 apical, radiating appendages; appendages 14-30.5(-38) μm long, 0.8-1.5 μm in diameter at the base, slightly tapering toward the tip, rigid, straight or slightly curved, diverging from a common base, 1.5-2 μm long, 1.5 μm in diameter.

Habitat: On seedlings, live and dead prop roots of *Rhizophora mangle*.
Discription: Based on Kohlmeyer and Kohlmeyer (1979).
Distribution: West Coast:-Kerala: On driftwood (Khan and Manimohan, 2011).

Key for identification of higher marine fungi of India:

Classification and orders according to that outlined by Hibbets et al. (2007) and Jones et al. (2009a). Genera and species of unknown affinities are referred to *incertae sedis.*

1 Meiospores produced endogenously, formed in asci: Phylum - Ascomycota (Sub-Phylum: Pezizomycotina) 2

1' Meiospores produced exogenously, borne on basidia or similar structures: Phylum- Basidiomycota 128

1" Mitospores exogenous, or in pycnidia : Anamorphic fungi 129

Key for identification of Ascomycetes:

2(1') Ascomata with bitunicate asci 3

2'(1') Ascomata with unitunicate asci 6

3(2) Lichen forming Ascomycetes 4

3'(2) Non-Lichenized Ascomycetes 5

4(3) Ascomata apothecial, marine species with no known phycobiont: .... Class: Lecanoromycetes: Ascospores 18-28 x 8-12 µm, spore wall with longitudinal striations .... *Dactylospora haliotrepha*

4'(3) Ascomata apothecial with various phycobionts: Class: Eurotiomycetes (No record from India)........

4"(3) Thallus varied, sometimes poorly developed or absent ....

Class: Arthoniomycetes (one species recorded from India) ....

Ascomata lirelliform, ascospores 1-septate, hyaline, lacking a sheath or appendages, 17-29 x 6-9 mm .... *Halographis runica*

5(3') Ascomata perithecial or apothecial ....

Class: Dothideomycetes 9

5'(3') Ascomata predominantly clestothecial, gymnothecial, globose,

asci evanescent .... Class: Eurotiomycetes (No record from India)

6(2') Parasitic on insects ....

....Class: Laboulbeniomycets (No record from India)

6'(2') Ascomycetes not parasitic on insects 7

7(6') Lichen forming ascomycetes with cyanophycean photobionts ....

.... Class: Lichinomycetes (No record from India)

7'(6') Non-lichenized ascomycetes 8

8(7') Ascomata perithecial ...... Class: Sordariomycetes 44

8'(7') Ascomata apothecial ......Class: Leotiomycetes, ...

ascospores 8-16 μm .... *Amylocarpu encephaloides*

9(5) Ascomata perithecial 10

9'(5) Ascomata apothecial, acospores hyaline, 4-9-septate, 24-48 x 6-24 μm .... *Patellaria atrata*

10(9) Ascospores 1-septate 11

10'(9) Ascospores multiseptate 25

10''(9) Ascospores muriform 39

11(10) Ascospores hyaline 12

11'(10) Ascospores brown 17

12(11) Ascospores with central septum 13

12'(11) Ascospores with eccentric septum 16

13(12) Ascospores lacking appendages/sheath, 16-25 x 10-16 µm....................
.... *Acrocordiopsis patilii*

13'(12) Ascospores with appendages/sheath 14

14(13') Ascospores with lenticular appendage on side, 17-28 x 8-12 µm...........
.... *Paraliomyces lentiferus*

14'(13') Ascospores lacking a lenticular appendage 15

15(14') Ascospores with tuberculate wall and surrounded by a sheath, 17- 25 x 10-15 µm .... *Belizeana tubercula*

15'(14) Ascospores not tuberculate, ascospores with sheathwhich drawn out at the poles to form tapering appendages, lackig a lenticular appendage, 28-39 x 7- 10 µm .... *Lophiostoma armatispora*

16(12') Ascospores with a sheath, on root and stem of *vicennia* sp., 25-32 x
10-15 µm .... *Didymella vicenniae*

16'(12') Ascospores without a sheath, on pneumatopores of *Avicennia* sp., 14-21 x 6-8.5 µm .... *Mycosphaerella neumatophorae*

17(11') Ascospores with eccentric septum 18

17'(11') Ascospores with centric septum 20

18(17) Ascomata superficial with stalk, ascopores 80-109 x 18-34 µm...............
.... *Manglicola* sp.

18'(17) Ascomata immersed 19

19(17') Ascomat multiloculate, ascspores with a delicate sheath, 36-51 x 18-26 µm .... *Helicascus kanaloanus*

19'(17') Ascomata unilocular, ascospores with bipolar germ tubes, no sheath, 59-72 x 24-30 µm .... *Salsuginea ramicola*

20(17') Ascomata mammiform, ascospores with bipolar germ tubes,................
22-30 x 10-12 µm .... *Caryosporella rhizophorae*

20'(17') Ascomata not mammiform and lacking bipolar germ tubes.............. 21

21(20') Ascomata semi-immersed, ascospores 14-21 x 5-8 µm ....
.... *Kirschsteiniothelia maritima*

21'(20') Ascomata immersed in substratum 22

22(21') Ascospores verrucose 23

22'(21') Ascospores not verrucose 24

23(22) Ascospores brown, 18-23 x 9-11 µm ....

.... *Didymosphaeria lignomaris*

23'(22) Ascospores dark brown, 16-23 x 7-11 µm .... *Verruculina enalia*

24(22') Ascospores striate, 23-33 x 9-12 µm .... *Lineolata rhizophorae*

24'(22') Ascospores not striate, 36-60 x 16-24 µm ....

.... *Coronopapilla mangrovei*

25(10') Ascospores 1-3-septate 26

25'(10') Ascospores more than 3-septate 36

26 (25) Ascospores hyaline 27

26'(25) Ascospores pale yellowish to brown 30

27(26) Ascomata superficial 28

27'(26) Ascomata immersed, ascospores with a sheath 29

28 (27) Ascomata superficial, crater-like, on mangrove wood, ascospores.........

46-62(-66) x 15-22 µm .... *Ascocratera manglicola*

28(27) Ascomata broadly oblong, on mangrove wood, asospores 37-55 x 7-11 µm .... *Rimora mangrovei*

29(27') Ascospores 45-56 x 14-19 µm .... *Morosphaeria velatospora*

29'(27') Ascospores 28-44 (-47) x 10-15 µm ...... *Halomassarina thalassiae*

30(26') Ascospores versicoloured 31

30'(26') Ascospores uniformly coloured 33

31(30) Ascospores striated, 31-38 x 6-9 µm ....

.... *Astrosphaeriella striataspora*

31'(30) Ascospores not striated 32

32(31') Asci with 4 ascospores, ascospores 64-88 x 24-28 µm .... *Passeriniella savoryellopsis*

32'(31') Asci with 8 ascospores, ascospores 44-62 x17-24 µm .... *Passeriniella mangrovei*

33 (30') Pseudoparaphyses trabeculate, ascospores 30-41 x 10-16 µm............ *Trematosphaeria mangrovei*

33'(30') Pseudoparaphyses cellular 34

34(33') Ascomata superficial, conoid, strongly thickened toward the base, with thick peridial wall, primarily occurring on Dicots, ascospores 19-27 x 6-9 µm .... Genus: *Leptosphaeria* 35

34'(33') Ascomata immersed, pyriform to ellipsoidal, with thin peridial wall, primarily occurring on Monocots, 17-29 (-32) x (4-) 5-8 µm........................ *Phaeosphaeria orae-maris*

34"(33') Ascomata on *Spartina*, also on driftwood, ascospores 23-35 x 5-8 µm yellow-brown .... *Phaeosphaeria halima*

35(34) Ascospores 19-27 x 6-9 µm .... *Leptosphaeria austaliensis*

35'(34) Ascospores with a sheath, 28-44 x 8-12 µm .... *Leptosphaeria pelagica*

35" (34) Ascospores with a sheath, 18-25 x 6-8 µm .... *Leptosphaeria avicenniae*

36(25') Ascospores hyaline, 51-80 x 14-20 µm.............. *Quintaria lignatilis*

36'(25') Asospores brown 37

37(36') Ascospores with septa at the ends, 55-82 x 16-25 µm .... *Biatriospora marina*

37'(36') Ascospores septa uniformly spread along ascospores 38

38(37') Ascomata superficial, ascospores with a sheath, 7-14-septate, 34-52 x 7-10 µm .... *Neomassariosphaeria typhicola*

38'(37') Ascomata immersed, ascospores versicolour with hyaline end-cells, ascospores 34-48x7-10 µm .... *Trematosphaeria lineolatispora*

39(10") Ascospores hyaline and thick-walled, 140-195 x 36-45 um.................
.... *Lautospora gigantea*

39'(10") Ascospores brown, ... 40

40(39') Ascospores with lighter end cells, only subapical cells covered by a gelatinous sheath .... Genus: *Aigialus* 41

40'(39') Ascospores uniformly brown with prominent sheath, 28-36 x 12-16 µm
.... *Julella avicenniae*

41(40) Ascospopres with 6 (rarely 8)-transverse-septate, 35-55 x 10-16 µm.....
.... *Aigialus mangrovei*

41'(40) Ascospores with more than 8-transverse-septate 42

42(41') Ascospores with (9)-10-11-(12)-transverse-septate, 49-72 x 19-26 µm
.... *Aigialus parvus*

42'(41') Ascospores with more than 13 transverse-septate 43

43(42') Ascospores rarely longi-septate in subapical cells, 67-101 x 18 - 29 µm,
.... *Aigialus grandis*

43'(42') Ascospores always long-septate in subapical cells, 70-95 x 18-25 µm,
.... *Aigialus rhizophorae*

44(8) Sardariomycetes, **three** subclasses with marine taxa (Perithecia or derived cleistothecial ascomata, unitunicate asci, basal or peripheral in ascoma with a wide range of anamorphs) 45

44'(8) Sardariomycetes *incertae sedis* 109

45(44) Stroma well developed, mostly consisting only fungal tissue, black with thick-wall, ostioles papillate, periphysate, ascomata perithecial, interascal tissue well developed, asci cylindrical with J+ apical apparatus, ascospores brown to black, with germ pores ... Sub-Class: Xalariomycetidae (**three** orders with marine taxa) 106

45'(44) Ascomata in a pseudostroma or coloured stroma or absent, lacking germ pores 46

46(45') Ascomata perithecial, rarely cleistothecial, sometimes stromatic and coloured, ostiole weakly to well developed, interascal tissue apical paraphyses or catenophyses or absent, asci thin-walled, often deliquescing, apical apparatus poorly developed, generally J-, ascospores septate, of variable morphology, hyaline to brown, anamorph may be present ..... Sub-Class: Hypocreomycetidae, **three** orders with marine taxa.. 47

46'(45') Ascomata perithecial, rarely cleistothe, cialrarely stromatic or in a pseudostroma, necks well developed, interascual tissue poorely developed or absent, asci cylindrical or clavate, often thick-walled but not fissitunicate, ascospores 0-1-septate, of varied morphology ....

.... Sub-Class: Sordariomycetidae 118

47(46) Ascomata in a stroma, perithecial, papillate or short necks, generally coloured, asci clavate to cylindrical, ascospores 1-septate, hyaline to pale brown, prominent anamorphs ....

.... Order: Hypocreales and Hypocreales *incertae sedis* 49

47'(46) Ascomata rarely in a stroma 48

48(47') Ascomata rarely stromatic, with well developed necks, asci generally clavate, ascospores hyaline, rarely coloured, 1-many-septate, rarely with anamorphs, Order: Halosphaeriales 56

48'(47') Ascomata dark, thick-walled, opening by an irregular lysigenous pore, asci clavate, long pedunculate, ascospores hyaline to brown, allantoid, Order: Coronophorales .... (No records from India).

49(47) Ascospores with polar appendages 50

49'(47) Ascospores without appendages 51

50(49) Ascospores 1-septate, 12-21 x 6-8 mm ....

.... *Etheirophora blepharospora*

50' (49) Ascospores 1-septate, 18-24 x 5-7 mm ....*Etheirophora unijubata*

50''(49) Ascospores 3-septate, 30.5-52 x 4-9 mm ....*Torpedospora radiata*

51(49') Ascospores unicellular, 6-11 x 4-7 μm .... *Payosphaeria minuta*

51'(49') Ascospores septate 52

52 (51') Ascospores 1-septate 53

52'(51') Ascospores 1-3-septate 55

53(52) Ascomata yellow to pale brown, ascospores 18-21 x 10-13 µm,............ *Trichothecium*-like anamorph .... *Heleococcum japonense*

53'(52) Ascomata orange, no anamorph, ascospores 17-26 x 8-13 µm 54

54(53') Ascospore wall smooth .... *Kallichroma glabrum*

54'(53') Ascospore wall with longitudinal ridges .... *Kallichroma tethys*

55(51') Ascospores 1-septate, 13-20x6-9 µm, ... *Swampomyces armeniacus*

55'(51') Ascospores 3-septate, weakly constricted at the septum, wall with granular ornamentations, 18-25 x 8-11 µm, ...*Swampomyces triseptatus*

56(48) Ascospores aseptate 57

56'(48) Ascospores septate 60

57(56) Ascospores globose to subglobose, 12.5-16.5 µm .... *Thalossogena sphaerica*

57'(56) Ascospores filiform 58

58(57') Ascospores thick-walled, a deep sea species, 50-72 x 4-6 µm .... *Bathyascus vermisporus*

58'(57') Ascospores thin-walled, intertidal species 59

59(58') Ascospores 90-145 x 2.5-4 µm, ... *Bathyascus avicenniae*

59'(58') Ascospores 205-300 x 9-15 µm, ... *Bathyascus grandisporus*

59"(58') Ascospores 125-160 x 2.5-5 µm, ... *Bathyascus mangrovei*

60 (56') Ascospores 1-septate 61

60'(56') Ascospores 1 to many-septate 92

61(60) Ascospores with no appendages 62

61'(60) Ascospores with polar / equatorial appendages or sheath 67

62(61) Asci deliquescing early, 70-100 x 8-9 µm .... *Bathyascus tropicalis*

62'(61) Asci persistent 63

63(62') Asci with no retraction of the plasmalemma at the ascus apex 64

63'(62') Asci with plasamalemma retraction at the ascus apex 65

64(63) Ascospores thick-walled, ascus with an apical pore, ascospores 22-36 x 12-16 µm .... *Lignincola tropica*

64'(63) Ascospores thin-walled, ascus without an apical pore and released through the neck, ascospores 13-24 x 5-8 µm, ... *Lignincola laevis*

64"(63) Ascospores 42-59 x 21- 31 µm, ... *Saagaromyces glitra*

65(63') Ascomata with short necks, ascospores without a faint sheath 66

65'(63') Ascomata with long necks (80-1125 x 13-40 mm), ascospores with a faint sheath, 13-19.5 x 4-7.5 mm .... *Neptunella longirostris*

66(65) Ascospores 20-25 x 14-18 µm, ... *Aniptodera haispora*

66' (65) Ascospores 10.5-13 x 7-8 µm, ... *Aniptodera intermedia*

66"(65) Ascospores 32-54 x 9-13 µm, ... *Aniptodera longispora*

66"'(65) Ascospores 21-37 x 7-15 µm, ... *Aniptodera chesapeakensis*

67(61') Ascospores with a single polar appendage 68

67'(61') Ascospores with bipolar hamate appendages or polar and equatorial appendages 72

68(67) Ascospores with a hamate polar appendage 69

68'(67) Ascospores with an ephemeral drop of polar mucilage, becoming 2- 4-septate on germination .... *Okeanomyces cucullatus*

69(68) Ascomata formed beneath a stroma, ascospores 6-21 x 6-8 µm............. .... *Ophiodeira monosemeia*

69'(68) Ascomata not stromatic 70

70(69') Ascospores oval 71

70'(69') Ascospores fusiform, 32-50 x 2(-3) µm .... *Oceanitis scuticella*

71(70) Asci with retracted plasamalemma at the apex, ascospores 24-32 x 8- 12 µm... *Tirispora unicaudata*

71'(70) Asci without retracted plasamalemma at the apex, ascospores 15-22 x 8-12 µm .... *Tirispora mandoviana*

72(67') Ascospores with polar unfurling appendages 73

72'(67') Ascospores appendages with a different morphology 81

73(72) Asci with plasamalemma retracted at the ascus apex 74

73'(72) Asci with no retraction of plasamalemma at the ascus apex 76

74(72) Ascospores with very thick wall, 21-37 x 7-15 µm ....

....*Aniptodera chesapeakensis*

74'(72) Ascospores wall less thick 75

75(74') Ascospores 14-20 x 4-7 µm .... *Aniptodera salsuginosa*

75'(74') Ascospores 37-45 x 12-14 µm .... *Aniptodera mangrovei*

76(73') Ascospores more than 48 µm long, 48-64(-72) x 22-28 µm.................

.... *Saagaromyces ratnagiriensis*

76'(73') Ascospores less than 48 µm long 77

77(76') Ascospores less than 14 µm wide 79

77'(76') Ascospores more than 14 um wide 78

78(77') Ascospores 33-47 x 14-22 µm, ... *Saagaromyces abonnis*

78'(77') Ascospores 32-44 x 18-24 µm, ... *Halosarpheia fibrosa*

78"(77') Ascospores 23-32 x 14-18 µm, ... *Halosarpheia trullifera*

79(77) Ascospores less than 15 µm long .... *Halosarpheia minuta*

79'(77) Ascospores more than 15 µm long 80

80(79') Ascospores 18-23 (-26) x 9-12 µm, appendages cap-like, then filamentous .... *Halosarpheia marina*

80'(79') Ascospores ellipsoidal, stain intensely around the mid-septum and at both apices, 21-31 x 8-11 µm, ... *Panorbis viscosus*

80"(79') Ascospores dufoid-ellipsoid, no staining as above, 20-33 x 7-11 µm.....

.... *Natantispora retorquens*

81(72') Ascospores with two types of appendages (polar hair-like and lateral sheath-like) 82

81'(72') Ascospores with only one type of appendages 84

82(81) Ascospores lacking hair-like appendages, sheath with many (8) folds, 22-29(-31) x 12-18 μm .... *Nimbospora octonae*

82'(81) Ascospores with lateral tuft of hair-like appendages 83

83(82') Ascospores with one lateral groop of appendages, 17-24 x 8-10 μm..... .... *Nimbospora effusa*

83'(82') Ascospores with 2 or more lateral groups of appendages, 18-27 x 8-12 μm .... *Nimbospora bipolaris*

84(81') Ascospores with polar or subpolar appendages 85

84'(81') Ascospores with an exosporic sheath 90

84"(81') Ascospores with polar and lateral appendages, polar appendage longer than equatorial appendages, ascospores 15-22 x 5-7 μm............................ .... *Ocostaspora apilongissima*

85(84) Ascospores with polar appendages formed by fragmentation of a sheath, spoon-shaped, delicate and fibrillar, ascospores 22-34 x 8-12 μm................ .... *Kochiella crispa*

85'(84) Ascospores with subpolar appendages 86

86(85') Ascospores with 2 subterminal appendages at each apex, these pairs arranged at right angles to one another, 20-35 x (6-)8-12 μm .... *Antennospora quadricornuta*

86'(85') Ascospores more than 2 subterminal appendages 87

87(86') Ascospores with 3-4 (rarely 5) sub-terminal spoon shaped appendages .... *Haiyanga salina*

87'(86') Ascospores with 3-4 thin, more or less rigid appendages at each apex 88

88 (87') Ascospores narrower than 6 μm wide, 16-18 x 3-4 μm .... *Arenariomyces parvulus*

88'(87') Ascospores more than 7 μm wide 89

89(88') Ascomata smooth, ascospores consistently with 3 appendages at each apex .... *Arenaromyces trifurcatus*

89'(88') Ascomata with spines on the peridium, ascospores with 3 or 4 appendages at each apex .... *Arenariomyces majusculus*

90(84') Ascospores with small equatorial and long polar appendages, ascospores (10-) 21-34 x 10-17 µm .... *Ceriosporopsis caduca*

90'(84') Ascospores lacking equatorial appendages 91

91(84') Ascospore appendages more than 750 µm long, ascospores (26-) 30-38 (-42) x 10-12 µm .... *Ceriosporopsis capillaceae*

91'(84') Ascospore appendages less than 750 µm long, 22.5-26 x 8-10 µm....... .... *Ceriosporopsis halima*

92(60') Ascospores hyaline or brown with polar and equatorial appendages....... .... Genus: *Corollospora* 94

92'(60') Ascospores hyaline, with only polar appendages 93

92"(60') Ascospores with sheath, lacking appendages, central cells dark brown, 27-42 x 6-24 µm .... *Carbosphaerella leptosphaerioides*

93(92') Ascospores 1-septate with an ephemeral drop of polar mucilage, becoming 2-4-septate on germination .... *Okeanomyces cucullatus*

93'(92') Ascospores 5-11-septste, 34-60 x 4-5 µm .... *Oceanities cincinnatula*

94(92) Ascospores light to dark brown .... *Corollospora cinnamomea*

94'(92) Ascospores hyaline 95

95(94') Ascospores 1-septate 96

95'(94') Ascospores with more than 1-septate 97

96(95) Ascospores wider than 8 µm, 22-33 x 8-10 µm .... *Corollospora maritima*

96'(95) Ascospores narrower than 8 µm, 26-45 x 3-7 µm, polar appendages 6.5-12 µm ....,,, *Corollospora gracilis*

97(95') Ascospores with 3-septate 98

97'(95') Ascospores with more than 3-septate 99

98(97) Ascospores wider than 7 µm, 25-34 x 7-12 µm, polar appendages 10-14 µm .... *Corollospora intermedia*

98'(97) Ascospores narrower than 7 µm, 35-57 x 3-7.5 µm, polar appendages 3-8 µm .... *Corollospora angusta*

99(97') Ascospores with polar appendages 103

99'(97') Ascospores without polar appendages 100

100(99') Ascospores 5-septate, 50-85 x 5-8 µm .... *Corollospora luteola*

100' (99') Ascospores predominately 7-septate 101

100"(99') Ascospores with more than 7-septate 102

101(100') Ascospores 60-108 x 13-26 µm .... *Corollospora colossa*

101'(100") Ascospores 52-122 x 7-16 µm .... *Corollospora pulchella*

102(101") Ascomata wall tuberculate, ascospores 7-11-septate, 65-98 x 8-12 µm .... *Corollospora pseudopulchella*

102'(101") Ascomata wall smooth, ascospores 13-septate, 73-120 x 5-8 µm.... .... *Corollospora filiformis*

103(99) Ascospores 3-5-septate 104

103'(99) Ascospores more than 5-septate .... *Corollospora indica*

104(103) Ascospores 39-63 x 10-19 µm, polar appendages 10-14 µm long...... .... *Corollospora lacera*

104'(103) Ascospores 38-59 x 8-10 µm, polar appendages 5-12 µm long......... .... *Corollospora quinqueseptata*

106(45) Ascospores 0-septate 107

106'(45) Ascospores 13-17-septate, 10-16x5-7 µm, yellow to brown, curved and hook shaped at one end .... *Pedumispora rhizophorae*

107(106) Ascospores lacking germ slits 108

107'(106) Ascospores with germ slits, 18-28x 7.5-13 µm, occurring on various mangrove wood .... *Halorosellinia oceanica*

108(107) Ascospores 6.5-8 x 1.2-16 µm, hyaline to pale yellow, occurring on mangrove wood .... *Eutypella naqsii*

108'(107) Ascospores 6-12 x 2.2-8 µm, olive-brown, occurring on mangrove wood .... *Eutypa bathurstensis*

108"(107) Ascospores 10-16 x 5-7 µm, light brown, occurring on intertidal wood of *Rhizophora apiculata* .... *Cryptosphaeria mangrovei*

109(44') Ascomata perthecial, non-stromatic, ostiolate, papillate with long necks, coriaceous to carbonaceous, paraphysate,asci variable, thin- walled, deliquescing early, ascospores variable, fusiform, ellipsoidal, filiform, generally with end-chambers containing mucilage, anamorphs generally dematious Hyphomycetes ... Order: Lulworhales 10

109'(44') Ascomata perithecial, immersed in the host tissue, subiculate or without subiculum, periphysate, paraphysate, asci clavate to fusiform, unitunicate, deliquescing, ascospores septate, hyaline, thick-walled, no apical chamber or sheaths ...... Order: Koralionastetales .... (No records from India).

109"(44') Ascomaypeat, perithecial, immersed in host tissue, often clypeate,ostiole perphysate, paraphyses wide thin-walled, asci cylindrical, persistent with inconspicuoua J-apical ring, ascosppores aseptate, anamorphs Coelomycetes ... Order: Phyllachorales (1 species from India) Ascospores 3-septate, not constricted at the septa, many oil globules per cell, no sheath or appendages, 26-31 x 6-8.5 mm .... *Marinosphaera mangrovei*

109"'(44') Ascomata perithecil, ostiolate, mmersed or superficial, often staining the substratum, paraphysate, paraphyses wide at the base tapering apically, asci cylindrical, unitunicate, ascospores septate, often filiform, hyaline, yellow to brown ... Order: Magnaporthales 114

110(109) Ascospores with apical chambers or appendages 111

110'(109) Ascomata on sea grasses, ascospores lacking appendages, 230-390 x 3-6 mm .... *Lindra thalassiae*

110"(109) Ascomata on wood, ascospores lacking appendages, 101-188 x 3.5-5.5 mm .... *Lindra hawaiiensis*

111(110) Ascospores 9-12-septate, 170-240x4 µm ... *Lulworthia lindroidea*

111'(110) Ascospores with no septa 112

112(111') Ascospores over 500 μm long, 500-756 x 3-5 μm .... *Lulworthia grandispora*

112'(111') Ascospores shorter than 500 μm 113

113(112') Ascomata dark brown, ascospores 230-320 x 3.6-5.7 μm .... *Lulworthia floridana*

113'(112') Ascomata purplish brown turning green in lactophenol, elongate, ascospores 185-265 x 3-4 μm .... *Lulworthia purpurea*

113"(112') Ascospores variable in dimensions .... *Lulworthia* sp.

114(109"') Ascomata perithecial 116

114'(109"') Ascomata cleisthothecial 115

115(114') Ascospores 12-15 x 4.5-5 μm .... *Dryosphaera tropicalis*

115'(114') Ascospores 8-12 x 3-5 μm .... *Dryosphaera navigans*

116 (114) Ascospores 0-septate, 20-28 x 7-10 μm ....*Rhizophila marina*

116'(114) Ascospores septate 117

117(116') Ascospores with (4-)5(-6)-septate, 19-26 x 6-8 μm .... *Saccardoella rhizophorae*

117'(116') Ascospores with 3-septa, 25-31 x 7.5-10 μm .... *Saccardoella marinosphora*

118(46') Ascomata stromatic, ascomata perithecial, long-nacked, interascal tissue absent, asci thick-walled, J- apical ring, ascospores vasried, no germ slit, anamorphs Coelomycetes .... Order: Diaporthals 121

118'(46') Ascomata not stromatic 119

119(118') Paraphyses filiform and septate, ascomata superficial or immersed, perithecial, ostolate or non-ostiolate, asci subglobose, clavate or cylindrical, thin-walled, ascospores unicellular, hyaline or pigmented, with germ slits, anamorphs Phialidic ... Order: Chaetosphaeriale..........(No record from India).

119'(118') Interascl tissue absent or thin-walled and inconspicuous 120

120(119') Ascomata perithecial, thick-walled, asci cylindrical or clavate, ascospores ith at least one dark cell with gern pore, appendaged, generally lacking anamorphs ... Order: Sordarials 122

120'(119') Ascomata perithecial, rarely cleistothecial, thin-walled, long nacled, ostiolar setae, asci small, evanescent, formed in chains, ascospores small, unicellular, appendaged or eith eccentric wall thickening, anamorphs Hyphomycetes .... Order: Ophiostomatales ... (No record from India). 121(118) Asci 64-spored, ascospores pale yellow- pale brown, verrucose, 5.6-11(-18) x 1.6-3.6 µm .... *Cryptovalsa mangrovei*

121'(118) Asci multi-spored, hyaline, 5-7 x 1-1.2 µm ....

.... *Cryptovalsa suaedicola*

122(120) Ascospores 1-septate ..... Genus *Zopfiella* 123

122'(120) Ascospores 3-septate 124

123(122) Ascospores with subapical germ pore, upper cell 16-22(-25) x 10-13(-15) mm, olivaceous to brown; basal cell 4-8(-9) x 3.5-7 mm, hyaline

.... *Zopfiella latipes*

123'(122) Ascospores with apical germ pore, upper cell 14-20 x 10-14 mm, olivaceous to dark broen, basal cell up to 4 mm diam., hyaline ....

.... *Zopfiella marina*

124(122') A deep water species, ascospores hyaline, with bipolar semiglobose cap-like polar appendages, 18-20 x 3.5-4 mm .... *Abyssomyces hydrozoicus*

124'(122') Ascospores versicolours 124

125(124') Asci 2-spored, 36-60 x 12-16.5 mm .... *Savoryella paucispora*

125'(124') Asci 8-spored 126

126(125') Ascospores appendaged, 26-35 x 11-16 mm ....

.... *Savoryella appendiculata*

126'(125') Ascospores not appendaged 127

127(1268') Ascospores 24-36 x 8-12 mm .... *Savoryella lignicola*

127'(126') Ascospores 33.5-46.5 x 7.5-12 m .... *Savoryella longispora*

Key for identification of Basidiomycota:

128(1) Basidioma sub-globose, puff ball-like, no stalk, basidiospores with appendages, 11-12 x 7-8 μm .... *Nia vibrissa*

128'(1) Basidioma cyphalloid, white, surface hairy smooth, basidiospres 8-10 x 8-9.5 μm .... *Halocyphina villosa*

128"(1) Basidioma yellowish to ochre yellow, cup-like, wall with surface hairs, brownish, dextrinoid, minutely encrusted, basidiospores 8- 12 x 6-10 μm .... *Calathella mangrovei*

Key for identification of Anamorphic Fungi:

129 (1") Conidia produced on hyphae .... **(Hyphomycetes)** 130

129'(1") Conidia produced in sporocarps .... **(Coelomycetes)** 155

MITOSPORIC FUNGI: Hyphomycetes

130 (129) Conidia hyaline 131

130'(129) Conidia coloured 134

131 (130) Conidia unicellular 132

131'(130) Conidia septate 133

132 (131) A deep water species, conidia produced in a superficial sporodochium on wood, 16-24 x 6.5-9 μm .... *Allescheriella bathygena*

132'(131) A intertidal species, conidia produced on subglobose conidiophores on hyphae, lateral or apical conidia, 2-5 μm .... *Botryophialophora marina*

133 (131') Conidia filamentous, 9-12(-19) -septate, end cells with cytoplasm, 150- 225(-312) x 2.5-4 μm .... *Anguillospora marina*

133'(131') Conidia elongate, obovoidal, ellipsoidal, clavate, 1-5-septate, predacious on nematodes in marine habitats, conidipophores up to 440 μm, conidia obovoid to pyriform, 17.5-28.8 x 10-16 μm, 10-15 conidia per conidiophore .... *Arthrobotrys oligospora*

134(130') Conidia coloured, unicellular 135

134'(130') Conidia coloured, septate 138

135 (134') Conidiphores penicilliate or on a terminal ampulla 136

135'(134') Conidiophores not as above, conidia acroblastosporic 137

136(135) Phialids borne on penicilliate conidiophores, conidia variously coloured .... *Penicillium* sp.

136'(135) Phialids on a terminal ampulla, conidia brown...................................

.... *Apergillus sydowii*

137(135') Conidia 6-13 µm diam., light to dark brown, littoral species..............

.... *Periconia prolifica*

137'(135') Conidia 16-20 µm diam., brown, deep sea species ....

.... *Periconia abyssa*

138(134') Conidia with transverse septa 139

138'(134') Conidia muriform 145

138"(134') Conidia helicoids 147

138"'(134) Conidia tetraradiate or staurospores, diverging upper arms 20-70 x 4-6(-8) µm, light brown (rarely 1 armed conidia occurs)...............................

.... *Clavatospora bulbosa*

139(138) Conidia with a prominent basal scar.......... *Cladosporim algorum*

139'(138) Conidia without a basal scar 140

140(139') Conidiophores in sporodochia, conidia obpyriform, 1-2-septate, 20-33.6 x 14.5-20.5 µm .... *Bactrodesmium linderi*

140'(139) Conidiophores not in sporodochia, conidial cells in straight row, constricted at the septa .... Genus: *Trichocladium* 141

141(140') Conidia 1-2(3)-septate 142

141'(140') Conidia more than 3-septate 143

142(141) Conidia 10-22(-38) x 8-18 µm, apical cell 8.5-15.5 x 7-12 µm, fuscous .... *Trichocladium alopallonellum*

142'(141) Conidia 6.5-14 x 3.5-9 μm, apical cell 6.5-13 x 4-9 μm, dark brown .... *Trichocladium melhae*

143(141') Conidia with 2-5(-6)-septate, 25-32 x 12-17 μm, slightly constricted at the septa, light brown .... *Trichocladium lignicola* 143'(141') Conidia 2-4-septate 144

144(145') Conidia 25-47 x 8-20 μm, subglobose, markedly constricted at the septa, reddish-brown ....*Trichocladium constrictum*

144'(145') Conidia (15-)20-34(-45) x (8-)10-24 μm, often sporodochial, less constricted at the septa, dark brown .... *Trichocladium achrasporum*

145(138') Conidia borne acropetally in chains .... *Alternaria* sp.

145'(138') Conidia single on conidiophores 146

146 (145') Conidia entire, clavate, 15-41x12-37 μm ......*Monodictys pelagica*

146'(145') Conidia irregular in shape and number of cells, 12-66 x 9-28(-36) μm, dark brown .... *Dictyosporium pelagicum*

147(138") Cells in conidia of equal width along the coil, conidia 4-27(-35)-septate, 1-5(-6) times coiled, conidial filament (4-)6-12 μm diam., fuscus or almost Black .... *Zalerion maritimum*

147'(138") Cells variable in width along the coil 148

148(147') Apical cell the largest in the coil .... Genus: *Cirrenalia* 149

148'(147') Apical cells variable in width 153

149(148) Conidia 6-12-septate, 1-1 &1/2 times ontorted, apical cell 10-20 μm diam., umber to reddish-brown .... *Cirrenalia tropicalis*

149'(148) Conidia 5-septate or less 150

150(149') Apical cell over 16 μm long 151

150'(149') Apical cell shorter than 16 μm 152

151(150) Apical cell hooked appearance, 16-23 μm, conidia 3-4-septate,, black to fuscous .... *Cirrenalia pygmea*

151'(150) Apical cell not hooked, 16-20 μm, 3-6-septate, fuscous to grey brown .... *Cirrenalia pseudomacrocephala*

151"(150) Apical cell not hooked, 11-22 μm, 2-4-septate, brown to dark brown .... *Cirrenalia fusca*

152(150') Apical cell 9-16 μm, 3-5-septate, pale brown ....

.... *Cirrenalia basiminuta*

152'(150') Apical cell 6-14(-17) μm, 2-6-septate, reddish-brown.....................

.... *Cirrenalia macrocephala*

153(148') Conidia spirally contorted into a ball of up to 100 cells, cells less than 10 μm wide, conidia 15-65 x 14-56 μm .... *Halenospora varia*

153'(148') Conidia initially a spiral/coil, cells over 10 μm wide, forming a ball of cells by division in several planes ... Genus: *Comulospora* 154

154(153') Conidia 52-91 x 40-71 μm, up to 20 cells, cells up to 30 μm wide .... *Cumulospora marina*

154'(153') Conidia 24-87 x 21-51 μm, up to 40 cells, cells up to 20 μm...........

.... *Cumulospora varia*

MITOSPORIC FUNGI: Coelomycetes

155(129') Conidioma eustromatic on wood, roots and seedlings of *Rhizophora* sp., conidia 3-6 x 1-1.5 μm .... *Cytospora rhizophorae*

155'(129') Conidioma pycnidial 156

156(155') Conidia hyaline 157

156'(155') Conidia coloured 159

157(156) Conidia produced endogenously in phialids, 2.5-4.5 x 1.5-2 μm.........

.... *Phialophorophoma litoralis*

157'(156) Conidia produced exogenously on conidiophores 158

158(159') On *Suaeda* sp., conidia 6-8 x 3-4 μm .... *Phoma suaedae*

158'(159') On *Rhizophora* sp., conidia 11-18 x 3-4 μm ....

.... *Phomopsis mangrovei*

158"(157') On other substrates .... *Phoma* and *Microphoma* sp.

159(158') Conidia 1-celled, 4-8 x 4-6 µm .... *Coniothyrium obiones*

159'(158') Conidia 1-septate, on Salicornia sp., olive to light brown, 4-8 x 4-6 µm .... *Ascochyta salicorniae*

159"(158') Conidia muriform ... Genus: *Camarosporium* 160

160(159") Conidia without sheath, 3-transverse-septate, 1-longi-septate, 10-20 x 7-3 µm ....*Camarosporium rumeguerii*

160'(159') Conidia with sheath, 5-transverse-septate, 1-longi-septate, 20-34 x 9-20 µm .... *Camarosporium palliatum*

Distribution and habitat range of marine fungi of India.

States and Union Territories:- DI: Diu Island, DA: Daman, GU: Gujarat, MS: Maharashtra, GO: Goa, KA: Karnataka, KE: Kerala, PM: Pondecherry (Mahe), LI: Lakshadeep Islands, TN: Tamil Nadu, AP: Andhra Pradesh, OR: Orissa, WB: West Bengal, AN: Andaman & Nicobar Islands, PN: Pondecherry. '+': Recorded, '-': Not-recorded, 'P': Present study.

SU: Substrates:- a: Intertidal Wood, b: Mangrove wood, c: Marine algae, d: Salt Marsh plants, e: Sea grasses, f: Animal substrates, g: Intertidal / Deep Sea sediments, h: Propagules in Sea foam samples,

S N	Name of the fungal species	SU	West Coast									East Coast					
	Phylum: Labyrinthulomycota		DI	DA	GJ	MS	GO	KA	KE	PM	LI	TN	AP	OR	WB	AN	PN
1	*Aplanochytrium minuta* (Watson & Raper) Leander & D. Porter	c e	-	-	-	-	+	-	-	-	+	-	-	-	-	-	-
2	*A. yorkensis* (Perkins) Leander & D. Porter	c f	-	-	-	-	+	-	-	-	+	-	-	-	-	-	-
3	*Aurantiochytrium mangrovei* (Raghukumar) R. Yokoyama & D. Honda	b c	-	-	-	-	+	-	-	-	-	-	-	-	-	-	-
4	*Labyrinthula* sp.	c e	-	-	+	-	+	+		-	+	+	-	-	-	-	-
5	*Schizochytrium aggregatum* S. Goldstein & Belsky	g	-	-	-	-	-	-	-	-	-	+	-	-	-	-	-
6	*Schizochytrium* sp.	c f	-	-	-	-	+	-	-	-	+	+	-	-	-	-	-
7	*Thraustochytrium aggregatum* A. Ulken	c	-	-	-	-	+	-	-	-	-	-	-	-	-	-	-
8	*T. gaertnerium* Bongiorni *et al.*	b	-	-	-	-	+	-	-	-	-	-	-	-	-	-	-

9	*T. motivum* S. Goldstein	b c e f	-	-	-	-	+	-	-	-	+	-	-	-	-	-	-
10	*T. multirudimentale* S. Goldstein	b	-	-	-	-	+	-	-	-	-	-	-	-	-	-	-
11	*T. striatum* J. Schneider	b	-	-	-	-	+	-	-	-	-	+	-	-	-	-	-
12	*Thraustochytrium* sp.	c f	-	-	-	-	+	+	-	-	+	+	-	-	-	-	-
13	*Ulkenia visurgensis* (A. Ulken) A. Gaertner	c f	-	-	-	-	+	-	-	-	+	+	-	-	-	-	-
14	*U. amoeboidea* (G. Bahnweg & F.K. Sparrow) A. Gaertner	c	-	-	-	-	+	-	-	-	-	-	-	-	-	-	-
	Phylum: Chytridiomycota																
15	*Chytridium polysiphoniae* Cohn	c	-	-	-	-	+	-	-	-	+	+	-	-	-	-	-
16	*Coenomyces* sp.	c	-	-	+	-	+	-	-	-	+	-	-	-	-	-	-
17	*Olpidium rostriferum* Tokunaga	c	-	-	+	-	+	-	-	-	+	+	-	-	-	-	-
18	*Olpidium sp.*	c	-	-	-	-	-	-	-	-	-	+	-	-	-	-	-
	Phylum: Oomycota																
19	*Lagenidium calllinectus* Couch	f	-	-	-	-	-	-	-	-	-	+	-	-	-	-	-
20	*Sirolpidium bryopsidis* (de Bruyne) H.E. Petersen	c	-	-	-	-	+	-	-	-	+	+	-	-	-	-	-
21	*Ostracoblabe implexa* E. Bornet & Flahault	f	-	-	-	-	-	-	-	-	+	-	-	-	-	-	-
22	*Pontisma lagenidioides* H.E. Petersen	c	-	-	+	-	+	-	-	-	-	+	-	-	-	-	-
	Phylum: Ascomycota																
23	*Abyssomyces hydrozoicus* Kohlm.	g	D	e	e	p		s	e	a		F	u	n	g	u	s
24	*Acrocordiopsis patilii* Borse & K.D. Hyde	b	-	-	-	+	-	-	-	-	-	-	+	-	+	+	-
25	*Aigialus grandis* Kohlm. & S. Schatz	a b	-	-	+	+	-	-	+	-	-	+	+	-	+	+	-
26	*A. mangrovei* Borse	a b	-	-	-	+	+	+	+	+	-	+	+	+	+	+	-
27	*A. parvus* S. Schatz & Kohlm.	a b	-	-	+	+	-	-	+	-	-	+	+	-	+	+	-
28	*A. rhizophorae* Borse	b	-	-	-	+	-	-	-	-	-	+		-	-	-	-
29	*Amylocarpus encephaloides* Curr.	a	-	-	-	-	-	+	-	-	-	-	-	-	-	-	-
30	*Aniptodera chesapeakensis* Shearer & M.A. Mill.	a b	-	-	-	+	+	+	+	+	-	+	+	+	+	+	-
31	*A. haispora* Vrijmoed, K.D. Hyde & E.B.G. Jones	a b	-	-	-	-	-	-	+	+	-	-	+	-	-	-	-
32	*A. intermedia* K.D. Hyde & Alias	b	-	-	-	-	-	-	-	-	-	-	-	-	+	-	-
33	*A. longispora* K.D. Hyde	a	-	-	-	-	+	+	+	-	-	-	-	-	+	-	-
34	*A. mangrovei* K.D. Hyde	a b	-	-	-	+	+	+	+	+	-	+	+	-	+	+	-
35	*A. salsuginosa* Nakagiri & Tad. Ito	a b	-	-	-	-	-	-	+	+	-	+	-	-	-	-	-
36	*Antennospora quadricornuta* (Cribb & J.W. Cribb) T.W. Johnson	a b	+	+	+	+	+	+	+	+	+	+	-	-	+	+	+
37	*Arenariomyces majusculus* Kohlm. & Volkm.-Kohlm.	a b h	-	-	-	+	-	+	+	+	+	+	-	-	+	-	+
38	*A. parvulus* Jorgen Koch	a b f h	-	-	+	+	+	+	+	+	+	-	-	+	+	-	+
39	*A. trifurcatus* Hohnk	a b h	-	-	+	+	+	+	+	+	+	+	-	+	+	-	-

40	*A. triseptatus* Kohlm.	a	-	-	-	-	-	-	+	-	-	-	-	-	-	-	-
41	*Anthostomella nypensis* K.D. Hyde *et al.*	a	-	-	-	-	-	-	+	-	-	-	-	-	-	-	-
42	*Ascocratera manglicola* Kohlm.	a b	-	-	-	+	-	+	+	-	+	+	+	-	+	+	-
43	*Astrosphaeriella striataspora* (K.D. Hyde) K.D. Hyde	a b	-	-	-	+	-	+	-	-	-	+	-	-	+	+	-
44	*Bathyascus avicenniae* Kohlm.	b	-	-	-	-	-	-	-	-	-	+	+	-	-	+	-
45	*B. grandisporus* K.D. Hyde & E.B.G. Jones	b	-	-	-	-	-	-	-	-	-	-	-	-	-	+	-
46	*B. mangrove* Ravikumar & B.P.R. Vittal	b	-	-	-	-	-	-	-	-	-	+	-	-	-	-	-
47	*B. tropicalis* Kohlm.	a b	-	-	-	+	-	-	+	+	+	+	+	-	+	-	-
48	*B. vermisporus* Kohlm.	g	D	e	e	p		S	e	a		F	u	n	g	u	s
49	*Belizeana tuberculata* Kohlm. & Volkm.-Kohlm.	b	-	-	-	-	-	-	-	-	-	+	-	-	-	+	-
50	*Biatriospora marina* K.D. Hyde & Borse	b	-	-	-	+	-	+	+	-	+	-	-	-	+	+	-
51	*Biconiosporella cornuculata* Schaumann	a	-	-	-	-	-	-	+	-	-	-	-	-	-	-	-
52	*Carbosphaerella leptosphaerioides* I. Schmidt	a	-	-	-	-	-	-	+	-	-	+	-	-	+	-	-
53	*Caryosporella rhizophorae* Kohlm.	a b h	-	-	-	-	+	+	+	-	+	-	-	-	-	+	-
54	*Ceriosporopsis caduca* E.B.G. Jones & Zainal	a b	-	-	-	+	-	-	-	-	-	+	-	-	-	-	-
55	*C. capillacea* Kohlm.	a	-	-	-	+	-	-	+	-	-	+	-	-	-	-	-
56	*C. halima* Linder	a b	-	+	+	+	+	+	+	-	-	+	-	-	+	-	-
57	*Corollospora angusta* Nakagiri & Tokura	a b f h	-	-	-	-	-	+	+	+	-	+	-	+	-	-	+
58	*C. cinnamomea* Jorgen Koch	a b f h	-	-	+	+	+	+	-	+	-	-	-	+	-	-	-
59	*C. colossa* Nakagiri & Tokura	a b h	-	-	-	+	+	+	+	+	-	+	-	+	+	-	+
60	*C. filiformis* Nakagiri	a h	-	-	-	+	-	+	+	-	-	+	-	+	+	-	-
61	*C. gracilis* Nakagiri & Tokura	a h	-	-	-	-	+	-	+	+	+	-	+	+	+	-	+
62	*C. indica* Prasannarai, Ananda & K.R. Sridhar	a f h	-	-	+	-	-	+	-	+	-	+	+	+	-	-	+
63	*C. intermedia* I. Schmidt	a f h	-	-	-	-	+	+	+	+	-	+	-	-	-	-	-
64	*C. luteola* Nakagiri & Tubaki	a	-	-	-	-	-	-	+	-	-	-	-	-	-	-	+
65	*C. maritima* Werderm.	a b f h	+	-	+	+	+	+	+	+	+	+	+	+	+	-	+
66	*C. pseudopulchella* Nakagiri & Tokura	b	-	-	-	-	-	-	+	-	-	+	-	-	-	-	-
67	*C. pulchella* Kohlm., I. Schmidt & N.B. Nair	a b h	-	+	+	+	+	+	+	+	-	+	+	+	+	-	+
68	*C. quinqueseptata* Nakagiri	a h	-	-	-	-	-	+	+	+	-	-	-	+	-	-	-
69	*Coronopapilla mangrovei* (K.D. Hyde) Kohlm. & Volkm.-Kohlm.	a b	-	-	-	-	-	+	-	-	-	-	-	-	-	+	-
70	*Cryptosphaeria mangrovei* K.D. Hyde	b	-	-	-	-	-	-	-	-	-	-	+	-	-	-	-
71	*Cryptovalsa mangrovei* Abdel-Wahab & Inderbitzin	b	+	-	-	+	-	-	-	-	-	-	-	-	-	-	-

72	*C. suaedicola* Spooner	b	-	-	-	-	-	-	-	-	-	-	+	-	-	-	-
73	*Dactylospora heliotrepha* (Kohlm. & E. Kohlm.) Hafellner	a b	+	+	+	+	+	+	+	+	+	+	+	+	+	+	-
74	*Didymella avicenniae* S.D.	b	-	-	-	+	-	+	-	-	-	+	-	-	-	-	-
	Patil & Borse																
75	*Didymospharia lignomaris* Strongman & Mill.	a b	-	-	-	-	-	+	+	-	+	-	-	-	-	-	-
76	*D. tropicalis* Kohlm. & Volkm.-Kohlm.	a b	-	-	-	+	-	+	+	+	-	+	-	-	+	-	+
77	*Etheirophora blepharospora* (Kohlm. & E. Kohlm.) Kohlm. & Volkm.-Kolhm.	b	-	-	-	-	-	-	-	-	-	-	-	-	-	+	-
78	*E. bijubata* Kohlm. & Volkm.-Kohlm.	a	-	-	-	-	-	-	+	-	-	-	-	-	-	-	-
79	*E. unijubata* Kohlm. & Volkm.-Kohlm.	b	-	-	-	-	-	+	-	-	-	-	-	-	-	-	-
80	*Eutypa bathurstensis* K.D. Hyde & Rappaz	b	-	-	+	+	+	+	+	-	-	+	+	-	+	-	-
81	*Eutypella naqsii* K.D. Hyde	b	-	-	-	-	-	-	-	-	-	-	-	-	+	-	-
82	*Haiyanga salina* (Meyers) K.L. Pang & E.B.G. Jones	a b	+	-	+	+	+	+	+	-	-	+	-	+	+	+	-
83	*Halographis runica* Kohlm. & Volkm.-Kohlm.	b	-	-	-	-	-	+	-	-	-	-	-	-	-	-	-
84	*Halomassarina thalassiae* (Kohlm. & Volkm.-Kohlm.) Suetrong *et al.*	a b	-	-	+	+	-	-	-	-	+	+	+	+	-	+	-
85	*Halorosselinia oceanica* (S. Schatz) Whalley *et al.*	a b	-	-	+	+	+	+	+	+	-	+	+	+	+	+	-
86	*Halosarpheia fibrosa* Kohlm.	a b	-	-	+	+	+	+	+	-	-	+	-	-	+	-	-
87	*H. marina* (Cribb & J.W. Cribb) Kohlm.	a b	-	-	-	+	-	+	+	+	+	+	+	-	-	-	-
88	*H. minuta* W.F. Leong	a b	-	-	-	+	-	-	+	+	-	+	+	-	-	+	-
89	*H. trullifera* (Kohlm.) E.B.G. Jones, S.T. Moss & Cuomo	b	-	-	-	-	-	-	-	-	-	+	-	-	-	-	-
90	*Halosphaeria appendiculata* Linder	a	-	-	-	-	-	-	+	-	-	-	-	-	-	-	-
91	*Heleococcum japonense* Tubaki	b	-	-	-	-	-	-	-	-	-	+	+	-	-	-	-
92	*Helicascus kanaloanus* Kohlm.	b	-	-	-	-	-	-	-	-	-	+	-	-	-	+	-
93	*Julella avicenniae* (Borse) K.D. Hyde	a b	+	-	+	+	+	+	+	-	-	+	+	+	+	+	-
94	*Kallichroma glabrum* (Kohlm.) Kohlm. & Volkm.-Kohlm.	b	-	-	+	+	-	-	-	-	-	-	-	+	-	+	-
95	*K. tethys* (Kohlm & E. Kohlm.) Kohlm. & Volkm.-Kohlm.	a b	-	-	+	+	+	+	+	-	+	+	+	-	+	+	+
96	*Kirschsteiniothelia maritima* (Linder) Hawksw.	b	-	-	-	-	-	-	+	-	-	-	+	-	-	-	-
97	*Kochiella crispa* (Kohlm.) Sakayaroj, *et al.*	a	-	-	-	-	-	-	-	-	-	-	-	+	-	-	-
98	*Lautospora igantean* K.D. Hyde & E.B.G. Jones	b	-	-	-	-	-	-	+	-	-	-	+	-	-	-	-
99	*Leptosphaeria australiensis* (Cribb & J.W. Cribb) G.C. Hughes	a b	-	-	+	+	+	+	+	+	+	+	+	+	+	+	-

100	*L. avicenniae* Kohlm. & E. Kohlm.	b	-	-	-	-	-	-	-	-	-	-	-	-	+	-	-
101	*L. pelagica* E.B.G. Jones	a b	-	-	-	-	+	+	-	-	-	-	-	-	-	-	-
102	*Lignincola laevis* Hohnk	a b	-	+	+	+	+	+	+	+	+	+	+	+	+	+	-
103	*L. tropica* Kohlm.	a b	-	-	-	-	-	+	+	-	-	+	+	-	-	-	-
104	*Lindra hawaiiensis* Kohlm. & Volkm.-Kohlm.	a b	-	-	-	-	-	-	+	-	-	-	-	-	-	-	-
105	*L. thalassiae* Orpurt, Meyers, Boral & Simins	a e	-	-	-	-	-	+	-	-	+	-	-	-	-	-	-
106	*Lineolata rhizophorae* (Kohlm. & E. Kohlm.) Kohlm. & Volkm.-Kohlm.	a b	-	-	+	+	-	+	+	-	-	+	+	-	+	+	-
107	*Lophiostoma armatispora* (K.D. Hyde *et al.*) Liew *et al.*	b	-	-	+	-	+	+	-	-	-	-	-	-	-	+	-
108	*Lulworthia floridana* Meyers	a b	-	-	-	+	-	-	+	-	-	-	-	-	-	-	-
109	*L. grandispora* Meyers	a b	+	+	+	+	+	+	+	+	+	+	+	-	+	+	+
110	*L. lindroidea* Kohlm.	b	-	-	-	-	-	+	+	-	+	-	-	-	-	-	-
111	*L. purpurea* (I.M. Wilson) T.W. Johnson	a b	-	-	-	+	-	-	-	-	-	-	-	-	-	-	-
112	*Lulwothia* sp.	a b	-	-	+	+	+	+	+	-	+	+	+	+	+	+	-
113	*Manglicola* sp.	b	-	-	-	-	-	-	-	-	-	-	+	-	-	-	-
114	*Marinosphaera mangrovei* K.D. Hyde	a b	-	-	-	-	-	-	+	+	+	+	+	-	+	+	-
115	*Morosphaeria velataspora* (K.D. Hyde & Borse) Suetrong *et al.*	a b	+	-	+	+	+	+	-	-	+	+	+	-	+	+	-
116	*Mycosphaerella pneumatophorae* Kohlm.	b	-	-	-	-	-	-	-	-	-	+	+	-	+	-	-
117	*Natantispora retorquens* (Shearer & J.L. Crane) J. Campb., *et al.*	a b	-	-	-	+	-	+	+	-	-	-	-	-	-	-	-
118	*Neomassariosphaeria typhicola* (Karst.) Yin *et al.*	b	-	-	-	-	-	-	-	-	-	-	+	-	-	-	-
119	*Neptunella longirostris* (Cribb & J.W. Cribb) K.L. Pang & E.B.G. Jones	a b	-	-	+	+	+	+	+	+	-	+	+	-	+	-	-
120	*Nimbospora bipolaris* K.D. Hyde & E.B.G. Jones	a	-	-	-	-	-	+	+	-	-	-	-	-	-	-	-
121	*N. effusa* Jorgen Koch	a	-	-	-	-	-	+	-	-	-	+	-	-	-	-	-
122	*N. octonae* Kohlm.	a	-	-	-	+	-	-	-	-	-	-	-	-	-	-	-
123	*Oceanitis cincinnatula* (Shearer & J.L. Crane) Dupont & E.B.G. Jones	a	-	-	-	+	+	+	+	-	-	-	-	+	-	-	-
124	*O. scuticella* Kohlm.	g	D	e	e	p	-	S	e	a	-	F	u	n	g	u	s
125	*O. viscidula* (Kohlm. & E. Kohlm.) Dupont & E.B.G. Jones	b	-	-	-	-	-	-	-	-	-	-	-	-	-	+	-
126	*Ocostaspora apilongissima* E.B.G. Jones, T.W. Johnson & S.T. Moss	a b	-	-	-	-	+	-	+	-	-	-	-	-	-	+	-
127	*Okeanomyces cucullatus* (Kohlm.) K.L. Pang & E.B.G. Jones	a b	-	-	-	+	-	-	+	-	-	+	-	-	-	-	-
128	*Ophiodeira monosemeia* Kohlm. & Volkm.-Kohlm.	b	-	-	-	-	-	-	-	-	-	-	+	-	-	-	-
129	*Panorbis viscosus* (I. Schmidt) J. Campb., *et al.*	a b	-	-	-	+	-	+	+	-	-	+	+	-	-	-	-
130	*Patellaria atrata* (Hedw.) Fr.	b	-	-	-	-	-	-	-	-	-	-	+	-	-	-	-
131	*Paraliomyces lentiferus* Kohlm.	a	-	-	-	-	-	-	-	-	-	+	-	-	-	-	-

132	*Passeriniella mangrovei* Maria & K.R. Sridhar	b	-	-	-	-	-	+	+	-	-	-	-	-	-	-	-
133	*P. savoryellopsis* K.D. Hyde & Mouzouras	b	-	-	-	-	-	-	-	-	-	-	-	-	-	+	-
134	*Payosphaeria minuta* W.F. Leong	a b	-	-	-	-	-	-	+	-	+	-	-	-	-	+	-
135	*Pedumispora rhizophorae* K.D. Hyde & E.B.G. Jones	b	-	-	-	-	-	-	-	-	-	-	+	-	-	-	-
136	*Phaeosphaeria halima* (T.W. Johnson) Shoemaker & Babcock	b	-	-	-	-	-	-	+	-	-	-	-	-	-	-	-
137	*Phaeosphaeria orae-maris* (Linder) Khashna. & Shearer	a b	-	-	-	+	+	+	-	-	-	+	-	-	-	-	-
138	*Quintaria lignatilis* (Kohlm.) Kohlm. & Volkm.-Kohlm.	a b	-	-	+	+	-	-	-	-	+	+	+	-	+	+	-
139	*Rhizophila marina* K.D. Hyde & E.B.G. Jones	b	-	-	-	+	+	+	+	-	-	+	+	-	-	+	-
140	*Remispora quadric-remis* (Hohnk) Kohlm.	a	-	-	-	-	-	-	+	-	-	-	-	-	-	-	-
141	*Rimora mangrovei* (Kohlm. & B.P.R. Vittal) Kohlm., *et al.*	a b	-	-	+	+	+	+	+	-	+	+	+	+	+	+	+
142	*Saagaromyces abonnis* (Kohlm.) K.L. Pang & E.B.G. Jones	a b	-	-	+	+	-	-	+	-	+	-	+	-	+	+	-
143	*S. glitra* (J.L. Crane &.Shearer) K.L. Pang & E.B.G. Jones	a b	-	-	-	-	+	-	+	-	-	+	+	-	-	+	-
144	*S. ratnagiriensis* (S.D. Patil & Borse) K.L. Pang & E.B.G. Jones	a b	-	-	+	+	+	+	+	-	-	+	+	-	+	+	+
145	*Saccardoella marinospora* K.D. Hyde	b	-	-	-	+	-	-	-	-	-	+	+	-	+	+	-
146	*S. rhizophorae* K.D. Hyde	b	-	-	-	-	-	-	-	-	-	+	+	-	-	+	-
147	*Salsuginea ramicola* K.D. Hyde	a b	-	+	+	+	+	+	+	+	-	+	-	+	+	+	-
148	*Savoryella appendiculata* K.D. Hyde & E.B.G. Jones	a h	-	-	-	-	-	-	+	-	-	+	+	+	+	-	-
149	*S. lignicola* E.B.G. Jones & R.A. Eaton	a b	-	+	+	+	+	+	+	+	+	+	+	+	+	+	+
150	*S. longispora* E.B.G. Jones & K.D. Hyde	b	-	-	-	-	-	+	+	-	-	-	-	+	-	-	-
151	*S. paucispora* (Cribb & J.W. Cribb) Jorgen Koch	a b	-	-	-	-	+	+	+	+	+	+	+	-	-	+	+
152	*Swampomyces armeniacus* Kohlm. & Volkm.- Kohlm.	a b	-	-	-	-	-	+	-	-	-	+	-	-	-	+	-
153	*S. triseptatus* K.D. Hyde & Nakagiri	a b	-	-	+	-	-	-	-	-	-	-	-	-	+	-	-
154	*Thalossogena sphaerica* Kohlm. & Volkm.- Kohlm.	b	-	-	-	-	-	-	-	-	-	-	-	-	+	-	-
155	*Tirispora unicaudata* E.B.G. Jones & Vrijmoed	a b	-	-	-	+	+	+	+	-	-	+	-	-	-	-	-
156	*T. mandoviana* V.V. Sarma & K.D. Hyde	b	-	-	-	-	+	+	-	-	-	+	-	-	-	-	-
157	*Torpedospora radiata* Meyers	a b h	-	-	+	+	+	+	+	+	+	+	-	+	+	-	-
158	*Trematosphaeria lineolatispora* K.D. Hyde	a b	-	-	-	-	-	-	+	-	-	-	-	+	+	-	-
159	*T. mangrovei* Kohlm.	a	-	-	-	-	-	+	-	-	-	-	-	-	-	-	-
160	*Verruculinia enalia* (Kohlm.) Kohlm. & Volkm.-Kohlm.	a b	+	+	+	+	+	+	+	+	+	+	+	+	+	+	+
161	*Zopfiella latipes* (Lundq.) Maloch & Cain	a b	-	-	+	+	-	+	+	+	-	+	+	+	+	-	-
162	*Z. marina* Furuya & Udagawa	b	-	-	-	-	-	-	-	-	-	-	+	-	-	-	-
	Phylum: Basidiomycota																
163	*Calathella mangrovei* E.B.G. Jones & Agerer	a b	-	-	-	+	-	+	-	-	-	-	-	-	-	-	-

164	*Halocyphina villosa* Kohlm. & E. Kohlm.	a b	-	-	+	+	+	+	+	+	+	+	+	+	+	+	+
165	*Nia vibrissa* R.T. Moore & Meyers	a b	-	-	+	+	+	+	+	-	-	+	+	+	+	-	-
	Mitosporic Fungi:			**H**	**y**	**p**	**h**	**o**	**m**	**y**	**c**	**e**	**t**	**e**	**s**		
166	*Alleschieriella bathygena* Kohlm.	g	D	e	e	p		S	e	a		F	u	n	g	u	s
167	*Alternaria* sp.	a b	-	-	-	-	-	-	+	+	-	-	-	+	+	-	-
168	*Anguillospora marina* Nakagiri & Tubaki	b	-	-	-	-	-	+	-	-	-	-	-	-	-	-	-
169	*Arthobotrys oligospora* Fresenius	a	-	-	-	-	-	+	-	-	-	-	-	-	-	-	-
170	*Aspergillus sydowii* (Bainier & Sartory) Thom & Church	b g	-	-	-	-	-	-	-	-	-	+	-	-	+	-	+
171	*Aspergillus* sp.	b	-	-	-	+	-	+	+	-	-	-	-	-	+	-	-
172	*Bactrodesmium linderii* (J.L. Crane & Shearer) Palm & Stewart	a b	-	-	-	+	+	+	+	-	-	-	+	-	+	-	-
173	*Blastobotrys* sp.	a	-	-	-	-	-	+	-	-	-	-	-	-	-	-	-
174	*Botryophialophora marina* Linder	b	-	-	-	-	-	-	-	-	-	-	-	-	+	-	-
175	*Cirrenalia basiminuta* Raghukumar & Zainal	a b	-	-	+	+	+	+	+	-	+	+	+	+	-	-	-
176	*C. macrocephala* (Kohlm.) Meyers & R.T. Moore	a b	-	-	+	+	-	+	+	-	-	+	-	-	-	-	+
177	*C. pseudomacrocephala* Kohlm.	a b	-	-	-	+	+	+	-	-	+	+	-	+	-	+	+
178	*Cladosporium algarum* Cooke & Massee	a	-	-	-	-	-	+	+	-	-	+	-	-	-	-	+
179	*Clavatospora bulbosa* (Anastasiou) Nakagiri & Tubaki	a b f h	-	-	+	+	+	+	+	+	+	+	+	+	-	+	-
180	*Cummulospora marina* I. Schmidt	a b	-	-	-	-	-	+	+	-	-	+	-	-	+	-	-
181	*Dictyosporium pelagicum* (Linder) G.C. Hughes	a b	-	-	-	+	+	+	+	-	-	+	-	+	-	-	+
182	*Fusarium* sp.	b c e	-	-	-	-	-	+	+	-	-	+	-	-	-	-	-
183	*Halazoon fuscus* (I. Schmidt) Abdel-Wahab & K.L. Pang	b	-	-	-	-	-	-	+	+	-	-	-	-	-	-	-
184	*Halenospora varia* (Anastasiou) Jones	a b h	-	+	+	+	+	+	+	+	+	+	+	+	+	+	+
185	*Hydea pygmea* (Kohlm.) K.L. Pang & E.B.G. Jones	a b	-	-	+	+	+	+	+	+	+	+	+	+	+	+	+
186	*Matsusporium tropicale* (Kohlm.) E.B.G. Jones & K.L. Pang	a b	-	-	-	+	+	+	+	-	+	+	+	+	+	+	+
187	*Monodictys pelagica* (T.W. Johnson) E.B.G. Jones	a b h	-	-	-	+	+	+	-	-	+	+	-	+	-	+	-
188	*Paecilomyces ramosus* Samson & Evans	a	-	-	-	-	-	+	-	-	-	-	-	-	-	-	-
189	*Periconia abyssa* Kohlm.	g	D	e	e	p		S	e	a		F	u	n	g	u	s
190	*P. prolifica* Anastasiou	a b	+	+	+	+	+	+	+	+	+	+	+	+	+	+	+
191	*Scolecobasidium* sp.	f	-	-	-	-	-	-	-	-	+	-	-	-	-	+	-
192	*Trichocladium achrasporum* (Meyers and R.T.Moore) Dixon	a b	-	-	+	+	+	+	+	+	+	+	+	+	+	+	-
193	*T. alopallonellum* (R.T. Moore & Meyers) Kohlm. & V.-Kohlm.	a b f	-	-	+	+	+	+	+	+	+	+	+	+	+	+	-
194	*T. constrictum* I. Schmidt	b	-	-	-	-	-	-	+	-	-	-	-	+	+	-	-
195	*T. lignicola* I. Schmidt	a b	-	-	+	-	-	-	-	-	-	-	-	-	-	-	-
196	*T. melhae* E.B.G. Jones, Abdel-Wahab & Vrijmode	a b	-	-	-	-	-	+	-	-	-	-	+	+	+	-	-
197	*Zalerion maritium* (Linder) Anastasiou	a b f h	+	+	+	+	+	+	+	+	+	+	-	+	+	-	-

	Mitosporic Fungi:			C	o	e	l	o	m	y	c	e	t	e	s		
198	*Ascochyta salicorniae* P. Magnus	d	-	-	+	+	-	-	-	-	-	-	-	-	-	-	-
199	*Camarosporium palliatum* Kohlm. & E. Kohlm.	a b	-	-	-	+	-	-	-	-	-	-	-	-	-	-	-
200	*C. roumeguerii* Sacc.	b d	-	-	+	+	-	-	-	-	-	+	+	-	+	+	-
201	*Coniothyrium obiones* Japp	b	-	-	-	-	-	-	-	-	-	-	-	+	-	-	-
202	*Cytospora rhozophorae* Kohlm. & E. Kohlm.	b	-	-	+	+	+	+	+	-	-	+	+	-	-	+	-
203	*Macrophoma* sp.	b	-	-	-	-	-	-	-	-	-	-	-	+	-	-	-
204	*Phialophorophoma litoralis* Linder	a	-	-	-	-	-	-	+	-	-	+	-	-	-	-	-
205	*Phoma suaedae* Japp	d	-	-	-	-	-	-	-	-	-	-	-	-	+	-	-
206	*Phoma* sp.	a b	-	-	+	+	-	+	+	+	-	+	+	+	+	+	+
207	*Phomopsis mangrovei* K.D. Hyde	b	-	-	+	+	-	-	-	-	-	-	+	-	+	+	-
208	*Robillarda rhizophorae* Kohlm.	a	-	-	-	-	-	-	+	-	-	-	-	-	-	-	-

Literature cited

Abdel-Wahab, M.A. (2011a) Lignicoulous marine fungi from Yokosuka, Japan. *Bot. Mar.*, **54:** 209-221.

Abdel-Wahab, M.A. (2011b) Marine fungi from Sarishima Island, Japan with a phylogenetic evaluation of the genus *Naufragella*. *Mycotaxon*, **115:** 443-456.

Abdel-Wahab, M.A. & Jones, E.B.G. (2000) Three new marine Ascomycetes from driftwood in Australian sand dunes. *Mycoscience,* **41:** 379-388.

Abdel-Wahab, M.A. & Nagahama, T. (2011a) *Halosarpheia japonica* sp. nov. (Halosphaerials, Ascomycota) from marine habitats in Japan. *Mycol. Progress*, **9:** DOI 10.1007/s11557-010-0731-0

Abdel-Wahab, M.A. & Nagahama, T. (2011b) *Gesasha* (Halosphaerials, Ascomycota) a new genus with three new species from Gasashi mangroves in Japan. *Nova Hedwigia*, **92:** 497-512.

Abdel-Wahab, M.A., Nagahama, T. & Abdel-Aziz, F.A. (2008) Two new *Corollospora* species and one anamorph based on morphological and molecular data. *Mycoscience*, **50:** 147-155.

Abdel-Wahab, M.A., Pang, KL, Nagahama, T., Abdel-Aziz, F.A. & Jones, E.B.G. (2010) Phylogetic evaluation of anamorphic species of *Cirrenalia* and *Cumulospora* with the description of eight new genera and four new species. *Mycol. Progress*, **9:** 537-558.

Abraham, S.P., Hoang, T.D., Alam, M. & Jones, E.B.G. (1994) Chemistry of the cytotaxic principals of the marine fungus *Lignincola laevis*. *Pure Appl. Chem.*, **66:** 2391-2394.

Agrawal, K.C. (1998). "*Biodiversity*", Agrobot. Publisher, Bikaner, pp. 150.

Ahearn, D.G. & Crow, S.A. (1986) Fungi and hydrocarbons in the marine environment. In: *The biology of marine fungi* (ed. Moss, S.T.), Cambridge Univ. Press, pp. 11-18.

Ahearn, D.G. & Roth, F.J. & Meyers, S.P. (1968) Ecology and characterization of yeasts from aquatic regions of South Florida. *Mar. Biol.*, **1:** 291-308.

Alam, M., Jones, E.B.G., Hossain, M. & Bilayet, H.D. (1996) Isolation and structure of isoculmorin from the marine fungus *Kallichroma tethys*. *J. Nat. Prod.*, **59:** 454-456.

Alderman, D.J. & Jones, E.B.G. (1967) Shell disease of *Ostrea edulis* L. *Nature*, **216:** 797-798.

Alderman, D.J. & Jones, E.B.G. (1971) Shell disease of oyster. *Fishery Invert.* Ser. II, **26:** 1-16.

Aleem, A.A. (1980) Distribution and ecology of marine fungi in Sierra Leone (Tropical west Africa). *Bot. Mar.,* **23:** 279-688.

Alias, S.A. & Jones, E.B.G. (2000a) Colonization of mangrove wood by marine fungi at Kuala Selangor mangrove stand, Malaysia. *Fungal Diversity*, **5:** 9-12.

Alias, S.A. & Jones, E.B.G. (2000b) Vertical distribution of marine fungi on *Rhizophora apiculata* at Morib mangroves Selangor mangrove stand, malaysia. *Mycoscience*, **41:** 431-436.

Alias, S. A. & Jones, E. B. G. (2009) *"Marine fungi from mangroves of Malaysia",* IOES monograph series 8, Institute of Ocean and Earth Scinces (IOES), University of Malaya, Kuala Lumpur. pp. 108.

Alias, S.A., Zianuddin, N. & Jones, E.B.G. (2010) Biodiversity of marine fungi in Malaysian mangroves. *Bot. Mar.*, **53:** 545-554.

Alker, A.P., Smith, W. & Kim, K. (2001) Characterization of *Aspergillus sydowii* Thom & Church, a fungus pathogen of Caribbean sea fan corals. *Hydrobiol.*, **460:** 105-111.

Alongi, G., Catra, M. & Cormaci, M. (1999) First record of *Haloguignardia cystoseira* (Ascomycota) parasitic on *Cystoseira elegans* (Fucophyceae) from the Mediteranean Sea. *Bot. Mar.*, **42:** 33-36.

Al-Saadoon, A.H. (2006) A new arenicolous species of *Corollospora* from Iraq. *Marsh Bull.*, **2:** 134-139.

Alva, P., McKenzie, E.H.C., Pointing, S.B., Pena-Murala, R. & Hyde, K.D. (2002) Do seagrasses harbour endophytes? In: *Fungi in marine environment* (ed. Hyde, K.D.), *Fungal Diversity*, **7:** 167-178.

Amon, J P. (1986) Growth of marine fungi at ambient nutrient levels. In: *The biology of marine fungi* (ed. Moss, S.T.), Cambridge Univ. Press, pp. 70-80.

Ananda, K. & Sridhar, K.R. (2001a) *Aniptodera indica*, a new species of mangrove inhabiting Ascomycete from West Coast of India. *Journal of Environmental Biology*, **22:** 283-286.

Ananda, K. & Sridhar, K.R. (2001b) Mycoflora on dead animal substrates of mangrove habitats of Karnataka coast, India. *Sri Lanka J. Aquat. Sci.*, **6:** 85-93.

Ananda, K. & Sridhar, K.R. (2002) Diversity of endophytic fungi in the roots of mangrove species on the West Coast of India. *Can. J. Micorbiol.*, **48:** 871-878.

Ananda, K. & Sridhar, K.R. (2003) Filamentous fungal Assemblage of Two Island Mangroves of India. In: *Prospects and Problems of Environment* (eds. Madhyastha, M.V. et al.,), Daya Publication House, Delhi, pp. **4:** 35-44.

Ananda, K. & Sridhar, K.R. (2004) Diversity of filamentous fungi on decomposing leaf and woody litter of mangrove forest in Southwest coast of India. *Current Science,* **10:** 1431-1437.

Ananda, K., Prasannarai, K. & Sridhar, K.R. (1998) Occurrence of higher marine fungi on marine animal substrates of some beaches along the west coast of India. *Indian Journal of Marine Sciences,* **27:** 233-236.

Anastasiou, C.J. (1961) Fungi from salt lakes. I: A new species of *Clavariopsis*. *Mycologia*, **53:** 11-16.

Anastasiou, C.J. (1963) Fungi from salt lakes. II: Ascomycetes and fungi imperfecti from the Saltan Sea. *Nava Hedwigia,* **6:** 243-276.

Aptroot, A. & van den Boom, P.P.G. (1995) *Strigula lateralis* sp. nov. with notes on the genus *Julella* (Ascomycetes). *Mycotaxon*, **56:** 1-8.

Bahnweg, G. & Sparrow, F.K. (1974) Occurrence, distribution and kinds of zoosporic fungi in subantarctic and Antarctic waters. *Veroff. Inst. Meeresforsch Bremarh.*, **5:** 149-157.

Bajpai, P., Bajpai, P.K. & Ward, O.P. (1991) Production of docosaheaxaenoic acid by *Thraustochytrium aureum. Appl. Microbiol. Biotechnol.*, **35:** 706-710.

Barata, M. (2002) Fungi on the halophyte *Spartina maritima* in salt marshes. In: *Fungi in marine environment* (ed. Hyde, K.D.), Fungal Diversity Res. Ser., **7:** 179-193.

Barghoorn, E.S. & Linder, D.H. (1944) Marine fungi: their taxonomy and biology. *Farlowia,* **1:** 395-467.

Barr, M.E. (1985) On *Julella*, *Delacourea*, and *Decaisnella*, three dictyosporous genera described by J.H. Fabre. *Sydowia*, **38:** 11-19.

Basitulo, A., Moslem, M.A. & Hyde, K.D. (2010) occurrence and distribution of marine fungi in a mangrove forest on Siargao Island, Philippines. *Bot. Mar.*, **53:** 535-543.

Bass, D.A., Howe, A., Brown, N., Barton, H., Demidova, M., Michelle, H., Li, L., Watkinson, S.C.C., Willcock, S. & Richards, T.A. (2007) Yeasts from dominate fungal diversity in the deep oceans. *Proc. R. Soc. B*, **274:** 3069-3077.

Becker, G. & Kohlmeyer, J. (1958 a) Holzzerstarung durch meerespilze in Indien and ihre besondere, Bedeutung a fur fischere fathrzeuge. *Arch. Fischere wiss,* **9:** 29-40.

Becker, G. & Kohlmeyer, J. (1958b) Deterioration of wood by marine fungi in India and its special significance for fishing crafts. *J. Timber Dryer's Presserv. Assoc. India,* **4:** 1-10.

Beena, K.R., Ananda, K. & Sridhar, K.R. (2000a) Fungal endophytes of three sand dune plant species of west coast of India. *Sydowia*, **52:** 1-9.

Beena, K.R., Raviraja, N.S. & Sridhar, K.R. (1997) Association of arbuscular mycorrhizal fungi with *Launaea sarmentosa* on maritime sand dunes of west coast of India. *Kavaka*, **25:** 53-60.

Beena, K.R., Raviraja, N S., Arun, A.B. & Sridhar, K.R. (2000b) Diversity of arbuscular mycorrhizal fungi on the coastal sand dunes of the west coast of India. *Curr. Sci.*, **79:** 1459-1465.

Bhat, D.J. (2010) "*Fascinating microfungi (Hyphomycetes) of Western Ghats - India*", Broadway Book Centre, Pangim, Goa. pp. 1-221.

Bilgrami K.S. (1995) "*Taxonomy and Biodiversity*". CBS Publishers and Distributors, Delhi.

Bilgrami, K.S., Jamaluddin, S. & Rizwi, M.A. (1979) "*Fungi of India*", Part I (List and References). Today and Tomorrow's Printers and Publishers, New Delhi. pp. 1- 467.

Bilgrami, K.S., Jamaluddin, S. & Rizwi, M.A. (1981) "*Fungi of India*" Part II. (Host index and Addenda), Today and Tomorrow's Printers and Publishers, New Delhi. pp. 268.

Bilgrami. K.S., Jamaluddin, S. & Rizwi M.A. (1991) "*Fungi of India – List and References*". Today and Tomorrow's Printers and Publishers, New Delhi. pp. 1-798.

Blackwell, M. (2011) The Fungi: 1, 2, 3 ... 5.1 million species? *Am. J. Bot.,* **98:** 426-438.

Bongiorni, G., Jain, R., Raghukumar, S. & Aggarwal, R.K. (2005) *Thraustochytrium gaertnerium* sp. nov.: a new thraustochytrid stramenopilan protest from mangroves of Goa, India. *Protist*, **156:** 303-315.

Boonmee, S., Ko, T.W.K., Chukeatirote, Chen, H., Cai, L., McKenzie, E.H.C., Jones, E.B.G., Hassan, B.A., & Hyde, K.D. (2011) Tow new *Kirschsteiniothelia* species with a *Dendryphiopsis* anamorph cluster in Kirschsteiniotheliaceae fam. nov. (in press).

Booth, T. (1969) Marine fungi from British Columbia: monocentric Chytrids and Chytridiaceous species from coastal and interior halomorphic soils. *Syesis*, **2:** 141-161.

Booth, T. (1971) Ecotypic responses of Chytrids and Chytridiaceous species to various salinity and temperature combinations. *Can. J. Bot.*, **49:** 1757-1767.

Booth, T. & Kenkel, N. (1986) Ecological studies of lignicolous marine fungi : A distribution model based on cordination and classification In: *The Biology of marine fungi* (ed. Moss, S.T.), Cambridge University Press, Cambridge, U.K. pp. 297 – 310.

Borse, B.D. (1983) *Marine fungi from Maharashtra*. Unpublished Ph. D. Thesis, University of Poona.

Borse, B.D. (1984) Marine fungi from India - I. *Ind. Bot. Reptr.*, **3:**156-157.

Borse, B.D. (1985a) Marine fungi from India - II. *Curr. Sci.*, **17:** 881-882.

Borse, B.D. (1985b) Two new marine Loculoascomycetes from Maharashtra, India. Abstract, In: *"The 4th International Marine Mycology Symposium"*, Portsmouth, England, United Kingdom, p.26.

Borse, B.D. (1987a) Marine fungi from India - IV. *Trans. Mycol. Soc. Japan*, **28:** 55-61.

Borse, B.D. (1987b) Marine fungi from India-V. *Curr. Sci.*, **21:** 1109-1111.

Borse, B.D. (1987c) New species of *Aigialus* from India. *Trans. Br. Mycol. Soc.*, **88:** 424-426.

Borse, B.D. (1987d) Marine fungi from India-VI. *Ind. Bot. Reptr.*, **6:** 108-110.

Borse, B.D. (1987e). Marine fungi from India - VII. *Indian Bot. Reptr.*, **6:** 62-64.

Borse, B.D. (1988). Frequencey of occurrence of marine fungi from Maharashtra coast, India. *Ind. J. Mar. Sci*,, **17:** 165-167.

Borse, B.D. (2000a) Marine fungi from India -X. *Savoryella* Jones and Eaton (Ascomycotina). In: *Ecology of fungi*. (eds. Bhat, D.J. and Raghukumar, S.), Goa University Press, Goa. pp. 163-165.

Borse, B.D. (2000b) Marine fungi from Maharashtra (India) - V. A check list. *J. Phytol. Res.*, **13:** 123- 128.

Borse, B.D. (2002) Marine fungi from India - XI: A check list . *J. Indian Bot. Soc.,* **81:** 83-92.

Borse, B.D. (2011) Biodiversity of Marine and Freshwater fungi from India. Abstract, In: *"Asian Mycological Congress 2011 & the 12th International Marine and Freshwater Mycology Symposium"*, Incheon, Korea, p. 451.

Borse, B.D. & Borse, K. N. (2001) New reports of marine Ascomycetes from Orissa, India. *Geobios,* **28:** 62-64.

Borse, B.D. & Hyde, K.D. (1989) Marine fungi from India- III. *Acrocordiopsis patilii* gen. *et* sp. nov., from mangrove wood. *Mycotaxon,* **34:** 535-540.

Borse, B.D. & Pawar, N.S. (2001) *Carbosphaerella* and *Dryosphaera*: Two new generic records of marine Ascomycetes from West Bengal (India). *Geobios,* **28:** 117-120.

Borse B.D. & Pawar, N.B. (2005) Marine fungi from Mahe (Pondicherry), India. *J. Adv. Sci. & Technology*, **8:** 43-48.

Borse, B.D. &. Shrivastava, A.D. (1988) Marine fungi from India - VIII. *Indian Bot. Reptr.,* **7:** 26-30.

Borse, B.D. & Shrivastava, A.D. (1994) Marine fungi from India - IX: Loculoascomycetes. *Indian Bot. Reptr.*, **13:** 1-3.

Borse, B.D. & Tuwar, A.R. (2006) Marine fungi from Goa coast (India) - I. *Bioinfolet*, **3:** 251-264.

Borse, B.D., Borse, K.N. & Pawar, N.S. (2005a) Marine fungi from Maharashtra: Diversity and Taxonomy. In: *Fungi: Diversity and Biotechnology* (eds. Rai M.K. and Deshmukh, S.K.), Scientific Publishers (India), Jodhpur. pp. 63-92.

Borse, B.D., Borse, K.N. & Pawar, N.S. (2012) Marine fungi from India-XII: A revised checklist.Ind. J. Mar. Sci., 41:(in press)

Borse, B.D., Kelkar, D.J. & Patil, A.C. (2000a) Frequencey of occurrence of marine fungi from Pirotan island (Gujarat), India. *Geobios.,* **27:** 145-148.

Borse, B.D., Nandan, S.N. & Shinde, D.N. (1999a) Higher marine fungi from Goa Coast (India). *J, Adv. Sci. Tech.,* **2:** 52-55.

Borse, B.D., Patil, R.V. & Kelkar, D.J. (1999b) Marine fungi from Div Island (India). *J. Adv. Sci. Tech.,* **11:** 1-8.

Borse, B.D., Patil, K.B., Patil, R.V. & Kelkar, D.J. (2000b) Marine fungi in foam, intertidal wood and dead *Avicennia marina* wood from Daman Coast, India. *Geobios,* **27:** 42-44.

Borse, B.D., Pawar, N.S. & Borse, K.N. (2001a) Marine fungi from Sundarbans (India) - I. sea foam spora. *J. Ind. Bot. Soc.,* **80:** 237-240.

Borse, B.D., Pawar, N.S., & Borse, K.N. (2002a) Marine fungi from Sunderbans (India) - II. The Genus *Arenariomyes* (Halosphaeriaceae). *J. Basic Appl. Mycol.,* **1:** 8-10.

Borse, B.D., Ramesh, Ch. & Shrivastava, A.D. (1988) Marine fungi from Maharashtra (India). *Indian Bot. Reptr.*, **7:** 18-25.

Borse, K.N. & Borse, B.D. (2005) Marine fungi from Orissa (India) III: Mitosporic fungi. In: *Frontiers in Plant Sciences* (eds. Mukherji, K.G. et al.), I.K. International pvt. Ltd., New Dehli. pp. 35-46.

Borse, K.N. Pawar, N.S. & Borse, B.D. (2002b) Marine fungi from Orissa (India) - I. The genus *Corollospora* (Ascomycotina). *Geobios,* **29:** 258-263.

Borse, K.N., Pawar, N.S. & Borse, B.D. (2001b) Marine fungi from Orissa (India) - II. Arenicolous group. *J. Adv. Sci. Tech.,* **4:** 17-22.

Borse, K.N., Pawar, N.B., Pawar, N.S. & Borse, B.D. (2005b) Marine fungi from Orissa (India) - IV. In: *Plant Diversity and Biotechnology*, (eds. Nandan et al.,), Pub. SSVPS's Sci. College, Dhule, Maharashtra. pp 102-106.

Borse, K.N., Pawar, N.S. & Borse, B.D. (2009) Marine fungi from Orissa (India) - VI. In: *Biodiversity, Sustainable Development and Human Welfare*, (eds. Nandan et al.,), Pub. SSVPS's Sci. College, Dhule, Maharashtra. pp 35-41.

Bower, S.M. (1987) *Labyrinthuloides haliotidis* (Protozoa: Labyrinthomorpha), a parasite of juvenile abalone in a British Columbia mariculture facility. *Can. J. Zool.*, **65:** 1996-2007.

Bowles, R.D., Hunt, A.E., Bremer, G.B., Duchars, M.G. & Eaton, R.A. (1999) Long-chain n-3 polyunsaturated fatty acid production by member of the marine protistan group the thraustochytrids; screening of isolates and optimization of docosahexaenoic acid production. *J. Biotecnol.*, **70:** 193-202.

Bremer, G.B. (2000) Isolation and culture of Thraustochytrids. In: *Marine Mycology - A Practical Approach* (eds. Hyde, K.D. and Pointing, S. B.), Fungal Diversity Research Series-1, Fungal Diversity Press, Hong Kong, China.

Brown, A.H.C. & Smith, G. (1957) The genus *Paecilomyces* and its perfect stage *Byssochlamys* Westling. *Trans. Br. Mycol. Soc.*, **40:** 17-89.

Bugni, T.S., & Ireland, C.M. (2004) Marine-derived fungi: a chemically and biologically diverse group of microorganisms. *Nat. Prod. Res.*, **21:** 143-163.

Butcher, V.V.C., Hyde, K.D., Pointing, S.B. & Reddy, C.A. (2004) production of wood decay enzymes, mass loss and lignin solubilization in wood by marine ascomycetes and their anamorphs. *Fungal Diversity*, **15:** 1-14.

Byrne, P. J. & Jones, E. B. G. (1975a) Effects of salinity on the reproduction of terrestrial and marine fungi. *Trans. Br. Mycol. Soc.,* **56:** 185-200.

Byrne, P.J. & Jones, E.B.G. (1975b) Effects of salinity on the reproduction of terrestrial and marine fungi. *Trans. Br. Mycol. Soc.,* **65:** 185-200.

Cai, L., Hyde, K.D. & Tsui, C.K.M. (2006) "*Genera of freshwater fungi*". Fungal Diversity Research Series (18), Fungal Diversity Press, Hong Kong, China. pp. 1-261.

Campbell, J., Anderson, J.L. & Shearer, C.A. (2003) Systematics of *Halosarpheia* based on morphological and molecular data. *Mycologia,* **95:** 530-552.

Campbell, J., Volkmann-Kohlmeyer, B., Grafenhan, T., Spatafora, J.W. & Kohlmeyer, J. (2005) A re-evaluation of Lulworthiales: relationship based on 18S rDNA. *Mycol. Res.*, **109:** 556-568.

Campbell, J., Inderbitzin, J., Kohlmeyer, J. & Volkmann-Kohlmeyer, B. (2009) Koralionastetales, a new order of marine Ascomycetes in the Sordariomycetes. *Mycol. Res.*, **113:** 373-380.

Cannon, P.F. (1997) Stratergies for rapid assessment of fungal diversity. *Biodiversity and conservation,* **6:** 669-680.

Cannon, P.F. and Kirk, P.M. (2007) Fungal Families of the world. CABI, Egham, UK.

Chakravarty, D.K. (1974) On the ecology of the infection of the marine diatom *Coscinodiscus granii* by *Lagenisma coscinodisci* in the Weser estuary. *Veroff. Inst. Meeresforsch. Bremerh.*, **5:** 115-122.

Chakravarty, D.L. (1979) *Schizochytrium aggregatum*, a marine fungus newly recorded from India. *Kavaka*, **6:** 37-40.

Chatmala, I., Sakayaroj, J., Somrithipol, S. and Phongpaichit, S. (2004) Marine hypomycetes of Thailand and *Cumulospora varia* sp. nov. *Fungal Diversity*, **17:** 1-9.

Chen, Y.S. Yanagida, F. & Chen, L.Y. (2009) Isolation of marine yeasts from coastal waters of northeastern Taiwan. *Aquatic Biol.*, **8:** 55-60.

Chinnaraj, S. (1992) Higher marine fungi of Lakshadweep islands and a note on *Quintaria lignalitis. Cryptogamie Mycol.*, **13:** 312-319.

Chinnaraj, S. (1993a) Higher marine fungi from mangroves of Andaman and Nicobar Islands. *Sydowia,* **45:** 109-115.

Chinnaraj, S. (1993b) *Studies on fungi associated with mangrove ecosystems of Indian subcontinent*, Ph. D. thesis, Goa University, Goa, India, 1993.

Chinnaraj, S. (1994) Higher marine fungi from mangroves (Manglicolous fungi). In: *Conservation of mangrove forest genetic resources- A training manual* (eds. Deshmukh, S.V. and Balaji, V.), CRSARD, pp. 271-277.

Chinnaraj, S. & Untawale, A.G. (1992) Manglicolous fungi from India. *Mahasagar*, **25:** 25-29.

Chowdhery, H.J. & Rai, J.N. (1980) Microfungi from mangrove swamps of West Bwngal, India. *Nova Hedwigia*, **32:** 229-236.

Clokie, J.J.P. (1974) Site selection by thraustochytrid zoospores in *Pinus* pollen. *Veroff. Inst. Meeresforsch. Bremerh.*, **5:** 159-174.

Clokie, J.J.P. & Dickinson, C.H. (1972) The use of pollen colonization for growth studies of thraustochytrids. *Veroff. Inst. Meeresforsch. Bremerh.*, **2:** 265-270.

Cohn, F. (1865) Chytridii species novae marinae. *Hedwigia,* **12:** 169-170.

Crane, J.L. & Shearer, C.A. (1978) Two new species of *Trichocladium* (Hyphomycetes) from submerged wood. *Mycologia*, **70:** 866-874.

Crane, J.L. & Shearer, C.A. (1986) *Nias glitra*, an Ascomycete from red mangrove in Everglands National Park, Florida. *Trans. Br. Mycol. Soc.*, **86:** 509-512.

Cribb, A.B. & Cribb, J.W. (1955) Marine fungi from Queensland. I. Papers, *University of Queensland, Department of Botany,* **3:** 78-81.

Cribb, A.B. & Cribb, J.W. (1956) Marine fungi from Queensland II. Papers, *University of Queensland, Department of Botany*, **3:** 97-105.

Crouan, P.L. & Crouan, H.M. (1867) "*Florule du Finistere*". Klincksieck, Paris and Brest

Cuomo, V., Vanzanella, F., Fresi, F., Cinelli, F., Mazzella, L. & Scipione, M.B. (1982) Micoflora delle fenerogame dell 'Isola d'Ischia: *Posidonia oceanica* (L.) Delile e *Cymodocea nodosa* (Ucria) Aschers. *Bull. Musea. Inst. Biol.* Univ. Genova, **50:** 162-166.

Cuomo, V., Vanzanella, F., Fresi, F., Cinelli, F. & Mazzella, L. (1985) Fungal flora of *Posidonia oceanica* and its ecological significance. *Trans. Br. Mycol. Soc.*, **84:** 35-40.

Cuomo, V. Jones, E.B.G. & Grasso, S. (1988) Occurrence and distribution of marine fungi along the coast of the Mediterranean Sea. In: *Aspects of marine microbiology* (eds. Jones, E.B.G. and Miller, J.D.), Pergamon Press, Oxford, Prog. Ocean, **21:** 189-200.

Damare, V. & Raghukumar, S. (2006) Morphology and physiology of the marine straminipilan fungi, the aplanochytrids isolated from the equatorial Indian Ocean. *Ind. J. Mar. Sci.*, **35:** 326-340.

Damare, S., Raghukumar, C. & Raghukumar, S. (2006a) Fungi in deep-sea sediments of the Central Indian Basin. *Deep-Sea Research,* **53:** 14-27.

Damare, S., Raghukumar, C., Muraleedharan, U.D. & Raghukumar, S. (2006b) Deep-sea fungi as a source of alkaline and cold-tolerent proteases. *Enzyme and Microbial Technology,* **39:** 172-181.

Danieles, R.J.R. (1993) The Nilgiri biosphere reserve and its role in conservating India's biodiversity. *Current Science,* **64:** 706-708.

Dargan, J.S., Arti, N.S. & Dhingra, G.S. (2005) *"The Fungi – Diversity and Conservation in India"*. Bishen Singh Mahendra Pal Singh, Dehra Dun, India. pp. 1-316.

Das, S., Lyla, P.S. & Khan, A. (2009) Filamentous fungal population and species diversity from the continental slope of bay of Bengal, India. *Acta Oceanologia,* **35:** 269-279.

Dayal, R. (1996) "*Advances in zoosporic fungi*". M.D. Publications Pvt. Ltd., New Delhi. pp. 1- 291.

Dayal, R. (1997) "*Chytrids of India*". M.D. Publications Pvt. Ltd. New Delhi. pp. 1- 316.

Dayal, R. (2000) "*Predaceous fungi*". Commonwealth Publishers, New Delhi. pp. 1-169.

Desmaziere, J.B.H.J. (1849) "*Plants cryptogams de france*", 2nd ed. No. 1778, Lille.

Devarajan, P.T., Suraynarayana, T.S. & Geetha, V. (2002) Endophytic fungi associated with the tropical seagrass *Halophila ovalis* (Hydrocharitaceae). *Ind. J. Mar. Sci.*, **31:** 73-74.

Dick, M.W. (2001) "*Straminipilous Fungi*". Kluwer Academic publishers, Doldrecht, Netherland. pp. 1-670.

D'Souza, D.T., Tiwari, R., Sah, A.K. & Raghukumar, C. (2006) Enhanced production of laccase by a marine fungus during treatment of coloured effluents and synthetic dyes. *Enzyme and Microbial Tecnology*, **38:** 504-511.

Du, L., Feng, S.A., Zhau, B., Li, B., Cai, S., Zhu, T., Wang, F., Xiao, X. & Gu, Q. (2010) Alkaloides from a deep ocean sediment-derived fungus *Penicillium* sp. and their antitumor activities. J. antibit., **63:** 165-170.

Duc, P.M., Hatai, K., Kurata, O., Tensha, K., Uchida, Y.,Yaguchi, T. & Udagawa, S.I. (2009) Fungal infection of mantis shrimp (*Oratasquilla oratoria*) by two anamorphic fungi found in Japan. *Mycopathol.*, **167:** 229-247.

Dupont, J., Magnin, S., Rousseau, F., Zbinden, M., Frebourh, G., Samadi, S., Richer da Forges, B. & Jones, E.B.G. (2009) Molecular and ultrastructural characterization of two ascomycetes found on sunken wood off Vanuatu Islands in the deep Pacific Ocean. *Mycol. Res.,* **113:** 1351-1364.

Durieu, de maisonneuve C. & Montagne, J.F.C. (1869) Pyrenomycetes Fr. In: "*Exploration Scientifique del. Algeric Botanique*" (eds. Bory de saint-Vincent, J. and G. Durieu maisonneuve), Paris, pp. 443-608.

Durve, V.S. & Bal, D.V. (1960) Shell disease in *Crossostrea gryphoides* (Schlotheim). *Curr. Sci.*, **29:** 489-490.

Ebel, R. (2006) Secondary metabolites from marine-derived fungi. In: *Frontiers in marine biotechnology* (eds. Proksch, P. and Muller, W.E.G.), Horizon Bioscience, Wymondham, UK. pp. 73-144.

Eckblad, F.E. & Kristinsen, G. (1989) *Ostracoblabe implexa*, a taxonomic reappraisal. *Mycol. Res.*, **94:** 706-708.

Ellis, M.B. (1971) *"Dematiaceous Hyphomycetes"*, Commomw. Mycol. Inst., Kew, Surry, England.

Ellis, M.B. (1976) "*More Dematiaceous Hyphomycetes*", Commomw. Mycol. Inst., Kew, Surry, England.

Fan, K.W., Chen, F., Jones, E.B.G. & vrijmoed, L.L.P. (2000) Utilization of food processing waste by Thraustochytrids. In: *Aquatic mycology: across the millennium* (eds. Hyde, K.D., Ho, W.H. and Pointing, S.B.), *Fungal Diversity*, **5:** 185-194.

Farrant, C.A., Hyde, K.D. & Jones, E.B.G. (1985) Further studies on lignicolous marine fungi from Danish sand dunes. *Trans. Br. Mycol. Soc.*, **85:** 164-167.

Fell, J.W. (1967) Distribution of yeasts in Indian Ocean. *Bull. Mar. Sci.*, **17:** 454-470.

Fell, J.W. (1974) Distribution of yeast in water masses of the southern oceans. In: *Effect of the ocean environment on microbial activities* (eds. Colwell, R.R. and Morita, R.Y.), Univ. Park Press, Baltimore, pp. 510-523.

Fell, J.W. (1976) Yeasts in oceanic regions. In: *Recent advances in Aquatic Mycology* (ed. Jones, E.B.G.), Elek Press, London, pp. 93-124.

Fell, J.W., Ahearn, D.G., Meyers, S.P. & Rith, F.J. (1960) isolation of yeasts from Biscayne Bay, Florida and adjacent benthic waters. *Limno. Oceanogr.*, **5:** 366-371.

Fell, J.W., Statzall-Tallmann, A. & Kurtzman, C.P. (2004) *Lachancea meyersii* sp. nov., an ascosporogenous yeast from mangrove region in the Bahama Islands. *Studies Mycol.*, **50:** 359-363.

Fell, J.W., Statzall-Tallmann, A., Scorzetti, G. & Gutierrez, M.H. (2010) Five new species of yeasts from fresh and marine habitats in the Florida Everglades. *Antonie Van Leeuwenhoek*, doi:10.1007/s10482-010-9521-6

Fisher, W.W., Nilson, E.H. & Schleser, B.A. (1975) Effect of the gungus *Halipthoros milfordensis* on the juvenile stages of the American lobster *Homarus americanus*. *J. Invert. Path.*, **26:** 42-45.

Frohlich, J. & Hyde, K.D. (2000) "*Palm microfungi*". Fungal Diversity press, Hong Kong.

Fletcher, A. (1980) Marine and Maritime lichens of rocky shores, their ecology, physiology and biological interactions. In: "*The shore Environment*", vol. 2, ecosystems (eds. Price J.H., Ir-vine D.E.G. and Farnham W.F.)., *Academic Press,* London & New York, pp. 789-842.

Fuller, M.S. & Jaworski, A. (1987) "*Zoosporic Fungi in teaching and research*". Southeastern Publishing Corporation, Athens, Georgia.

Furuya, K. & Udagawa, S. (1975) Two new species of clestothecial Ascomycetes. *J. Jpn. Bot.*, **50:** 249-254.

Gachon, M.M., Kupper, H., Kupper, F.C. & Setlik, I. (2006) Single-cell chlorophyll fluorescence kinetic microscopy of *Pylaiella littorais* (Phaeophyceae) infected by *Chytridium polysiphoniae* (Chytridiomycota). *Eur. J. Phycol.*, **41:** 395-403.

Gachon, M.M., Strittmatter, M., Muller, D.G., Kleinteich, J. & Kupper, F.C. (2009) Detection of differential host susceptibility to the marine Oomycete pathogen *Eurychasms dicksonii* by real-time PC: Not all algae are equal. *Appl. Environ. Microbiol.*, **73:** 322-328.

Gachon, M.M., Sime-Ngando, T., Strittmatter, M., Chambouvet, A. & Kim, G.H. (2010) Algal disease: spotlight on a black box.*Trends Plant Sci.*, **15:** 633-640.

Gaertner, A. (1968) (ed.) Marine Mrkologie. Symposium uber Niedere Pilze im Kustenbereich in Bremerhaven vom 17. Bis 19 Oktober 1966, *Veroff. Inst. Meeresforschung Bremerhaven*, pp. 519.

Gaertner, A. (1972) Characters used in the classification of thraustochtriaceous fungi. *Veroff. Inst. Meeresforschung Bremerhaven,* **13:** 183-194.

Gaertner, A. (1974) (ed.) Marine Mrkologie. 2. Internationales Symposium in Bremerhaven vom 11. bis 16. September 1972. *Veroff. Inst. Meeresforschung Bremerhaven*, pp. 159.

Gessner, R.V. & Kohlmeyer, J. (1976) Geographical distribution and taxonomy of fungi from salt marsh *Spartina. Can. J. Bot.*, **54:** 2023-2037.

Ginns, J. & Malloch, D. (1977) *Halocyphina*, a marine basidiomycete (Aphyllophorales). *Mycologia*, **69:** 874-880.

Gleason, F.H., Kupper, F.C., Amon, J.P., Picard, K., Gachon, S.M.M., Marano, A.V., Sime-Ngando, T. & Lilje, O. (2011) Zoosporic true fungi in marine ecosystems: a review. *Mar. Freshwater Res.*, **62:** 383-393.

Goldstein, S. (1963) Studies of a new species of *Thraistochytrium* that displays light stimulated growth. *Mycologia*, **55:** 799-811.

Goldstein, S. (1973 Zoosporic marine fungi (Thraustochytriaceae and Dermaocystidiaceae). *Ann. Rev. Microbiol.*, **27:** 13: 26.

Gonzales, M.C. & Hanlin, R.T. (2010) Potential use of marine arenicolous ascomycetes as bioindicators of ecosystem distrurbances on sandy Canadian beaches: *Corollospora maritima* as a candidate species. *Bot. Mar.*, **53:** 577-580.

Gopalan, U.K., Meenakshikunjamma, P.P., Purushan, K.S. (1980) Fungal infection in the tiger prawn (*Penaeus monodon*) and in other crustaceans from the Cochin backwaters. *Mahasagar Bull. Nat. Inst. Oceanogr.*, **13:** 359-365.

Grasso, S., La Ferela, R. & Jones, E.B.G. (1985) Lignicolous marine fungi in a harbour environment (Milazzo). *Bot. Mar.*, **28:** 259-164.

Grasso, S., Panebianco, C. & La Ferala, R. (1990) Lignicolous marine fungi in the Straits of Messina, Italy. *Hyrobiol.*, **206:** 149-154.

Guerriero, A., D'Ambrosio, M., Cuomo, V., Vanzanella, F. & Pietra, F. (1988) Dendryphiellin A, the first fungal trinor-ermophilane isolated from the marine deuteromycete *Dendryphiella salina* (Sutherland) Pugh et Nicot. *Helvetica Chim. Acta.*, **71:** 57-61.

Hale, M.S. & Eaton, R.A. (1984) Soft rot cavity widening, a consideration of the kinetics. *Inter. Res. Group Wood Preservation.* No IRG/WP/1227.

Hale, M.S. & Eaton, R.A. (1985a) The ultrastructure of soft rot fungi. II. Cavity-forming hyphae in the wood cell walls. *Mycologia*, **77:** 594-605.

Hale, M.S. & Eaton, R.A. (1985b) The oscillatory growth of fungal hyphae on wood cell walls. *Trans. Br. Mycol. Soc.*, **84:** 227-288

Harder, R. & Uebelmesser, E.R. (1955) Uber marine saprophytische Chytridiales und einiege andere Pilze vom Meeresboden und Meeresstand. *Arch. Mikrobiol.*, **22:** 87-144.

Harrison, J.L. & Jones, E.B.G. (1974) Pattern of salinity tolerance displayed by the lower fungi. *Veroff. Inst. Meeresforsch. Bremerh. Suppl.*, **5:** 197-220.

Harvey, J.B.J. & Golf, L.J. (2010) Genetic covariation of the marine fungal symbiot *Haloguignardia irritans* (Ascomycota, Pezizomycotina) with its algal host *Cystoseira* and *Halidrys* (Phaeophyceae, Fucales) along the west coast of North America. *Mycol. Res.,* **114:** 82-95.

Hawksworth, D.L. (1987a) *Kirschsteiniothelia*, a new genus for the *Microthelia incrustans*-group (Dothideales). *Bot. J. Linn. Soc.*, **91:** 181-202.

Hawksworth, D.L. (1987b) Observations on three algicolous micro-fungi. *Transactions of the Royal Botanic Garden, Edinburgh,* **44:** 549-560.

Hawksworth, D.L. (1988) The variety of fungal–algal symbiosis, their evolutionary significance and the nature of lichens. *Bot. J. Linn. Soc.,* **96:** 3-20.

Hawksworth, D.L. (1991) The fungal dimension of biodiversity, Mannitude, significance and conservation. *Mycol. Res.*, **95:** 641-665.

Hawksworth, D. L. (2001) The mangnitude of fungal diversity: the 1.5 million species estimate revisited. *Mycol. Res.*, **105:** 1422-1432.

Hawksworth, D.L. & Hop-Negger, R. (1994) The Lichen thallus: a symbiotic phenotype of nutritionally specialize fungi and its response to gall producers. In: *Plant Galls* (ed. Wiliams M.A.J.), Alarendon Press, Oxford. pp. 77-98.

Hawksworth, D.L., Kirk, P.M., Sutton, B.C. & Pegler, D.N. (1995) *Ainsworth and Bisby's Dictionary of the fungi* (8th Eds). CAB International, Wallingford, U. K. pp. 1- 616.

Hibbett, et al. (2007) A higher-level phylogenetic classification of the fungi. *Mycol. Res.*, **111:** 509-547.

Hohnk, W. (1952) Studien Zur Breackund seewasser mykologie - I. *Veroff. Inst. Morresforsch. Bremerhaven,* **1:** 115-125.

Hohnk, W. (1954) Studien Zur Brackand See wassermy Kologie V. Hohere pilze des submersen Holzes. *Veroff. Inst. Morresforsch. Brremerhaven* **3:** 199-227.

Hohnk, W. (1955a) Studien zur Brack-und Seewassermykologie V. Hohere Pilze des submersem Holzes. *Veroff. Inst. Morresforsch. Bremerhaven,* **3:** 199-227.

Hohnk, W. (1955b) Marine Pilze vom watt und meeresgrund (Chytridiales und Traustochytriaceae). *Natwissen,* **42:** 348-349.

Hohnk, W. (1961) Afurther contribution to the oceanic mycology. *Cons. Inter. Explor. Mer.,* **12:** 202-208.

Hohnk, W. & Aleem, A.A. (1953) Ein Brackwasserpilze: Olpidium maritimum nov. spec. *Veroff. Inst. Morresforsch. Bremerhaven,* **2:** 224-229.

Hughes, S.J. (1958) Revisions Hyphomycetum atiquot cum appendice de nominibus rejiciendis. *Can. J. Bot.,* **36:** 727-836.

Hughes, G.C. (1974) Geographical distribution of the higher marine fungi. *Veroff. Inst. Meeresforsch. Bremerhaven supplement,* **5:** 419-441.

Hughes, G.C. (1986) Biogeography of marine fungi In: *The Biology of marine fungi* (ed. Moss, S.T.) Cambridge University Press, Cambridge, U. K., pp. 275-295.

Huhndorf, S.M. (1994) Neotropical ascomycetes. 5. Hypostromataceae, a new family of Loculoascomycetes and *Manglicola samuelsii*, a new species from Guyana. *Mycologia,* **86:** 266-269.

Hyde, K.D. (1988a) A study of vertical zonation of intertidal fungi on *Rhizophora apiculata* at Kampong Kapok mangrove, Brunei. *Aquat. Bot.,* **36:** 255-262.

Hyde, K.D. (1988b) Studies on the tropical marine fungi of Brunei. *Bot. J. Linn. Soc.,* **98:** 135-151.

Hyde, K.D. (1989) Intertidal mangrove fungi from north Sumatra. *Can. J. Bot.,* **67:** 3078-3082.

Hyde, K.D. (1990a) Vertical zonation of intertidal mangrove fungi. In: *Recent advances in microbial ecology* (eds. Hattori, T., Ishida, Y., Maruyuama, Y., Morita, R.Y. and Echida, A.), Japan Scientific Societies Press, Tokyo. pp. 302-306.

Hyde, K.D. (1990b) A new marine ascomycete from Brunei. *Aniptodera longispora* sp. nov. from intertidal mangrove wood. *Bot. Mar.,* **33:** 335-338.

Hyde, K.D. (1991a) *Helicascus kanaloanus*, *Helicascus nypae* sp. nov. and *Salsuginea ramicola* gen. et sp. nov. from intertidal mangrove wood. *Bot. Mar.*, **34:** 311-318.

Hyde, K.D. (1991b) *Massarina velatispora* and a new mangrove-inhabiting species, *M. ramunculicola* sp. nov. *Mycologia,* **83:** 839-845.

Hyde, K.D. (1991c) *Phomopsis mangrovei*, from intertidal prop roots of *Rhizophora* spp. *Mycol. Res.*, **95:** 1149-1151.

Hyde, K.D. (1992a) *Julella avicenniae* (Borse) comb. Nov. (Thelenellaceae) from intertidal mangrove wood and miscellaneous fungi from the NE coast of Queensland. *Mycol. Res.*, **96:** 939-942.

Hyde, K.D. (1992b) The genus *Saccardoella* from intertidal mangrove wood. *Mycologia*, **84:** 803-810.

Hyde, K.D. (1992c) Intertidal mangrove fungi from the west coast of Mexico, including one new genus and two new species. *Mycol. Res.*, **96:** 25-30.

Hyde, K.D. (1993a) *Cryptosphaeria mangrovei* sp. nov. (Diatrypaceae) from intertidal mangrove wood. *Trans. Mycol. Soc. Japan,* **34:** 311-314.

Hyde, K. D. (1993b) *Cryptovalsa halosarceicola* sp. nov. an intertidal saprotroph of *Halosarceia halocnemoides. Mycol. Res.*, **97:** 779-800.

Hyde, K.D. (1995) *Eutypella naqsii* sp. nov. from intertidal *Avicennia. Mycol. Res.*, **97:** 1462-1464.

Hyde, K.D. (ed.) (2002) "*Fungi in marine environment*", Fungal Diversity Press, Hong Kong. pp. 1-397.

Hyde, K.D. & Borse, B.D. (1986a) Marine fungi from Seychelles -V. *Biatriospora marina* gen. *et* sp. nov., from mangrove wood. *Mycotaxon,* **26:** 263-270.

Hyde, K.D. & Borse, B.D. (1986b) Marine fungi from Seychelles -VI. *Massarina velataspora*. A new marine ascomycotina from mangrove wood. *Mycotaxon,* **27:** 161-167.

Hyde, K.D. & Jones, E.B.G. (1984) Marine fungi from Seychelles -II. *Nimbospora effuse* and *Nimbospora bipolaris* sp. nov. from drift wood. *Can. J. Bot.*, **63:** 611-615.

Hyde, K.D. & Jones, E.B.G. (1987) Marine fungi from the Seychelles –VII. *Bathyascus grandisporous* sp. nov. from Mangrove wood. *Bot. Mar.*, **30:** 413-416.

Hyde, K.D. & Jones, E.B.G. (1988) Marine Mangrove Fungi. P. S. Z. N. I.: *Marine Ecology,* **9:** 15-33 .

Hyde, K.D. & Jones, E.B.G. (1989a) Intertidal fungi from Brunei. *Lautospora gigantea* gen. et sp. nov., a new Loculoascomycetes from prop roots of Rhizophora spp. *Bot. Mar.*, **32:** 479-482.

Hyde, K.D. & Jones, E.B.G. (1989b) Marine fungi from Seychelles - VIII. *Rhizophila marina*, a new Ascomycete from mangrove prop roots. *Mycotaxon*, **34**: 527-533.

Hyde, K.D. & Jones, E.B.G. (1992) Intertidal mangrove fungi: *Pedumispora gen. nov*. (Diaportahales). *Mycol. Res.*, **96:** 87-80.

Hyde, K.D. & Mouzouras, R. (1988) *Passeriniella savoryllopsis* sp. nov., a new Ascomyceye from intertidal mangrove wood. *Trans. Br. Mycol. Soc.*, **91:** 179-185.

Hyde, K.D. & Nakagiri, A. (1992) Intertidal fungi from Australia. The genus *Swampomyces* including *S. triseptatus* sp. nov. *Sydowia*, **44:** 122-130.

Hyde, K.D. & Pointing, S.B. (eds.) (2000) *Marine Mycology: A Practical Approach*. Fungal Diversity Research Series 1, Fungal Diversity Press, Hong Kong, China, pp. 1-377.

Hyde, K.D. & Rappaz, F. (1993) *Eutypa bathurstensis* sp. nov. from intertidal *Avicennia Mycol. Res.*, **97:** 861-864.

Hyde, K.D. & Sarma, V.V. (2000) Pictorial key to higher marine fungi. In: *Marine Mycology– A Practical Approach* (eds. Hyde, K.D. and Pointing, S.B.), Fungal Diversity Research Series 1, Fungal Diversity Press, Hong Kong. pp. 205 -270.

Hyde, K.D. & Sarma, V.V. (2006) Biodiversity and ecological observations on filamentous fungi of mangrove palm *Nypa fruticans* Wurumb. (Liliopsida, Arecales) along the Tutong River, Brunei. *Ind. J. Mar. Sci.*, **35:** 297-307.

Hyde, K.D., Sarma, V.V. & Jones, E.B.G. (2000) Morphology and taxonomy of higher marine fungi. In: *Marine Mycology– A Practical Approach* (eds. Hyde, K.D. and Pointing, S.B.), Fungal Diversity Research Series 1, Fungal Diversity Press, Hong Kong. pp. 172- 204.

Hyde, K.D., Farrant, C.A. & Jones, E.B.G. (1986) Marine fungi from Seychelles. III. *Aniptodera mangrovii* sp nov. from mangrove wood. *Can. J. Bot.*, **64:** 2989-2992.

Hyde, K.D., Vrijmoed, L.L.P., Chinnaraj, S. & Jones, E.B.G. (1992) *Massarina armatispora* sp. nov., a new intertidal Ascomycete from mangroves. *Bot. Mar.,* **35:** 325-328.

Hyde, K.D., Goh, T.K., Lu, B.S. & Alias, S.A. (1999) Seven new intertidal fungi from *Nypa fruiticans*. *Mycol. Res.*, **103:** 1409-1422.

Inderbitzin, P. & Desjardin, D.E. (1999) A new halotolerant species of *Physalacria* from Hong Kong. *Mycologia*, **91:** 666-668.

Inderbitzin, P., Abdel-Wahab, M.A., Jones, E.B.G. & Vrijmoed, L.L.P. (1999) A new species of *Cryptovalsa* from Mai Po mangroves in Hong Kong. *Mycol. Res.*, **103:** 1628-1630.

Inderbitzin, P., Landvic, S., Abdel-Wahab, M.A. & Berbee, M.L. (2001) Aliquandostipitaceae, a new family for two new tropical ascomycetes with unusually wide hyphae and dimorphic ascomata. *Amer. J. Bot.,* **88:** 52-61.

Ingold, C.T. (1942) Aquatic Hyphomycetes of decaying alder leaves. *Trans. Br. Mycol. Soc.*, **25:** 339-417.

Jain, R., Raghukumar, S. & Chandramohan, D. (2004) Enhanhancement of the production of the polyunsaturated fatty acid, docosahexaenoic acid in thraustchytrid protist. *Mar. Biotechnol.* (Suppl), **6:** 59-65.

Jain, R., Raghukumar, S., Tharanathan, R. & Bhosale, N.B. (2005) Extra-cellular polysaccharide production by thraustchytrid protist. *Mar. Biotechnol*, 7**:** 184-192.

Jamaluddhin, S., Goswami, M.G. & Ojha, B.M. (2004) " *Fungi of India*", 1989-2001, Scientific Publishers, Jodhpur (India), Pp. 326.

Janardhanan, K K., Rajendran, C., Natrajan, K. & Hawksworth, D.L. (eds.) (1997) "*Tropical Mycology*". Oxford & IBH Publishers Co. Pvt. Ltd, New Delhi, pp. 315.

Janson, R., Bernan, V.S,. Greenstein, M., Bugni, T.S. & Ireland, C.M. (2005) *Penicillium dravuni*, a new marine-derived species from an alga in Fiji. *Mycotaxon,* **97:** 444-453.

Jennings, D.H. (1986a) Some aspects of the physiology and biochemistry of marine fungi. *Biol. Rev.*, **58:** 423-459.

Jennings, D.H. (1986b) Fungal growth in the sea. In: *The biology of marine fungi* (ed. Moss, S.T.), Cambridge Univ. Press, pp. 1-10.

Jiang, G.C., Lin, Y.C., Zhou, S.N., Vrijmoed, L.L.P. & Jones, E.B.G. (2000) Studies on the secondary metabolites of mangrove fungus no. 1403 from South China Sea. *J. Sunb. Yet-sen Univ. (Nat. Sci.),* **39:** 68-72.

Johnson, T.W. Jr. (1956) Marine fungi II. Ascomycetes and Deutero-mycetes from submerged wood. *Mycologia,* **48:** 841-851.

Johnson, T.W. Jr. & Sparrow, F.K. Jr. (1961) "*Fungi in oceans and esturies*", Cramer Wcinheim, New York. pp. 668.

Jones, E.B.G. (1962) Marine fungi. *Trans. Br. Mycol. Soc.*, **45:** 93-114.

Jones, E.B.G. (1968a) The distribution of marine fungi on wood submerged in the sea. In: *Biodeterioration of materials* (eds. Walters, A.H. and Elphick, J.J.), Elsevier, Amsterdam, pp. 237-258.

Jones, E.B.G. (1968b) Marine fungi, *Current Science*, **37:** 378-379.

Jones, E.B.G. (1971) The ecology and rotting ability of marine fungi. In: *Marine borers, fungi and fouling organisms of wood* (eds. Jones, E.B. G. and Eltringham, S.K.), OECD, Paris, pp. 237-258.

Jones, E.B.G. (Ed.) (1976) "*Recent advances in Aquatic Mycology*". Elek Science, London. pp. 1-749.

Jones, E.B.G. (1985) Wood-inhabiting marine fungi from San Juan Island with special reference to ascospore appendages. *Bot. J. Lin. Soc.*, **91:** 219-231.

Jones, E.B.G. (1988) Do fungi occure in the Sea? *The Mycologist,* **2:** 150-157.

Jones, E.B.G. (1993) Tropical marine fungi In: *Aspects of tropical mycology* (eds. Issac, S., J.C. Frankland, R. Walling and WhalleyJ.S.), Cambridge University Press, Cambridge. pp. 73-89.

Jones, E.B.G. (2000) marine fungi: some facters influencing biodiversity. *Fungal Diversity*, **4:** 53-73.

Jones, E.B.G. (2008) Marine compounds in marine organisms. *Bot. Mar.,* **51:** 161-162.

Jones, E.B.G. (2010) Fungi. In: *Checklist of the flora and fauna in Italian seas. Boil. Mar. Mediterr.*, 17 (suppl. 1), pp. 681-684.

Jones, E.B.G. (2011a) Fifty years of marine mycology. *Fungal Diversity*, DOI 10.1007/s13225-011-0119-8

Jones, E.B.G. (2011b) Are there more marine fungi to be described? *Bot. Mar.*, (In Press).

Jones, E.B.G. & Agerer, R. (1992) *Calathella mangrovei* sp. Nov. and observation on the mangrove fungus *Halocyphina villosa. Bot. Mar.*, **35:** 259-265.

Jones, E. B. G. & Alias, S. A. (1997) Biodiversity of mangrove fungi. In : *Biodiversity of tropical microfungi* (ed. Hyde, K.D.), Hong Kong University Press, Hong Kong, pp. 71-92.

Jones, E.B.G. & Eaton, R.A. (1969) *Savoryella lignicola* gen. et sp. nov. from water cooling towers. *Trans. Br. Mycol. Soc.*, **52:** 161-174.

Jones, E.B.G. & Jennings, D.H. (1964) The effect of salinity on the growth of marine fungi in comparison with non-marine species. *Trans. Br. Mycol. Soc.,* **47:** 619-625.

Jones, E.B.G. & Hyde, K.D. (1992) Taxonomic studies on *Savoryella* Jones *et* Eaton (Ascomycotina). *Bot. Mar.*, **35:** 83-91.

Jones, E.B.G. & Le Campion-Alsumand (1970a) Marine fungi on polyurethane plates submerged in the sea. *Nava Hedwigia*, **19:** 567-590.

Jones, E.B.G. & Le Campion-Alsumand (1970b) The biodeterioration of polyurethane by marine fungi. *Int. Biodeterior Bull.,* **6:** 119-124.

Jones, E.B.G. & Mitchell, J.I. (1996) Biodiversity of marine fungi. In: Biodiversity: International Biodiversity Seminar. (eds. A. Cimerman and Gunde-Cimerman, N.), National Institute of Chemistry and Slovenia National Commission for UNESCO, Ljubljana, Slovenia. pp. 31-42.

Jones, E.B.G. & Vrijmoed, L.L.P. (2003) Biodiversity of marine fungi in Hong Kong coastal waters. In: *Perspective on marine environment change in Hong Kong & Southern China 1977-2001* (ed. Morton, B.), Proc. Inter. Workshop Reunion Conference. Hong Kong Univ. press, Hong Kong, pp. 75-92.

Jones, E.B.G. & Zainal, A. (1988) *Ceriosporopsis caduca*, a new lignicolous marine ascomycotina from submerged wood. *Mycotaxon*, **32:** 237-244.

Jones, E.B.G., Byrne, P. & Alderman, D.J. (1971) The response of fungi to salinity. *Viet et Milieu supplement,* **22**: 265-280.

Jones, E.B.G., Johnson, R.G. & Moss, S.T. (1983a) Taxonomic study of the Halosphaeriaceae: *Corollospora* Werdermann. *Bot. J. Linn. Soc.*, **87:** 193-212.

Jones, E.B.G., Johnson, R.G. & Moss, S.T. (1983b) *Ocostaspora apilongissima* gen. et sp. nov: A new marine Pyrenomycete from wood. *Bot. Mar.*, **26:** 353-360.

Jones, E.B.G., Vrijmoed, L.L.P., Read, S.J. & Moss, S.T. (1994) *Tirispora,* a new Ascomycetous genus in the Halosphaeriales. *Can. J. Bot.*, **72:** 1373-1378.

Jones, E.B.G., Koch, J., McKeown, T.A. & Moss, S.T. (1995) *Limacospora* gen. nov. and a reevaluation of *Ceriosporopsis cambrensis*. *Can. J. Bot.,* **73:** 1010-1018.

Jones, E.B.G., Vrijmoed, L. L. P. & Alias, S.A. (1998) Intertidal marine fungi from San Juan Island and comments on temperate water species. *Bot. J. Scotl.*, **50:** 177-184.

Jones, E.B.G., Abdel-Wahab, M.A. & Vrijmoed, L.L.P. (2001) *Trichocladium melhae* sp. nov., a new tropical marine fungus. *Fungal Diversity*, **7:** 49-52.

Jones, E.B.G., Chatmala, I. & Pang, K.L. (2006) Two new genera of the Halosphaeriaceae isolated from marine habitats in Thailand: *Pseudolignicola* and *Thalespora. Nova Hedwigia*, **83:** 219-232.

Jones, E.B.G., Stanley, S.J. & Pinruan, U. (2008a) Marine endophytes sources of new chemical natural products: a review. *Bot. Mar.,* **51:** 163-170.

Jones, E.B.G., Chatmala, I., Klasuban, A. & Pang, K.L. (2008b) Ribosomal DNA phylogeny of marine anamorphic fungi: *Cumulospora varia, Dendryphiella* species and *Orbimyces spectabilis. The Raffles Bull. Zool. Suppl.*, **19:** 11-18.

Jones, E.B.G., Sakayaroj, J., Suestrong, S., Somrithipol, S. & Pang, K.L. (2009a) Classification of Ascomycota, anamorphic taxa and Basidiomycota. *Fungal Diversity*, **35:** 1-187.

Jones, E.B.G., Zuccaro, A., Nakagiri, A., Mitchell, J.L. & Pang, K.L. (2009b) Phylogenetic relationships of the genus *Sigmoidea* and a new genus *Halosigmoidea* gen. nov. *Bot. Mar.*, **52:** 349-359.

Kagami, M. de Bruin, A., Ibelinbgs, B.W. & Van Donk, E. (2007) Parasitic chytrids: their effect on phytoplankton communities and food-web dynamics. *Hydrobiol.*, **578:** 113-129.

Kamat, T., Rodrigues, C. & Naik, G. (2008) Marine-derived fungi as a source of proteases. *Ind. J. Mar. Sci.*, **37:** 326-328.

Kamble, V.M., Borse, B.D. & Borse, K.N. (2008) Marine fungi From Kutcch (Gujarat Coast). *Research Link,* **6:** 67-68.

Kamble, V.M., Borse, K.N. & Borse, B.D. (2011) Marine fungi From Gujarat Coast. *Bioinfolet,* **8:** 124-127.

Karling, J.S. (1977) Chytridiomycetarum Iconographia, Lubrecht and Cramer, Vaduz.

Karling, J.S. (1981) *Predominatly holocarpic and eucarpic simple biflagellate Phycomycetes*. J. Cramer, Vaduz. pp. 1-252.

Karunsagar, I., Karunsagar, I. & Umesha, R.K. (2004) Microbial diseases in Shrimp aquaculture. In: *Marine Microbiology: Facets and Opportunities.* (ed. Ramaiah, N.), National Institute of Oceanography, Goa, pp. 121-134.

Kato, H., Yoshida, T., Tokue, T., Nojiri, Y., Hirota, H., Ohta, T., Williams, R.M. & Tsukamoto, S. (2007) Notoamides A-D; prenylated indole alkalods isolated from marine-derived fungus *Aspergillus* sp. Agnew. *Chem. Int. Ed. Engl.*, **46:** 2254-2256.

Kendrick, B., Risk, M.J., Michaelids, J. & Bergman, K. (1982) Amphibious microborers: bioeroding fungi isolated from live corals. *Bull. Mar. Sci.*, **32:** 862-867.

Khan, S.S. & Manimohan, P. (2011) Diversity and abun ance of marine fungi on driftwood collected from Kerala State and Lakshadweep Island, India. *Mycosphere*, **2:** 223-229.

Khao, L.V., Hatai, K. & Aoki, T. (2004) *Fusarium incarnatum* isolated from black tiger shrimp *Panaeus mondon* with black gill disease cultured in Vietnam. *J. Fish Dis.*, **27:** 507-515.

Khao, L.V., Hatai, K., Yuasa, A. & Sawada, K. (2005) Morphology and molecular phylogeny of *Fusarium solani* isolated from kuuruma prawn *Panaeus japonicas* with black gills. *Fish Pathol.*, **40:** 103-109.

Khulbe, R.D. (2001) *"A manual of aquatic fungi (Chytridiomycetes and Oomycetes)"*, Daya Publishing House, New Delhi. pp. 1- 255.

Kirk, P.W. Jr. (1969) Isolation and Culture of lignicolous marine fungi. *Mycologia,* **61:** 174 – 177.

Kirk, P.W. Jr. (1983) Direct enumeration of marine arenicolous fungi. *Mycologia,* **75:** 670-682.

Kirk, P.W. Jr. & Brandit, J.M. (1980). Seasonal distribution of lignicouls marine fungi in the lower Chesapeake Bay. *Bot. Mar.,* **23:** 629-638.

Kirk, P.W. Jr. & Gorden, A.S. (1988) Hydrocarbon degradation by filamentous marine higher fungi, *Mycologia,* **80:** 776-786.

Kirk, P. W. Jr. & Schatz, S. (1980) Higher fungi affected by declining salinity and seasonal factors in a coastal embayment. *Bot. Mar.,* **23:** 629-638.

Kirk, P.M., Cannon, P.F., David, J.C. & Stalpers, J.A. (2001) *Ainsworth and Bisby's Dictionary of fungi*, 9th edition,. CAB Internation, Wallingford, pp. 1-655.

Kirk, P.M., Cannon, P.F., Minter, D.W. & Stalpers, J.A. (2008) *Ainsworth and Bisby's Dictionary of fungi*, 10 th edition,. CAB Internation, Wallingford.

Kitancharoen, N., Nakamura, K., Wada, S. & Hatai, K. (1994) *Atkinsiella awabi* sp. nov. isolated from stocked abalone, *Haliotis sieboldii*. *Mycoscience*, **35:** 265-270.

Koch, J. (1974) Marine fungi of drift wood from the West coast of Jutland, Denmark, *Freesia,* **10:** 209-250.

Koch, J. (1982) Some lignicolous marine fungi from Sri Lanka. *Nordic. J. Bot.,* **2:** 163-169.

Koch, J. (1986) Some lignicolous marine fungi from Thailand, including two new species. *Nord. J. Bot.*, **6:** 497-499.

Koch, J. & Jones, E.B.G. (1985) *Ceriosporopsis sundica*, a new lignicolous marine ascomycete from Denmark. *Nord. J. Bot.*, **6:** 339-344.

Koch, J. & Jones, E.B.G. (1989) The identity of *Crinigera maritima* and three new genera of marine cleistothecial ascomycetes. *Can. J. Bot.*, **67:** 1183-1197.

Koch, J. & Petersen, K.R.L. (1996) A check list of higher fungi on wood from Danish coasts. *Mycotaxon*, **60:** 397-414.

Koch, J., Jones, E.B.G. & Moss, S.T. (1983) *Goenhiella bivestia*, gen et sp. nov., a lignicolous marine fungus from Denmark. *Bot. Mar.*, **26:** 265-270.

Koch, J., Pang, K.L. & Jones, E.B.G. (2007) *Rostrupiella danica* gen. et sp. nov., a *Lulworthia*-like marine lignicolous species from Denmark and the USA. *Bot. Mar.*, **50:** 1-8.

Kohlmeyer, J. (1959) Neufunde hazberrede Inder meerspilze. *Nova Hedwigia,* **1:** 77-99.

Kohlmeyer, J. (1963) Zwei neue Ascomyceten-Gattungen auf *Posidonia*-Rhizomen. *Nova Hedwigia*, **6:** 5-13.

Kohlmeyer, J. (1966a) Ecological observations on arenicolous marine fungi. *Z. Alig. Mikrobiol.*, **6:** 95-106.

Kohlmeyer, J. (1966b) Nene Meerespilse an mangroven. *Ber. Dtsch. Bot. Ges.,* **79:** 27-37.

Kohlmeyer, J. (1967) Intertidal and phycobilous fungi from Tenerife (Canary Island). *Trans. Br. Mycol. Soc.*, **50:** 137-147.

Kohlmeyer, J. (1968) Marine fungi from the tropics. *Mycologia*, **60:** 252-270.

Kohlmeyer, J. (1969a) The role of marine fungi in the penetration of calcareous substances. *Am. Zool.*, **9:** 741-746.

Kohlmeyer, J. (1969b) Marine fungi of Hawaii including the new genus *Helicascus. Can. J. Bot.*, **47:** 1469-1487.

Kohlmeyer, J. (1972a) Parasitic *Haloguignandia oceanica* (Ascomycetes) and hyperparasitic *Sphaceloma cecidii* sp. nov. (Deuteromycetes) in drift *Sargassum* in North Carolina. *J. Elisha Mitchell. Soc.*, **88:** 255-259.

Kohlmeyer, J. (1972b) Marine fungi deteriorating chitin of hydrozoa and keratin-like annelid tubes. *Inter, J. Oceans Coastal Waters*. **12:** 277-284.

Kohlmeyer, J. (1973) Fungi on marine algae. *Bot. Mar.*, **16:** 201-215.

Kohlmeyer, J. (1977) New genera and species of higher fungi from the deep sea (1615-5315 m). *Rev. Mycol.*, **41:** 189-206.

Kohlmeyer, J. (1979) Marine fungal pathogens among Ascomycetes and Deuteromycetes. *Experientia*, **35:** 437-439.

Kohlmeyer, J. (1980) Tropical and subtropical filamentous fungi of the Western Atlantic ocean. Bot. *Mar.*, **23:** 529-544.

Kohlmeyer, J. (1981) Marine fungi from Martinque. *Can. J. Bot.*, **59:** 1314-1321.

Kohlmeyer, J. (1984) Tropical marine fungi. *P.S.Z.N.I.: Marine Ecology*, **5**: 329-378.

Kohlmeyer, J. (1985a) *Caryosporella rhizophorae* gen. *et* sp. nov. (Massariaceae), a marine Ascomycete from *Rhizophora mangle*. *Proc. Indian Acad. Sci.* (Plant Sci.), **94:** 355-361.

Kohlmeyer, J. (1985b) *Nimospora octonae*, a new marine ascomycete from Hawaii. *Can. J. Bot.*, **63:** 1122-1125.

Kohlmeyer, J. (1986) *Ascocratera manglicola* gen. et. sp. nov., and a key to the marine Loculoascomycetes on mangroves. *Can. J. Bot.*, **64:** 3036-3042.

Kohlmeyer, J. & Demoulin, V. (1981) Parasitic and symbiotic fungi on marine algae. *Bot. Mar.*, **24:** 9-18.

Kohlmeyer, J. & Hawkes, M.W. (1983) A suspected case of mycophycobiosis between *Mycosphaerella apophlaeae* (Ascomycetes) and *Apophylaea* spp. (Rhodophyta). *J. Phycol.*, **19:** 257-260.

Kohlmeyer, J. & Kohlmeyer, E. (1964-1969) "*Icones Fungorum Maris*" J. Cramer. Weinhein Fasc. 1-7, pls. 1-90.

Kohlmeyer, J. & Kohlmeyer, E. (1965) New marine fungi from mangroves and trees along eroding shorelines. *Nova Hedwigia*, **9:** 89-104.

Kohlmeyer, J. & Kohlmeyer, E. (1971) Marine fungi from tropical America and Africa. *Mycologia*, **63:** 831-861.

Kohlmeyer, J. & Kohlmeyer, E. (1977) Bermuda marine fungi. *Trans. Br. Mycol. Soc.*, **68:** 207-219.

Kohlmeyer, J. & Kohlmeyer, E. (1979) *Marine Mycology: The Higher Marine fungi,*. Academic press, New York, pp. 689.

Kohlmeyer, J. & Schatz, S. (1985) *Aigialus* gen. nov. (Ascomycetes), with two new marine species from mangroves. *Trans. Br. Mycol. Soc.*, **85:** 699-707.

Kohlmeyer, J. & Vittal, B.P.R. (1986) *Lophiostoma mangrovie*, a new marine ascomycete from the tropics. *Mycologia,* **78:** 485-489.

Kohlmeyer, J. & Volkmann-Kohlmeyer, B. (1987a) Marine fungi from Aldabra, the Galopagosn and other tropical Islands. *Can. J. Bot.*, **65:** 571-582.

Kohlmeyer, J. & Volkmann-Kohlmeyer, B. (1987b) Koralionastetaceae fam. Nov. (Ascomycetes) from coral rock. *Mycologia*, **79:** 764-778.

Kohlmeyer, J. & Volkmann-Kohlmeyer, B. (1987c) *Thalassogena*, a new Ascomycetous genus in the Halosphaeriaceae. *Systema Ascomycetum*, **6:** 223-228.

Kohlmeyer, J. & Volkmann-Kohlmeyer, B. (1988a) *Halographis* (Opegraphales), a new endolithic lichenoid from corala and snails. *Can. J. Bot.*, **66:** 1138-1141.

Kohlmeyer, J. & Volkmann-Kohlmeyer, B. (1988b) *Ophiodeira* gen. nov. (Halosphaeriales) a survey of higher marine fungi from Saint Croix (Virgin Island) *Can. J. Bot.*, **66:** 2062-2067.

Kohlmeyer, J. & Volkmann-Kohlmeyer, B. (1989a) A new *Lulworthia* (Ascomycotina) from Corals. *Mycologia,* **81:** 289-292.

Kohlmeyer, J. & Volkmann-Kohlmeyer, B. (1989b) Hawaiian marine fungi, including two genera of Ascomycotina. *Mycol. Res.*, **92:** 410-421.

Kohlmeyer, J. & Volkmann-Kohlmeyer, B. (1990a) New species of *Koralionastes* (Ascomycotina) from the Caribbean and Australia. *Can. J. Bot.,* **68:** 1554-1559.

Kohlmeyer, J. & Volkmann-Kohlmeyer, B. (1990b) Revision of marine species of *Didymosphaeria* (Ascomycotina). *Mycol. Res.*, **94:** 685-690.

Kohlmeyer, J. & Volkmann-Kohlmeyer, B. (1991) Illustrated key to the filamentous Higher marine fungi. *Bot. Mar.*, **34:** 1-61.

Kohlmeyer, J. & Volkmann-Kohlmeyer, B. (1992 a) *Corallicola nana* gen. *et.* sp. nov. and other Ascomycetes from coral reefs. *Mycotaxon,* **44:** 417-424.

Kohlmeyer, J. & Volkmann-Kohlmeyer, B. (1992 b) Two Ascomycotina from Coral Reefs in the Caribbean and Australia. *Cryptogame Botany,* **2:** 367-374.

Kohlmeyer, J. & Volkmann-Kohlmeyer, B. (1993a) A new marine *Dryosphaera* (Ascomycotina) from the tropics. *Can. J. Bot.*, **71:** 992-995.

Kohlmeyer, J. & Volkmann-Kohlmeyer, B. (1993b) Observations on *Hydronectria* and *Kallicroma* gen. nov. *Mycol. Res.*, **97:** 753-761.

Kohlmeyer, J. & Volkmann-Kohlmeyer, B. (2000) Fungi on *Juncus roemerianus.* 14. Three new Coelomycetes, including *Floricola,* anam. Gen. nov. *Bot. Mar.,* **43:** 385-392.

Kohlmeyer, J. & Volkmann-Kohlmeyer, B. (2001a) The biodiversity of fungi on *Juncus roemerianus. Mycol. Res.*, **105:** 1411-1412.

Kohlmeyer, J. & Volkmann-Kohlmeyer, B. (2001b) Fungi on *Juncus roemerianus*: new Coelomycetes with notes on *Dwayaangam. Mycol. Res.,* **105:** 500-505.

Kohlmeyer, J. & Volkmann-Kohlmeyer, B. (2001c) Fungi on *Juncus roemerianus*. 16. More new Coelomycetes, including *Tetranacriella* gen nov. *Bot. Mar.,* **44:** 147-156.

Kohlmeyer, J. & Volkmann-Kohlmeyer, B. (2003) Marine Ascomycetes from Algae and Animal hosts. *Bot. Mar.*, **46:** 285-306.

Kohlmeyer, J., Schmidt, I. & Nair, N.B. (1967) *Eine neye Corollospora (Ascomycetes) qus dern Indischen ozeam and der ostsee Bericheten den denstschen bosanisctien Gesellschaft,* **80:** 98-102.

Kohlmeyer, J., Volkmann-Kohlmeyer, B. & Eriksson, O.E. (1995) Fungi on *Juncus roemerianus*. 2. New dictyosporous ascomycetes. *Bot. Mar.,* **38:** 165-174.

Kohlmeyer, J., Volkmann-Kohlmeyer, B. & Eriksson, O.E. (1997) Fungi on *Juncus roemerianus*. 9. New obligate and facultative marine ascomycotina. *Bot. Mar.,* **40:** 291-300.

Kolhmeyer, J., Spataflora, J.W. & Volkmann-Kolhmeyer, B. (2000) Lulworthales, a new order of marine Ascomycota. *Mycologia*, **2:** 453-458.

Kohlmeyer, J., Volkmann-Kohlmeyer, B. & Tsui, C.K.M. (2005) Fungi on *Juncus roemerianus*. 17. New Ascomycetes and Hyphomycetes genus *Kolletes* gen. nov. *Bot. Mar.,* **48:** 306-317.

Kovacik, L. & Pereira, A.B. (2001) Green alga *Prasiola crispa* and its licherized from *Mastodia tessellate* [Sic!] in Antarctic environment: general aspects. *Nova Hedwigia, Beiheft,* **123:** 465 -478.

Kumaran, K.P.N., Shindikar, M. & Limaye, R.B. (2004a) Mangrove associated lignite beds of Malvan, Konkan: Evidence for higher sea-level during the late Tertiary (Neogene) along the west coast of India. *Curr. Sci.*, **86:** 335-340.

Kumaran, K.P.N., Shindikar, M. & Limaye, R.B. (2004b) Fossil record of marine manglicolous fungi from Malvan (Konkan) west coast of India. *Ind. J. Mar.Sci.*, **33:** 257-261.

Kumaresen, V. & Suryanarayanan, T.S. (2001) studied occurrence and distribution of endophytic fungi in a mangrove community. *Mycol. Res.*, **105:** 1388-1391.

Kumaresen, V. & Suryanarayanan, T.S. (2002) Enophyte assemblage in young, mature and senescent leaves of *Rhiziphora apiculata*: evidence for the role of endophytes in mangrove litter degradation. *Fungal Diversity*, **9:** 82-91.

Kumaresan, V., Suryanarayana, T.S. and Johnson, J.A. (2002) Ecology of mangrove endophytes. In: *Fungi in Marine Environment* (ed. Hyde, K.D.), *Fungal Diversity,* **7:** 45-66.

Kupta, J., Anke, T., Stenglich, W. & Zechlin, L. (1981) Antibitics from basidiomycetes. XI. The biological activity of siccayne from the marine fungus *Halocyphina villosa* J. & E. Kohlmeyer. *J. Nat. Prod.*, **34:** 298-304.

Kurtzman, C.P. & Robnett, C.J. (2007) Multigene phylogenetic analysis of the *Trichomonascus*, *Wickerhamiella* and *Zygoascus* yeast clades, and the proposal of *Sugiyamaella* gen. nov. and 14 new combinations. *FEMS Yeast Res.*, **7:** 141-151.

Kutty, S.N. & Philip, R. (2008) Marine yeasts, a review. *Yeast*, **7:** 465-483.

Lachance, M.A. & Starmer, W.T. (1998) Aquatic habitats, ecology and yeasts. In: *The Yeasts- A Taxonomic study* (eds. Kurtzman, C.P. and Fell, F.W.), 4th edn. Elsevier, Amsterdam, pp. 21-30.

Lakhanpal, T.N. & Mukerji, K.G. (1981) *"Taxonomy of the Indian Myxomycetes"*. J. Cramer, Vaduz. pp. 1-530.

Lai, X., Cao, L., Tan, H., Fang, S., Huang, Y. & Zhou, S. (2007) Fungal communities from methane hydrate-bearing deep-sea marine sediments in South China Sea. *ISME J.* **1:** 75-76.

Leander, C.A. & Porter, D. (2000) Redefining the genus *Aplanchytrium* (Phylum Labyrinthulomycota). *Mycotaxon*, **76:** 339-444.

Le Calvez, T., Burgaud, G., Mahe, S., Barbier, G. & Vanden-Koornhuyse, P. (2009) Fungal diversity in deep-sea hydrothermal ecosystems. *Appl. Environ. Microbiol.*, **75:** 6415-6421.

Le Campion-Alsumard, T., Golubic, T. & Priess, K. (1995) Fungi in corals symbiosis or disease? Interactions between polyps and fungi causes pearl-like skeleton biominerilization. *Mar. Ecol. Prog. Ser.*, **117:** 137-147.

Leightly, L.E. (1980) Wood decay activities of marine fungi. *Bot. Mar.*, **23:** 387-395.

Leightly, L.E. & Eaton, R.A. (1979) *Nia vibrissa*, a marine white rot fungus. *Trans. Br. Mycol. Soc.*, **73:** 35-40.

Leong, W.F., Tan, T.K., Hyde, K.D. & Jones, E.B.G. (1990) *Payosphaeria minuta gen. et sp. nov.*, an Ascomycete on mangrove wood. *Bot. Mar.*, **33:** 551-514.

Leong, W.F., Tan, T.K., Hyde, K.D. & Jones, E.B.G. (1991) *Halosarpheia minuta* sp. nov., an ascomycete from submerged mangrove wood. *Can. J. Bot.*, **69:** 883-886.

Lin, Y.C., Shao, Z., Jiang, G.C., Zhou, S.N., Cai, J.W., Vrijmoed, L.L.P. & Jones, E.B.G. (2000) Penicilazine, a unique quinolone derivative with 4H-5,6-dihydro-1,2-oxazine ring system from the marine fungus *Penicillium* sp. (strain #386) from South China Sea. *Tetrahedron*, **56:** 9607-9609.

Lin, Y.C., Wu, X. Y., Feng, S. A., Jiang, G.C., Zhou, S.N., Vrijmoed, L.L.P. & Jones, E.B.G. (2001a) A novel N-cinnamoylcyclopeptide containg an allenic ether from the fungus *Xylaria* sp. (strain #2508) from the South China Sea. *Tetrahedron Lett.*, **42:** 449-451.

Lin, Y.C., Wu, X.Y., Feng, S.A., Jiang, G.C., Luo, J.H., Zhou, S.N., Vrijmoed, L.L.P., Jones, E.B.G., Krohn, K., Steingrover, K. & Zsila, F. (2001b) Five unique compounds: Xyloketols from mangrove fungus *Xylaria* sp. from South China Sea. *J. Org. Chem.*, **66:** 6252-6256.

Lin, Y.C., Wang, J., Wu, X. Y., Zhou, S.N., Vrijmoed, L.L.P. & Jones, E.B.G. (2002) A Novel compound, enniatin G, from the mangrove fungus *Halosarpheia* sp. (strain 731) from the South China Sea. *Aust. L. Chem.*, **55:** 225-227.

Lincoln, R., Boxshell, G., & Clark, P. (1998) "*A dictionary of Ecology, Evolution and Systematics*", 2nd ed., Cambridge University Press, Cambrige, U.K., **253:** 758-762.

Lintott, W.H. & Lintott, E.A. (2002) Marine fungi from New Zealand. In: *Fungi in marine environment* (ed. Hyde, K. D.), Fungal Diversity Res. Ser., **7:** 285-292.

Liu, J.K., Jones, E.B.G., Chukeatirote, E., Bahkali, A.D.H. & Hyde, K.D. (2011) *Lignincola conchicola* from palms with key to the species of *Lignincola*. *Mycotaxon*, **117:** 343-349.

Lud, D., Huiskes, A.H.L. & Ott, S. (2001) Morphological evidence for the symbiotic character of *Turgidosculum complicatium* Kohlm. & Kohlm. (*Mastodia tessellata* Hook F. and Harvey). *Symbiosis,* **31:** 141-151.

Lopez-Garcia, P., Rodriguez-Valera, F., Pedros-Allo, C. & Moreira, D. (2001) Unexpected diversity of small eukaryotes in dee-sea Antarctic plankton. *Nature*, **409:** 603-607.

Lopez-Garcia, P., Philip, H., Gall, F. & Moreira, D. (2003) Autochthonous eukaryotic diversity in hypothermal sediments and experimental microcolonizers at the Mid-Atlantic Ridge. *Proc. Natl. Acad. Sci.* USA, **100:** 697-702.

Loque, C.P., Medeiros, A.O., Pellizzari, F.M., Oliveira, E.C., Rosa, C.A. & Rosa, L.H. (2009) Fungal community associated with marine macroalgae from Antarctica. *Polar Biol.*, **33:** 641-648.

Maekawa, N., Suhara, H., Kinijo, K., Kondo, R. & Hashi, Y. (2005) *Haloaleurodiscus mangrovei* gen. sp. nov. (Basidiomycota) from mangrove forest in Japan. *Mycol. Res.*, **109:** 825-832.

Manimohan, P., Amritha, M. & Sairabanu, N.K. (2011) A comparison of diversity of marine fungi on three co-habiting mangrove plants. *Mycosphere*, **2:** 533–538.

Manoharachary, C. (2001) Biodiversity, Conservation and Biotechnology of fungi. *Presidential Address, Section –Botany, 89th Session of Indian Science Congress*, Lucknow.

Manoharachary, C., Sridhar, K.R, Singh Reena, Adholeya, A., Suryanarayanan, T.S., Rawat Seema & Johri, B.N. (2005) Fungal biodiversity: Distribution, conservation and prospecting of fungi from India. *Current science*, **89:** 58-60.

Mantel, P.G., Hawksworth, D.L., Pazoutova, S., Collinson, L.M. & Rassing, B.R. (2006) *Amorosia littoralis* gen. sp. nov., a new genus and species name for the scorpinone and caffeine-producing hyphomycetes from the littoral zone in the Bahamas. *Mycol. Res.*, **110:** 1371-1378.

Margules, C.R., Pressey, R.L. & Williams, P.H. (2002) Representing biodiversity: Data and producers for identifying priority areas for conservation. *J. Biosci.*, **27:** 309-326.

Maria, G.L. & Sridhar, K.R. (2002a) Richness and diversity of filamentous fungi on woody litter of mangroves along the west coast of India. *Current Science*, **83:** 1573-1580.

Maria, G.L. & Sridhar, K.R. (2002b) A new ascomycete, *Passeriniella mangrovei* sp. nov. from the mangrove forest of India. *Indian J. Forestry*, **25:** 319-322.

Maria, G.L. & Sridhar, K.R. (2003a) Diversity of filamentous fungi on woody litter of five mangrove plant species from the south west coast of India. *Fungal Diversity*, **14:** 109-126.

Maria, G.L. & K.R. Sridhar (2003b) Endophytic fungi assemblage of two halophytes from west coast mangrove habitatas, India. *Czech Mycol.*, **55:** 241-251.

Maria, G.L., Sridhar, K.R. & Raviraja, N.S. (2005) Antimicrobial and enzyme activity of mangrove endophytic fungi of southwest coast of India. *J. Agri. Techno.*, **1:** 67-80.

Maria, G.L. & Sridhar, K.R. (2004) Fungal Colonization of immersed wood in mangroves of the southwest coast of India. *Can . J. Bot.,* **82:** 1409-1418.

Maria, G.L., Sridhar, K.R. & Raviraja, N.S. (2005) Antimicrobial and enzyme activity of mangrove endophytic fungi of southwest coast of India. *J. Agri. Techno.*, **1:** 67-80.

Matondkar, S.G.P., Mahtani, S. & Mavinkurve, S. (1980a) Fungal flora of the mangrove swamps of Goa. *Mahasagar Bull. Nat. Inst. Oceanogr.,* **13:** 5.

Matondkar, S.G.P., Mahtani, S. & Mavinkurve, S. (1980b) Seasonal variation in microflora from mangrove swamps. *Ind. J. Mar. Sci*., **9:** 119-120.

Meyers, S.P. (1957) Taxonomy of marine Pyrenomycetes. *Mycologia*, **49:** 475-528.

Meyers, S.P. (1953) Marine fungi in Biscanyne Bay, Florida. *Bull. Mar. Sci. Gulf Caribb.*, **2:** 590-601.

Meyers, S. P. (1968) Observations on the physiological ecology of marine fungi. *Bull. Misaki. Mar. Biol. Inst. Kyoto Univ.*, **12:** 207-225.

Meyers, S.P. (1971a) Isolation and identification of marine fungi. In: *Marine borers, fungi and fouling organisms of wood* (eds. Jones, E.B.G. and Eltringham, S.K.), OECD, Paris, pp. 89-113.

Meyers, S.P. (1971b) Developments in the biology of filamentous marine fungi. In: *Marine borers, fungi and fouling organisms of wood* (eds. Jones, E.B.G. and Eltringham, S.K.), OECD, Paris, pp. 217-258.

Meyers, S.P. (1996) Fifty years of marine mycology: highlights of the past, projections for the coming century. *SIMS News*, **46:** 119-127.

Meyers, S.P., Ahearn, D.G., Gunkel, W. & Roth, F.J. (1967) Yeasts from the North Sea. *Mar. Biol*., **1:** 118-134.

Meyers, S.P. and Moore, R.T. (1960) Thalassiomycetes. II. New genera and species of Deuteromycetes. *Am. J. Bot*., **47:** 345-349.

Molitoris, H.P. & Schaumann, K. (1986) Physiology of marine fungi. In: *The biology of marine fungi* (ed. Moss, S.T.), Cambridge Univ. Press, pp. 35-47.

Moore, R.T. & Meyers, S.P. (1959) Thalassiomycetes I. Principals of delimitation of the marine mycota with the descriptions of a new aquatically adapted deuteromycete genus. *Mycologia*, **51:** 871-876.

Moro, I., Negrisolo, E., Callogaro, A. & Andreoli, C. (2003) *Aplanochytrium stocchinoi*, a new Labyrinthulomycota from the Southern Ocean (Ross Sea, Antarctica). *Protist*, **154:** 331-340.

Morrison-Gardiner, S. (2002) Dominant fungi from Australian reefs. *Fungal Diversity*, **9:** 105-121.

Moss, S.T. (1985) An ultrastructural study of taxonomically significant character of the Thrausthochytriales Labyrinthulales. *Bot. J. Linn. Soc.*, **91:** 329-257.

Moss, S.T. (ed.) (1986) *"The Biology of marine fungi"*. Cambridge University Press, Cambridge. pp. 1-382.

Mouzouras, R. (1986) Pattern of timber decay caused by marine fungi. In: *"The Biology of marine fungi"* (ed. Moss, S.T.), Cambridge University Press, Cambridge. pp. 341-353.

Mouzouras, R. (1989) Soft rot decay of wood by marine microfungi. *J. Inst. Wood Sci.*, **11:** 393-401.

Mouzouras, R., Jones, E.B.G., Venkatsamy, R. & Holt, D.M. (1988) Microbial decay of lignocellulose in the marine environment. In: *Marine Biodeterioration* (eds. Thompson, M.F., Sarojini, R. and Nagabhushanam, R.), Oxford and JBH Publishing, New Delhi, pp. 329-354.

Mugambi, G.K. & Huhndorf, S.M. (2009) Molecular phylogenetics of Pleosporales: Melanommataceae and Lophiostomataceae re-circumscribed (Pleosporomycetidae, Dothideomecetes, Ascomycota). *Studies Mycology*, **64:** 103-121.

Muller, G., Kupper, F. C. & Kupper, H. (1999) Infection experiments reveal broad host range of *Eurychama dicksonii* (Oomycota) and *Chytridium polysiphoniae* (Chytridiomycota), two eukaryotic parasites in marine brown algae. *Phycol. Soc.*, **4:** 2-23.

Nagno, Y., Nagahama, T., Hatada, Y., Numoura, T., Takami, H., Miyazaki, J., Takai, K. & Horikoshi, K. (2010) Fungal diversity in deep-sea sediments- the presence of novel fungal groups. *Fungal Ecol.*, **3:** 316-325.

Nair, L.N., Rao, V.P. & Chaudhuri, S. (1991) Microflora of Avicennia officinalis Linn. In: *Proceedings of the Symposium on significance of mangroves* (eds. Agate, A.D., Bonde, S.D. and Kumaran, K.P.), Ahgarkar Research Institute, Pune, pp. 52-55.

Nair, N.B. (1970) The problem of timer destroying organisms along the Indian coast. *Proc. Int. Cong. Mari. Coross. Foul.* 2nd 1968, pp. 1-7.

Nair, N.B., Fernandez, T.V., Salim, M. & Balakrishnan, M. (1987a) Ecology of estuarine boring organisms along the south west coast of India. In: Proceedings of the National Seminar on Estuarine Management (ed. N.B. Nair), ST Joseph Press, Trivandraum, Kerala, India. pp. 492-500.

Nair, N.B., Chandran Shoba R., Ammal Rathi, S., Maya, S. & Suryanarayanan, H. (1987b) Vertical distribution of certain algae and other organisms in brackish water lagoon in Kerala. In: Proceedings of the National Seminar on Estuarine Management (ed. N.B. Nair), ST Joseph Press, Trivandraum, Kerala, India. pp. 296-300.

Nakagiri., A. (1989) Marine fungi in sea foam from Japanese coast. *IFO Res. Commu.,* **14:** 52-79.

Nakagiri, A. & Ito, T. (1994) *Aniptodera salsuginosa*, a new mangrove-inhabiting ascomycete, with observation on the effect of salinity on ascospore appendage morphology. *Mycol. Res.*, **98:** 931-936.

Nakagiri, A. & Tokura, R. (1987) Taxonomic studies of the genus *Corollospora* (Halosphaeriaceae, Ascomycotina) with description of seven new species. *Trans. Mycol. Soc. Japan*, **28:** 413-436.

Nakagiri, A. & Tubaki, K. (1982) A new marine ascomycete and its anamorph from Japan. *Trans. Mycol. Soc. Japan*, **23:** 101-110

Nakagiri, A. & Tubaki, K. (1983) *Lindra obtuse*, a new marine Ascomycete and its *Anguillospora* anamorph. *Mycologia*, **75:** 487-497.

Nakagiri., A., Okane, I. & Ito., T.(1999) Geogrophical and seasonal distribution of arenicolous marine fungi along the pacific coast of the Bouson peninsula. *IFO Res. Commun.*, **19:** 22-33.

Nambiar, G.R. & Raveendran, K. (2006) A comparative account of Pokkali and mangrove associated marine and manglicolous marine fungi from Valapattanam estuary, Kannur District (Kerala). *Extended Abstract*, 18th. *Kerala Science Congress*, CESS, Akkulam, Thiruvananthpuram, pp. 559-561.

Nambiar, G.R. & Raveendran, K. (2007) Estuaries marine mycoflora of North Malabar (Kerala). *J. Mar. Atmos. Res.*, **3:** 29-31.

Nambiar, G.R. & Raveendran, K. (2008a) Diversity of Mangrove fungi of North Malabar Kerala. *The Indian Forester*, **134:** 1658-1662.

Nambiar, G.R. & Raveendran, K. (2008b) Marine and manglicolous fungal diversity in the coastal wetlands of Kerala. *Seaweed Res. Utiln., (Special Issue)*, **30:** 107-111.

Nambiar, G.R. & Raveendran, K. (2008c) Marine mycoflora of Pondicherry and Mahe. *Eco- Chronical,* **3:** 47-50.

Nambiar, G.R. & Raveendran, K. (2008d) Impact of coir retting on manglicolous marine fungi of Kerala coastal waters. *Poll. Res.*, **27:** 481-483.

Nambiar, G.R. & Raveendran, K. (2008e) Biodiversity of marine mangrove fungi of Valapattanam and Pichavaram mangrove forest (South India). *Eco-Chronicle*, **3:** 137-140.

Nambiar, G.R. & Raveendran, K. (2009a) Influence of coir retting on mangrove ecosystem. In: *Proceeding of National Seminar on Conservation of Biodiversity in protected areas-opportunities and challenges* (ed. Antoney, P.V.), Chist University Bangalore (India), pp.34-36.

Nambiar, G.R. & Raveendran, K. (2009b) Lignicolus marine fungi in selected wetland of Nort Malabar (Kerala). In: *Compendium on wetland Biodiversity and Conservation* (eds. Bijoy Nandan S., Salim M., Somanathan Pillai G. and Reeny Varghese), Published by LAK, *Irinjalakuda, Kerala,* Pp. 136-140.

Nambiar, G.R. & Raveendran, K. (2009c) Frequency of arenicolous marine fungi on sand buried wood substrates along Kerala beaches. In: *Biodiversity, Sustainable Development and Human Welfare*, (eds. Nandan, S.N. et al.,), Sripushpa Press, Dhule, Maharashtra, India. pp. 394-399.

Nambiar, G.R. & Raveendran, K. (2009d) Manglicolous marine fungi on *Avicennia* and *Rhizophora* along Kerala coast (India). *Middle-East J.Sci. Res.,* **4:** 48-51.

Nambiar, G.R. & Raveendran, K. (2009e) Exploration of untrapped potentiality of coastal paddy fields of Kerala (India) - A case study. *Middle East J. Sci. Res.,* **4:** 44-47.

Nambiar, G.R. & Raveendran, K. (2009f) A checklist of marine fungi Kerala state, India. *American-Eurasian J. Botany*, **2:** 57-61.

Nambiar, G.R. & Raveendran, K. (2009g) Frequency and abundance of marine mycoflora in mangrove ecosystem of North Malabar, Kerala (India). *Acad. J. Pl. Sci.*, **2:** 65-68.

Nambiar, G.R. & Raveendran, K. (2009h) Marine mycoflora on *Avicennia* and *Bruguiera* along Kerala coast (India). *Acad. J. Pl. Sci.*, **2:** 78-81.

Nambiar, G.R. & Raveendran, K. (2009i) Marine mycoflora in backwater ecosystem of Kerala, India. *J. Environ. Biol.*, **30:** 30-32.

Nambiar, G.R. & Raveendran, K. (2009j) Marine mycoflora of Andhra Pradesh, India. *Eco-Chronicle*, **3:** 13-16.

Nambiar, G.R. & Raveendran, K. (2009k) Marine mycoflora of south India with special emphasis to lignicolous marine fungi. *Front. Biol.China,* **4:** 436-441.

Nambiar, G.R. & Raveendran, K. (2009l) Eco-biodiversity of marine mycoflora of Gulf of mannar biosphere reserve (India). *Ecology, Environ. & Cons.*, **15:** 739-743.

Nambiar, G.R. & Raveendran, K. (2009m) Manglicolous marine fungi of Kerala (South India). *Bot. Res. Internat.*, **2:** 206-210.

Nambiar, G.R. & Raveendran, K. (2010a) Frequency and abundance of arenicolous marine fungi along South Indian Beaches. *J. Sci. Res.*, **2:** 138-143.

Nambiar, G.R. & Raveendran, K. (2010b) Impact of hospital waist on mangrove fungi of Kerala. *Poll. Res.*, **29:** 59-61.

Nambiar, G.R. & Raveendran, K. (2010c) Frequency and abundance of marine fungi of South India. *Asian J. Micro. Biotech. & Environ.Sci.,* **12:** 103-108.

Nambiar, G.R. & Raveendran, K. (2010d) Arenicolous marine fungi of Kerala. *Asian J. Micro. Biotech. & Environ. Sci.,* **12:** 117-120.

Nambiar, G.R. & Raveendran, K. (2010e) Frequency and abundance of algicolous marine fungi along Indian coast. *Seaweed Res. Utlin.*, **30:** 1-5.

Nambiar, G.R. & Raveendran, K. (2011a) *Arenariomycec triseptatus* and *Halosphaeria appendiculata*: Two new record of marine fungi from India. *Geobios*, **41:** 17-19.

Nambiar, G.R. & Raveendran, K. (2011b) *Anthostomella nypensis* and *Biconiosporella corniculata*: Additions to the marine mycoflora of India. *Geobios,* **43:** 24-27.

Nambiar, G.R., Pratheesh, P.V. & Raveendran, K. (2006) Pokkali associated marine fungi of north Malabar of Kerala coastal waters. *Seaweed Res. Util.*, **28:** 55-61.

Nambiar, G.R., Jaleel, C. & Raveendran, K. (2008a) A comparative account of backwater and brackish water marine mycoflora of North Malabar Kerala (India), *C. R. Biologies*, **331:** 294-297.

Nambiar, G.R., Raveendran, K., Zhao, C. & Jaleel, C.H. (2008b) A glimpse of lignicolous marine fungi occurring in coastal water bodies of Tamil Nadu (India). *C. R. Biologies*, **331:** 475-480.

Nandan, S.N., Shinde, D.N. & Borse, B.D. (1993) Marine fungi from Goa Coast (India). *Biol. Ind.,* **4:** 29-34.

Newell, S.Y. & Fell, J.W. (1980) Mycoflora of turtlegrass (*Thalassia testudinum*) as recorded after seawater incubation. *Bot. Mar.*, **23:** 165-275.

Norkrans, B. (1966) Studies on marine occurring yeasts. Growth related to pH, NaCI concentration and temperature. *Arch. Mikrobiol.*, **54:** 374-392.

Norkrans, B. & Kylin, A. (1969) Regulation of potassium to sodium and of the osmotic potential in relation to salt tolerance in yeasts. *J. Bacteriol.*, **100:** 836-845.

Nilsson, S. (1964) Freshwater Hyphomycetes. Taxonomy, morphology and ecology. *Symb. Bot. Ups.*, **18:** 1-130.

Okane, I., Nakagiri, A. & Ito, T. (2001) Assemblage of endophytic fungi on *Bruguiera gymnorrhiza* in the Shiira River Basin, Irimote Island. *IFO Research*, **20:** 41-49.

Olive, L.S. (1975) "*The* Mycetozooans". Academic press, New York.

Pan, J.H., Jones, E.B.G., She, Z.G., Pang, J.Y. & Lin, Y.C. (2008) Review of bioactive compounds from fungi in the South China Sea. *Bot. Mar.*, **51:** 179-190.

Pande, Alaka (2008) "*Ascomycetes of Peninsular India*". Scientific Publishers (India), Jodhpur, India. pp. 1-568.

Panebianco, C., Tam, W.Y. & Jones, E.B.G. (2002) The effect of preinoculation of balsa wood by selected marine fungi and their effect on subsequent colonization in the sea. In: *Fungal Succession* (eds. Hyde, K.D. & Jones, E.B.G.), *Fungal Diversity*, **10:** 77-88.

Pang, K.L. (2002) Systematics of the Halosphaeriales which morphological characters are important? In: *Fungi in marine environment* (ed. Hyde, K.D.), *Fungal Diversity Res. Ser.,* **7:** 35-57.

Pang, K.L., Abdel-Wahab, M.A., El-Sharouney, H.M., Sivichai, S. & Jones, E.B.G. (2002) Jahnulales (Dothideomycetes, Ascomucota) a new order of lignicolous freshwater ascomycetes. *Mycol. Res.*, **106:** 1013-1042.

Pang, K.L., Vrijmoed, L.L.P., Kong, R.Y.C. & Jones, E.B. (2003) *Lignincola* and *Nais*, polyphyletic genera of the Halosphaeriales (Ascomycota). *Mycologycal Progress*, **2:** 29-36.

Pang, K.L., Jones, E.B.G., Vrijmoed, L.L.P. & Vikineswary, S. (2004) *Okeanomyces*, a new genus to accommodate *Halosphaeria cucullata* (Halosphaeriales, Ascomycota). *Bot. J. Linn. Soc.*, **146:** 223-229.

Pang, K.L., Jones, E.B.G., She, Z.Y. & Ling, Y.C. (2008a) Review of bioactive compounds from fungi in the South China Sea. *Bot. Mar.*, **51:** 179-190.

Pang, K.L., Vrijmoed, L.L.P., Goh, T.K., Plaingam, N. & Jones, E.B.G. (2008b) Fungal endophytes associated with *Kandelia candel* (Rhizophoraceae) in Mai Po Nature Reserve, Hong Kong. *Bot. Mar.*, **51:** 171-178.

Pang, K.L., Cheng, J.S. & Jones, E.B.G. (2010a) *Marine mangrove fungi of Taiwan.* National Taiwan Ocean University, Chilung, pp. 1-131.

Pang, K.L, Alias, A.S., Chlang, M.W.L., Vrijmoed, L.L.P. & Jones, E.B.G. (2010b) *Sedecimlella taiwanensis* gen. et sp. nov., a marine mangrove fungus in the Hypocreales (Hypocreomycetidae, Ascomycota). *Bot. Mar.*, **53:** 493-498.

Pang, K.L, Sharuddin, S.S., Alias, A.S., Nor, N.A. M. & Awaluddin, H.H. (2010c) Diversity and abundance of lignicolous marine fungi from the east and west coasts of Pennisular Malaysia and Sabah (Borneo Island). *Bot. Mar.*, **53:** 515-523.

Pang, K.L., Jheng, J.S., Koch, J. & Jones, E.B.G. (2011) Taxonomy and phylogeny of *Arenariomyces* (Halosphaeriaceae, Ascomycota). Abstract for 12 th International marine and freshwater mycology symposium, Incheon, Republic of Korea, pp. 437.

Patersen, R.A. (1958) Parasitic and saprophytic phycomycetes which invade planktonic organisms 1. New taxa and records of Chytridiaceous fungi. *Mycologia*, **20:** 85-96.

Patersen, K.R.L. & Koch, J. (1997) Substrate preference and vertical zonation of lignicolous marine fungi on mooring posts of oak (*Quercus* sp.) and larch (*Larix* sp.) in Svanemollen harbour, Denmark. *Bot. Mar.*, **40:** 451-463.

Patil, K.B. & Borse, B.D. (2001) Studies on higher marine fungi from Gujarat coast (India). *Geobios*, **28:** 41-44.

Patil, S.D. & Borse, B.D. (1982) Marine fungi from Maharashtra (India) - I. The genus *Halosarpheia* J. and E. Kohlm. *Indian Bot. Reptr.*, **1:** 102-106.

Patil, S.D. & Borse, B.D. (1983a) Marine fungi from Maharashtra (India) - II. Some Arenicolous fungi . *Indian Bot. Reptr.,* **2:** 86-87.

Patil, S.D. and Borse, B.D. (1983b) Marine fungi from Maharashtra (India) - III. Some fungi from mangroves. *Indian Bot Reptr.*, **2:** 56-58.

Patil, S.D. & Borse, B.D. (1985) Marine fungi from Maharashtra (India) - IV : Some Loculoascomycetes. *Trans. Mycol. Soc. Japan,* **26:** 271-276.

Patil, S.D. & Borse, B.D. (1986) Marine fungi from Indian mangroves. In : *The mangroves* (ed. Bhosale, L.J.), Shivaji University Press, Kolhapur, India, pp. 151-152.

Pawar, N.S. & Borse, B.D. (2004) Marine fungi from Sundarbans (India) -IV. *J. Adv. Sci. & Tech.*, **7:** 17-28.

Pawar, N.S. & Borse, B.D. (2005a) Marine fungi from Sunderbans (India) - III. In: *Emerging Trends in Mycology, Plant Pathology and Microbial Biotechnology* (eds. Bhagyanarayana et al.), B.S. Publication Hyderabad, PP. 170-179.

Pawar, N.S. & Borse, B.D. (2005b) Marine fungi from Sundarbans (India) - V. *J. Phytol. Res.,* **18:** 1-9.

Pawar, N.S., Pagar, D.V., Borse, K.N. & Borse, B.D.(2005) Marine fungi from Sundarbans (India) - VI. In: *Plant Diversity and Biotechnology* (eds. Nandan et al.), Pub. SSVPS's Sci. College, Dhule, maharashtra, pp. 107-115.

Pawar, N.S., Borse, K.N. & Borse, B.D. (2009) Marine fungi from Sundarbans (India) -VII. In: *Biodiversity, Sustainable Development and Human Welfare*, (eds. Nandan et al.,), Pub. SSVPS's Sci. College, Dhule, Maharashtra. pp. 91-103.

Pawar, V.H. & Thirumalachar, M.J. (1966) Studies on halophilic soil fungi from Bombay. *Nova Hedwigia*, **12:** 497-508.

Pawar, V.H., Mathur, P.N. & Thirumalachar, M.J. (1967) Species of *Phoma* isolated from marine soils in India. *Trans. Br. Mycol. Soc.*, **50:** 259-265.

Phongpaichit, S., Preedana, S., Rungiondamai, N., Sakayaroj, J., Benzies, C., Chuaypat, J. & Plathong, S. (2006) Aspergillosis of the gorgonian sea fan *Annella* sp. after the 2004 tsunami at Mu Ko Similan National Park, Andaman Sea, Thailand. *Coral Reefs.*, doi: 1007/s00338-006-0104-y

Pilantanapak, A., Jones, E.B.G. & Eaton, E.A. (2005) Marine fungi on *Nypa fruticans* in Thailand. *Bot. Mar.*, **48:** 1-9.

Pivikin, M.V. (2000) Filamentous fungi associated with holthurians from the Sea of Japan, off the Primorye coast of Russia. *Biol. Bull.*, **198:** 101-109.

Pivikin, M.V., Afyatullov, S.S. & Elyakov, G.B. (1999) Biodiversity of marine fungi and new bioactive substances from them. In: *From organisms to ecosystems in the Pacific* (eds. Chou, C.H., Walker, G.R. and Reinhardt, C.), Biodiversity Alleoathy, pp. 91-99.

Poch, G.K. & Gloer, J.B. (1989) Helicascoides A and B: new lactone from the marine fungus *Helicascus kanaloanus*. *J. Nat. Prod.*, **52:** 257-260.

Pointing, S.B. (2001) Feasability of bioremidation by white-rot fungi. *Appl. Microbiol. Biotechnol.*, **57:** 20-33.

Pointing, S.B., Vrijmoed, L.L.P. & Jones, E.B.G. (1998) A qualitative assessment of lignocellulose degrading enzyme activity in marine fungi. *Bot. Mar.*, **41:** 293-298.

Pokorny, K.S. (1967) Labyrinthula. *J. Protozool.*, **14:** 697-708.

Poon, M.O.K. & Hyde, K.D. (1998a) Biodiversity of intertidal estuarine fungi on *Phragmites* at Mai Po marshes, Hong Kong. *Bot. Mar.*, **41:** 141-155.

Poon, M.O.K. & Hyde, K.D. (1998b) Evidence for the vertical distribution of saprophytic fungi on senescent *Phragmites australis* culms at Mai Po Marshes, Hong Kong. *Bot. Mar.*, **41:** 285-292.

Porter, D. (1990) Phylum Labyrinthulomycota. In: *Handbook of Protoctista* (eds. Margulis, L., Corliss, J.O., Melkonian, M. and Chapman, D.J.), Jones and Bartlett Publishers, Boston.

Porter, D. & Farnham, W.F. (1986) *Mycaureola dilseae* a marine basidiomycete Parasite on the red alga, *Dilsea carnosa. Trans. Br. Mycol. Soc.,* **85:** 575-582.

Porter, D. & Lingle, W.F. (1982) Endolithic thraustochytrid marine fungi from planted shell fragments. *Mycologia*, **84:** 289-299.

Postlethwait, J.H. & Hoson, J.L. (1989) "*The Nature of Lice*". Mc.-Graw Hill Publishing Co., New York. pp. 1-820.

Prabhakaran, N., Gupta, R. & Krishnakutty, M. (1987) Fungal activity in Mangalvan: an estuarine ecosystem. In: Proceedings of the National seminar on Estuarine Management (ed. Nair, N.B.), STEC, Kerala. pp. 458-463.

Prasannarai, K. (1999) *Studies on the marine fungi from the west coast of India*. Unpublished Ph. D. Thesis, Mangalore University, Karnataka, India. pp. 1-170.

Prasannarai, K. & Sridhar, K.R. (1997) Effect of incubation period of drift wood on the occurrence of marine fungi. *Indian Journal of Marine Sciences,* **26:** 380-382.

Prasannarai, K. &. Sridhar, K.R. (2001) Diversity and abundance of higher marine fungi on woody substrates along the west Coast of India. *Current Science,* **81:** 303-311.

Prasannarai, K. & Sridhar, K.R. (2000-2001) Observations on Intertidal marine fungi of islands adjacent to the west coast of India. *Kavaka,* **28** & **29:** 27-33.

Prasannarai, K., Ananda, K. & Sridhar, K.R. (1999) Intertidal fungi in Magalore Harbour, Southern India. *Bot. Mar.,* **42:** 117-122.

Prasannarai, K. Ananda, K. & Sridhar, K.R. (2000) *Corollospora indica*, a new marine fungal species from west coast at India. *Journal of Environmental Biology,* **21:** 235-239.

Proksch, P., Ebel, R., Edrada, R., Riebe, F., Liu, H., Diesel, A., Bayer, M., Li, X., Lin, W.H., Grebenyuk, V., Mueller, W.E.G., Draeger, S., Zuccaro, A. & Schulz, B. (2008) Sponge-associated fungi and their bioactive compounds: the suberites case. *Bot. Mar.*, **51:** 209-218.

Pueschel, C.M. & Vandermee, J.P. (1985) Ultrastructure of the fungus *Petersenia palmariae* (Oomycota) parasitic on algae *Palmaria mollis*: (Rhodophyceae). *Can. J. Bot.*, **63:** 409-418.

Purvetkar, R.R., D'Souza, C., Tripathi, A. & Naik, C.G. (2009) Aspernoilides A and B and Butenolides from marine-derived fungus *Aspergillus terreus*. *Phytochemistry*, **70:** 128-132.

Raghukumar, C. (1986a) Fungal Parasites of the Marine Green Algae, *Cladophora* and *Rhizoclonium. Bot. Mar.,* **29:** 289-297.

Raghukumar, C. (1986b) The occurrence of *Chytridium polysiphoniae,* a fungal pathogen on the red alga, *Centroceras clavulatum* (C. Agardh) Montagne from Goa. *Ind. J. Mar. Sci.,* **15:** 42-44.

Raghukumar, C. (1987a) Fungal parasites of marine algae from Mandapam (Sauth India). *Dis. Aquat. Organ.,* **3:** 137-145.

Raghukumar, C. (1987b) Fungal parasites of marine algae, *Cladophora* and *Rhizoclonium. Bot. Mar.*, **29:** 289-297.

Raghukumar, C. (1996) Zoosporic fungal parasites of marine biota. In: *Advances in zoosporic fungi* (ed. Rayal, R.), M.D. Publications Pvt. Ltd. New Delhi. pp. 61-83.

Raghukumar, C. (1998) Barotolerance of fungi isolated from deep sea sediments of the Indian ocean. *Aquatic Microbial Ecology*, **15:** 153-163.

Raghukumar, C. (2000) Fungi from marine habitats: an application in bioremediation. *Mycol. Res.*, **104:** 1222-1226.

Raghukumar, C. (2002) Bioremediation of colored pollutants by terrestrial versus facultative marine fungi. In: *Fungi in Marine Environment* (ed. K.D. Hyde), *Fungal Diversity,* **7:** 317-344.

Raghukumar, C. (2004) Marine fungi and their enzymes for decolorization of coloured effluents. In: *Marine Microbiology: Facets and Opportunities.* (ed. Ramaiah, N.), National Institute of Oceanography, Goa. pp. 203-224.

Raghukumar, C. (2006) Algal-fungal interactions in the marine ecosystem: symbiosis to parasitism. In: *Recent advances on applied aspects of Indian marine algae wih reference to global scenario.* Vol. I (ed. Tewari, A.), Central salt and marine chemicals research institute, Bhavnagar. pp. 366-385.

Raghukumar, C. (2008) Marine fungal Biotechnology: an ecological perspective. *Fungal Diversity,* **31:** 19-35.

Raghukumar, C. & Lande, V. (1988) Shell disease of rock oyster *Crassostrea cucullata. Dis. Aquat. Org.*, **4:** 77-81.

Raghukumar, C. & Raghukumar, S. (1991) Fungal invasion of massive corals. *P.S.Z.N.I: Marine Ecology*.**12:** 251-260.

Raghukumar, C. & Raghukumar, S. (1994) Investigation of fungal parasites of marine Algae in India: An appraisal. In: *Recent Advances in Phycology* (eds. Kashyap, A.K. and Kumar, H.D.), Rastogi Publications, Meerut, India. pp. 40-50.

Raghukumar, C. & Raghukumar, S. (1998) Barotolerance of fungi isolated from deep-sea sediments of the Indian Ocean. *Aquatic Microbial Ecology,* **15:** 153-163.

Raghukumar, C., Damare, S. & Singh, P. (2010) A review of deep sea fungi: occurrence, diversity and adaptations. *Bot. Mar.*, **53:** 479-492.

Raghukumar, C., D'Souza, T.M., Thom, R.G. & Reddy, C.A. (1999) Lignin modifying enzymes of *Flavodon flavus*, a basidiomycete isolated from a coastal marine environment. *Appl. Environ. Microbiol.*, **65:** 2100-2111.

Raghukumar, C. Nagarkar, S. & Raghukumar, S. (1992a) Association of Traustochytrids and fungi with living marine algae. *Mycological Reasearch*, **96:** 542-546.

Raghukumar, C., Chandramohan, D., Michel, F.C. & Reddy, C.A. (1996) Degradation of lignin and decolourization of paper mill bleach plant effluent (BPE) by marine fungi. *Biotechnology letters*, **18:** 105-106.

Raghukumar, C., Raghukumar, S., Sharma, S. & Chandramohan, D. (1992b) Endolithic fungi from deep-sea calcareous substrata: Isolation and laboratory studies. In: *Oceanography of the Indian Ocean* (ed. B.N. Desai), Oxford IBH Publication., New Delhi. pp. 3-9.

Raghukumar, C., Raghukumar, S., Chinnaraj, S., Chandramohan, D., D'zouza, T.M. & Reddy , C.A. (1994) Laccase and other lignocelluslose modifying enzymes of marine fungi isolated from the coast of India. *Bot. Mar.,* **37:** 515-523.

Raghukumar, C., Muraleedharan, U.D., Goud, V.R. & Mishra, R. (2004a) Xylanases of marine fungi of potential use for biobleaching of paper pulp. *Journal of Industial Microbiology and Biotechnology,* **31:** 433-441.

Raghukumar, C., Raghukumar, S., Sheelu, G., Gupta, S.M., Nath, B.N. & Rao, B.R. (2004b) Buried in time: culturable fungi in a deep-sea sediment core from the Chagos Trench, Indian ocean. *Deep-Sea Research,* **51:** 1759-1768.

Raghukumar, C., Shailaja, M. S., Parameswaran, P.S. & Singh, S.K. (2006) Removal of polycyclic aromatic hydrocarbons from aqueous media by the marine fungus NIOCC #312: involment of lignin-degrading enzymes and exopolysaccharides. *Ind. J. Mar. Sci.*, **35:** 380-387.

Raghukumar, S. (1971) *Studies on fungi from marine and brackish environment*. Unpublished Ph. D. Thesis, University of Madras.

Raghukumar, S. (1973) Marine lignicolous fungi from India. *Kavaka,* **1:** 73-85.

Raghukumar, S. (1985) Enumeration of thraustochytrids (heterotrophic microorganisms) from the Arabian sea. *Mahasagar, Bull. Natl. Inst. Oceanogr.*, **18:** 457-465.

Raghukumar, S. (1987) Occurrence of the thrastochytrid, *Corallochytrium limacisporum* gen. *et* sp. nov., in the coral reef lagoons of the Lakshadweep islands in the Arabian sea. *Bot. Mar.*, **30:** 83-89.

Raghukumar, S. (1988a) *Schizochytrium mangrovei* sp. nov., a Thraustochytrid from mangroves in India. *Trans. Br. Mycol. Soc.,* **90:** 627- 631.

Raghukumar, S. (1988b) Detection of the thraustochytrid protests *Ulkenia visurgensis* in a hydroid, using immunofuorescence. *Mar. Biol.*, **97:** 253-258.

Raghukumar, S. (1988c) Some interesting lignicolous marine fungi from the West Coast of India. *Kavaka,* **16:** 77-80.

Raghukumar, S. (1990) Speculations on niches occupied by fungi in the sea with relation to bacteria. *Proc. Indian Acad. Sci. (Pl. Sci.).* **100:** 129-145.

Raghukumar, S. (1996a) Morphology, Taxonomy and Ecology of Thraustochytrids and Labyrinthulids. The marine counterparts of zoosporic fungi. In: *Advances in zoosporic fungi* (ed. Dayal, R.), M.D. Publishers, New Delhi. pp. 35-45.

Raghukumar, S. (1996b) Fungi in the marine realm: Status Challenges and prospects. *Kavaka,* **22:** 25-34.

Raghukumar, S. (2002) Ecology of the marine protests, the Labyrinthulomycetes (Thraustochytrids and Labyrinthulids). *Europ. J. Protistol.,* **38:** 127-145.

Raghukumar, S. (2004) The role of marine detrital processes. In: *Marine Microbiology: Facets and Opportunities.* (ed. Ramaiah, N.), National Institute of Oceanography, Goa. pp.125-140.

Raghukumar, S. (2006) Marine Microbial eukaryotic diversity, with particular reference to fungi: lessons from prokaryotes. *Indian J. Mar. Sci.*, **35:** 388-398.

Raghukumar, S. (2008) Thraustochytrid marine protests: production of PUFAs and other emerging technologies. *Mar. Biotechnol.,* **10:** 631-640.

Raghukumar, S. & Balasubramanian, R. (1991) Occurrence of thraustochytrid fungi in corals and coral mucus. *Ind. J. Mar. Sci.*, **20:** 176-181.

Raghukumar, S. & Damare, V.S. (2011) Increasing evidence for the important role of Labyrinthulomycetes in marine ecosystems. *Bot. Mar.*, **54:** 3-11.

Raghukumar, S. & Raghukumar, C. (1988) A report on the occurrence of Thraustochytrid species in Indian waters. *Kavaka,* **16:** 40-44.

Raghukumar, S. & Raghukumar, C. (1999) Thraustochytrid fungoid protests in faecal pellets of the tunicate *Pegea confoederata*, their tolerance to deep-sea conditions and implication in degradation processes. *Mar. Ecol. Progr. Ser.*, **190:** 133-140.

Raghukumar, S., Raghukumar, C. & Rajendran, A. (1990) noticed abundance of thraustochytid fungi in the Arabian Sea. *East. Coastal & Shelf Sci.*, **31:** 351-358.

Raghukumar, S., Ramaiah, N. & Raghukumar, C. (2001) Dynamics of thraustochytrid protests in the water column of the Arabian Sea. *Aquat. Microbial. Ecol.*, **24:** 175-186.

Raghukumar, S., Santhakumaran, L.N. & Chandramohan, D. (1988b) Association of marine fungi on timber panels treated with preservatives In: *Marine Biodeterioration* (eds. M. F. Thompson., R. Sarojini and Nagabhushanam, R.), Oxford & IBH Publishing Co. Pvt. Ltd., New Delhi, India. pp. 651-656.

Raghukumar, S., Zainal, A.K. & Jones, E.B.G. (1988a) *Cirrenalia basiminuta* : A new lignicolous marine Deuteromycete from the tropics. *Mycotaxon,* **31:** 163-170.

Raghukumar, S., Anil, A.C., Khandeparkar, L. & Patil, J.S. (2000) Thraustochytrid protests as a component of marine microbial films. *Mar. Biol.*, **136:** 603-609.

Raghukumar, S., Sathe-Pathak, V., Sharma, S. & Raghukumar, C. (1995) Thraustochytrids and fungal components of marine detritus. III. Field studies on decomposition of leaves of the mangrove *Rhizophora apiculata* Blume. *Aquat. Microb. Ecol.*, **9:** 117-125.

Raghukumar, S., Sharma, S., Raghukumar, C., Sathe-Pathak, V & D Chandramohan (1994) Thraustochytrid and fungal component of detritus. IV. Laboratory studies on decomposition of leaves of the mangrove *Rhizophora apiculata* Blume. *J. Exp. Mar. Biol. Ecol.*, **183:** 113-131.

Rai, M.K. & Deshmukh, S.K. (2005) *"Fungi: Diversity and Biotechnology"*. Scientific Publishers (India), Jodhpur. pp. 1-510.

Rai, J.N., Tewari, J. P. & Mukerji, K.G. (1969) Mycoflora of mangrove mud. *Mycopath. Et Mycol. Appl.*, **38:** 17-31.

Raikar, M.T., Raghukumar, S., Vani, V., David, J.J. & Chandramohan, D. (2001) Thraustochytrid protests degrade hydrocarbons. *Ind. J. Mar. Sci.*, **30:** 139-145.

Ramaiah, N. (2006) A review on fungal disease of algae, marine fishes, shrimps and corals. *Ind. J. Mar. Sci.*, **35:** 380-387.

Ramasamy, P., Rajan, P.R., Jayakumar, R., Rani, S. & Brenner, G.P. (1996) *Lagenidium callinectes* (Couch, 1942) infection and its control in cultured larval Indian tiger prawn *Penaeus monodon* Fabricus. *J. Fish Dis.*, **19:** 75-82.

Ramesh, Ch. &. Borse, B.D. (1989) Marine fungi from Maharashtra Coast (India). *Acta Botanica Indica,* **17:** 143-146.

Ramesh, S., Jayaprakashvel, M.& Mathivanan, N. (2006) Microbial status in seawater and coastal sediments during pre- and post-tsunami periods in the Bay of Bengal, India. *Marine Ecology*, **27:** 198-203.

Rao, G.P., Manoharachary, C., Bhat, D.J., Rajak, R.C., & Lakhanpal, T.N. (eds.) (2003) *Frontiers of Fungal Diversity in India* - Prof. Kamal Festschrift Volume, International Book Distributing Co. Lucknow, India.

Rao, V., Manoharachary, C., Sureshkumar, G. & Subodh, K. (2004). *Fungi: Around some Aquatic Bodies in Andhra Pradesh. India*, B.S. Publisher Hydrabad, pp. 1-167.

Raveendran, K. (2002) *Marine Fungi of Kerala: A Preliminary Floristic and Ecological Study*, Unpublished Ph. D. Thesis, University of Calicat, Kerala, India.

Raveendran, K. & Manimohan, P. (Eds.), (2007) *Marine Fungi of Kerala, A Preliminary Floristic and Ecological Study*, Malbar Natural History Society, Calicat. Kerala, India. pp. 1-270.

Ravikumar, D.R. (1991) *Studies on fungi from mangroves of the east coast of India.* Ph. D. thesis, University of Madras, India.

Ravikumar, D.R. & Purushothaman (1988a) *Cirrenalia tropicalis* an addition to Indian marine fungi. *Curr. Sci.,* **57:** 674.

Ravikumar, D.R. & Purushothaman (1988b) *Corollospora intermedia,* a lignicolous marine fungi from India. *Curr. Sci.,* **57:** 898-899.

Ravikumar, D.R. & Suryanarayanam, T.S. (1998) Influence of salinity on the activity of polyol metabolism enzymes and peroxidase in the marine fungus *Cirrenalia pygmea* (*Hyphomycetes*). *Indian J. Mar. Sci.,* **27:** 237-238.

Ravikumar, D.R. & Vittal, B.P.R. (1987) Studies on mangrove fungi of India. Notes on some Ascomycetes. *Kavaka,* **15:** 99-103.

Ravikumar, D.R. & Vittal, B.P.R. (1991a) A new species of *Bathyascus* from Indian mangrove. *Mycol. Res.,* **95:** 370-371.

Ravikumar, D.R. & Vittal, B.P.R. (1991b) Two *Corollospora* species new to India. *Kavaka,* **19:** 17-18.

Ravikumar, D.R. & Vittal, B.P.R. (1996) Fungal diversity on decomposing biomass of mangrove plant *Rhizophora* in pichavaram estuary, east coast of India. *Indian Journal of Marine Sciences,* **25:** 142-144.

Ravikumar, M., Sridhar, K.R., Sivakumar, T., Karamchand, K.S., Sivakumar, N. & Vellaiyan, R. (2009) Diversity of filamentous fungi on coastal woody debris after tsunami on the southeast coast of India. *Czech. Mycol.*, **61:** 1-7-115.

Ravindran , J., Raghukumar, C. & Raghukumar, S. (2001) Fungi in *Porites lutea,* association with healthy and deseased corals. *Dis. Aquat. Org.,* **47:** 219-228.

Ravishanker, J.P., Suryanarayanan, T.S. & Murugandam, V. (2006) Strategeis for osmoregulations in the marine fungus *Cirrenalia pygmea* Kohlm. (Hyphomycetes). *Ind. J. Mar. Sci.*, **35:** 351-358.

Read, S.J., Haieh, H.-Y., Jones, E.B.G., Moss, S.T. & Chang, H.S. (1992) *Paraliomyces lentiferus*: an ultrastructural study of a little-known marine Ascomycete. *Can. J. Bot.*, **70:** 2223-2232.

Reeb, D., Best, P.B., Botha, A., Cloete, K.J., Thornton, M. & Mouton, M. (2011) Fungi associated with the skin of a southern right whale (*Eubalaena australis*) from South Africa. *Mycology*, **1:** 155-162.

Rees, G., Johnson, R.G. & Jones, E.B.G. (1979) Lignicolous marine fungi from Danish sand dunes. *Trans. Br. Mycol. Soc.*, **72:** 99-106.

Rossello, M.A., Descals, E. & Cabrer, B. (1993) *Nia epidermoidea*, a new marine gastromycete. *Mycol. Res.*, **97:** 68-70.

Saccardo, P.A. (1880) Fungi gallici, Series II. *Michelia*, **2:** 39-135.

Sadaba, R.B. & Sarinas, B.G.C. (2010) Fungal communities in bunker C oil-impacted sites off southern Guimaras, Philippines: a post-spill assessment of Solar 1 oil spill. *Bot. Mar.*, **53:** 565-576.

Sadaba, R.B., Vrijmoed, L.L.P., Jones, E.B.G. & Hodgkiss, I.J. (1995) Observations on vertical distribution of fungi associated with standing senescent *Acanthus ilicifilius* stems at Mai Po mangroves, Hong Kong. *Hydrobiol.*, **295:** 119-126.

Salique, S.M., Krishnamurty, K. & Palaniappan, R. (1985) Ecology of fungal populations in the Vellar edtuary and Pichavaram mangroves, South India. In: *The Mangroves* (ed. Bhosale, L.J.), Shivaji University Press, Kolhapur, Maharashtra. pp. 153-162.

Sakayaroj, J., Jones, E.B.G., Chatmala, I. & Phongpaichit, S. (2004) In: *Thai fungal diversity* (eds. Jones, E.B.G, Tantichareon, M. and Hyde, K.D.), BIOTEC, Thailand, pp. 107-117.

Sakayaroj, J., Pang, K.L., Phongpaichit, S. & Jones, E.B.G., (2005a) A phylogenetic study of the genus *Haligena* (Halosphaeriales, Ascomycota). *Mycologia*, **97:** 804-811.

Sakayaroj, J., Pang, K.L., Jones, E.B.G., Vrijmoed, L.L.P. & Abdel-Wahab, M.A. (2005b) A systematic reassessment of marine ascomycetes *Swampomyces* and *Torpedospora. Bot. Mar.*, **48:** 395-406.

Sakayaroj, J., Pang, K.L. & Jones, E.B.G. (2010a) Multi-gene phylogeny of the Halosphaeriaceae: its ordinal status, relationships between genera and morphological character evolution. *Fungal Diversity*, **46:** 87-109.

Sakayaroj, J., Preedanon, S., Supaphon, O., Jones, E.B.G. & Phongpaichit (2010b) Phylogenetic diversity of endophyte assemblage associated with the tropical seagrass *Enhalus acoroides* in Thialand. *Bot. Mar.*, **42:** 27-45.

Sarbhoy, A.K., Agrawal, D.K. & Varshey, J.L. (1986) "*Fungi of India* (1986)". Association Publishing company, New Delhi. pp. 274.

Sarbhoy, A.K., Vershey, J.L. & Agrawal, D.K. (1996) "*Fungi of India* (1982-1992)", CBS Publishers and Distributors, New Delhi. pp. 1-350.

Sarma, V.V. (1998) *Studies on the biodiversity of fungi colonizing litter of mangrove plants in Godavari and Krishna deltas, east coast of India.* Unpublished Ph. D. Thesis, University of Madras, India.

Sarma, V.V. & Hyde, K.D. (2000a) A synoptic key to higher marine fungi. In: *Marine Mycology* – A practical approach (eds. Hyde, K.D. and Pointing, S.B.). *Fungal Diversity Press, Research series 1.* The University of Hong Kong, Hong Kong. pp. 271-307.

Sarma, V.V. & Hyde, K.D. (2000b) *Tirispora mandoviana* sp. nov. from Chorao mangroves, Goa, the West coast of India. *Australian Mycologist*, **19:** 52-56.

Sarma, V.V. & Hyde, K.D. (2001) A review on frequency of occurrence of fungi in mangrove. *Fungal Diversity*, **8:** 1-34.

Sarma, V.V. & Vittal, B.P.R. (1998-1999) Ecological studies on manglicolous fungi from Godavari and Krishna deltas, East coast of India–observations on the seasonal occurrence. *Kavaka,* **26-27:** 105-120.

Sarma, V.V. & Vittal, B.P.R. (2000) Biodiversity of mangrove fungi on different substrata of *Rhizophora apiculata & Avicennia sp.* from Godavari & Krishna deltas, East coast of India In : *Aquatic mycology across the Millenium* (eds K.D. Hyde., W.H. Ho and Pointing, S.B.), *Fungal Diversity,* **5:** 23-41.

Sarma, V.V. & Vittal, B.P.R. (2001) Biodiveristy of manglicolous fungi on selected plants in the Godavari & Krishna deltas, East Coast of India. *Fungal Diversity,* **6:** 115-130.

Sarma, V.V. & Vittal, B.P.R. (2002) A preliminary study on vertical distribution of manglicolous fungi on prop roots of *Rhizophora apiculata* Blume at Krishna delta, East coast of India. *Kavaka,* **30:** 21-29.

Sarma, V.V. & Vittal, B.P.R. (2004) Manglicolous fungi recorded from Godavary and Krishna deltas, A. P., East coast of India along with dichotomous key and notes on some taxa. *Kavaka,* **32:** 65-111.

Sarma, V.V., Hyde, K.D. & Vittal, B.P.R. (2001) Frequency of occurrence of mangrove fungi from the east coast of India. *Hydrobiologia,* **455:** 41-53.

Sarma, V.V., Newell, S.Y. & Hyde, K.D. (2001) *Koorchaloma spartinicola* sp. nov., a new marine sporodochial fungus from *Spartina alterniflora. Bot. Mar.,* **44:** 321-326.

Sarmiento-Ramirez, J.M., Abella, E., Martin, M.P., Telleria, M.T., Lopez-Jurado, L.F., Marco, A. & Dieguez-Uribeondo, J. (2010) *Fusarium solani* is responsible for mass mortalities in nests of loggerhead sea turtle, *Caretta caretta*, in Boavista, Cape Verde. *Res. Lett.*, **312:** 192-200.

Sathe-Pathak, V., Raghukumar, S., Raghukumar, C. & Sharma, S. (1993) Thraustochytrid and fungal component of marine detritus. I. Field studies on decomposition of the brown alga *Sargassum cinereum. Indian J. Mar. Sci.,* **22:** 159-167.

Sati, S.C., Saxena, J. & Dubey, R.C. (eds.) (1997a) "*Himalayan Microbial Diversity*", Part- I. Today & Tomorrow's Printers and Publishers, New Delhi. pp. 1-315.

Sati, S.C., Saxena, J. & Dubey, R.C. (eds.) (1997b) "*Himalayan Microbial Diversity*", Part - II. Today & Tomorrow's Printers and Publishers, New Delhi. pp. 1-210.

Schatz, S. (1980a) Taxonomic revision of the two Pyrenomycetes associated with littoral-marine green algae. *Mycologia,* **72:** 110-117.

Schatz, S. (1980b) The life history, developmental morphology, and taxonomy of *Lautitia danica* gen. nov., comb. Nov. *Can. J. Bot.*, **62:** 28-32.

Schatz, S. (1988) *Hypoxylon oceanicum* sp. nov. from mangroves. *Mycotaxon*, **33**: 413-418.

Schaumann, K. (1968) Marine hohere Pilze (Ascomycetes und Fungo-imperfect aus dem Weser-Astuar. *Veroff. Inst. Meeresforsch. Brremerhaven,* **11:** 93-117.

Schaumann, K. (1969) Uber marine hohere Pilze von Holzsubstraten der Nordsee-Insel Helgoland. *Ber. Dtsch. Botan Ges.*, **82:** 307-327.

Schaumann, K. (1975) marine Pilzfunde von der Norwegischen Rinne, der Parents-See und von denkusten Westafrikas und deer Kanarischen Inseln. *Veroff. Inst. Meeresforsch. Brremerhaven,* **15:** 183-194.

Schaumann, K., Mulach, W. & Molitoris, H.P. (1986) Comparative studies on growth and exoenzyme production of different *Lulworthia* isolates. In: *The biology of marine fungi* (ed. Moss, S.T.), Cambridge Univ. Press, pp. 49-67.

Schmit, J.P. & Shearer, C.A. (2003) A checklist of mangrove associated fungi. *Mycotaxon*, **80:** 423-477.

Schmit, J.P. & Shearer, C.A. (2005) Geographical and host distribution of lignicolous mangrove microfungi. *Bot. Mar.*, **47:** 496-500.

Schoch, C.L., Sung, G.H., Volkmann-Kohlmeyer, B., Kohlmeyer, J. & Spatafora, J.W. (2006a) Marine fungal lineages in the Hypocreomycetidae. *Mycol. Res.*, **110:** 257-263.

Schoch, et al. (2006b) The Ascomycota tree of life: a phylum-wide phylogeny clarifies the origin and evolution of fundamental reproductive and ecological traits. *Syst. Biol.*, **58:** 211-223.

Schulz, B., Draeger, S., Cruz, T.E. dela, Rheinheimer, J., Siems, K., Loesgen, S., Bitzer, J., Schloerke, O., Zeeck, A., Kock, I., Hussain, H., Dai, J. & Krohn, K. (2008) Screening strategies for obtaining novel, biologically active, fungal secondary metabolites from marine habitats. *Bot. Mar.*, **51:** 219-234.

Sekimoto, S., Yoko, K., Kawamura, Y. & Honda, D. (2008a) Taxonomy, molecular phylogeny, and ultrastucture of *Olpidiopsis porphyrae* sp. nov. (Oomycetes, Straminipiles), a unicellular obligate endoparasite of *Bangia* and *Porphyra* spp. (Bangiales, Rhodophyta). *Mycol. Res.*, **112:** 361-374.

Sekimoto, S., Beakes, W., Gachon, C.M.M., Muller, D.G., Kupper, F.C. & Honda, D. (2008b) The developmental, ultrastructural cytology, and molecular phylogeny of the basal Oomycete *Eurychasma dicksonii*, infecting the filamentous

Phaeophyta algae *Ectocarpous stiliculosus* and *Pylaiella littoralis*. *Protist.*, **159:** 299-318.

Sharmȧ, J. R., Mudgal, V. & Hajra, P.K. (1997) Floristic diversity- Review, scope and perpective. In: *Floristic Diversity and conservation strategies in India.* (eds. Magdal, V. and Hajra, P.K.). Vol. I, BSI, Calcutta, pp. 1-45.

Sharma, S., Raghukumar, S., Raghukumar, C. & Sathe-Pathak, V. (1994) Thraustochytrid and fungal component of marine detritus. II. Laboratory studies on decomposition of the brown alga *Sargassum cinereum*. J. Ag. *J. Exp. Mar. Biol. Eco.*, **175:** 227-242.

Shearer, C.A. & Burgos, J. (1987) Lignicolous marine fungi from Chile. *Bot. Mar.*, **30:** 455-458.

Shearer, C.A. & Crane, J.L. (1980) Fungi of the Chesapeak Bay and its tributaries VIII. Ascomycetes with unfurling appendages. *Bot. Mar.*, **23:** 607-615.

Shearer, C.A. & Miller, M. (1977) Fungi of the Chesapeak Bay and its tributaries. V. *Aniptodera chesapeakensis*. *Mycologia*, **69:** 887-898.

Shearer, C.A., Descals, E., Kohlmeyer, B., Kohlmeyer, J., Marvanova, L., Padgett, D., Porter, D., Raja, H.A., Schmit, J.P., Thorton, H.A. & Voylymayer, H. (2007). Fungal Biodiversity in aquatic habitats. *Biodiversity Conservation,* **16:** 49-67.

Shindikar, M. & Borse, B.D. (2002) Marine fungi from mangrove swamps of Vikaroli, Maharashtra, India. *J. Adv. Sci. Tech.,* **5:** 69-74.

Shini, K., Sridhar, K.R. & Karamchand, K.S. (2009-10) Assemblage and diversity of fungi in two under explored mangroves of India. *Kavaka*, **37** & **38**: 79-85.

Shrivastava, A.D. (1989) Marine fungi from Bombay - III. some Loculoascomycetes. *Indian Bot. Reptr.*, **8:** 16-19.

Shrivastava, A.D. (1994) Marine fungi from from Bombay - IV. *Indian Bot. Reptr.*, **13:** 92-93.

Shrivastava, A.D. (1995) Marine fungi from Bombay - V. *Indian Bot. Reptr.*, **14:** 74-77.

Singh, A., Wilson, S. & Ward, O.P. (1996) Docosahexaenoic acid (DHA) production by *Thraustochytrium* sp. ATCC 20892. *World Microbiol. Biotechnol.*, **12:** 76-81.

Singh, P., Raghukumar, C., Verma, P. & Scouche, Y. (2010) Phylogenetic diversity of culturable fungi from the deep sea sediments of the Central Indian Basin and their growth characteristics. *Fungal Diversity*, **40:** 89-102.

Somrithipol, S., Chatmala, I. & Jones, E.B.G. (2002) *Cirrenalia nigropora* sp. nov. and *C. tropicalis* from Thailand. *Nova Hedwigia*, **75:** 477-485.

Sparks, A.K. (1982) Observations on the histopathology and possible progression of the disease caused by *Trichomaris invadens*, an invasive ascomycete, in the tanner crab, *Chionoecetes baindi*. *J. Invert. Path.*, **40:** 242-254.

Sparrow, F.K. Jr. (1934) Observations on marine Phycomycetes collected in Denmark. *Densk. Bot.*, **8:** 1-24.

Sparrow, F.K. Jr. (1936) Biological observations of the marine fungi of woods hole waters. *Biological Bulletin,* **70:** 236-263.

Sparrow, F.K. Jr. (1937) The occurrence of Saprophytic fungi in marine muds. *Biol. Bull.,* (Woods Hok, Mass.) **73:** 242-248.

Sparrow, F. K. Jr. (1960) "*Aquatic Phycomycetes*". Univ. Michigan Press, Ann Arbor, Michigan. pp. 1-1187.

Sparrow, F.K. Jr. (1969) Zoosporic marine fungi from the Pacific Northwest (USA). *Arch Mikrobiol.*, **66:** 129-146.

Spooner, B.M. (1981) New records and species of British microfungi. *Trans. Br. Mycol. Soc.*, **76:** 205-301.

Spatafora, J.W. & Blackwell, M. (1994) The polyphyletic origin of ophiostoatoid fungi. *Mycol. Res.*, **98:** 1-9.

Spatafora, J.W., Volkmann-Kohlmeyer, B. & Kohlmeyer, J. (1998) Independent terrestrial origin of the Halosphaeriales (marine ascomycota). *Amer. J. Bot.*, **85:** 1569-1580

Sridhar, K.R. (2004) Mangrove fungi from India. *Curr. Sci.,* **86:** 1586-1587.

Sridhar, K.R. (2005) Diversity of fungi in mangrove ecosystems In: *Microbial Diversity: Current Perspectives and Potential Applications* (eds. Satyanarayana, T. and B.N. Johri), I.K. Internatinal Pvt. Ltd., New Delhi. pp. 129-147.

Sridhar, K.R. (2009a) Mangrove fungi of the Indian Peninsula. In: *Frontiers in Fungal ecology, Diversity and Metabolites* (ed. Sridhar, K.R.), I. K. International Publishing House Pvt. Ltd., New Delhi. pp. 28-50.

Sridhar, K.R. (2009b) Fungal diversity of Pichavaram Mangroves, Southeast coast of India. *Nature & Science*, **7:** 67-75.

Sridhar, K.R. & Kaveriappa, K.M. (1988) Occurrence and survival of Aquatic Hyphomycetes in brackish and seawater. *Archiv fur Hydrobiologie*, **113:** 153-160.

Sridhar, K.R. & Kaveriappa, K.M. (1991) A note on marine fungi from mangalore coast. *Mahasagar,* **24:** 63-66.

Sridhar, K.R. and Maria, G.L. (2006) Fungal diversity on mangrove woody litter *Rhizophora mucronata* (Rhizophoraceae). *Ind. J. Mar. Sci.*, **35:** 318-325.

Sridhar, K.R. & Prasannarai, K. (1993) Marine fungi on Decaying wood from mangalore coast. *Mahasagar,* **26:** 33-38.

Sridhar, K.R. & Prasannarai, K. (2001) Biodiversity of higher marine fungi in tropics. A Review. *Env. & Cons.,* **7:** 219-234.

Sridhar, K.R., Karamchand, K.S. & Prabhu, Sumathi (2010) Fungal colonization and breakdown of sedge (*Cyperus malaccensis* Lam.) in an Indian mangroves. *Bot. Mar.*, **53:** 525-533.

Statzell-Tallman, A., Belloch, C. & Fell, J. W. (2008) *Kwoniella mangroviensis* gen. nov., sp. nov. (Tremallales, Basidiomycota), a teleomorphic yeast from mangrove habitats in the Florida Everglades and Bahamas. *FEMS Yeast Res.*, **8:** 103-113.

Statzell-Tallman, A., Scorzetti, G. & Fell, J.W. (2010) *Candida spencermartinsiae* sp. nov., *Candida taylorii* sp. nov. and *Pseudozyma abaconensis* sp. nov., novel yeasts from mangrove and coral reef ecosystems. *Inter. J. Syst. Evol. Microbiol.*, **60:** 1978-1984.,

Steinke, T.D. & Jones, E.B.G. (1993) Marine and mangrove fungi from the Indian Ocean coast of South Africa. *South Africa J. Bot.*, **59:** 385-390.

Stevens, R.L. (1990) New and noteworthy Puerto Rican fungi. *Bot. Gaz.,* **70:** 399-402.

Stork, N. (1993) "How many species are there?" *Biodvr. & Conserv.* **2:** 215-232.

Strittmatter, M., Gachon, C.M.M. & Kapurr, F.C. (2009) Ecology of lower Oomycetes. In: *Oomycetes genetics and genomics: Diversity, interactions and research tools* (eds. Lamour, K. and Kamoun, S.), Wiley, pp. 25-46.

Strongman, D., Miller, J.D. & Whitney, N. (1985) Lignicolous marine fungi from Prince Edward Island with a description of *Didymosphaeria lingnomaris* sp. nov. *Proc. N. S. Inst. Sci.*, **35:** 99-105.

Subramanian, C.V. (1971) *"Hyphomycete, an account of Indian species except Cercosporae"*. ICAR, New Delhi. pp. 1-930.

Subramanian, C.V. (1979) Fungi from Indian mangroves. Paper presented in the 3rd International Marine Mycology Symposium, Morehead city, North Carolina, USA.

Subramanian, C.V. (1983) "*Hyphomycetes, Taxonomy & Biology*". Academic Press, London and New York. pp. 1-502.

Suetrong S., Schoch, C.L., Spatafora, J.W., Kolhmeyer, J., Volkmann-Kolhmeyer, B., Sakayaroj, J., Phongpaichit, S., Tanaka, K., Hirayama, K. & Jones, E.B.G. (2009) Molecular systematics of the marine Dothideomycetes. *Stud. Mycol.,* **64:** 155-173.

Suetrong, S., Hyde, K.D., Zhang, Y., Bahkali, A.H. & Jones, E.B.G. (2011) Trematosphaeriaceae fam. nov. *Mycologia*, (in press).

Sundari, R. Vikineswary, S., Yusoff, M. & Jones, E.B.G. (1996a) Observations on tropical arenicolous marine fungi on driftwood from Malaysia and Singapore. *Bot. Mar.*, **39:** 327-333.

Sundari, R., Vikineswary, S., Yusoff, M. & Jones, E.B.G. (1996b) *Corollospora besarispora*, a new arenicolous marine fungus from Malaysia. *Mycol. Res.*, **100**: 1259-1262.

Suryanarayana, T.S. & Kumaresan, V. (2000) Endophytic fungi of some halophytes from an estuarine mangrove forest. *Mycol. Res.*, **104:** 1465-1467.

Suryanarayana, T.S., Ambayeram, V., Thirunavukkarasu, N., Ravishankar, P., Doble, M. & Venkatachalam, Geetha (2010) Internal mycobiota of marine macroalgae from the Tamilnadu coast: distribution, diversity and biotechnology potential. *Bot. Mar.*, **53:** 457-468.

Sutherland, G.K. (1915) New marine fungi on Pelvetia. *New Phytol.* **14:** 33-42.

Sutherland, G.K. (1916a) Marine Fungi Imperfecti. *New Phytol.* **15:** 35-48.

Sutherland, G.K. (1916b) Addition notes on marine Pyrenomycetes. *Trans. Br. Mycol. Soc.*, **5:** 257-263.

Sutton, B.C. (1980) "*The Coelomycetes, fungi Imperfect with pycnidia Acervuli and stromata*", Common wealth Mycological Institute, Kew survey, England, pp. 1-696.

Swart, H.J. (1958) An invesigatin of the mycoflora in the soil of some mangrove swamps. *Acta Bot. Neerland.*, **7:** 741-768.

Swart, H.J. (1963) Further investigation of the mycoflora in the soil of some mangrove swamps. *Acta Bot. Neerland.*, **12:** 98-111.

Takishita, K., Tsuchiya, M., Reimer, J.D. & Maruyama, T. (2011) Molecular evidence demonstrating the basidiomycetous fungus *Cryptococcus curvatus* is the dominant microbial eukaryote in sediment at the Kuroshima Knoll methane seep. *Extremophiles*, **10:** 165-169.

Takuma, D., Sano, A., Wada, S., Kurata, O. & Hatai, K. (2011) *Aphanomyces sinensis*, sp. nov., isolated from juvenile soft-shelled turtle, *Pelodiscus sinensis*, in Japan. *Mycoscience*, **52:** 119-131.

Thind, K.S. (1977) *"The Myxomycetes"*. ICAR, New Delhi.

Thongkantha, S., Jeewon, R., Vijaykrishna, D., Lumyong, S., McKenzie, E. H.C. & Hyde, K.D. (2009) Molecular phylogeny of Magnaporthaceae (Sordariomycetes) with a new species *Ophiocera chinadaoense* from *Dracaena loueriri* in Thailand. *Funlgul Diversity,* **34:** 157-173.

Tillmann, U., Hesse, K.J. & Tillmann, A. (1999) Large-scale parasitic infection of diatoms in the Northfrisian Wadden Sea. *J. Sea Res.*, **42:** 255-261.

Tokura, R. (1982) Arenicolous marine fungi from Japanese beaches. *Trans Mycol. Soc. Japan,* **23:** 423-433.

Tomlinson, P.B. (1986) *"The Botany of mangroves"*. Cambridge Univ. Press, Cambridge, London. pp. 1-414.

Tsui, C.K.M., Fan, K.W., Chow, R.K.K.., Jones, E.B.G. & Vrijmoed, L.L.P. (2011) Zoospore production and motility of mangrove thraustochytrids from Hong Kong under various salinities. Mycoscience, doi:10.1007/s 10267-011-0127-2.

Tubaki, K. (1967) An undescribed species of *Heleococcum* from Japan. *Trans. Mycol. Soc. Japan*, **8:** 5-10.

Tubaki, K. (1969) Studies on the Japanese marine fungi Lignicolous group (III), algicolous group and general consideration. *Annu. Rep. Inst. Ferment.* Osaka, **4:** 12-41.

Tuwar, A.R. (2011) Biodiversity of Marine fungi from Goa Coast (India). Abstract, In: *"Asian Mycological Congress 2011 & the 12th International Marine and Freshwater Mycology Symposium"*, Incheon, Korea, p. 448.

Udea, S. (1980) A mycoflora study on brackish water sediments in Nagasaki, Japan. *Trans. Mycol. Soc. Japan*, **21:** 103-112.

Udea, S. & Udagawa, S.I. (1983) A new Japaneses species of Neocospora from marine sludge. *Mycotaxon*, **16:** 387-395.

Ulken, A. (1967) Einige beobachtungen uber das Vorkmmen von Phycomycete aus der Reihe der Chytridiales im brackigen und marinen Wasser. *Veroff. Inst. Meeresforsch. Brremerhaven,* **10:** 167-172.

Ulken, A. (1968) Einige beobachtungen uber das Vorkmmen von uniflaggelaten Phycomycete (Chytridiales) inder Wassermundung. *Veroff. Inst. Meeresforsch.Brremerhaven,* Suppl. **3:** 59-66.

Ulken, A. (1969) Uber das Vorkomann niederer saprophytischer Phycomycete (Chytridiales im Bassin d'Arcachon (Frankreich). *Veroff. Inst. Meeresforsch. Brremerhaven,* **11:** 303-308.

Ulken, A. (1974) Chytridineen im Kustenbereich. *Veroff. Inst. Meeresforsch. Brremerhaven,* Suppl. **5:** 27-36.

Ulken, A. (1990) "*Marine thraustochytrids and chytridiomycetes in the North Sea area and in selected other regions*". J. Cramer, Berlin.

Van Uden, N. & Casttelo-Branco, R. (1963) Distribution and population densities of yeast species in Pacific water, air, animals, and kelp off southern California. *Limnol. Oceanogr.*, **8:** 323-3329.

Van Uden, N. & Fell, J.W. (1968) Marine yeasts. *Adv. Microbiol* Sea, **1:** 167-201.

Van Uden, N. & ZoBell, C.E. (1962) *Candida marina* nov. spec., *Torulopsis torresii* nov. spec. and *T. maris* nov. spec., three yeasts from the Torres Strait. *Antonie van Leeuenhoek*, **28:** 275-283.

Vasilyeva, L.N. & Stephenson, S.L. (2005) Pyrenomycetes of the Great Smoky Mountain National Park. II. *Cryptovalsa* Ces. Et De Not. and *Diatrypella* (Ces. Et De Not.) Nitcshke (Diatrypaceae). *Fungal Diversity*, **19:** 189-200.

Vasilyeva, L.N. & Stephenson, S.L. (2006) Pyrenomycetes of the Great Smoky Mountain National Park. III. *Cryptosphaeria*, *Eutypa* and *Eutypella* (Diatrypaceae). *Fungal Diversity*, **22:** 243-254.

Venkatesan, T. & Natarajan, R. (1985) Intertidal mycoflora of Pichavaram mangroves near porto Novo In: *The Mangroves: proc. Nat. Symposium Biol. Util. Cons. Mangroves* (ed. Bhosale, L.J.), Shivaji Uni. Press, India. pp. 163-169.

Verbist, J-F., Sallenave, C. & Pouchus, Y-F. (2000) Matine fungal substances. In: *Studies in natural products chemistry* (ed. Rahman, A.), *Elsevier Sci.*, **24:** 979-1009.

Vishwakiran, Y., Thalur, N. L., Raghukumar, S., Yennawar, P.L. & Anil, A. C. (2001) Spatial and temporal distribution of fungi and wood-borers in the coastal tropical waters of Goa, India. *Bot. Mar.*, **44:** 47-56.

Vittal, B.P.R. & Sarma, V.V. (2005) Fungal Diversity on mangroves. In: *The fungi-Diversity and conservation in India* (eds. Dargan, J.S., Arti, N.S. and Dhingara, G.S.), Bishen Singh Mahendra Pal Singh, Dehra Dun (India), pp. 33-45.

Vittal, B.P.R. & Sarma, V.V. (2006) Diversity and ecology of fungi of mangroves of Bay of Bengal - An overview. *Ind. J. mar. Sci.,* **35:** 308-317.

Volkmann-Kohlmeyer, B. & Kohlmeyer, J. (1996) How to prepare truly permanent microscopic slides, *Mycologist,* **10:** 107-108.

Vrijmoed, L.L.P., Hyde, K.D. & Jones, E.B.G. (1994) Observation on mangrove fungi from Macau and Hong Kong, with the description of two new ascomycetes: *Diaporthe salsuginosa* and *Aniptodera haispora*. *Mycol. Res.*, **98:** 699-704.

Vrijmoed, L.L.P., Hodgkiss, I.J. & Thrower, L.B. (1982) Factors affecting the distribution of lignicolous marine fungi in Hong Kong. *Hydrobiologia,* **87:** 143-160.

Vrijmoed, L.L.P., Hyde, K.D. & Jones, E.B.G. (1994) Observation on mangrove fungi from Macau and Hong Kong, with the description of two new ascomycetes: *Diaporthe salsuginosa* and *Aniptodera haispora*. *Mycol. Res.*, **98:** 699-704.

Walsh, J.J. (1983) Death in the sea – Enigmatic phytoplankton losses. *Prog. Oceanogr.*, **12:** 1-86.

Webber, F.C. (1967) Observation on the structure, life history and biology of *Mycosphaerella ascophylli*. *Trans. Br. Mycol. Soc.*, **50:** 583-601.

Whalley, A.S.J., Jones, E.B.G., Hyde, K.D. & Lassoe (2000) *Halorosellinia* gen. nov. to accommodate *Hypoxylon oceanicum*, a common mangrove species. *Mycol. Res.*, **104:** 368-374.

Williams, P.H., Margules, C.R. & Hilbert, D.W. (2002) Data requirements and data sources for biodiversity priority areas selection. *J. Biosci.*, **27:** 327-338.

Wilson, I.M. (1951) Notes on some marine fungi. *Trans. Br. Mycol. Soc.*, **34:** 540-543.

Wilson, W.I. (1998) Isolation of endophytes from seagrasses from Bermuda. The University of New Brunswick, Canada, M.Sc. Thesis.

Wong, M.K.M., Poon, M.O.K. & Hyde, K.D. (1998) *Phragmitensis marina* gen. et sp. nov., an intertidal saprotraph from *Phragmites australis* in Hong kong. *Bot. Mar.*, **41:** 379-382.

Wu, Q.X., Crews, M.S., Draskovic, M., Sohn, J., Johnson, T.A., Tenney, K., Valeriote, F.A., Yao, X.J., Bjeldanes, L.F. & Crews, P. (2010) Azonazin, a novel dipeptide from a Hawaiian marine sediment-derived fungus, *Apergillus insilicola*. *Org. Lett.*, **12:** 4458-4461.

Yaguchi, T., Tanaka, S., Nakahara, T. & Higashuhara, T. (1997) Production of docosahexaenoic ascid production by *Schozochytrium* sp. *JAOCS* **74:** 1431-1434.

Yokoyama, R. & Honda, D. (2007) Taxonomic rearrangement of the genus *Schizochytrium sensu lato* based on morphology, chemotaxonomy characteristics, and 18S rRNA gene phylogeny (Thraustochytrium, labyrinthulomycetes): emendation for *Schizochytrium* and erection of *Aurantiochytrium* and *Oblongichytrium* gen. nov. *Mycoscience*, **48:** 199-211.

Yokoyama, R., Salleh, B. & Honda, D. (2007) Taxonomic rearrangement of the genus *Schizochytrium sensu lato* based on morphology, chemotaxonomy characteristics, and 18S rRNA gene phylogeny (Thraustochytrium, labyrinthulomycetes): emendation for *Ulkenia* and erection of *Botryochytrium*, *Perietichytrium* and *Sicyoidiochytrium* gen. nov. *Mycoscience*, **48:** 329-341.

Zhang, Y., Wang, H.K., Crous, P.W., Pointing, S.B. & Hyde, K.D. (2009a) Towards a phylogenetic clarification of Lophiostoma / Massarina and morphological similar genera in the Pleosporales. *Fungal Diversity*, **38:** 225-251.

Zhang, Y., Schoch, C.L., Fourneor, J., Crous, P.W., de Gruyter, J., Woudenberg, J.H.C., Hirayama, K., Tanaka, K., Pointing, S.B. & Hyde, K.D. (2009b) Mult-locus phylogeny of Pleosporales: a taxonomic, ecological and evolutionary re-evaluation. *Studies Mycology*, **64:** 85-103.

Zellers, S.M. (1918) Fungi found on *Codium mucronatum*. *Publications of the Puget sound Biology Station,* **2:** 121-125.

Zhao, G.Z. & Liu, X.Z. (2005) A review of *Cirrenalia* (Hyphomycetes) and a new species. *Fungal Diversity*, **18:** 201-209.

Zobell, L.E. (1968) Bacterial life in the Deep Sea. *Bull, Misaki Mar. INST, Kuota. University,* **12:** 77-96.

Zuccaro, A., Schulz, B. & Mitchell, J.I. (2003) Molecular detection of ascomycetes associated with *Fucus serratus*. *Mycol. Res.*, **107:** 1451-1466.

Zuccaro, A., Summerbell, R.C., Gams, W., Schroers, H. F. & Mitchell, J.I. (2004) A new *Acremonium* species associated with *Fucus* sp., and its affinity with a phylogenetically distinct marine *Emericellopsis* clade. *Stud. Mycol.*, **50:** 283-292.

Zuccaro, A., Schoch, C.L., Spatafora, J.W., Kohlmeyer, J., Draeger, S. & Mitchell, J.I. (2008) Detection and identification of fungi intimately associated with the brown seaweed *Fucus serratus*. *Appl. Environ. Microbiol.*, **74:** 931-941.

Fungus index

Italicized names are synonyms; the names of doubtfully recorded taxa are followed by a question mark; boldface page numbers indicate technical description.

Other titles on Goa

Goa n Pioneers in Bombay
Teresa Albuquerque Rs. 295. Pp.197 Pb. ISBN 9789380739236

Being a Goan Christian
Victor Ferrao Rs. 195. Pp. 120. Pb. ISBN 9789380837352

Will Bombay be at Peace
Pradosh Karapurkar Rs. 195. Pp 218. Pb. ISBN 9789380837277

Epistemology Search for Truth is made easy
Dr. Antonio Rodrigues C.Ss.R. Rs. 150. Pp. 186. Pb. ISBN 9789380837369

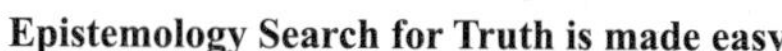

Undra Muja Mama-Jose Pereira, Micael Martins &
Antonio da Costa Rs. 295. Pp. 210.ISBN 9789380739267

Infinite Articulation
Perin Ilavia Rs 495. Pp 81. Hb. ISBN 9789380837338

Goa Folklore studies
Dr. Pandurang Phaldesai. Rs 295. Pp 280. Pb. ISBN 9789380739229

Inhuman Vivek Nayak. Rs 495. Pp 496. Pb. ISBN 9789380837284

Friendship Factor
Dr. U. G. Barad. Rs 395. Pp 368. Pb. ISBN 9789380837307

Take Control of Diabetes
Dr. U. G. Barad. Rs 295. Pp 265. Pb. ISBN 9789380837291

The Tulsi Edila Gaitonde. Rs 195. Pp 160. Pb. ISBN 9789380739250

Goa Traffic
Marissa De Luna. Rs 350. Pp 260. Pb. ISBN 9789380837260

The Parish Churches of Goa: A Study of Façade Architecture
Jose Lourenço. Rs 595. Pp 220. Pb. ISBN 9789380837192

Konkani-Italiano Italian-Konknni dictionary
Ave Cleto Afonso. Rs 995. Pp 398. Hb. ISBN 9789380837215

Inside Out
Dr. U.G.Barad Rs 400. Pp 406. Pb ISBN 9789380837086.

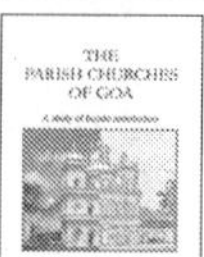

Glimpses From India's Natural World
A. K. Sahay. Rs 595. Pp 179. Pb. ISBN 9789380837123

Song of Goa:Crown of Mandos- Jose Pereira, Micael Martins &
Antonio da Costa Rs. 550. Pp 485. ISBN 9789380739038

Veni, Vidi... Goa
Luis S. R. Vas Rs 350. Pp 370. Hb. Rs. 295. Pp. 370 Pb. ISBN 9789380837222

Skin
Margaret Mascarenhas. Rs 295. Pp 263. Pb ISBN 9789380739052

Green Tops in Goa
A. K. Sahay Rs 250. Pp 124. Pb

Cozinha de Goa: History & Tradition of Goan Food
Fatima da Silva Gracias. Rs 295. Pp 278. ISBN 9789380739090

The Hindus of Goa and the Portuguese Republic
Antonio de Noronha. Ave Cleto Afonso (transl.) Rs 395. Pp 233.ISBN 9788190571609

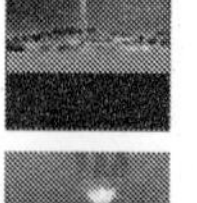

The Kadambas of Goa
S.G. Kadamb Rs 595. Pp 322. Hb ISBN 9789380837109.

The 'Portuguese' Law of Goa: Succession and Inventory
Ave Cleto Afonso. Rs 995. Pp 424. Hb ISBN 9789380837062.

The Goan Voice of Reason
Juino De Souza. Rs 350. Pp 267. Pb ISBN 9789380837116.

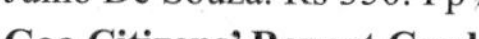

Goa Citizens' Report Card
Aureliano Fernandes. Rs 395. Pp 157. Pb ISBN 9789380837086.

Delights of Goa
Alda Figueiredo. Rs 195. Pp 162. ISBN 9789380739212

Heartbreak Passage
Lambert Mascarenhas Rs 350. Pp 200. Pb. ISBN 9789380837017

Kornelia's Kitchen
Kornelia Santoro. Rs 295. Pp 233. Pb ISBN 9789380739076.

Employment & Development in Goa
Silvia M. de Mendonça e Noronha Rs 395. Pp 125. Hb ISBN 9789380837093.

English-Konkani Dictionary
Angelus F.X. Maffei Rs 695. Pp 545. Hb ISBN 9789380837161

Aroma
Shilpa Sinari. Rs 150. Pp 130. ISBN 9789380837079

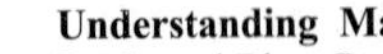

Understanding Mahabharata
Raghupati Bhatt Rs 195. Pp 146. Pb ISBN 9789380837154.

Fascinating Microfungi (Hyphomycetes) of Western Ghats - India
Dr. Jayarama Bhat Rs. 1800. Pp 221.Pb ISBN 9789380837055.

Plant And Fungal Biodiversity &Bioprospecting
S. Krishnan Dr. D. J. Bhat Rs. 795. Pp 195. Pb ISBN 9789380837000.

The Cultural History of Goa
A. R. S. Dhume Rs 295. Pp 402. Hb ISBN 9788190571678.

Operation Vijay: The Ultimate Solution
Shrikant Y Ramani. Rs 495. Pp 402. Hb ISBN 9788190571661.

Information on the Right to Information
Anil A. Parulekar Rs. 250. Pb. ISBN 9788190571661

From Goa to Patagonia
Alfredo F. de Mello. Rs. 295. Pp. 215 ISBN 9788190571616 Hb.

Passionate and Unrestrained
Alexandre Barbosa (transl.) Rs. 225. Pb ISBN 9788190571647

Life is Beautiful
Tony Martin Rs 200. Pp 254. Pb. ISBN 9788190571623.

Mirror to Goa
Donna J. Young. Rs 195. Pp 169. Pb ISBN 9788190568210.

Practical English Teacher
Tony Martin. Rs. 125. Hb. ISBN 9788190571630

Into the Diaspora Wilderness
Selma Carvalho. Rs 345. Pp 270. ISBN 9789380739021

Modern Goan Literature: An anthology
Peter Nazareth (Ed.) Rs 395. Pp 420. Pb. ISBN 9788190568258.

The Sting of Peppercorns
Antonio Gomes. Rs 295. Pp 268. Pb. ISBN 978819056296.

Sur June Pan Noor Nava
Laxman Krishna Pitre. Rs. 99 Pp. 71 Hb. ISBN 9789380837031

Medieval Goa: A Socio-Economic History
Teotonio R. de Souza, Rs 495 (hb), Rs 395 (pb). Pp 265. ISBN 9788190568265.

Goa Masala
Rs.195 Pp.260 Pb. ISBN 9789380739045

Laxmanresha
Laxman Krishna Pitre.
Rs. 180 Pp. 176 Hb. ISBN 9788190571623

The Cannons of Goa
Abhijit Ambekar, Rohini Pande and Tejas Garge. Rs. 795. Pp 80. Hb. ISBN 9789380837130

When the Curtains Rise
Andre Rafael Fernandes. Rs. 195 Pp. 215 Hb. ISBN 9789380739014

Abbé Faria
Luis S. R. Vas
Rs. 295 Hb ISBN 9788190571609

Patriotism in Action
Valmiki Faleiro
Rs. 600 Hb. ISBN 9789380739069

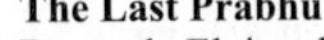

Inside Out : New Writing from Goa
Rs. 195 Pp. 244
ISBN 97893807391131

The Last Prabhu
Bernardo Elvino de Sousa
Rs.195 Pp. 163 ISBN 9789380739151

Francisco Luís Gomes 1829-1869
Luís De Assis Correa
Rs. 350 Pp., Rs 500 Hb 456

The Art of Coconut Craft
Vijaydatta Lotlikar
Rs. 195 Pp. 156. ISBN 9788190568234